Handbook of High-Temperature Superconductor Electronics

APPLIED PHYSICS

A Series of Professional Reference Books

Series Editor

ALLEN M. HERMANN

University of Colorado at Boulder
Boulder, Colorado

1. Hydrogenated Amorphous Silicon Alloy Deposition Processes, *Werner Luft and Y. Simon Tsuo*
2. Thallium-Based High-Temperature Superconductors, *edited by Allen M. Hermann and J. V. Yakhmi*
3. Composite Superconductors, *edited by Kozo Osamura*
4. Organic Conductors: Fundamentals and Applications, *edited by Jean-Pierre Farges*
5. Handbook of Semiconductor Electrodeposition, *R. K. Pandey, S. N. Sahu, and S. Chandra*
6. Bismuth-Based High-Temperature Superconductors, *edited by Hiroshi Maeda and Kazumasa Togano*
7. Handbook of High-Temperature Superconductor Electronics, *edited by Neeraj Khare*

Additional Volumes in Preparation

Handbook of High-Temperature Superconductor Electronics

edited by
Neeraj Khare
National Physical Laboratory
New Delhi, India

MARCEL DEKKER, INC. NEW YORK • BASEL

Library of Congress Cataloging-in-Publication Data
A catalog record for this book is available from the Library of Congress.

ISBN: 0-8247-0823-7

This book is printed on acid-free paper.

Headquarters
Marcel Dekker, Inc., 270 Madison Avenue, New York, NY 10016, U.S.A.
tel: 212-696-9000; fax: 212-685-4540

Distribution and Customer Service
Marcel Dekker, Inc., Cimarron Road, Monticello, New York 12701, U.S.A.
tel: 800-228-1160; fax: 845-796-1772

Eastern Hemisphere Distribution
Marcel Dekker AG, Hutgasse 4, Postfach 812, CH-4001 Basel, Switzerland
tel: 41-61-260-6300; fax: 41-61-260-6333

World Wide Web
http://www.dekker.com

The publisher offers discounts on this book when ordered in bulk quantities. For more information, write to Special Sales/Professional Marketing at the headquarters address above.

Current printing (last digit):

10 9 8 7 6 5 4 3 2 1

PRINTED IN THE UNITED STATES OF AMERICA

Preface

The discovery of high-temperature superconductors (HTS) exhibiting superconductivity above liquid nitrogen temperature has led to rapid growth in the development of many special-purpose electronics devices that can be broadly grouped under the umbrella term of "superconductor electronics."

Superconductor electronics promises particular advantages over conventional electronics: higher speed, less noise, lower power consumption, and much higher upper-frequency limit. Such characteristics are advantageous in communication technology, high-precision and high-frequency electronics, magnetic field measurement, superfast computers, etc. The potential of several superconductor electronics devices has already been established using low-T_c conventional superconductors. The discovery of cuprate superconductors with higher transition temperature and higher energy gap extends the capability of superconductor electronics considerably. Rapid advancement in the synthesis of HTS thin films and artificial grain boundary HTS Josephson junctions has elicited considerable interest in the development of electronic devices found to be very promising for future applications, such as superconducting quantum interference devices (SQUIDs) small microwave, and digital devices. Some of the HTS devices are already on the market.

Advances in the physics and material aspects of HTS have been well documented in the form of books and monographs, serving as a starting block for general readers and beginners. However, the literature was scattered. Thus, this book is vital, bringing together contributions from leaders in different areas of research and development in HTS electronics.

The contents are organized to be self-explanatory, comprehensive, and useful to both general reader and specialist. In each chapter care has been taken to introduce basic terminology so that the readers in other fields interested in high-temperature superconductor electronics will find no difficulty in reading it. Professionals will find it an easily available collection of valuable and relevant information. The chapters are sequentially organized for use as a text for the study of high-T_c devices at the graduate and advanced undergraduate level.

Chapter 1 is an introduction to high-T_c superconductors, presenting the developments in the discovery of various HTS compounds, its structure, preparation, various properties, and comparison to low-T_c superconductors. The developments of various techniques for high-T_c thin-film fabrication are described in Chapter 2. Readers interested in knowing the advancements in high-T_c film fabrication will find it very interesting and informative.

Chapters 3 and 4 present fabrication details and characteristics of multilayer edge junctions and step-edge junctions in high-T_c superconducting films.

It is not easy to prepare S/I/S Josephson junctions in high-T_c as it is usually done in low-T_c superconductors (LTS), due to the short coherence length of HTS. Natural grain boundaries in high-T_c materials are found to behave as Josephson junctions. Detailed studies of these grain boundaries have led to the development of several techniques for realizing artificial grain boundaries and junctions whose behavior is similar to that of Josephson junctions. Grain boundaries in HTS are of central importance in numerous applications, such as electronic circuits and sensors and SQUIDs. Also, for many experiments elucidating the physics of high-T_c superconductivity, grain boundaries have been used with outstanding success.

Chapter 5 discusses the progress in understanding the conduction noise in high-T_c superconductors. Chapter 6 reviews noise mechanisms in HTS junctions, experimental techniques, and quantitative data on the noise properties of a range of junctions and devices.

Noise in electronic systems sets limits the sensitivity of devices. Superconducting devices offer levels of performance that are difficult or impossible to achieve by conventional methods, but are also subject to limitations due to intrinsic noise. A full understanding of the noise mechanism remains one of the outstanding tasks in the way of successful high-T_c applications. Intrinsic noise is in orders of magnitude greater than the limits imposed by quantum mechanics, and it becomes important to understand the mechanism that causes the excess noise.

In recent years, progress in the development of the high-T_c SQUID has been remarkable. It is among the first HTS devices to reach the market. The field sensitivity achieved in HTS SQUIDs is sufficiently high for several applications including biomagnetism measurement, nondestructive evaluation, and geophysical measurement. Progress in high-T_c rf-SQUIDs and SQUID magnetometer are presented in Chapters 7 and 8.

Chapter 9 presents an overview of progress in HTS digital circuits. Chapter 10 reviews the progress in the development of several HTS microwave devices such as filters, delay lines, low loss resonators, and antennas etc. Chapter 11 describes the principles and characteristics of high-T_c IR detectors.

HTS digital circuits are more suitable for use in single-flux quantum (SFQ) circuits than in LTS ones, because HTS Josephson junctions are naturally overdamped, which means that their I-V curves do not show hysteresis, and the junctions in SFQ circuits must be overdamped junctions. The I_cR_n product of HTS junctions can also be expected to be larger than that of LTS junctions because it intrinsically depends on the gap voltage of the superconductor.

For a widespread application of HTS electronics, a package of high-T_c components in closed-cycle cryocoolers is required. Chapter 12 presents advances in the area of cryocoolers and high-T_c devices. In order to make this chapter more comprehensive for beginners, the principles and details of various closed-cycle methods such as the Joule-Thomson, Brayton, Claude, Stirling, Gifford-McMahon, and pulse tube cryocoolers along with their relative merits, are discussed. Finally, the last Chapter 13 presents a summary of the status and future of HTS electronics.

This book would have never been possible without the support of all the contributors. I am grateful to all of them for their contributions. In spite of their own busy schedules and commitments, they spared the time to prepare an exhaustive and critical review. The idea of preparing a book on HTS electronics came after a thought-provoking discussion with Prof. Allen M. Hermann. I am grateful to him for the enthusiasm he created and for his support during the entire course of preparation of the book. I am thankful to the publisher, Marcel Dekker, Inc., for inviting me to edit this book, which indeed proved to be a very interesting and rewarding experience. I am also thankful to my production editor, Brian Black, for his editorial support.

I have greatly benefited from the experienced advice of Prof. S. Chandra on several occasions and I am grateful to him for all the encouragement and support. Encouragement and guidance received from Prof. S. K. Joshi, Dr. K. Lal, Dr. Praveen Chaudhari, Prof. G. B. Donaldson, Prof. O. N. Srivastava, Prof. E. S. Rajagopal, Prof. A. K. Raychaudhuri, and Dr. A. K. Gupta are gratefully acknowledged. I am thankful to Dr. N. D. Kataria and Dr. Vijay Kumar for their help and cooperation.

Concern and words of appreciation of Prof. O. P. Malviya have been a great source of encouragement for me. Emotional support from my well-wishers particularly came from Priyadarshan Malviya, Pankaj Khare, and Alka Wadhwa. I wish to express my gratitude to my wife, Sangeeta, for her untiring help, cooperation, and patience, without which it would not have been possible to complete this book. The smiling face and shining eyes of my little son, Siddharth have been a great source of stress relief for me and always inspired me to devote more time to completing the book.

Neeraj Khare

Contents

Contributors

John C. Brasunas NASA's Goddard Space Flight Center, Greenbelt, Maryland, U.S.A.

L. Hao* Department of Physics and Applied Physics, University of Strathclyde, Glasgow, Scotland

Mutsuo Hidaka NEC Corporation, Ibaraki, Japan

Q. X. Jia Superconductivity Technology Center, Los Alamos National Laboratory, Los Alamos, New Mexico, U.S.A.

Neeraj Khare National Physical Laboratory, New Delhi, India

László Béla Kish Texas A&M University, College Station, Texas, U.S.A.

F. Lombardi Chalmers Institute of Technology and Göteborg University, Göteborg, Sweden

J. C. Macfarlane Department of Physics and Applied Physics, University of Strathclyde, Glasgow, Scotland

David P. Norton University of Florida, Gainesville, Florida, U.S.A.

C. M. Pegrum Department of Physics and Applied Physics, University of Strathclyde, Glasgow, Scotland

**Current affiliation:* Centre for Basic Metrology, National Physical Laboratory, Teddington, England

Ray Radebaugh National Institute of Standards and Technology, Boulder, Colorado, U.S.A.

V. I. Shnyrkov Institute for Low Temperature Physics and Engineering, Academy of Sciences, Kharkov, Ukraine

Shoji Tanaka Superconductivity Research Laboratory, ISTEC, Tokyo, Japan

A. Ya. Tzalenchuk National Physical Laboratory, Middlesex, England

Handbook of High-Temperature Superconductor Electronics

1

Introduction to High-Temperature Superconductors

Neeraj Khare
National Physical Laboratory, New Delhi, India

1.1 INTRODUCTION

The discovery of superconductivity in copper oxide perovskite (1) has opened a new era of research in superconducting materials. This class of materials not only show high-temperature superconductivity but also show properties that are different from classical superconductors. This offers a great challenge to understanding the basic phenomenon that causes superconductivity in these materials and to developing the appropriate preparation methods so that these can be exploited for a wide range of applications. During the last one and half decades after the discovery of high-T_c materials, several high-T_c superconductors have been discovered which show superconductivity at temperatures higher than liquid-nitrogen temperature (77 K). There has also been great progress in understanding the properties of these materials, developing different methods of preparation, and realizing superconducting devices which use these superconductors.

This chapter will give a brief description of the historical developments in raising the transition temperature (T_c) of the superconductors, preparation, and structure of the material. Different properties of the high-T_c materials such as critical magnetic field, penetration depth, coherence length, critical current density, weak link, and so forth are also discussed.

1.2 RAISING THE TRANSITION TEMPERATURE

Superconductivity is the phenomenon in which a material loses its resistance on cooling below the transition temperature (T_c). Superconductivity was first discovered in mercury by Onnes (2) in 1911. The temperature at which mercury becomes superconducting was found to be close to the boiling point of liquid helium (4.2 K). Subsequently, many metals, alloys, and intermetallic compounds were found to exhibit superconductivity. The highest T_c known was limited to 23.2 K (*3*) in the Nb_3Ge alloy; however, in September 1986, Bednorz and Muller (1) discovered superconductivity at 30 K in La–Ba–Cu–O. The phase responsible for superconductivity was identified to have nominal composition of $La_{2-x}Ba_xCuO_{4-y}$ ($x = 0.2$). The discovery of high-temperature superconductivity in ceramic cuprate oxides by Bednorz and Muller led to unprecedented effort to explore new superconducting oxide material with higher transition temperatures. The value of T_c in $La_{2-x}Ba_xCuO_4$ was found to increase up to 57 K with the application of pressure (4). This observation in $La_{2-x}Ba_xCuO_4$ material raised the hope of attaining even higher transition temperatures in cuprate oxides. This, indeed, turned out to be true when Chu and co-workers (5) reported a remarkably high superconductivity transition temperature (T_c) of 92 K on replacing La by Y in nominal composition $Y_{1.2}Ba_{0.8}CuO_{4-y}$. Later, different groups identified (6–8) that the superconducting phase responsible for 90 K has the composition $YBa_2Cu_3O_{7-y}$.

The discovery of superconductivity above the boiling point of liquid nitrogen led to extensive search for new superconducting materials. Superconductivity at transition temperatures of 105 K in the multiphase sample of the Bi–Sr–Ca–Cu–O compound was reported by Maeda et al. (9) in 1988. The highest T_c of 110 K was obtained in the Bi–Sr–Ca–Cu–O compound having a composition $Bi_2Sr_2Ca_2Cu_3O_{10}$ (10,11). Sheng and Hermann (12) substituted the nonmagnetic trivalent Tl for R in R-123, where R is a rare-earth element. By reducing the reaction time to a few minutes for overcoming the high-volatility problem associated with Tl_2O_3, they detected superconductivity above 90 K in $TlBa_2Cu_3O_x$ samples in November 1987. By partially substituting Ca for Ba, they (13) discovered a $T_c \sim 120$ K in the multiphase sample of Tl–Ba–Ca–Cu–O in February 1988. In September 1992, Putillin et al. (14) found that the $HgBa_2CuO_x$ (Hg-1201) compound with only one CuO_2 layer showed a T_c of up to 94 K. It was, therefore, rather natural to speculate that T_c can increase if more CuO_2 layers are added in the per unit formula to the compound. In April 1993, Schilling et al. (15) reported the detection of superconductivity at temperatures up to 133 K in $HgBa_2Ca_2Cu_3O_x$. The transition temperature of $HgBa_2Ca_2Cu_3O_x$ was found to increase to 153 K with the application of pressure (16).

Figure 1.1 depicts the evolution in the transition temperature of superconductors starting from the discovery of superconductivity in mercury. The slow but steady progress to search for new superconductors with higher transition temper-

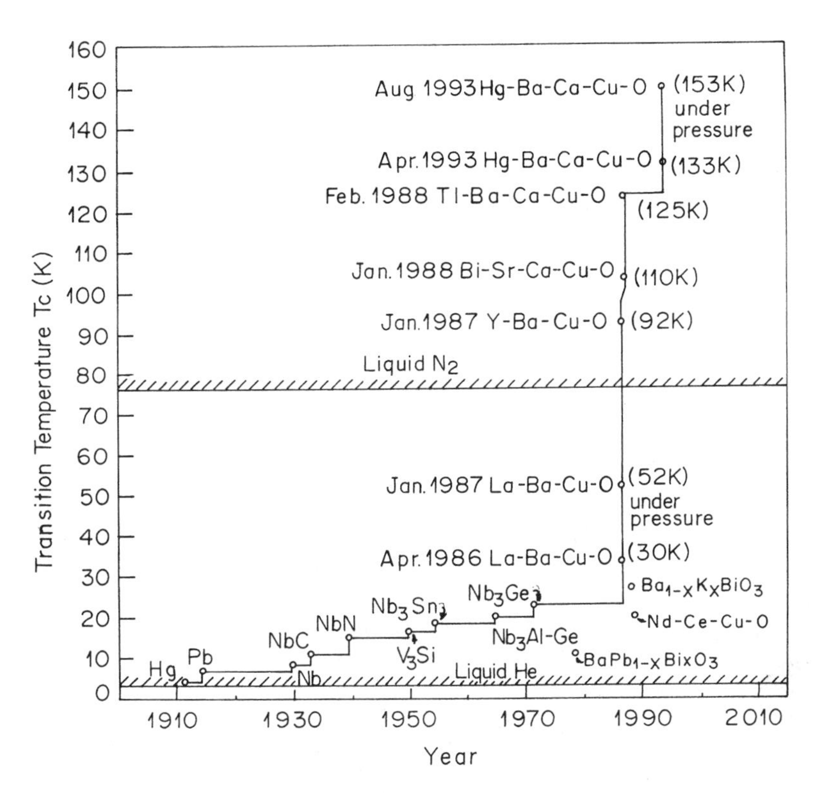

FIGURE 1.1 The evolution of the transition temperature (T_c) subsequent to the discovery of superconductivity.

atures continued for decades until superconductivity at 30 K in La–Ba–Cu–O oxide was discovered in 1986. Soon after this, other cuprate oxides such as Y–Ba–Cu–O, Bi–Sr–Ca–Cu–O, Tl–Ba–Ca–Cu–O with superconductivity above the liquid-nitrogen temperature were discovered.

Table 1.1 gives a list of some of high-T_c superconductors with their respective transition temperature, crystal structure, number of Cu–O layers present in unit cell, and lattice constants. Transition temperature has been found to increase as the number of Cu–O layer increases to three in Bi–Sr–Ca–Cu–O, Tl–Ba–Ca–Cu–O, and Hg–Ba–Ca–Cu–O compounds. In all of the cuprate superconductors described so far, the superconductivity is due to hole-charge carriers, except for $Nd_{2-x}Ce_xCuO_4$ ($T_c \sim 20$ K), which is an n-type superconductor (17). The superconductor $Ba_{0.6}K_{0.4}BiO_3$, which does not include Cu, was reported by Cava et al. (18) in 1988 exhibiting $T_c \sim 30$ K. A homologous series of compounds $(Cu,Cr)Sr_2Ca_{n-1}Cu_nO_y$ [Cr12($n-1$)n] has been synthesized under high pressure.

TABLE 1.1 Transition Temperature (T_c), Crystal Structure and Lattice Constants of Some High-T_c Superconductors

High-T_c superconductors					
Formula	Notation	T_c (K)	n^a	Crystal structure	Lattice constants (Å)
$La_{1.6}Ba_{0.4}CuO_4$	214	30	1	Tetragonal	$a = 3.79$, $c = 13.21$
$La_{2-x}Sr_xCuO_4$	214	38	1	Tetragonal	$a = 3.78$, $c = 13.23$
$YBa_2Cu_3O_7$	123	92	2	Orthorhombic	$a = 3.82$, $b = 3.89$, $c = 11.68$
$YBa_2Cu_4O_8$	124	80	2	Orthorhombic	$a = 3.84$, $b = 3.87$, $c = 27.23$
$Y_2Ba_4Cu_7O_{14}$	247	40	2	Orthorhombic	$a = 3.85$, $b = 3.87$, $c = 50.2$
$Bi_2Sr_2CuO_6$	Bi-2201	20	1	Tetragonal	$a = 5.39$, $c = 24.6$
$Bi_2Sr_2CaCu_2O_8$	Bi-2212	85	2	Tetragonal	$a = 5.39$, $c = 30.6$
$Bi_2Sr_2Ca_2Cu_3O_{10}$	Bi-2223	110	3	Tetragonal	$a = 5.39$, $c = 37.1$
$TlBa_2CuO_5$	Tl-1201	25	1	Tetragonal	$a = 3.74$, $c = 9.00$
$TlBa_2CaCu_2O_7$	Tl-1212	90	2	Tetragonal	$a = 3.85$, $c = 12.74$
$TlBa_2Ca_2Cu_3O_9$	Tl-1223	110	3	Tetragonal	$a = 3.85$, $c = 15.87$
$TlBa_2Ca_3Cu_4O_{11}$	Tl-1234	122	4	Tetragonal	$a = 3.86$, $c = 19.01$
$Tl_2Ba_2CuO_6$	Tl-2201	80	1	Tetragonal	$a = 3.86$, $c = 23.22$
$Tl_2Ba_2CaCu_2O_8$	Tl-2212	108	2	Tetragonal	$a = 3.86$, $c = 29.39$
$Tl_2Ba_2Ca_2Cu_3O_{10}$	Tl-2223	125	3	Tetragonal	$a = 3.85$, $c = 35.9$
$HgBa_2CuO_4$	Hg-1201	94	1	Tetragonal	$a = 3.87$, $c = 9.51$
$HgBa_2CaCu_2O_6$	Hg-1212	128	2	Tetragonal	$a = 3.85$, $c = 12.66$
$HgBa_2Ca_2Cu_3O_8$	Hg-1223	134	3	Tetragonal	$a = 3.85$, $c = 15.78$
$(Nd_{2-x}Ce_x)\ CuO_4$	T	30	1	Tetragonal	$a = 3.94$, $c = 12.07$
(Nd, CeSr) CuO_4	T*	30	1	Tetragonal	$a = 3.85$, $c = 12.48$

[a] *n* represents the number of Cu-O planes in the unit cell.

In the Cr series, the value of n can be changed from 1 to 9, with a maximum T_c of 107 K at $n = 3$. The $Pr(Ca)Ba_2Cu_3O_y$ compound has also been synthesized under high pressure, showing a transition temperature of 97 K (19).

1.3 CRYSTAL STRUCTURE OF HIGH-T_c SUPERCONDUCTORS

The structure of a high-T_c superconductor is closely related to perovskite structure. The unit cell of perovskite consists of two metal (A, B) atoms and three oxygen atoms, with the general formula given as ABO_3. The ideal perovskite structure is shown in Fig. 1.2a. Atom A, sitting at the body-centered site, is coordinated by 12 oxygen atoms. Atom B occupies the corner site and the oxygen atom occupies the edge-centered position.

Figure 1.2b shows the unit cell of $La_{2-x}Ba_xCuO_4$, which has a tetragonal symmetry and consists of perovskite layers separated by rock-salt-like layers made of La (or Ba) and O atoms. This compound is often termed 214 because it has two La, one Cu, and four O atoms. The 214 compound has only one CuO_2 plane. Looking at the exact center of Fig. 1.2b, the CuO_2 plane appears as one copper atoms surrounded by four oxygen atoms, with one LaO plane above the CuO_2 plane and one below it. The entire structure is layered. The LaO planes are said to be intercalated. The CuO_2 plane is termed the conduction plane, which is responsible for superconductivity. The intercalated LaO planes are called "charge-reservoir layers." When the intercalated plane contains mixed valence atoms, electrons are drawn away from the copper oxide planes, leaving holes to form pairs needed for superconductivity.

The structure of $YBa_2Cu_3O_7$ is shown in Fig. 1.2c. The unit cell of $YBa_2Cu_3O_7$ consists of three pseudocubic elementary perovskite unit cells (8). Each perovskite unit cell contains a Y or Ba atom at the center: Ba in the bottom unit cell, Y in the middle one, and Ba in the top unit cell. Thus, Y and Ba are stacked in the sequence [Ba–Y–Ba] along the *c*-axis. All corner sites of the unit cell are occupied by Cu, which has two different coordinations, Cu(1) and Cu(2), with respect to oxygen. There are four possible crystallographic sites for oxygen: O(1), O(2), O(3), and O(4). The coordination polyhedra of Y and Ba with respect to oxygen are different. The tripling of the perovskite unit cell (ABO_3) leads to nine oxygen atoms, whereas $YBa_2Cu_3O_7$ has seven oxygen atoms accommodat-

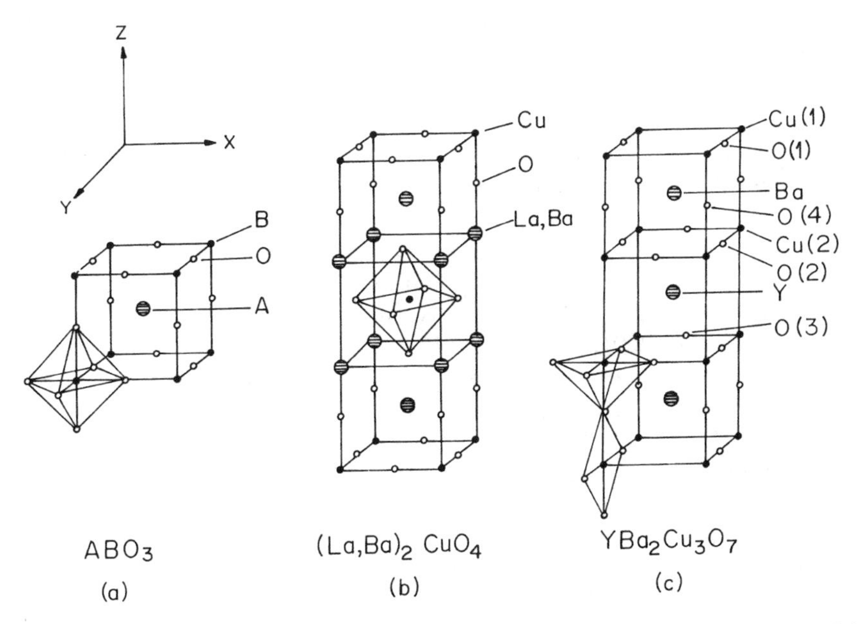

FIGURE 1.2 Structure of (a) perovskite ABO_3, (b) $(La,Ba)_2CuO_4$, and (c) $YBa_2Cu_3O_7$.

ing the deficiency of two oxygen atoms. Thus, the structure of the 90 K phase deviates from the ideal perovskite structure and, therefore, is referred to as an oxygen-deficient perovskite structure. Oxygen atoms are missing from the Y plane (i.e., $z = 1/2$ site); thus, Y is surrounded by 8 oxygen atoms instead of the 12 if it had been in ideal perovskite structure. Oxygen atoms at the top and bottom planes of the $YBa_2Cu_3O_7$ unit cell are missing in the [100] direction, thus giving (Cu–O) chains in the [010] direction. The Ba atom has a coordination number of 10 oxygen atoms instead of 12 because of the absence of oxygen at the (1/2 0 z) site. The structure has a stacking of different layers: (CuO)(BaO)(CuO_2)(Y)(CuO_2)(BaO)(CuO). One of the key feature of the unit cell of $YBa_2Cu_3O_{7-\delta}$ (YBCO) is the presence of two layers of CuO_2. The role of the Y plane is to serve as a spacer between two CuO_2 planes. In YBCO, the Cu–O chains are known to play an important role for superconductivity. T_c maximizes near 92 K when $\delta \approx 0.15$ and the structure is orthorhombic. Superconductivity disappears at $\delta \approx 0.6$, where the structural transformation of YBCO occurs from orthorhombic to tetragonal.

The crystal structure of Bi-, Tl-, and Hg-based high-T_c superconductors are very similar to each other. Like YBCO, the perovskite-type feature and the presence of CuO_2 layers also exist in these superconductors. However, unlike YBCO, Cu–O chains are not present in these superconductors. The YBCO superconductor has an orthorhombic structure, whereas the other high-T_c superconductors have a tetragonal structure (see Table 1.1).

The Bi–Sr–Ca–Cu–O system has three superconducting phases forming a homologous series as $Bi_2Sr_2Ca_{n-1}Cu_nO_{4+2n+y}$ (n = 1, 2, and 3). These three phases are Bi-2201, Bi-2212, and Bi-2223, having transition temperatures of 20, 85, and 110 K, respectively (10,11). The structure of Bi-2201 together with Bi-2212 and Bi-2223 is shown in Fig. 1.3. All three phases have a tetragonal structure which consists of two sheared crystallographic unit cells. The unit cell of these phases has double Bi–O planes which are stacked with a shift of (1/2 1/2 z) with respect to the origin. The stacking is such that the Bi atom of one plane sits below the oxygen atom of the next consecutive plane. The Ca atom forms a layer within the interior of the CuO_2 layers in both Bi-2212 and Bi-2223; there is no Ca layer in the Bi-2201 phase. The three phases differ with each other in the number of CuO_2 planes; Bi-2201, Bi-2212, and Bi-2223 phases have one, two, and three CuO_2 planes, respectively. The c axis of these phases increases with the number of CuO_2 planes. The lengths of the c axis are 24.6 Å, 30.6 Å, and 37.1 Å respectively for the Bi-2201, Bi-2212, and Bi-2223 phases. The coordination of the Cu atom is different in the three phases. The Cu atom forms an octahedral coordination with respect to oxygen atoms in the 2201 phase, whereas in 2212, the Cu atom is surrounded by five oxygen atoms in a pyramidal arrangement. In the 2223 structure, Cu has two coordinations with respect to oxygen: one Cu atom is bonded with four oxygen atoms in square planar configuration and another Cu atom is coordinated with five oxygen atoms in a pyramidal arrangement.

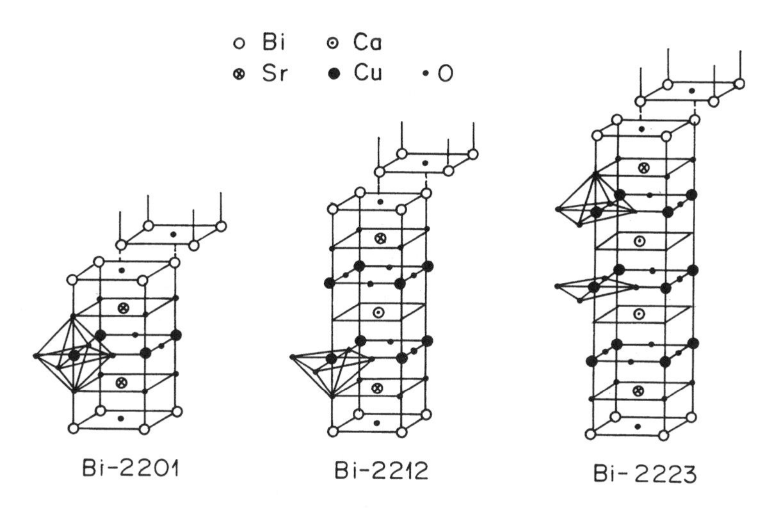

FIGURE 1.3 Unit cells of the $Bi_2Sr_2Ca_{n-1}\ Cu_nO_x$ compound with $n = 1, 2,$ and 3. (Adapted from Ref. 11.)

Figure 1.4 shows the unit cells of two series of the Tl–Ba–Ca–Cu–O superconductor (20). The first series of the Tl-based superconductor containing one Tl–O layer has the general formula $TlBa_2Ca_{n-1}Cu_nO_{2n+3}$, whereas the second series containing two Tl–O layers has a formula of $Tl_2Ba_2Ca_{n-1}Cu_nO_{2n+4}$ with $n =$ 1, 2, and 3. In the structure of $Tl_2Ba_2CuO_6$, there is one CuO_2 layer with the stacking sequence (Tl–O) (Tl–O) (Ba–O) (Cu–O) (Ba–O) (Tl–O) (Tl–O). In $Tl_2Ba_2CaCu_2O_8$, there are two Cu–O layers with a Ca layer in between. Similar to the $Tl_2Ba_2CuO_6$ structure, Tl–O layers are present outside the Ba–O layers. In $Tl_2Ba_2Ca_2Cu_3O_{10}$, there are three CuO_2 layers enclosing Ca layers between each of these. In Tl-based superconductors, T_c is found to increase with the increase in CuO_2 layers. However, the value of T_c decreases after four CuO_2 layers in $TlBa_2Ca_{n-1}Cu_nO_{2n+3}$, and in the $Tl_2Ba_2Ca_{n-1}Cu_nO_{2n+4}$ compound, it decreases after three CuO_2 layers.

The crystal structure of $HgBa_2CuO_4$ (Hg-1201), $HgBa_2CaCu_2O_6$ (Hg-1212), and $HgBa_2Ca_2Cu_3O_8$ (Hg-1223) is similar to that of Tl-1201, Tl-1212, and Tl-1223 (Fig. 1.4) with Hg in place of Tl (21). It is noteworthy that the T_c of the Hg compound (Hg-1201) containing one CuO_2 layer is much larger as compared to the one-CuO_2-layer compound of thallium (Tl-1201). In the Hg-based superconductor, T_c is also found to increase as the CuO_2 layer increases. For Hg-1201, Hg-1212, and Hg-1223, the values of T_c are 94, 128, and 134 K respectively, as shown in Table 1.1. The observation that the T_c of Hg-1223 increases to 153 K under high pressure (16) indicates that the T_c of this compound is very sensitive to the structure of the compound.

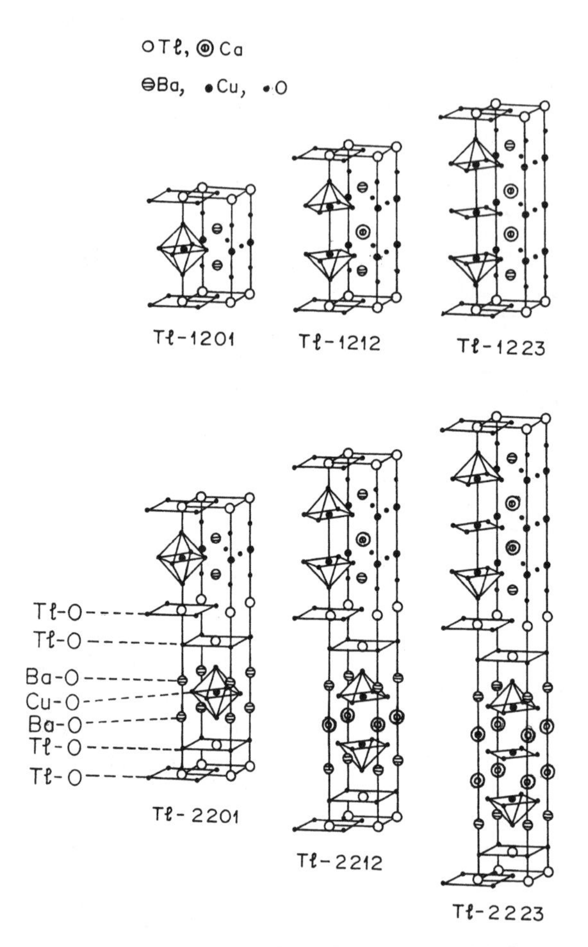

FIGURE 1.4 Unit cells of the $Tl_1Ba_2Ca_{n-1}Cu_nO_{2n+3}$ compound containing one Tl–O layer and the $Tl_2Ba_2Ca_{n-1}Cu_nO_{2n+4}$ compound containing two Tl–O layers for $n = 1, 2,$ and 3. (Adapted from Ref. 20.)

1.4 PREPARATION OF HIGH-T_c SUPERCONDUCTORS

High-T_c superconductors are prepared in the form of bulk, thick films, thin films, single crystals, wires, and tapes. Fabrication in the form of wires and tapes are required for high-current applications. On the other hand, thick and thin films are needed for electronic application. Strict control of the stoichiometry of the composition is very much required for preparing high-T_c superconductors with desirable characteristics. Even a small change in oxygen content or a small change in cation doping level can transform the material from a superconductor to a low-carrier-density metal or even to an insulator. The following paragraphs give a brief description of high-T_c superconductors in the form of bulk and thick films. The preparation of high-T_c thin films is given in more detail in the other chapters of this book.

The simplest method for preparing high-T_c superconductors is a solid-state thermochemical reaction involving mixing, calcination, and sintering. The appropriate amounts of precursor powders, usually oxides and carbonates, are mixed thoroughly using a ball mill. Solution chemistry processes such as coprecipitation, freeze-drying, and sol–gel methods are alternative ways for preparing a homogenous mixture. These powders are calcined in the temperature range from 800°C to 950°C for several hours. The powders are cooled, reground, and calcined again. This process is repeated several times to get homogenous material. The powders are subsequently compacted to pellets and sintered. The sintering environment such as temperature, annealing time, atmosphere, and cooling rate play a very important role in getting good high-T_c superconducting materials. The $(La_{1-x}Ba_x)_2CuO_{4-\delta}$ high-T_c superconductor is prepared by heating a mixture of La_2O_3, $BaCO_3$, and CuO in a reduced oxygen atmosphere at 900°C. After regrinding and reheating the mixtures, the pellet is prepared and sintered at 925°C for 24 h. The $YBa_2Cu_3O_{7-\delta}$ compound is prepared by calcination and sintering of a homogenous mixture of Y_2O_3, $BaCO_3$, and CuO in the appropriate atomic ratio. Calcination is done at 900–950°C, whereas sintering is done at 950°C in an oxygen atmosphere. The oxygen stoichiometry in this material is very crucial for obtaining a superconducting $YBa_2Cu_3O_{7-\delta}$ compound. At the time of sintering, the semiconducting tetragonal $YBa_2Cu_3O_6$ compound is formed, which, on slow cooling in oxygen atmosphere, turns into superconducting $YBa_2Cu_3O_{7-\delta}$. The uptake and loss of oxygen are reversible in $YBa_2Cu_3O_{7-\delta}$. A fully oxidized orthorhombic $YBa_2Cu_3O_{7-\delta}$ sample can be transformed into tetragonal $YBa_2Cu_3O_6$ by heating in a vacuum at temperature above 700°C.

The preparation of Bi-, Tl-, and Hg-based high-T_c superconductors is difficult compared to YBCO. Problems in these superconductors arise because of the existence of three or more phases having a similar layered structure. Thus, syntactic intergrowth and defects such as stacking faults occur during synthesis and it becomes difficult to isolate a single superconducting phase. For Bi–Sr–Ca–Cu–O, it is relatively simple to prepare the Bi-2212 ($T_c \sim 85$ K) phase, whereas it is very difficult to prepare a single phase of Bi-2223 ($T_c \sim 110$ K). The Bi-2212 phase appears only after few hours of sintering at 860–870°C, but the larger fraction of the Bi-2223 phase is formed after a long reaction time of more than a week at 870°C (11). Although the substitution of Pb in the Bi–Sr–Ca–Cu–O compound has been found to promote the growth of the high-T_c phase (22), a long sintering time is still required.

Toxicity and low vapor pressure of Hg–O and Tl–O make fabrication of Hg- and Tl-based high-T_c superconductors much more difficult and one has to follow special precautions and stringent control on the preparation atmosphere. The Tl-based superconductor is prepared by thorough mixing of Tl_2O_3, BaO, CaO, and

CuO in appropriate proportions and pressing the powders into a pellet. The pellet is wrapped in a gold foil and fired at 880°C for 3h in a sealed quartz tube containing 1 atm oxygen to reach superconductivity (20).

For the preparation of a Hg-based high-T_c superconductor (15), first a precursor material with the nominal composition $Ba_2CaCu_2O_5$ is obtained from a homogenous mixture of the respective metal nitrates by sintering at 900°C in oxygen. Dry boxes are used for grinding and mixing of the powders. After regrinding and mixing with HgO powder, the pressed pellet is sealed in an evacuated quartz tube. This tube is placed horizontally in a tight steel container and sintered at 800°C for a few hours.

Several techniques such as screen printing (23–27), spin-coating (28) and spray pyrolysis (29–33) are used in preparing high-T_c thick films. For the screen printing or spin-coating method, the first step is to prepare homogenous powders of high-T_c materials; this is accomplished by solid-state reaction or by a chemical route involving mixing, calcination, and sintering of appropriate powders in the form of oxides or carbonates. After sintering the powders are sieved through a screen woven from stainless steel or nylon wire. The diameter of the screen wire and the size of the opening can vary depending on the process requirement. The opening size is usually given in terms of a standard mesh number that varies from 100 to 400. The fine sieved powders are converted into thick paste by mixing with an organic solvent such as propylene glycol, octyl alcohol, heptyl alcohol, triethanolamine, or cyclohexagonal. In the screen-printing technique, thick paste is used for printing the substrate through the mesh screen and dried at an appropriate temperature. In the spin-coating method, one drop of the paste is put on the substrate and the substrate is spun to get a uniform coating of the material. The resultant films are fired at a suitable annealing temperature. In general, single-crystal and polycrystalline substrates of magnesium oxide (MgO), strontium titnate ($SrTiO_3$), lanthanum aluminate ($LaAlO_3$), yattria-stabilized zirconia (YSZ), and aluminum oxide (Al_2O_3) are used for the high-T_c thick-film preparation.

For YBCO thick films, the sintering temperature is kept between 940°C and 970°C followed by slow cooling in an oxygen atmosphere (23). In order to achieve YBCO films with a larger grain size and higher current density, the firing temperature is increased to 1000°C (24). Bi-2212 high-T_c films are prepared by firing the films at 880–885°C. It has been found that partial melting and quenching of the Bi-2212 films from 885°C to room temperature leads to a T_c as high as 96 K (25). For high-T_c films with a Bi-2223 phase, the films are fired at ~880°C for a few minutes and then annealed at 864°C for a duration of 70–80 h (26). The preparation of Tl–Ba–Ca–Cu–O thick films requires a two-step process (27). In the first step, a film of Ba–Ca–Cu–O is prepared, and in the second step, this precursor film is heated in Tl_2O_3 vapor followed by slow cooling to room temperature.

Spray pyrolysis is another simple and inexpensive technique for preparing high-T_c films (30–33). For YBCO film, an aqueous solution for the spray is prepared by dissolving $Y(NO_3)\cdot 6H_2O$, $Ba(NO_3)$, and $Cu(NO_3)\cdot 3H_2O$ in triple-distilled water in a 1 : 2 : 3 stoichiometric ratio (29). The solution is sprayed on a single-crystal YSZ or $SrTiO_3$ substrate through a glass nozzle using oxygen for few minutes and then slowly cooled to room temperature. The starting solution for depositing Bi-2212 film is prepared by mixing aqueous solution of Bi_2O_3, $SrCO_3$, $CaCO_3$, and CuO in dilute nitric acid (30). A two-step process is used for preparing Tl- and Hg-based high-T_c films by the spray pyrolysis technique (31–33). The first step involves preparation of Ba–Ca–Cu–O precursor films by spraying an aqueous solution of Ba, Ca, and Cu nitrates on a single-crystal substrate. In the second step, Tl or Hg is incorporated in the precursor films by annealing the film in a controlled Tl–O or Hg–O vapor atmosphere.

Different techniques such as sputtering, evaporation, molecular beam epitaxy, laser ablation, chemical vapor deposition, and so forth have been used successfully to prepare thin films of high-T_c superconductors. A detailed account of these techniques is given in Chapter 2. Most of these techniques work in a vacuum environment and the oxygen partial pressure near the substrate is controlled to obtain a superconducting film. This can be done during the film deposition (in situ process) or by postdeposition oxygen annealing. The substrate temperature during the deposition is a crucial parameter that determines microstructural details such as texture and the degree of epitaxy of the film. Substrate–film interaction such as interdiffusion can affect the quality of the films. Thus, it is desirable to develop processes that allow a low substrate temperature.

1.5 PROPERTIES OF HIGH-TEMPERATURE SUPERCONDUCTORS

A superconducting state is defined by the transition temperature (T_c) at which material exhibits zero resistance on cooling. Apart from the transition temperature, other properties characterizing the high-T_c superconductors are critical magnetic field, penetration depth, coherence length, critical current density and weak link, energy gap, and so forth. A brief description of these is presented here.

1.5.1 Anisotropy

As described in Section 1.3, the crystal structure of high-T_c superconductors is highly anisotropic. This feature has important implications for both physical and mechanical properties. In high-T_c superconductors, electrical currents are carried by holes induced in the oxygen sites of the CuO_2 sheets. The electrical conduction is highly anisotropic, with a much higher conductivity parallel to the CuO_2 plane than in the perpendicular direction. Other superconductivity properties such as coherence length (ξ), penetration depth (λ), and energy gap (Δ) are also anisotropic. The mechanical properties of high-T_c materials are also very anisotropic. For example, in YBCO, upon cooling, the lattice contracts far more along *a-b* planes than along *c* axis. Torque magnetometry measurements have been made for several high-T_c superconductors for studying anisotropy (34,35). For Tl-2212, an anisotropy of $\sim 10^5$ is found for the ratio of the mass along the *c* axis to that of *a-b* plane. A similar large ratio is obtained for the Bi-2212 compound. In Y-123, the value of this ratio is found to be $\sim$25, which is much smaller compared to Bi and Tl compounds. The anisotropy factor of a high-T_c superconductor at the optimally doped composition is related to the interlayer spacing between CuO_2 layers in the unit cell. It has been also noted that increasing carrier doping or substituting ions on the blocking layer for certain other ions such as Pb in Bi-2212 reduces anisotropy without changing the interlayer spacing significantly.

1.5.2 Critical Magnetic Field

The abrupt transition from the normal to superconducting state occurs at a boundary defined not only by the transition temperature (T_c) but also by the magnetic field strength. There is a critical value of magnetic field, H_c, above which the superconductivity is destroyed. If a paramagnetic material is placed in a magnetic field, then the magnetic lines of force penetrate through the material. However, when the same material is made superconducting by cooling to a low temperature below T_c, then the magnetic lines of force are completely expelled from the interior of the material. This effect is called the Meissner effect. Based on the Meissner effect, the superconducting materials are classified as type I and type II superconductors. If there is a sharp transition from the superconducting state to the normal state, then this type of material is called a type I superconductor. This kind of behavior is shown, in general, by pure metals. In type II superconductors, there are two values of the critical field: the lower critical field, H_{c1}, and the upper critical field, H_{c2}. For $H < H_{c1}$, the field is completely expelled from the superconductor. However, for $H > H_{c1}$, the magnetic field penetrates the material slowly and continues up to H_{c2}, beyond which the material transforms completely from the superconducting state to the normal state. The state between H_{c1} and H_{c2} is called the vortex or mixed state. Figure 1.5a shows the *H–T* phase diagram for conventional low-T_c superconductors. At low fields, there is Meissner state, and at high fields, vortices enter the material and form a vortex lattice. Superconductivity is completely destroyed at H_{c2}, for which the density of vortices is such that the normal cores fill the entire material. For low-T_c superconductors, this behavior is exhibited, in general, by alloys and compounds. On the other hand, all high-T_c superconductors behave as type II superconductors. For high-T_c superconductors, the value of $H_{c1}(0)$ is low ($\sim$100 Oe), whereas the value of $H_{c2}(0)$ is quite high (about few hundred tesla). The value of the critical field is anisotropic for these materials. For a YBCO single crystal, values of $H_{c1}(0)$ in the direction parallel to the *c* axis and in the *a-b* plane are estimated as 850 and 250 Oe,

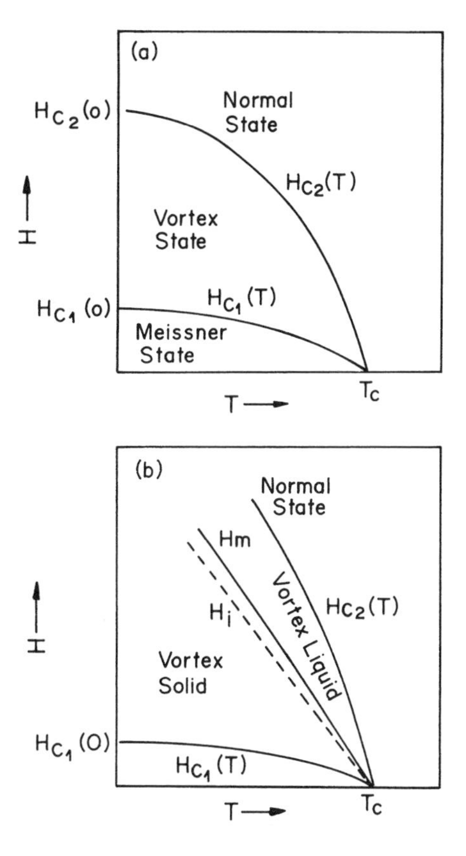

FIGURE 1.5 *H*–T phase diagram of (a) low-T_c type II superconductors and (b) high-T_c superconductor.

respectively (36), whereas the value of $H_{c2}(0)$ is estimated to be 670 T and 120 T in the *a-b* plane and along the *c* axis, respectively (37).

Cuprate high-T_c superconductors display a complex *H–T* phase diagram (Fig. 1.5b) due to their high-T_c, short coherence length, layered structure and anisotropy (38). Apart from H_{c1} and H_{c2}, there are irreversibility (H_i) and melting (H_m) lines. The melting line separates a vortex lattice and a vortex liquid state. The irreversibility line occurs in the vicinity of the melting line. This line provides a boundary between the reversible and irreversible magnetic behavior of a superconductor.

The structure of the vortex line in high-T_c superconductors is different from the conventional type II superconductors. The individual two-dimensional "pancake" vortices on neighboring layers couple to form three-dimensional vortex lines. A weakness of attractive interaction between the "pancakes" from different layers results in a strong reduction of the shear modulus of the vortex lattice along the layers as well as a strong influence from thermal fluctuations. The phase diagram of such flexible vortices in the presence of thermal fluctuations and pinning is a topic of intense study. A better understanding of the dynamics of the vortices will help to increase the transport critical current density in the material and also to control the flux noise for electronic applications.

1.5.3 Penetration Depth

Below H_{c1}, the external magnetic field is excluded from the bulk of a superconducting material by a persistent supercurrent in the surface region, which induces a field that exactly matches the applied field. The depth of this surface is called the penetration depth (λ) and the external field penetrates the superconductor in an exponentially decreasing manner. To be more precise, penetration depth is the distance over which an applied magnetic field decays to $1/e$ of its value at the surface. For an isotropic superconductor, the lower critical field (H_{c1}) is related to the penetration depth by

$$H_{c1} \approx \frac{\Phi_0}{\lambda^2} \tag{1}$$

where Φ_0 is the flux quantum. The value of the penetration depth can be obtained from magnetization of the thin superconducting crystal, muon spin rotation, kinetic inductance, or microwave measurements. Anisotropy in λ can be estimated from flux decoration and magnetic torque experiments. For high-T_c superconductors, the penetration depth along the c axis is different than that along the a-b plane. For the YBCO single crystal, the value of $\lambda ab(T \to 0)$ is obtained as 1400 Å (39). There has been much interest in studying the temperature dependence of λ because it is expected to provide information about the symmetry of the order parameter of high-T_c superconductors. The two-fluid model describes the temperature dependence as

$$\lambda(\mathrm{T}) = \lambda(0)\left[1 - \left(\frac{T}{T_c}\right)^4\right]^{-1/2} \tag{2}$$

For a weak coupling BCS superconductor, the $\Delta\lambda$ [$=\lambda(\mathrm{T}) - \lambda(0)$] varies exponentially with temperature (40). If there are nodes in the energy gap, then $\lambda(\mathrm{T})$ varies linearly with T (41). The results of the temperature dependence of λ for high-T_c superconductors are discussed in Section 1.5.8.

1.5.4 Coherence Length

One of the important parameters determining the performance of a superconductor is the coherence length (ξ). It is a measure of the correlation distance of the superconducting charge carriers. Coherence length represents the size of the Cooper pair. In terms of the Fermi velocity (v_F) and transition temperature T_c, coherence length is given as

$$\xi = \frac{hv_F}{2\pi^2 k_B T_c} \tag{3}$$

where k_B is the Boltzman constant and h is Planck's constant. The higher value of T_c in copper oxide superconductors is expected to lead to a low value for the coherence length. Direct measurement of the coherence length is difficult. However, the value of the coherence length can be extracted from fluctuation contributions to the specific heat, susceptibility, or conductivity. The value of the coherence length can also be obtained via measurement of H_{c2} using

$$H_{c2} \approx \frac{\Phi_0}{2\pi\xi^2} \tag{4}$$

The value of the coherence length is found to be highly anisotropic for high-T_c materials. The coherence length parallel to the c axis is typically 2–5 Å, and in the a-b plane, the value is typically 10–30 Å. Thus, perpendicular to the a-b plane, the superconducting wave function is essentially confined to the few adjacent unit cells. In conventional low-T_c, type I superconductors, the coherence length is 1000 Å, which is several orders of magnitude larger than that in high-T_c superconductors. The low value of the coherence length in high-T_c superconductors means that the coherence volume contains only a few Cooper pairs, implying that the fluctuations may be much larger in the high-T_c superconductors than in the conventional superconductors. The low values of the coherence length make these materials very sensitive to the presence of local defects such as oxygen vacancies, dislocations, and deviation from the stoichiometry.

1.5.5 Flux Quantization

In the classical low-T_c superconductors, magnetic flux (Φ) trapped in a closed superconducting ring is always an integral multiple of a flux quantum, Φ_0:

$$\Phi = n\Phi_0 \tag{5}$$

where n is an integer, $\Phi_0 = h/2e = 2 \times 10^{-7}$ G/cm^2, h is Planck's constant, and e is the electronic charge; the factor 2 in the denominator shows that the superconducting ground state is composed of paired electrons.

Soon after the discovery of light-T_c superconductors, various experiments were performed to find out if the superconducting state in high-T_c superconductors consisted of paired electrons or of something else. One way to find out is by measuring a trapped flux in the high-T_c superconducting ring. Gough and coworkers (42) performed an experiment to measure the flux through a sintered

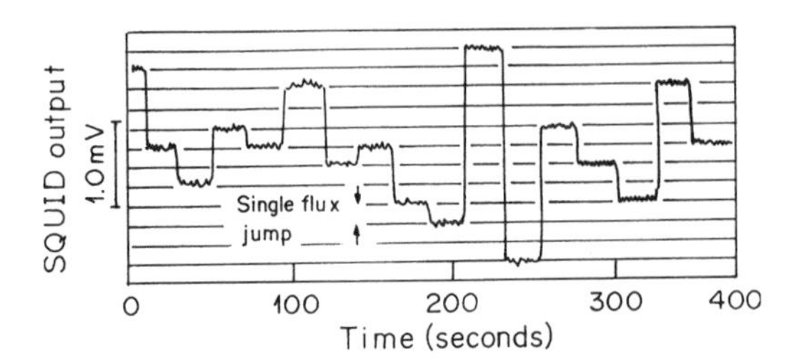

FIGURE 1.6 Flux jumps as a function of time when the YBCO ring at 4.2 K was exposed to a local source of electromagnetic noise, causing the ring to jump between quantized flux states. (Adapted from ref. 42.)

YBCO ring using a weakly coupled superconducting quantum interference device (SQUID) magnetometer. A small source of noise was applied to induce flux jumps. The quantized nature of the flux passing in and out of the ring was clearly observed in the experiment, as shown in Figure 1.6. The value of flux quantum was obtained as $\Phi_0 = (0.97 \pm 0.04)h/2e$. In another experiment, while studying flux line arrangement in YBCO single crystals through the magnetic decoration technique, Gammel et al. (43) found the value of flux quantum as $h/2e$, which is similar to that obtained by Gough et al. (42). These observations of flux quantization in high-T_c superconductor clearly indicate that the paired electrons are responsible for the superconducting state.

1.5.6 Critical Current Density and Weak Links

Similar to the transition temperature (T_c) and the critical magnetic field (H_c), the critical current density (J_c) is another important parameter which determines the boundary between superconducting and normal states. Following the discovery of high-T_c superconductors, it was found that the critical current density in bulk high-T_c materials is remarkably small ($\sim$10–100 A/cm^2) at 77 K and it is strongly dependent on the preparation condition. Very soon, it was realized that natural grain boundaries behaving as Josephson weak links (44–46) are responsible for the low value of J_c in high-T_c superconductors. Weak links are essentially localized regions in the superconductor where various superconducting properties are degraded. The role of weak links in J_c is quite different in high-T_c superconductors as compared to low-T_c superconductors. In conventional low-T_c superconductors, the defect such as grain boundaries increases the pinning and, therefore, enhances the J_c. On the other hand, the grain-boundary weak links in polycrystalline high-T_c superconductors limit the critical current density.

In epitaxial films, where grain boundaries are completely absent, the J_c value is found to be as large as 10^6 A/cm^2 at 77 K (47). The weak-link nature of the grain boundaries was more clearly established by growing YBCO high-T_c epitaxial films on a $SrTiO_3$ bicrystal substrate (48). The bicrystal substrates are fabricated by fusing two single-crystal substrates. When high-T_c film is grown epitaxially on the bicrystal substrate, a single grain boundary is realized. It has been demonstrated that grain boundaries are, indeed, weak links, which are Josephson coupled. It has also been found that the critical current across the grain boundary is a function of misorientation angle between the two crystal (48). A 45° misorientation angle formed by rotation about the *c* axis can reduce the critical current density by four orders of magnitude from that of the best films. The understanding of the weak-link nature of grain boundary in high-T_c superconductors has necessitated the development of single-crystal film technology for electronics application. Artificial weak links can be created in an epitaxial high-T_c film using bicrystal substrate, edge junction, and so forth. The details of the fabrication and properties of these artificially prepared weak links in high-T_c superconductors will be dealt in more detail in other chapters of this volume.

The weak-link nature of the grain boundaries in high-T_c superconductors makes processing of the wires more complicated. The grains have to be aligned such that the *c* axis is parallel and the spread of nearest-neighbor orientation is preferably less than 10° to obtain acceptably high current densities in finite magnetic fields. This has been accomplished by following special techniques in wire preparation (49). For bulk applications where different topological shapes such as rodes, sheets, blocks, and cylinders are required, the melt processing technique (50) is used, which minimizes the effect of the weak link and results in samples with high current densities.

1.5.7 Energy Gap

One of the important features of superconductivity is the existence of a gap in low-energy excitation. In a superconductor, the external energy ($E \geq 2\Delta$, where Δ is the energy gap) has to be supplied for creating an electron–hole pair close to the Fermi surface. For a weakly coupled BCS superconductor, the energy gap Δ at 0 K is related to T_c by

$$\frac{2\Delta(0)}{k_B T_c} = 3.52 \tag{6}$$

where k_B is Boltzman constant.

Tunneling spectroscopy is a widely used technique to study the superconducting gap. Apart from this, there are many other measurements such as infrared spectroscopy, photoelectron spectroscopy, inelastic light scattering, nuclear magnetic resonance (NMR), nuclear quadrople resonance (NQR), and so forth, which also provide information on the magnitude and the temperature dependence of energy gap. Several approaches have been applied to perform tunneling measurements in high-T_c superconductors, such as point-contact tunneling, break junction tunneling, and planar junction tunneling.

TABLE 1.2 Values of Energy Gap-to-Transition Temperature Ratio ($2\Delta/k_BT_c$) for Some High-T_c Superconductors

High-T_c superconductors	T_c (K)	Δ (mV)	$2\Delta/k_BT_c$	Ref.
$YBa_2Cu_3O_{7-\delta}$	85	20	6	51
$Bi_2Sr_2CaCu_2O_{8+\delta}$	62	20	7.5	53
$Bi_{1,7}Pb_{0,3}Sr_2CaCu_2O_x$	96	26	6.3	54
$Tl_2Ba_2CuO_6$	86	25	6.7	55
$Tl_2Ba_2Ca_2Cu_3O_x$	114	30	6.1	56
$HgBa_2Ca_2Cu_3O_{8+\delta}$	132	48	8.5	57
$(Nd_{2-x}Ce_x)CuO_{4-y}$	22	3.7	3.9	58
$Ba_{1-x}K_xBiO_3$	24.5	4.6	3.9	58

For high-T_c superconductors, a higher value of the energy gap-to-T_c ratio is observed as compared to the weakly coupled BCS superconductor. Anisotropy in the gap value along the *c* axis and in the *a-b* plane is also noticed. For YBCO, the energy gap-to-T_c ratio [$2\Delta(0)/k_BT_c$] has been found to be 3.5 for tunneling perpendicular to the Cu–O plane and a value of ~6 has been found for tunneling in the Cu–O plane (51). A similarly higher value of the energy gap-to-T_c ratio is also observed for tunneling in the Cu–O plane in Bi–Sr–Ca–Cu–O (52–54), Tl–Ba–Ca–Cu–O (55,56), and Hg–Ba–Ca–Cu–O (57) superconductors. However, Nd–Ce–Cu–O and Ba–K–Bi–O showed smaller values of this ratio (~3.9) (58). Table 1.2 shows values of the energy gap-to-T_c ratio [$2\Delta(0)/k_BT_c$] for some of high-T_c superconductors. Angle-resolved photoelectron spectroscopy has been used to investigate the energy gap in different *k* directions in Bi-2212, and an anisotropy of the gap value in the *a-b* plane was noted (59), which indicates the possibility of the existence of nodes in the energy gap. Low-temperature scanning tunneling microscopy in Hg-1201 showed the presence of different gaps to different crystallographic faces, implying a non-BCS electron–electron pairing mechanism (60).

Nuclear magnetic resonance and photoemission measurements in underdoped high-T_c cuprates indicated the presence of a gap in the spin excitation spectrum. This pseudogap opening occurs below to a characteristic temperature T^*, well above the T_c. This spin gap is not found in optimally doped material. The existence of the spin gap in underdoped samples is found to be a fundamental feature of high-T_c superconductors, and two-dimensional charge dynamics, reduced Drude spectral weight results from the spin gap (61). The existence of the pseudogap also implies that there must be some developing electronic order. However, the real importance of the existence of the pseudogap and its relation with superconducting gap has not yet properly understood.

1.5.8 Symmetry of the Order Parameter

Flux quantization measurements (42) and observation of the ac–Josephson effect (44,45,62) in high-T_c materials have established that the pairing is formed in the condensed state of superconductors. However, the nature of the pairing still remains to be understood. An important step toward understanding the coupling mechanism is to know the symmetry of superconducting order parameter.

The order parameter of a superconductor is described by the wave function of a Cooper pair. It is given as

$$\psi(r_1 - r_2) = \sum_k g(k) \exp[ik(r_1 - r_2)] \tag{7}$$

where r_1 and r_2 are position coordinates of the two electrons and $g(k)$ is the Fourier transform of the pair amplitude and it is proportional to the energy gap Δ_k.

Figure 1.7 shows the variation of the energy gap function corresponding to isotropic s, anisotropic s, d_{x2-y2}, and extended s-wave symmetry of the order parameter in the momentum space. The thick line represents the Fermi surface and the thin line shows the variation of gap function. The distance from the Fermi surface gives the amplitude: a positive value for a line lying outside the Fermi surface and a negative value inside the Fermi surface. The zero crossing points are called nodes. The gap surfaces are represented by the dashed line in Figure 1.7.

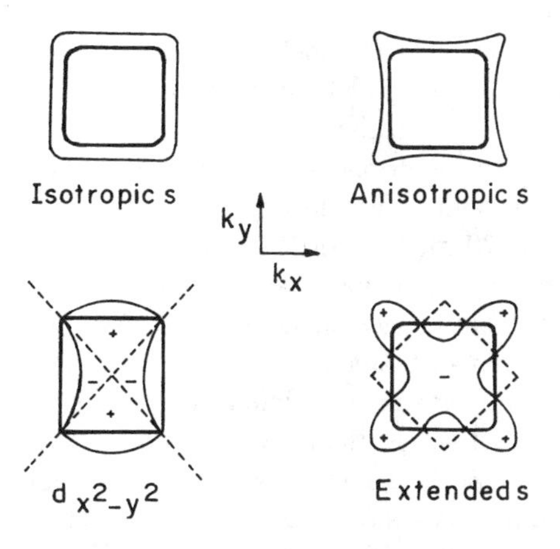

FIGURE 1.7 Variation of gap function in the momentum space for isotropic s, anisotropic s, d_{x2-y2}, and extended s-wave symmetry of the order parameter. The Fermi surface is represented by thick lines and the variation of the gap function is shown by thin lines. The gap nodes surfaces are represented by the dashed line.

For isotropic *s*-wave symmetry, Δ_k is constant along all directions of *k*. In the case of anisotropic *s*-wave symmetry, Δ_k does not remain constant. For d_{x2-y2} symmetry, the gap function Δ_k varies as $k_{x2} - k_{y2}$ which passes through zero along the $|k_x| = |k_y|$ directions. The corresponding real-space-pair wave function of the *d* wave has a $x^2 - y^2$ spatial symmetry with nodes and sign change upon rotation of 90°. This symmetry function can also be written $r^2 \cos 2\theta$. In the case of extended *s*-wave symmetry, the variation of gap function is expressed as $\Delta_k \approx [\cos(k_x a) + \cos(k_y a)]$. It is evident from Figure 1.7 that for extended *s*-wave symmetry, the variation of gap function over a 2π rotation in the momentum space exhibits eight nodes.

The order parameter has a magnitude and a phase. The magnitude of the order parameter for isotropic *s*-wave symmetry remains constant, whereas for the other three symmetries, it is different along different directions of *k*. The sign of the phase of the order parameter for isotropic and anisotropic *s* waves remains same, whereas for d_{x2-y2} symmetry and an extended *s* wave, the sign of the phase changes. In the case of d_{x2-y2}, the sign changes four times, whereas for the extended *s* wave, the sign changes eight times over a rotation of 2π in the momentum space.

Knowledge about the underlying symmetry is important in finding out which class of theory describes these materials. The BCS theory, which assumes phonon-mediated coupling, favors an order parameter with isotropic *s*-wave symmetry. Another class of theory assumes that exchange of antiferromagnetic spin fluctuation can provide a pairing mechanism leading to a pairing state with d_{x2-y2} symmetry (63,64). Chakravarty et al. (65) proposed a theoretical model which assumes electron–phonon interaction as a dominant mechanism and the interlayer tunneling leading to an anisotropic *s*-wave symmetry. Various other theoretical models argue for generalized *s*-wave symmetry (66) or for order-parameter symmetry to be a complex mixture of *s* and *d* waves (67) or a complex mixture of d_{x2-y2} and d_{xy} (68).

Several experimental techniques (69,70) such as NMR studies, temperature dependence of penetration depth, angle-resolved photoemission, Josephson junction, and SQUID studies have been carried out to understand the symmetry of the order parameter in high-T_c superconductors. NMR studies in YBCO which measure the relaxation rates and Knight shift supports *d*-wave symmetry (64). A paramagnetic effect (71) was observed in some Bi-2212 field-cooled samples which was interpreted in terms of π-junction and *d*-wave symmetry. For *s*-wave symmetry, the penetration depth is expected to vary exponentially with temperature, whereas for *d*-wave symmetry, λ varies linearly with temperature. Earlier measurements of temperature dependence of λ supported the *s* wave (39,72), whereas several others found that $\lambda(T)$ is proportional to T (41) or T^2 (73). An angle-resolved photoemission experiment of a Bi-2212 single crystal indicated anisotropy in superconducting gap in the *a-b* plane (59). The temperature dependence of the penetration depth and angular-resolved photoemission studies showed anisotropy in the energy gap. However, the existence of nodes could not be established because the magnitude of the order parameter is measured in these experiments. In order to find the existence of nodes, several phase-sensitive experiments based on the Josephson junction and SQUID have been performed (55,74–88). Experiments based on a specially designed tricrystal geometry have shown the existence of nodes by demonstrating the observation of spontaneously generated $\phi_0/2$ flux (81–83). The results of a majority of these experiments supported *d*-wave symmetry; however, an unanimous view about the symmetry of the order parameter has not yet been accepted.

The symmetry of the order parameter may have implications on the practical applications of high-T_c superconductors. For example, the *d*-wave model predicts a lower limit to the surface resistance and such a lower limit may constrain those applications seeking to maximize the Q of superconducting microwave circuits. Similarly, the presence of nodes may constrain the design of Josephson junction devices.

1.6 CONCLUSION

During the last one and half decades after the discovery of the high-T_c superconductor, more than 100 high-T_c compounds have been made which exhibit superconductivity above 23 K. Several of theses have a T_c higher than the liquid-nitrogen temperature. The highest T_c of 133 K is observed in $HgBa_2Ca_2Cu_3O_y$ at ambient pressure. The superconductivity in high-T_c cuprates is due to the presence of CuO_2 planes and the T_c of the material is found to depend on the number of CuO_2 planes in the unit cell. High-T_c superconductors are, in several ways, different from low-T_c superconductors such as short coherence length, large anisotropy, grain-boundary weak links, and layered structure. A better understanding of the grain boundary has enabled one to improve the quality of high-T_c films and superconducting wires and led to the development of artificial grain boundaries for electronic applications. High-T_c superconductors are extreme type II superconductors. More understanding of the flux dynamics is required, as these material exhibit a complex H–T phase diagram.

Similar to low-T_c superconductors, superconductivity in high-T_c materials is due to pairing of electrons, but the mechanism of pairing is still not clear. The energy gap-to-T_c ratio in several high-T_c superconductors show a value larger than the value predicted for weakly coupled BCS superconductors. Anisotropy in the gap in the *a-b* plane is also noted. Several studies on the measurement of penetration depth, NMR, angle-resolved photoemission, Josephson junction, and SQUID have been performed to explore the symmetry of the order parameter of high-T_c superconductors. The majority of these studies revealed *d*-wave symmetry of the order parameter. The observation of a spin gap in under doped high-T_c compounds at temperature much above the T_c is very interesting; however, its relationship with the superconducting energy gap needs to be understood.

REFERENCES

1. JG Bednorz, KA Muller. Possible high-T_c superconductivity in Ba–La–Cu–O system. Z Phys B: Condensed Matter 64:189–193, 1986.
2. HK Onnes. Akad van Wetenschappen (Amsterdam) 14:818, 1911.
3. JR Gavaler. Superconductivity in Nb–Ge films above 22K. Appl Phys Lett 23:480–482, 1973.
4. CW Chu, PH Hor, RL Meng, L Gao, ZJ Huang. Superconductivity at 52.5 K in the lanthanum–barium–copper–oxide system. Science 235:567–569, 1987.
5. MK Wu, JR Ashburn, CJ Torng, PH Hor, RL Meng, L Gao, ZJ Huang, YQ Wang, CW Chu. Superconductivity at 93 K in a new mixed-phase Y–Ba–Cu–O compound system at ambient pressure. Phys Rev Lett 58:908–910, 1987.
6. RJ Cava, B Batlogg, RB VanDover, DW Murphy, S Sunshine, T Siegrist, JP Remeika, EA Rietman, S Zahurak, GP Espinosa. Bulk superconductivity at 91 K in single-phase oxygen-deficient perovskite $Ba_2YCu_3O_{9-\delta}$. Phys Rev Lett 58:1676–1679, 1987.
7. CNR Rao, P Ganguly, AK Raychaudhuri, RAM Ram, K Sreedhar. Identification of the phase responsible for high-temperature superconductivity in Y–Ba–Cu oxides. Nature 326:856–857, 1987.
8. RM Hazen, LW Finger, RJ Angel, CJ Prewitt, NL Ross, HK Mao, CG Hadidiacos, PH Hor, RL Meng, CW Chu. Crystallographic description of phases in the Y–Ba–Cu–O superconductor. Phys Rev B 35:7238–7241, 1987.
9. H Maeda, Y Tanaka, M Fukutomi, T Asano. A new high-T_c oxide superconductor without a rare earth element. Jpn J Appl Phys 27:L209–L210, 1988.
10. RM Hazen, CT Prewitt, RJ Angel, NL Ross, LW Finger, CG Hadidiacos, DR Veblen, PJ Henaey, PH Hor, RL Meng, L Gao, J Bechtold, CW Chu. Superconductivity in the high-T_c Bi–Ca–Sr–Cu–O system: Phase identification. Phys Rev Lett 60:1174–1177, 1988.
11. JM Tarascon, WR McKinnon, P Barboux, DM Hwang, BG Bagley, LH Greene, GW Hull, Y LePage, N Stoffel, M Giroud. Preparation structure and properties of the superconducting compound series $Bi_2Sr_2Ca_{n-1}Cu_nO_y$ with $n=1$, 2 and 3. Phys Rev B 38:8885–8892, 1988.
12. ZZ Sheng, AM Hermann. Superconductivity in the rare-earth free Tl–Ba–Cu–O system above liquid nitrogen temperature. Nature 332:55–58, 1988.
13. ZZ Sheng, AM Hermann, A EI Ali, C Almasan, J Estrada, T Datta, RJ Matson. Superconductivity at 90 K in the Tl–Ba–Cu–O system. Phys Rev Lett 60:937–940, 1988.
14. SN Putillin, EV Antipov, O Chmaissem, M Marezio. Superconductivity at 94 K in $HgBa_2CuO_{4+\delta}$. Nature 362:226–228, 1993.
15. A Schilling, M Cantoni, JD Guo, HR Ott. Superconductivity above 130 K in the Hg–Ba–Ca–Cu–O system. Nature 363:56–58, 1993.
16. CW Chu, L Gao, F Chen, ZJ Huang, RL Meng, YY Xue. Superconductivity above 150 K in $HgBa_2Ca_2Cu_3O_{8+\delta}$ at high pressure. Nature 365:323–325, 1993.
17. Y Tokura, H Takagi, S Uchida. A superconducting copper oxide compound with electrons as the charge carriers. Nature 337:345–347, 1989.

18. RJ Cava, B Batlogg, JJ Krajewski, R Farrow, LW Rupp, AE White Jr, K Short, T Kometani. Superconductivity near 30 K without copper $Ba_{0.6}K_{0.4}BiO_3$ Perovskite. Nature 332:814–816, 1988.
19. YS Yao, YF Xiong, D Jin, JW Li, JL Luo, ZX Zhao. The high-pressure synthesis and superconductivity of $Pr_{0.5}Ca_{0.5}Ba_2Cu_3O_z$. Physica C 282–287:49–52, 1997.
20. SSP Parkin, VY Lee, AI Nazzal, R Savoy, R Beyers, SJ LaPlaca. $Tl_1Ca_{n-1}Ba_2Cu_nO_{2n+3}$ (n=1,2,3): A new class of crystal structures exhibiting volume superconductivity at up to $\simeq$ 110 K. Phys Rev Lett 61:750–753, 1988.
21. SN Putillin, EV Antipov, M Marezio. Superconductivity above 120 K in $HgBa_2CaCu_2O_{6+\delta}$. Physica C 212:266–270, 1993.
22. D Shi, MS Boley, JG Chen, M Xu, K Vandervoort, YX Liao, A Zangvil, J Akujieze, C Segre. Origin of enhanced growth of the 110 K superconducting phase by Pb doping in the Bi–Sr–Ca–Cu–O system. Appl Phys Lett 55:699–701, 1989.
23. J Tabuchi, K Utsumi. Preparation of superconducting Y–Ba–Cu–O thick films with preferred c-axis orientation by a screen-printing method. Appl Phys Lett 53:606–608, 1988.
24. A Bailey, G Alvarez, GJ Russell, KNR Taylor. High current capacity textured thick films of YBCO on YSZ obtained by melt processing. Cryogenics 30:599–602, 1990.
25. N Khare, S Chaudhry, AK Gupta, VS Tomar, VN Ojha. Superconducting thick films of $Bi_2Sr_2CaCu_2O_{8+x}$ with zero resistivity at 96 K. Supercond Sci Technol 3:514–516, 1990.
26. S Chaudhry, N Khare, AK Gupta. Growth kinetics of high T_c phase in Bi–Sr–Ca–Cu–O thick films. J Mater Res 7:2027–2033, 1992.
27. V Zeng, Z Zhao, Li Zhang, H Chen, Z Qian, D Wu. Preparation of the T1–Ba–Ca–Cu–O thick film by processing a screen-printed Ba–Ca–Cu–O film in Tl_2O_3 vapor. Appl Phys Lett 56:1573–1575, 1990.
28. CE Rice, RB van Dover, GJ Fisanick. Preparation of superconducting thin films of $Ba_2YCu_3O_7$ by a novel spin-on pyrolysis technique. Appl Phys Lett 51:1842–1844, 1987.
29. A Gupta, G Koren, EA Giess, NR Moore, EJM O'Sullivan, EI Cooper. $Y_1Ba_2Cu_3O_{7+\delta}$ thin films grown by a simple spray deposition technique. Appl Phys Lett 52:163–165, 1988.
30. DK Walia, AK Gupta, GSN Reddy, ND Kataria, N Khare, VN Ojha, VS Tomar. Growth of oriented superconducting films of Bi–Sr–Ca–CuO by spray pyrolysis technique. Mod Phys Lett B 4:393–397, 1990.
31. JA DeLuca, MF Garbauskas, RB Bolon, JG McMullen, WE Balz, PL Karas. The synthesis of superconducting Tl–Ca–Ba–Cu-oxide films by the reaction of spray deposited Ca–Ba–Cu-oxide precursors with Tl_2O vapor in a two-zone reactor. J Mater Res 6:1415–1423, 1991.
32. N Khare, AK Gupta, AK Saxena, KK Verma, ON Srivastava. Studies on the RF SQUID effect in Tl–Ca–Ba–Cu–O thin films prepared by the spray pyrolysis technique. Supercond Sci Technol 7:402–406, 1994.
33. N Khare, AK Gupta, HK Singh, ON Srivastava. Rf-SQUID effect in Hg(Tl)–Ba–Ca–Cu–O high-T_c thin film up to 121 K. Supercond Sci Technol 11:517– 519, 1998.

34. VG Kogan. Uniaxial superconducting particle in intermediate magnetic fields. Phys Rev B 38:7049–7050, 1988.
35. DE Farrell, RG Beck, MF Booth, CJ Allen, ED Bukowski, DM Ginsberg. Superconducting effective-mass anisotropy in $Tl_2Ba_2CaCu_2O_x$. Phys Rev B 42:6758–6761, 1990.
36. DH Wu, S Sridhar. Pinning forces and lower critical fields in $YBa_2Cu_3O_y$ crystals: Temperature dependence and anisotropy. Phys Rev Lett 65:2074–2077, 1990.
37. U Welp, M Grimsditch, H You, WK Kwok, MM Fang, GW Crabtree, JZ Liu. The upper critical field of untwinned $YBa_2Cu_3O_{7-\delta}$ crystals. Physica C 161:1–5, 1989.
38. G Blatter. Vortex matter. Physica C 282–287:19–26, 1997.
39. L Krusin-Elbaum, RL Greene, F Holtzberg, AP Malozemoff, Y Yeshurun. Direct measurement of the temperature-dependent magnetic penetration depth in Y–Ba–Cu–O crystals. Phys Rev Lett 62:217–220, 1989.
40. B Muhlschlegel. Die Thermodynamischen Funktionen des Supraleiters. Z Phys 155: 313–327, 1959.
41. WN Hardy, DA Bonn, DC Morgan, R Liang, K Zhang. Precision measurements of the temperature dependence of λ in $YBa_2Cu_3O_{6.95}$: Strong evidence for nodes in the gap function. Phys Rev Lett 70:3999–4002, 1993.
42. CE Gough, MS Colclough, EM Forgan, RG Jordan, M Keene, CM Muirhead, AI M Rae, N Thomas, J S Abell, S Sutton. Flux quantization in a high-T_c superconductor. Nature 362:855–857, 1987.
43. PL Gammel, DJ Bishop, GJ Dolan, JR Kwo, CA Murray, LF Scheemeyer, JV Waszczak. Observation of hexagonally correlated flux quanta in $YBa_2Cu_3O_7$. Phys Rev Lett 59:2592–2595, 1987.
44. JT Chen, LE Wenger, CJ McEwan, EM Logothetis. Observation of the reverse ac Josephson effect in Y–Ba–Cu–O at 240 K. Phys Rev Lett 58:1972–1975, 1987.
45. AK Gupta, SK Agarwal, B Jayaram, A Gupta, AV Narlikar. Observation of inverse A.C. Josephson effect in bulk Y–Ba–Cu–O possessing high T_c. Pramana-J Phys 28:L705–L707, 1987.
46. P Chaudhari, J Mannhart, D Dimos, CC Tsuei, J Chi, MM Oprysko, M Scheuermann. Direct measurement of the superconducting properties of single grain boundaries in $Y_1Ba_2Cu_3O_{7-\delta}$. Phys Rev Lett 60:1653–1656, 1988.
47. P Chaudhari, RH Koch, RB Laibowitz, TR McGuire, RJ Gambino. Critical-current measurements in epitaxial films of $YBa_2Cu_3O_{7-x}$ compound. Phys Rev Lett 58:2684– 2686, 1987.
48. D Dimos, P Chaudhari, J Mannhart. Superconducting transport properties of grain boundaries in $YBa_2Cu_3O_7$ bicrystals. Phys Rev B 41:4038–4049, 1990.
49. A Otto, LJ Masur, J Gannon, E Podtburg, D Daly, GJ Yurek, AP Malozemoff. Multifilamentary Bi-2223 composite tapes made by a metallic precursor route. IEEE Trans Appl Supercond 3:915–922, 1993.
50. M Murakami, M Morita, N Koyama. Magnetization of a $YBa_2Cu_3O_7$ crystal prepared by the quench and melt growth process. Jpn J Appl Phys 28:L1125–L1127, 1989.
51. JS Tsai, I Takeuchi, J Fujita, S Miura, T Terashima, Y Bando, K Iijima, K Yamamoto. Tunneling study of clean and oriented Y–Ba–Cu–O and Bi–Sr–Ca–Cu–O surfaces. Physica C 157:537–550, 1989.

52. M Boekholt, M Hoffmann, G Guntherodt. Detection of an anisotropy of the superconducting gap in $Bi_2Sr_2CaCu_2O_{8+\delta}$ single crystals by Raman and tunneling spectroscopy. Physica C 175:127–134, 1991.
53. L Ozyuzer, JF Zasadzinski, C Kendziora, KE Gray. Quasiparticle and Josephson tunneling of overdoped $Bi_2Sr_2CaCu_2O_{8+\delta}$ single crystals. Phys Rev B 61:3629–3640, 2000.
54. Q Huang, JF Zasadzinski, KE Gray, JZ Liu, H Claus. Electron tunneling study of the normal and superconducting states of $Bi_{1.7}Pb_{0.3}$ $Sr_2CaCu_2O_x$. Phys Rev B 40:9366–9369, 1989.
55. L Ozyuzer, Z Yusof, JF Zasadzinski, L Wei-Ting, DG Hinks, KE Gray. Tunneling spectroscopy of $Tl_2Ba_2CuO_6$. Physica C 320:9–20, 1999.
56. I Takechi, JS Tsai, Y Shimakawa, T Manako, Y Kubo. Energy gap of Tl–Ba–Ca–Cu–O compounds by tunneling. Physica C 158:83–87, 1989.
57. GT Jeong, JI Kye, SH Chun, S Lee, SI Lee, ZG Khim. Energy gap of the high-T_c superconductor $HgBa_2Ca_2Cu_3O_{8+\delta}$ determined by point-contact spectroscopy. Phys Rev B 49:15,416–15,419, 1994.
58. Q Huang, JF Zasadzinski, N Tralshawala, KE Gray, DG Hinks, JL Peng, RL Greene. Tunnelling evidence for predominantly electron-phonon coupling in superconducting Ba_{1-x} K_x BiO_3 and $Nd_{2-x}Ce_xCuO_{4-y}$. Nature 347:369–372, 1990.
59. Z-X Shen, DS Dessau, BO Wells, DM King, WE Spicer, AJ Arko, D Marshall, LW Lombardo, A Kapitulinik, P Dickinson, S Doniach, J DiCarlo, AG Loeser, CH Park. Anomalously large gap anisotropy in the a-b plane of $Bi_2Sr_2CaCu_2O_{8+\delta}$. Phys Rev Lett 70:1553–1556, 1993.
60. P Jess, U Hubler, HP Lang, HJ Guntherodt, K Luders, EV Antipov. Energy gap distribution of $HgBa_2CuO_{4+x}$ investigated by scanning tunneling microscopy/spectroscopy. J Low Temp Phys 105:1243–1248, 1996.
61. S Uchida. Spin gap effects on the c-axis and in-plane charge dynamics of high-T_c cuprates. Physica C 282–287:12–18, 1997.
62. J Niemeyer, MR Dietrich, C Politis. AC-Josephson effect in $YBa_2Cu_3O_8$/PbSn point contact junctions. Z Phys B: Condensed Matter 67:155–159, 1987.
63. P Monthoux, AV Balatsky, D Pines. Weak-coupling theory of high-temperature superconductivity in the antiferromagnetically correleated copper oxides. Phys Rev B 46:14,803–14,817, 1992.
64. N Bulut, DJ Scalapino. Analysis of NMR data in the superconducting state of $YBa_2Cu_3O_7$. Phys Rev Lett 68:706–709, 1992.
65. S Chakravarty, A Sudbo, PW Anderson, S Strong. Interlayer tunneling and gap anisotropy in high-temperature superconductors. Science 261:337–340, 1993.
66. DJ Scalapino. The case for $d_x2\text{-}_y2$ pairing in the cuprate superconductors. Phys Rep 250:329, 1995.
67. G Kotliar. Resonating valence bonds and d-wave superconductivity. Phys Rev B 37: 3664–3666, 1988.
68. DS Rokhsar. Pairing in doped spin liquids: Anyon versus d-wave superconductivity. Phys Rev Lett 70:493–496, 1993.
69. N Khare. Symmetry of Order Parameter of High-T_c Superconductors. In: A Narlikar, ed. Studies of High Temperature Superconductor vol 20 New York: Nova Science Publishers, 1996, pp 187–215.

70. RC Dynes. The order parameter of high T_c superconductors; experimental probes. Solid State Commun 92:53–62, 1994.
71. W Braunisch, N Knauf, V Kataev, S Nehuhausen, A Grutz, A Kock, B Roden, D Khomskii, D Wohlleben. Paramagnetic Meissner effect in Bi high-temperature superconductors. Phys Rev Lett 68:1908–1911, 1992.
72. SM Anlage, BW Langley, G Deutscher, J Halbritter, MR Beasley. Measurements of the temperature dependence of the magnetic penetration depth in $YBa_2Cu_3O_{7-\delta}$ superconducting thin films. Phys Rev B 44:9764–9767, 1991.
73. MR Beasley. Recent penetration depth measurements of the high-T_c superconductors and their implications. Physica C 209:43–46, 1993.
74. AG Sun, DA Gajewski, MB Maple, RC Dynes. Observation of Josephson pair tunneling between a high-T_c cuprate ($YBa_2Cu_3O_{7-\delta}$) and a conventional superconductor (Pb). Phys Rev Lett 72:2267–2270, 1994.
75. P Chaudhari, SY Lin. Symmetry of the superconducting order parameter in $YBa_2Cu_3O_{7-\delta}$ epitaxial films. Phys Rev Lett 72:1084–1087, 1994.
76. DA Wollman, DJ Van Harlingen, J Giapintzakis, DM Ginsberg. Evidence for d_{x2-y2} pairing from the magnetic field modulation of $YBa_2Cu_3O_7$–Pb Josephson junctions. Phys Rev Lett 74:797–800, 1995.
77. DA Brawner, HR Ott. Evidence for an unconventional superconducting order parameter in $YBa_2Cu_3O_{6.9}$. Phys Rev B 50:6530–6533, 1994.
78. JH Miller, Jr, QY Ying, ZG Zou, NQ Fan, JH Xu, MF Davis, JC Wolfe. Use of tricrystal junctions to probe the pairing state symmetry of $YBa_2Cu_3O_{7-\delta}$. Phys Rev Lett 74: 2347–2350, 1995.
79. A Mathai, Y Gim, RC Black, A Amar, FC Wellstood. Experimental proof of a time-reversal-invariant order parameter with a π shift in $YBa_2Cu_3O_{7-\delta}$. Phys Rev Lett 74:4523–4526, 1995.
80. P Kleiner, AS Katz, AG Sun, R Summer, DA Gajewski, SH Han, SI Woods, E Dantsker, B Chen, K Char, MB Maple, RC Dynes, J Clarke. Tunneling from c-axis $YBa_2Cu_3O_{7-x}$ to Pb: Evidence for s-wave component from microwave induced steps. Phys Rev Lett 76:2161–2164, 1996.
81. CC Tsuei, JR Kirtley, CC Chi, LS Yu-Jahnes, A Gupta, T Shaw, JZ Sun, MB Ketchen. Pairing symmetry and flux quantization in a tricrystal superconducting ring of $YBa_2Cu_3O_{7-\delta}$. Phys Rev Lett 73:593–596, 1994.
82. JR Kirtley, CC Tsuei, JZ Sun, CC Chi, LS Yu-jahnes, A Gupta, M Rupp, MB Ketchen. Symmetry of the order parameter in the high-T_c superconductor $YBa_2Cu_3O_{7-\delta}$. Nature 373:225–228, 1995.
83. CC Tsuei, JR Kirtley, M Rupp, JZ Sun, A Gupta, MB Ketchen, CA Wang, ZF Ren, JH Wang, M Bhushan. Pairing symmetry in single-layer tetragonal $Tl_2Ba_2CuO_{6+\delta}$ superconductors. Science 271:329–332, 1996.
84. JR Kirtley, P Chaudhari, MB Ketchen, N Khare, SY Lin, T Shaw. Distribution of magnetic flux in high-T_c grain boundary junctions enclosing hexagonal and triangular areas. Phys Rev B 51:12,057–12,059, 1995.
85. ZG Ivanov, EA Stepantsov, T Claeson, F Wenger, SY Lin, N Khare, P Chaudhari. Highly anistoropic supercurrent transport in $YBa_2Cu_3O_{7-\delta}$ bicrystal Josephson junctions. Phys Rev B 57:602–607, 1998.

86. L Alff, S Kleefisch, S Meyer, U Schoop, A Marx, H Sato, M Naito, R Gross. Determination of the order parameter symmetry in hole and electron doped cuprate superconductors. Physica B 284–288:591–592, 2000.
87. Q Li, YN Tsay, M Suenaga, GD Gu, N Koshizuka. A direct probe of superconducting order parameter symmetry in $Bi_2Sr_2CaCu_2O_{8+\delta}$ using bicrystal c-axis twist Josephson Junctions. Physica C 341–348:1665–1660, 2000.
88. WK Neils, BLT Blouride, DJV Harlingen. Search for superconducting phases with broken time reversal symmetry in *d*-wave grain boundary junctions and interfaces. Physica C 341–348:1705–1706, 2000.

2

Epitaxial Growth of Superconducting Cuprate Thin Films

David P. Norton
University of Florida, Gainesville, Florida, U.S.A.

2.1 INTRODUCTION

In 1986, Bednorz and Müller reported a superconducting transition temperature greater than 30 K in a multicomponent oxide compound, namely $La_{2-x}Ba_{x}CuO_{4-\delta}$ (1). The discovery of other layered copper oxide materials with superconducting transition temperatures, T_c, exceeding the boiling point of liquid nitrogen (77 K) soon followed. Today, numerous high-temperature superconducting (HTS) cuprate phases have been uncovered with transition temperatures as high as 135 K. Many of these materials have been synthesized as epitaxial thin films. A fundamental understanding of both the superconducting properties, as well as the materials science of these complex oxide materials, is still emerging. Although much is known about the synthesis and properties of HTS films, there remain significant challenges in this area, particularly in producing thin-film materials suitable for HTS technologies. Potential applications involving HTS films include high-frequency electronics for radio-frequency (RF) microwave communications, superconducting quantum interference devices (SQUIDs) for the detection of minute magnetic fields, and superconducting wires for energy-efficient delivery and use of electrical energy. This chapter provides an overview of the science and technology of HTS thin-film synthesis, focusing on the growth of epitaxial films.

In order to address the materials-related issues most relevant for HTS cuprate thin films, one must first discuss the generic structure for these materials. The layered crystal structure inherent to the HTS compounds yields highly anisotropic materials in terms of both the electronic properties and crystal-growth characteristics. A comprehensive overview of the various multielement crystal structures for HTS cuprates has been given elsewhere (2). A unit cell that is conceptually applicable to all of the HTS cuprates can be constructed from two distinct chemical blocks, as illustrated in Figure 2.1. The first block consists of one or more CuO_2 planes. The common feature of all cuprate phases that exhibit high-temperature superconductivity is the presence of two-dimensional CuO_2 sheets within their layered structure. Each Cu atom in the CuO_2 layer is surrounded by four O atoms in a square-planar configuration. For structures with more than one CuO_2 sheet per unit cell, the individual sheets are separated by a layer of divalent alkaline earth or trivalent rare-earth atoms. The CuO_2 sheets defines the *a-b* planes in all of the HTS crystal structures with the *c* axis of the crystal structure perpendicular to the sheets. The second block in the generic unit cell is referred to as the charge reservoir and can be used to define specific homologous HTS families of compounds. Within the HTS structure, this block appears to be largely responsible for providing charge carriers to the CuO_2 planes. It also determines the degree of anisotropy in the individual HTS compounds, as *c*-axis transport is primarily determined by this layer. Within a homologous series, the specific phases are dis-

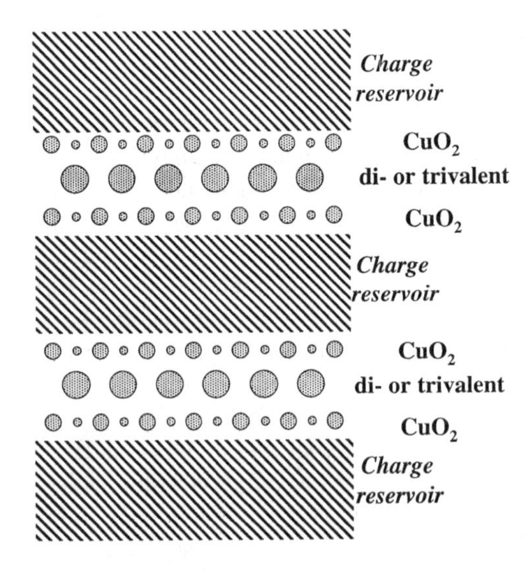

FIGURE 2.1 Generic structure of the superconducting cuprates showing the CuO_2 planes separated by the charge-reservoir blocks. The schematic illustrates the specific case of the $n=2$ structures.

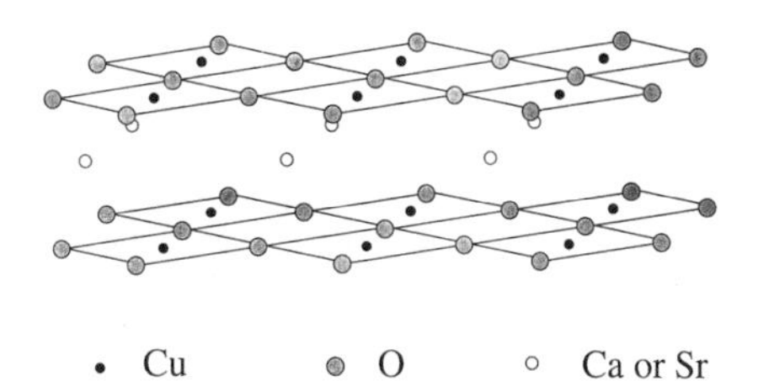

FIGURE 2.2 Schematic of the $(Ca,Sr)CuO_2$ crystal structure.

tinguished by the number, *n*, of CuO_2 planes per unit cell. For most of the HTS compounds, $n \leq 3$. The various HTS compounds can then be characterized by the number of CuO_2 planes contained in each unit cell and by the specific chemical block that separates these CuO_2 blocks and completes the structures.

The simplist HTS structure is the so-called "infinite-layer" $(Ca,Sr)CuO_2$ material. This compound, illustrated schematically in Figure 2.2, consists of four-fold coordinated CuO_2 sheets separated by alkaline earth atoms. It is distinct from the other HTS compounds in that it contains only CuO_2–alkaline earth blocks with no charge-reservoir layer. Hence, it is referred to as the "infinite-layer" ($n = \infty$) compound. As described, this structure is insulating. Carries are introduced by replacing some of the alkaline earth atoms with trivalent earth ions. In contrast, consider the $(La,Sr)_2CuO_4$ compound shown schematically in Figure 2.3. In this material, each CuO_2 plane is separated along the *c* axis by two (La,Sr)–O planes in a

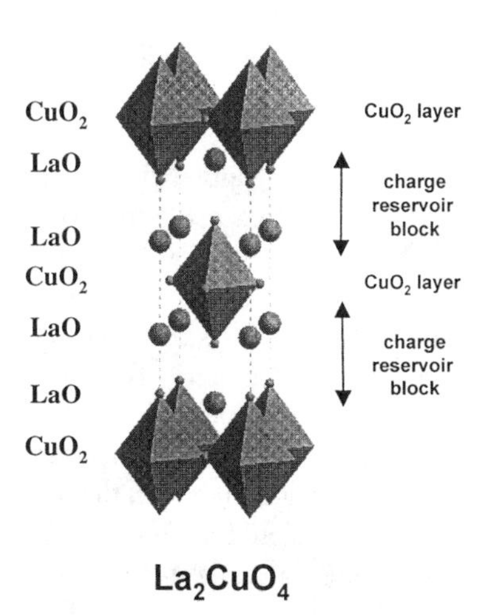

FIGURE 2.3 Schematic drawing illustrating the crystal structure for the La_2CuO_4 compounds.

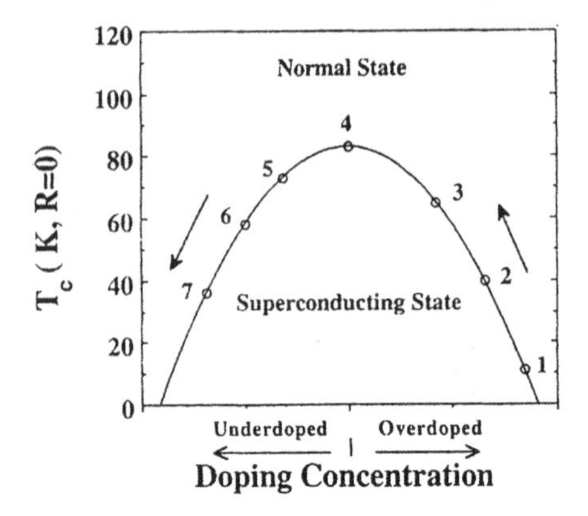

FIGURE 2.4 Variation of T_c with carrier density for the $Tl_2Ba_2CuO_6$ compounds. The carrier density is adjusted by varying the oxygen content. (From Ref. 3.)

rock salt structure. This particular compound is classified as a $n = 1$ structure, as there is only one CuO_2 plane in each unit cell. Other HTS structures include more complex charge-reservoir layers. For example, in the $T1Ba_2Ca_{n-1}Cu_nO_y$ homologous series, each unit cell contains a single T1–O layer sandwiched between two Ba–O layers. This comprises the charge-reservoir chemical block. The CuO_2 planes are adjacent to the Ba–O layers. For the $n = 2$ and 3 members of the series, the multiple CuO_2 planes are separated by Ca atoms. Other HTS compounds can be similarly constructed.

Carrier doping plays a critical role in determining the superconducting properties in all of the HTS cuprates. The charge carriers are holes (p-type) in most structures, with only two structure types supporting superconductivity with n-type doping. The hole-doped superconductors are characterized by either fivefold or sixfold coordinated bonding of the Cu atoms to oxygen. In this case, the additional coordination is provided by apical oxygen atoms above and/or below the CuO_2 planes. The electron-doped HTS compounds always contain only fourfold coordinated bonding of the Cu to oxygen atoms. Carrier concentration is controlled either by chemical substitution or changes in the oxygen stoichiometry. The transport properties of the cuprates can be varied from metallic and superconducting to insulating, with each compound possessing an optimum doping. For instance, La_2CuO_4 is an insulator that is driven metallic and superconducting with the partial substitution of a divalent alkaline earth (i.e., Sr) for trivalent La. Figure 2.4 illustrates the transition-temperature dependence on doping for the $n = 1$ T1 compound (3). In a similar manner, $YBa_2Cu_3O_{7-\delta}$ is a 90 K superconductor only when the oxygen content is near 7 ($\delta \sim 0$). As oxygen is removed, T_c decreases, with $YBa_2Cu_3O_6$ displaying semiconducting behavior.

The HTS cuprates possess other distinctive properties that contribute either to the difficulties or advantages associated with these materials. The superconducting coherence length in HTS cuprates is anisotropic and quite small, with typical values on the order of the atomic spacing. This presents difficulties in the fabrication of junction devices. As with other oxide materials, the HTS cuprates are brittle ceramics prone to fracture with applied stress. This introduces challenges in developing a flexible conductor from HTS materials. Another issue involves the ability of HTS materials to carry significant currents and remain superconducting in the presence of a magnetic field. As with any type II superconductor, magnetic fields penetrate the HTS cuprates in the form of quantized magnetic field lines. In the presence of an electrical current, microscopic defects are needed to immobilize or "pin" these flux lines against energy-dissipative motion. For some HTS materials, such as $YBa_2Cu_3O_7$, strong magnetic flux pinning has been demonstrated at 77 K (4). For other more anisotropic compounds, such as $Bi_2Sr_2Ca_2Cu_3O_{10}$, strong pinning has been realized only at much lower temperatures. The ability to pin magnetic flux lines at temperatures near T_c varies significantly among the HTS compounds and appears to correlate with the degree of anisotropy in the material.

One detrimental aspect in HTS materials is the effect of grain boundaries on transport. The density of current that can flow through the material is severely limited by the presence of grain boundaries in all of the HTS materials. This is particularly evident for boundaries with misorientation angles greater than 10°, as is shown in Figure 2.5 (5). As a result, the capacity to carry superconducting current in polycrystalline materials with large-angle grain boundaries is significantly less than that for single-crystal-like material. Studies of transport through individual

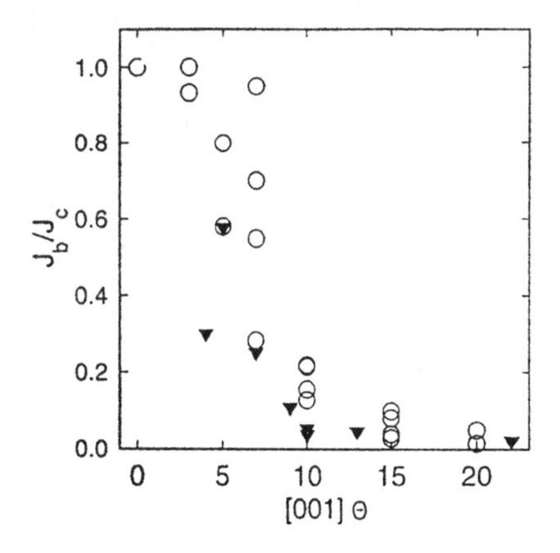

FIGURE 2.5 Relative drop in the grain boundary (J_c) as the misorientation angle increases. (From Ref. 5.)

grain boundaries in HTS bicrystals showed that large-angle grain boundaries act as weak links in the superconductor (5,6). This effect has proven fortuitous in the fabrication of Josephson junction device structures. However, this profound influence of grain boundaries in the HTS materials makes it necessary to utilize epitaxial films on single-crystal substrates in order to realize optimal material properties and device performance. It also implies that HTS wires with very high current-carrying capability will require fabrication techniques that result in highly oriented material with virtually no high-angle grain boundaries. Thus, a significant effort has been devoted to studying the epitaxial growth and properties of HTS films.

2.2 TECHNIQUES FOR HTS FILM GROWTH

The unique promise held by HTS materials in many applications has driven significant efforts in exploring their formation as thin films. The general requirements for the synthesis of HTS films with little or no impurity phase include stringent control of the composition during the deposition process, because each compound is a multication oxide with a rather complex crystal structure. Even with the correct cation composition, the formation of a specific HTS oxide phase requires an optimization of both the temperature and the partial pressure of the chosen oxidizing species consistent with the thermodynamic phase stability of the compound. Because the electronic properties of the superconducting cuprates show a significant dependence on oxygen content, specific oxidation conditions after film growth are generally required in order to achieve optimal doping for superconductivity. Control of film surface morphology is a key issue for the synthesis of multilayer device structures. This is particularly true for junction devices due in large part to the short, anisotropic superconducting coherence lengths in these materials. These collective requirements prove challenging for nearly all techniques presently employed in thin-film processing.

Numerous film-growth techniques have been investigated for the epitaxial growth of HTS films. These include in situ growth techniques, in which the correct crystallographic phase is formed as the material is being deposited, as well as ex situ techniques, where a film that is either amorphous or an assemblage of polycrystalline phases is deposited and subsequently annealed to form the desired HTS phase. For in situ growth, the kinetics of epitaxial film growth, along with the thermodynamic requirements for proper phase formation, typically require deposition at elevated temperatures (650–800°C) in an oxidizing ambient. The ability to produce relatively smooth film surfaces and synthesize multilayer film structures are obvious advantages with in situ film growth.

In situ film-growth techniques that have been successfully employed in the synthesis of epitaxial HTS materials include physical deposition techniques, such coevaporation (7,8), molecular beam epitaxy (9,10), pulsed laser deposition (11), and sputtering (12). With the physical deposition of HTS cuprates, the phase constituents are delivered as a flux of individual atoms or simple oxide species. Atomic-level control of the film-growth process is possible with most physical deposition approaches, thus enabling the formation of novel multilayer structures (13,14). Other techniques that have proven useful in obtaining epitaxial HTS films are metalorganic chemical vapor deposition (15) and liquid-phase epitaxy (16).

2.2.1 Coevaporation and Molecular Beam Epitaxy

In the growth of HTS films by coevaporation or molecular beam epitaxy (MBE), the flux is delivered by electron beam (e-beam) or thermal evaporation sources. A separate source is required for each element due to differences in vapor pressures for various elements or oxides. The flux from each source must be precisely controlled to ensure proper stoichiometry of the film. In situ monitoring of the flux from each source can be accomplished with the use of multiple crystal-quartz monitors. Optical techniques have also been developed in which the optical absorption coefficient of each element is used to monitor the flux (17,18). Film deposition by evaporation typically takes place in a background pressure less than 10^{-4} torr. This is lower than what is thermodynamically required for the in situ growth of HTS films; most of these compounds require molecular oxygen pressures much higher. To overcome this limitation, highly oxidizing gases, such as NO_2 (19) or O_3 (20), as well as atomic oxygen created by a plasma source (21), can be utilized. Oxidation of the HTS films can also be enhanced by irradiating the growing film with ultraviolet light (22). The ultraviolet (UV) photons produce excited-state O and O_2 species, thereby increasing the activity significantly. With these highly oxidizing species, background pressures less than 10^{-4} torr can often be maintained while growing epitaxial HTS films.

One approach developed to overcome this limitation to coevaporation with molecular oxygen utilizes a molecular oxygen pocket that is maintained at a higher pressure than that of the deposition chamber (7). The substrates are placed on a rotating disk and alternate between a zone of the metal vapor and a pocket into which oxygen is introduced. A partial pressure drop of 1:100 can be maintained between the oxygen pocket and vacuum chamber with the proper design of the rotating disk and oxygen pocket.

Film growth by evaporation can occur by the simultaneous coevaporation of all the components or by sequentially shuttering the delivery of each component. The latter is often associated with molecular beam epitaxy. This technique offers atomic-level control of the film-growth process and has proven useful in the formation of novel multilayered structures (23–25). For some HTS compounds, MBE can be used to tailor the formation of specific phases through layer-by-layer growth of the various components of the layered HTS compounds. Molecular beam epitaxy also permits the so-called "block-by-block" approach, illustrated in

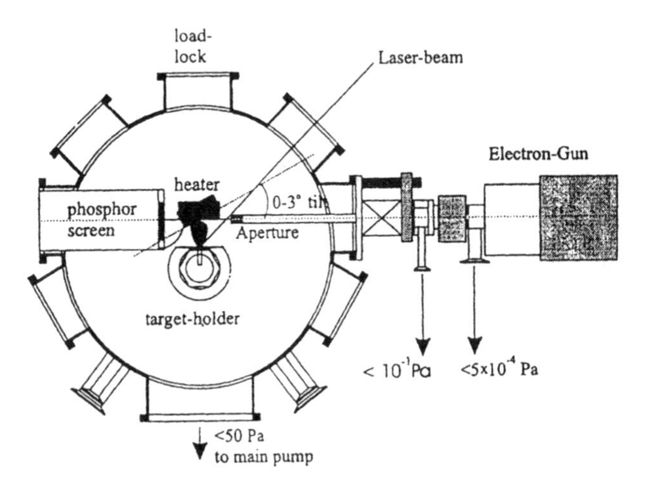

FIGURE 2.8 Schematic of a conventional pulsed-laser deposition system equipped with a differentially pumped RHEED system. (From Ref. 32.)

the plume. Plume energies must be moderated by controlling laser energy density and/or by using a background gas to thermalize the plume species.

As an electron-probe technique, RHEED has generally been restricted to in situ monitoring of film growth under background gas pressures less than 10^{-4} torr. This is unfortunate, as the most favorable film-growth conditions for many HTS materials using PLD are at much higher oxygen pressures. Recently, a modified RHEED system capable of operating under standard PLD film-growth conditions (100–300 mtorr O_2) has been demonstrated (32). In this system, illustrated in Figure 2.8, the electron beam entry and phosphor screen are placed in close proximity to the substrate. Using this approach, RHEED intensity oscillations for conventional PLD growth at oxygen pressures up to 300 mtorr have been observed (8).

2.2.3 Sputtering

Several sputtering techniques have been used in the growth of HTS films, including on-axis dc magnetron sputtering (33), cylindrical magnetron sputtering (34), ion beam sputtering (35), and off-axis sputtering (12). In sputter deposition, energetic ions created in a plasma bombard a metal or oxide target surface. This process ejects atoms from the target that subsequently deposit on a nearby substrate surface. In an on-axis configuration, the substrate and target are facing one another. Although this is the optimal geometry for the maximum deposition rate, the on-axis configuration can result in film damage due to the bombardment of the film surface with energetic species from the plasma. An alternative is the off-axis approach, in which the substrate surface is oriented perpendicular to the surface of the sputter target. This removes the film from the plasma region, eliminates sputter damage, and, generally, results in better films. Unfortunately, the off-axis approach also significantly reduces the growth rate that can be achieved by sputter deposition. One disadvantage with sputter deposition is that the stoichiometry of a multicomponent film is not necessarily that of the target material due to differences in sputtering yields for different elements.

2.2.4 Metal–Organic Chemical Vapor Deposition

For large-scale production of thin films, metal–organic chemical vapor deposition (MOCVD) is very attractive (15). It is routinely utilized in the electronics industry and is quite amenable to large-area deposition with high throughput. It is independent of line-of-site deposition and can be used for in situ growth at oxygen pressures near 1 atm. With MOCVD, the cations necessary for film growth are delivered as constituents of organometallic molecules. If the organometallic molecules are sufficiently volatile, they can be delivered to the heated substrate via a carrier gas. For nonvolatile precursors, the reactants are delivered as a condensed phase. The molecules thermally decompose at the heated substrate surface, resulting in film growth. For oxide film growth, oxygen is included within the gas flow. One key challenge in the synthesis of HTS films using MOCVD has been the development and reproducible synthesis of volatile precursor molecules that are stable in storage and transport and that decompose at elevated temperatures to yield good films with no contamination from the organic ligand.

2.2.5 Liquid-Phase Epitaxy

One of the more recent developments in HTS film growth has been progress in the use of liquid-phase epitaxy (LPE). In LPE, film growth occurs from a melt in contact with a substrate surface. For many years, this technique has proven to be quite useful in growing relatively thick semiconductor films with near-perfect crystallinity. Superior crystallinity is possible with LPE, as film growth from the melt takes place very near thermodynamic equilibrium. The structural and chemical complexities of the HTS materials have made it difficult in determine conditions for HTS film growth using LPE. Nevertheless, the epitaxial growth of HTS films with near-single-crystal-like properties has been achieved using this technique (16).

2.2.6 Ex Situ Postannealing

Despite success in growing epitaxial HTS films using in situ techniques, there are limitations to these approaches. In situ growth requires significant and uniform substrate heating in an oxygen ambient during film deposition. For most HTS phases, the temperature range at a given oxygen pressure for achieving optimal film properties is rather narrow, on the order of 20–40°C. This proves challenging for large-area, double-sided, or continuous-length deposition of HTS films. In contrast, ex situ processing requires no substrate heating during precursor deposition, greatly simplifying the film-growth apparatus. The precursor film can be deposited either by vacuum deposition or by wet-chemistry approaches. The desired crystallographic phase is formed through bulk diffusion and solid-phase epitaxy by annealing the "precursor" film at elevated temperatures. Annealing can also be performed as a batch process of multiple substrates. In addition, HTS compounds consisting of cations with high vapor pressures, such as Tl or Hg, are not easily grown by in situ film-growth techniques. These compounds are typically synthesized by ex situ annealing of precursor films, where substantial overpressure of the volatile component can be easily achieved. However, the use of solid-phase epitaxy places severe restrictions on the fabrication of multilayered thin-film structures.

2.3 SUBSTRATES FOR HTS FILMS

Multiple considerations are involved when evaluating the usefulness of a particular substrate for HTS film growth. A comprehensive review of substrate selection for HTS film growth has been published elsewhere (36). Film/substrate lattice match, thermal expansion match, and chemical compatibility are the most relevant factors when the singular consideration is film properties. Because the HTS cuprates are nearly tetragonal in their crystal structure, oxides with a square-planar surface orientation, such as the (001) face of cubic oxide crystal, are ideal for *c*-axis-oriented HTS films. Typically, the in-plane lattice spacing of the HTS film should closely match that of the substrate either aligned with or rotated 45° with respect to the principle axes. Significant differences in the thermal expansion coefficient should be avoided, as this will lead to cracking of the film. Of course, any chemical reaction between the substrate and film will likely inhibit good epitaxy and may prevent the formation of the HTS phase. The substrate material should also be stable against thermal cycling with no significant phase transitions.

A large array of oxide and nonoxide materials has been investigated as single-crystal substrates for epitaxial growth of HTS films (36). In many cases, attractive substrate materials can be prepared that possess smooth surfaces with only unit-cell-high steps, as revealed by scanning force microscopy (37,38). The substrates used for HTS film growth can generally be categorized into three distinct groups. The first is the perovskite-related materials (39), such as $SrTiO_3$, $LaAlO_3$, and $NdGaO_3$. These material, illustrated in Figure 2.9, are cubic or pseudocubic with lattices parameters very close to the *a–b* lattice spacing of the HTS cuprates. Because the alkaline earth–CuO_2 subunit block in the HTS materials can be viewed as a defect perovskite structure, perovskite crystals are generally the most chemically and structurally compatible substrates for growing high-quality epi-

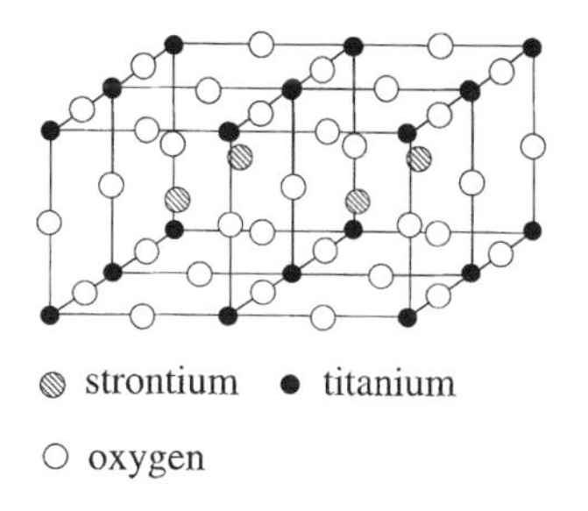

FIGURE 2.9 Crystal structure for $SrTiO_3$.

taxial HTS films. Recent developments in understanding the surfaces of perovskites have enabled the reproducible termination of several perovskite crystalline surfaces with specific cation species (40,41). For example, a simple aqueous treatment, etch, and annealing procedure yields (001) $SrTiO_3$ surfaces that are singularly TiO_2 terminiated. Figure 2.10 shows an atomic force microscopy (AFM) image of an atomically flat (001) $SrTiO_3$ surfaces that possesses (a) a singular TiO_2 termination and (b) the corresponding AFM line scan. These capabilities greatly enhance the ability to control phase nucleation and mulilayer structure formation in HTS epitaxy.

Next are the nonperovskite oxides, such as MgO and Al_2O_3. Nonperovskite oxides are of interest as HTS substrates if they possess advantageous physical properties for specific applications. For instance, the electronic properties of the substrate, including dielectric constant, conductivity, and loss tangent, are critically important for high-frequency applications of HTS films. MgO has a cubic NaCl structure with a significant lattice mismatch with most HTS compounds. Yet, its availability as an inexpensive substrate with a temperature-independent dielectric constant, ϵ, of 10 and a low dielectric tangent loss, tan δ, of 10^{-5} at 77 K

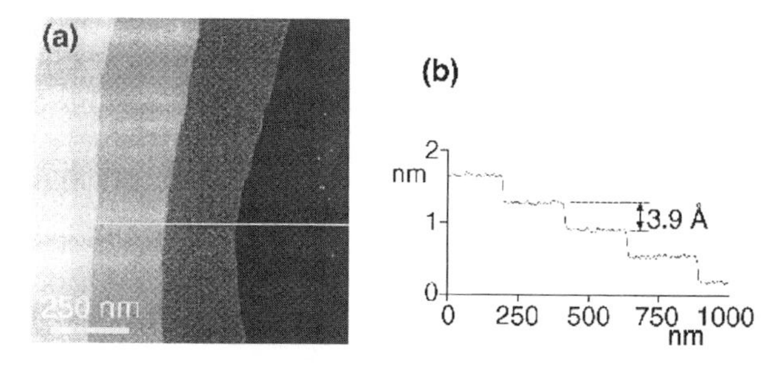

FIGURE 2.10 Atomic force microscopy image of (a) a TiO_2-terminated $SrTiO_3$ surface and (b) the associated line scan. (From Ref. 40.)

and 10 GHz make MgO an attractive HTS substrate for microwave applications. Al_2O_3 is also an attractive HTS substrate material for microwave applications despite significant problems with chemical reactivity. Single-crystal yttria-stabilized zirconia (YSZ) is attractive due to its low cost, mechanical strength, and chemical stability. However, the lattice constant of this cubic fluorite structure provides a relatively poor lattice match to the HTS films.

Third, one can consider nonoxide substrates that are of interest for specific applications. This last group, which includes metals and semiconductors, presents significant film-growth challenges due to chemical, thermal, and lattice-matching incompatibilities. In these cases, the use of a chemically compatible oxide buffer layer is necessary to achieve epitaxy.

2.4 EPITAXIAL GROWTH OF SPECIFIC HTS MATERIALS

2.4.1 $YBa_2Cu_3O_7$

The epitaxial growth and characterization of $YBa_2Cu_3O_7$ thin films has received significantly more attention than any other HTS compound. Compared to the other HTS materials, epitaxial $YBa_2Cu_3O_7$ films are the easiest to synthesize and achieve a T_c for the film that is near the bulk value. This is due, in part, to the relative stability of the $YBa_2Cu_3O_7$ phase, as there are no $n = 1$ or $n = 3$ members to compete with in phase formation. The structure of $YBa_2Cu_3O_{7-\delta}$, shown schematically in Figure 2.11, can be derived by stacking three oxygen-deficient perovskite unit cells ($ACuO_y$) in the layered sequence BaO–CuO_2–Y–CuO_2–BaO–CuO. $YBa_2Cu_3O_7$ contains two CuO_2 planes per unit cell separated by an Y atom. CuO chains lie between the BaO layers. The oxygen content can be varied from $\delta = 0$ to $\delta = 1$ through removal of oxygen from the CuO chain layer. Fully oxygenated $YBa_2Cu_3O_7$ is a hole-doped superconductor with $T_c = 92$ K. The

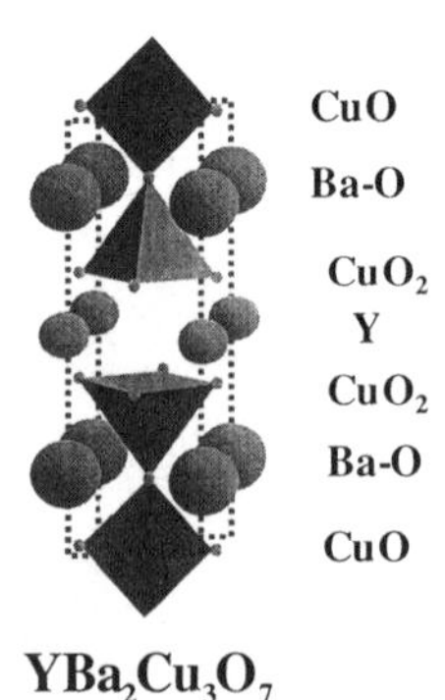

FIGURE 2.11 Crystal structure of $YBa_2Cu_3O_7$.

crystal structure is orthorhombic with $a = 3.82$ Å, $b = 3.88$ Å, and $c = 11.68$ Å, resulting in twinning for c-axis-oriented films. Microwave surface resistance lower than 200 μΩ for 10 GHz at 77 K has been measured in epitaxial $YBa_2Cu_3O_7$ films (42). A critical current density of ~2–5 MA/cm^2 at 77 K in a zero magnetic field (H = 0) is typical for high-quality films. $YBa_2Cu_3O_{7-\delta}$ is less anisotropic than other hole-doped HTS materials. This appears responsible for the strong magnetic flux pinning observed in $YBa_2Cu_3O_7$. The magnetic field dependence of J_c is anisotropic, due to intrinsic pinning from the layered structure, with the J_c highest for H parallel to the a-b planes. With few exceptions, near-optimal flux pinning for H parallel to the c axis is also observed in epitaxial $YBa_2Cu_3O_7$ films due to a fortuitous array of growth defects. Recently, Dam et al. made use of a sequential etching technique in an attempt to identify these defects (43). They suggest that edge and screw dislocations, which can be mapped quantitatively by this technique, are the linear defects that provide the strong pinning centers responsible for the high critical currents observed in these $YBa_2Cu_3O_7$ films. These collective properties make $YBa_2Cu_3O_7$ films quite attractive for many applications.

The most successful ex situ synthesis route for the epitaxial growth of $YBa_2Cu_3O_7$ is the so-called BaF_2 process (44–47). With this approach, a stoichiometric precursor film of Y, Cu, and BaF_2 is deposited at room temperature with minimal oxygen background pressure on a lattice-matched oxide surface, such as $SrTiO_3$. Other substrate materials with a large lattice mismatch between film and substrate results in $YBa_2Cu_3O_7$ films with a large fraction of polycrystalline grains. BaF_2 is used instead of Ba metal or BaO, as it is stable in air. Annealing the stoichiometric precursor film at a high temperature in oxygen and water vapor results in the epitaxial growth of $YBa_2Cu_3O_7$ by a solid-phase epitaxy process. Water vapor is necessary for the decomposition of BaF_2 and complete removal of fluorine from the film during the high-temperature anneal. When annealed at an oxygen pressure of 1 atm, the formation of c-axis-oriented $YBa_2Cu_3O_7$ films is limited to annealing temperatures $T > 830°C$ and film thickness less than ~0.4 μm (44). Lower annealing temperatures and/or thicker deposits result in significant a-axis-oriented nucleation. However, if the annealing process is performed at lower oxygen partial pressures, c-axis-oriented $YBa_2Cu_3O_7$ film growth is maintained for significantly lower temperatures and thick precursor film deposits (46). For instance, a thick Y–BaF_2–Cu–O precursor film processed in an oxygen partial pressure of 2.6×10^{-4} atm at 740°C yielded a 1-μm-thick c-axis-oriented epitaxial film with a T_c~90 K and J_c(77 K) ~ 1.9 MA/cm^2 (45). It is interesting to note that the $P(O_2)$–T conditions for ex situ synthesis of $YBa_2Cu_3O_7$ using the BaF_2 process are consistent with the $P(O_2)$–T phase space described by Hammond et al. (48) for in situ films. In addition to precursor films deposited using evaporation, a related process involves the use of meta-triflouroacetates as the precursor film (49). Conversion of this precursor proceeds much the same as with the e-beam evaporated case, with $J_c(77\text{ K}) > 1$ MA/cm^2 reported for 1-μm-thick films deposited on $LaAlO_3$

Films synthesized under the above-described conditions have high critical current densities, indicating strong flux pinning due to the presence of microstructural defects. In contrast, films that are processed at very high temperatures (~900°C) in 1 atm oxygen can exhibit cation alignment similar to that of single crystals, with low defect densities and subsequent reduced flux pinning and low critical current densities (44). The ability to produce $YBa_2Cu_3O_7$ films with low defect densities has not been demonstrated for in situ films obtained by vapor deposition techniques.

Epitaxial *c*-axis-oriented $YBa_2Cu_3O_7$ films with $T_c > 90$ K and $J_c(77\text{ K})$ $> 1\text{ MA/cm}^2$ can be routinely synthesized by a number of in situ physical and chemical deposition techniques, including coevaporation (7,50), MBE (10), PLD (51), sputtering (12), and MOCVD (15). A survey of the various growth conditions for synthesizing epitaxial $YBa_2Cu_3O_7$ films by different deposition techniques was used by Hammond et al. (48) to develop a $P(O_2)$–T "phase diagram," shown in Figure 2.12, for in situ epitaxial growth. For each technique, $P(O_2)$–T region can be defined for the optimized synthesis of high-quality $YBa_2Cu_3O_7$ films. In most cases, these regions lie just above the $YBa_2Cu_3O_7$ thermodynamic stability line. In general, in situ $YBa_2Cu_3O_{7-\delta}$ films are oxygen deficient ($\delta \neq 0$) at the growth temperatures and oxygen pressures typically used. This is consistent with the bulk $P(O_2)$–T phase diagram for $YBa_2Cu_3O_{7-\delta}$ (52) and with quenching experiments involving thin films (53). Real-time measurements of the film resistance at typical growth temperatures show that the oxidation of $YBa_2Cu_3O_7$ films is rapid when exposed to high oxygen pressures at elevated temperatures. In practice, the introduction of 300–760-torr oxygen during cooling of the film after growth is sufficient to achieve fully oxidized $YBa_2Cu_3O_7$ films.

High-quality $YBa_2Cu_3O_7$ films have been obtained by MBE and coevaporation using NO_2 (19), O_3 (20,50), or atomic oxygen (10,21,48). Molecular oxygen is not effective at the pressures compatible with these techniques. Significant control of the film growth process has been demonstrated with the formation of superconducting $YBa_2Cu_3O_7$ layers as thin as a single unit cell (50). Large-area $YBa_2Cu_3O_7$ films with excellent superconducting properties have also been realized. For instance, double-sided $YBa_2Cu_3O_7$ films on 4-in. $LaAlO_3$ substrates with a surface resistance, R_s, of 500 μΩ at 77 K and 10 GHz, $J_c(77\text{ K}) > 2$ MA/cm^2, and $T_c > 88$ K over nearly the entire area have been grown by the reactive thermal evaporation technique involving the rotating substrate and oxygen pocket (7).

Both pulsed-laser deposition and sputtering have been used for single-source deposition of high-quality $YBa_2Cu_3O_7$ films. With both techniques, molecular oxygen (O_2) is used as the oxidizing gas during film growth. Large-area deposition has also been realized for both PLD and sputtering through a combination

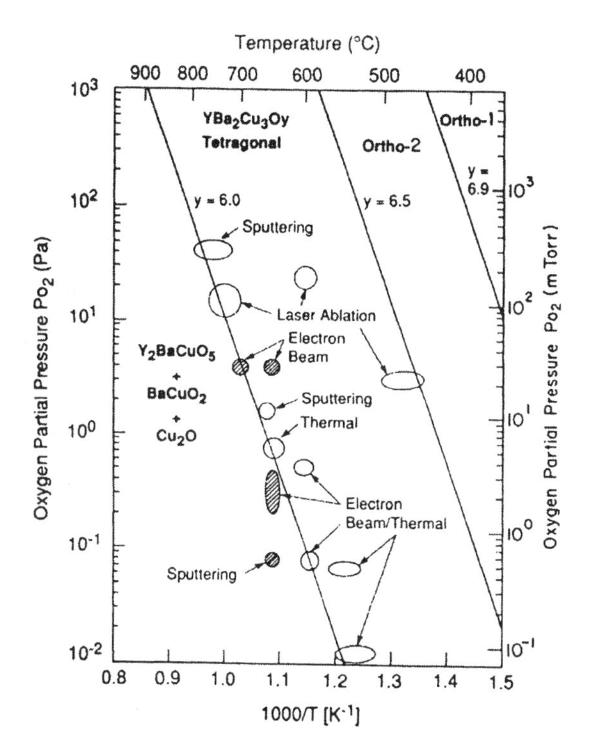

FIGURE 2.12 A $P(O_2)$–T phase diagram showing regions in which epitaxial $YBa_2Cu_3O_7$ can be obtained. (From Ref. 48.)

of substrate and target rotation. Double-sided $YBa_2Cu_3O_7$ films on 3-in.-diameter sapphire wafers with $J_c(77\ K) > 3\ MA/cm^2$ has been demonstrated with PLD (54). Using in situ off-axis magnetron sputtering, double-sided, 2-in.-diameter $YBa_2Cu_3O_7$ films on $LaAlO_3$ substrates with R_s (77 K) < 400 μΩ at 10 GHz have also been reported (55). Thickness uniformity of ±5% with $J_c(77\ K) > 1\ MA/cm^2$ and $T_c > 87$ K has been demonstrated for $YBa_2Cu_3O_7$ films over an 8-in.-diameter area using off-axis sputtering (56). In addition, atomic-level control of the growth process has been demonstrated with both techniques. Epitaxial $YBa_2Cu_3O_7$ films with a surface roughness less than one unit cell have also been achieved by conventional magnetron sputtering with an oscillating substrate configuration (57). Using PLD, $YBa_2Cu_3O_7/PrBa_2Cu_3O_7$ superlattice structures with $YBa_2Cu_3O_7$ layers as thin as one unit cell have been synthesized (58,59). In these studies, the superconducting properties of single-unit-cell-thick $YBa_2Cu_3O_7$ layers were examined.

Progress has been made in the development of better precursors for the deposition of $YBa_2Cu_3O_7$ by MOCVD. The properties of $YBa_2Cu_3O_7$ films deposited by MOCVD are approaching that of films grown by physical deposition techniques. For example, 3-in.-diameter double-sided $YBa_2Cu_3O_7$ films on $LaAlO_3$ have been deposited by MOCVD with $T_c \sim 87$ K and a microwave surface resistance as low as 260 μΩ at 10 GHz, 77 K (60). One interesting modification for MOCVD of $YBa_2Cu_3O_7$ films involves a photo-assisted technique (61,62). In this approach, a tungsten halogen lamp is used to irradiate the substrate surface with UV photons during growth, providing both substrate heating and photostimulation of the chemical processes involved in the reaction. The growth of high-quality $YBa_2Cu_3O_7$ films with $J_c(77\text{ K}) > 1$ MA/cm^2 at remarkably high growth rates (>800 nm/min) has been reported.

One of the more recent developments in $YBa_2Cu_3O_7$ film growth has been progress in the use of liquid-phase epitaxy (LPE). Recent results for $YBa_2Cu_3O_7$ films grown by this technique show near-perfect crystallinity that is far superior to that achieved with vapor deposition techniques. Although vapor-deposited HTS films typically have screw dislocation densities of $\sim 10^9$/cm^2 with unit-cell-step distances of 30 nm, films grown by LPE have microspiral densities on the order of $10^3/10^4$/cm^2 and interstep distances of up to 3 μm (63). However, substrate selection for growth by LPE is more restrictive. In addition to requirements of small lattice and thermal mismatch between the film and substrate, one must choose substrates that can withstand the high-temperature solutions of $YBa_2Cu_3O_7$ at temperatures as high as 1000°C. The growth temperature for liquid-phase epitaxy of $YBa_2Cu_3O_7$ can be reduced by the addition of BaF_2 to the growth flux, as seen in Figure 2.13 (64). The temperature of YBCO formation can be reduced to

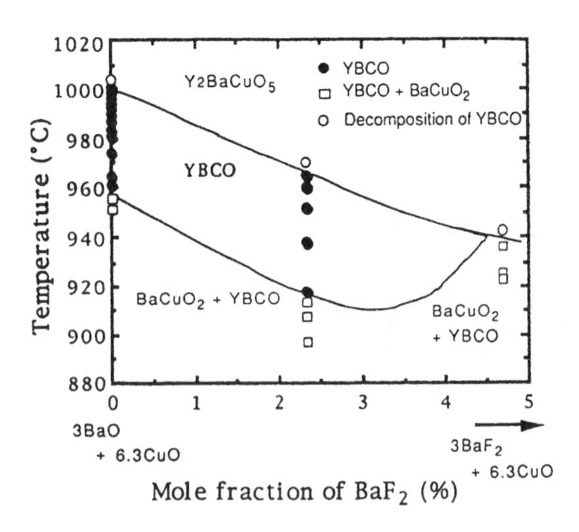

FIGURE 2.13 Relations between growth temperature and deposited phases as a function of fluoride concentration in liquid-phase epitaxy of $YBa_2Cu_3O_7$.

FIGURE 2.14 Illustrations of the three basic modes observed in film growth.

920°C, thereby enabling the LPE on a wide range of substrates. A seed layer of epitaxial $YBa_2Cu_3O_7$, typically deposited by physical vapor deposition techniques, remains an important determinant in the formation of films. In LPE, cation stoichiometry must be precisely controlled. In addition to smoother surfaces, other properties of LPE-grown $YBa_2Cu_3O_7$ films differ from that of vapor-deposited films. Oxidation of *c*-axis-oriented films progresses more slowly due to a lower density of dislocations, grain boundaries, and other defects that would enhance oxygen diffusion along the *c* axis. An absence of structural defects in highly perfect LPE-grown films results in weaker flux pinning, with typical J_c(77 K) values of 5×10^2 to 10^4 A/cm^2. Additional pinning can be introduced by growing on a substrate possessing a large lattice mismatch, such as MgO (65). The defects introduced by the lattice mismatch increase the pinning, with J_c(77 K) $\sim 10^5$ A/cm^2 in zero field and increased pinning evident at high fields.

2.4.1.1 $YBa_2Cu_3O_7$ Growth Mode and Microstructure

A microscopic understanding of the in situ epitaxial growth of $YBa_2Cu_3O_7$ has emerged through the use of various surface-sensitive probes, such as RHEED and scanning force microscopy, in monitoring and characterizing the film surface both during and after film growth. In general, epitaxial growth of thin films proceeds within the context of three basic modes, as illustrated in Figure 2.14 (66): layer by layer (Frank–van der Merwe), island formation (Volmer-Weber), and layer by layer followed by island formation (Stranski–Krastanov). True layer-by-layer growth is typically reserved for lattice-matched film/substrate systems in which no stress is imposed on the nucleating film. Oscillations in the RHEED specular intensity are a consequence of the nucleation of two-dimensional (2D) islands and their cyclical growth into flat terraces during layer-by-layer growth or the initial stages of Stranski–Krastanov growth. For film-growth experiments involving coevaporation (67) or PLD (68), strong RHEED intensity oscillations have been observed during the initial nucleation of *c*-axis-oriented $YBa_2Cu_3O_7$ on lattice-matched substrates, as shown in Figure 2.15 (67). These studies suggest that the minimum growth unit for $YBa_2Cu_3O_7$ is the 11.7-Å unit cell, as this satisfies both chemical composition and electrical neutrality considerations. Similar RHEED studies on the growth of $YBa_2Cu_3O_7$ on (100) MgO show no oscillations. This is

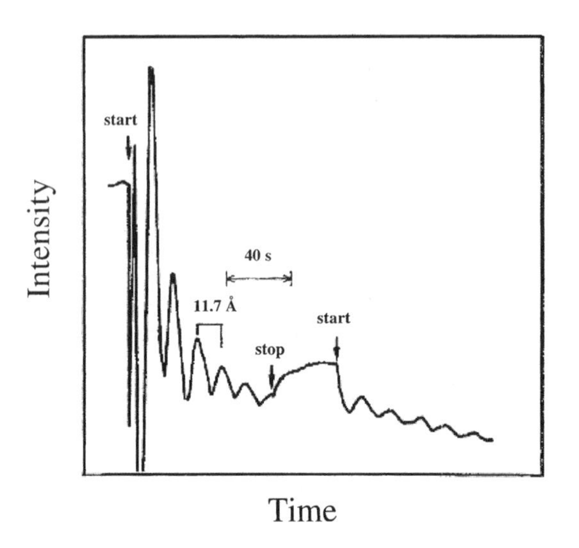

FIGURE 2.15 RHEED oscillations observed in the initial growth of $YBa_2Cu_3O_7$ on (001) $SrTiO_3$. (From Ref. 67.)

consistent with a 3D island nucleation for a film/substrate system with significant lattice mismatch.

Although RHEED oscillations suggest that layer-by-layer growth occurs during initial nucleation, scanning probe microscopy studies show that $YBa_2Cu_3O_7$ grows by an anisotropic Stranski–Krastanov mode on substrates with a low lattice mismatch. Scanning tunneling microscope (STM) images of $YBa_2Cu_3O_7$ films that are only a few unit cells thick show that the epitaxial growth of $YBa_2Cu_3O_7$ on $SrTiO_3$ proceeds by a Stranski–Krastanov mechanism, with a transition from layer-by-layer to island growth occurring at a film thickness between 8 and 16 unit cells (69). Stranski–Krastanov growth of $YBa_2Cu_3O_7$ is observed on other lattice-matched substrates, including $NdGaO_3$ (70). In contrast, STM studies confirm that deposits of $YBa_2Cu_3O_7$ on (100) MgO nucleate and grow by a Volmer–Weber island mechanism due to the lattice mismatch between the film and substrate.

Scanning tunneling microscopy images of the surface microstructure for thicker *c*-axis-oriented $YBa_2Cu_3O_7$ films grown on both (100)-oriented MgO and $SrTiO_3$ single crystals often show a predominance of growth spirals consisting of atomically flat terraces with growth steps one unit cell high, as shown in Figure 2.16 (71–73). These growth spirals presumably originate from screw dislocations in the film. Subsequent studies showed that the presence of a spiral-growth surface microstructure is a function of film-growth conditions, substrate–film lattice mismatch, and substrate miscut, although a terraced microstructure with unit-cell-high steps is a common feature for *c*-axis-oriented films (73). In some cases, high-quality films with a surface morphology more reminiscent of 2D terrace growth is observed, particularly for films deposited by laser ablation (74). It has been argued that the non-steady-state growth conditions of pulsed-laser deposition can inhibit the formation of spirals due to the absence of steady-state diffusion of adatoms to the growing step edge. One consequence of this is that the density of spiral morphological features does not necessarily correlate with the density of linear defects, such as dislocations, in the film. This is important when attempting to identify potential magnetic flux pinning centers in these films.

In recent years, there has been significant progress in understanding the initial nucleation of $YBa_2Cu_3O_7$ on single-crystal oxide surfaces, particularly for growth on the (001) $SrTiO_3$ surface. Studies have focused on the chemistry of $YBa_2Cu_3O_7$ formation on specific cation-terminated surfaces. Initial studies suggested that $YBa_2Cu_3O_7$ always nucleates as a complete unit cell. This conclusion was based largely on AFM, STM, and RHEED studies of relatively thick $YBa_2Cu_3O_7$ films. In the scanning probe microscopy studies, step heights of precisely one $YBa_2Cu_3O_7$ unit cell were generally observed. RHEED oscillations also correlated with unit cell by unit-cell film growth. However, recent efforts indicate that the nucleation of $YBa_2Cu_3O_7$ may proceed by sub-unit cell formation and that this process is highly dependent on substrate surface termination. The most studied case is that for nucleation on a $SrTiO_3$ (001) surface. $SrTiO_3$ (001) can be terminated either with the TiO_2 or the SrO atomic plane. For SrO-terminated surfaces, $YBa_2Cu_3O_7$ can nucleate with a Cu–O plane at the interface. In this case, the stacking sequence can be CuO–BaO–CuO–Y–CuO–BaO or CuO–Y–CuO–Ba–CuO–BaO. In either sequence, this leaves the cuprate (001) terminated with BaO, which is generally accepted to be the case, and incorporates all of the available cations for unit-cell growth. However, for TiO_2-terminated sur-

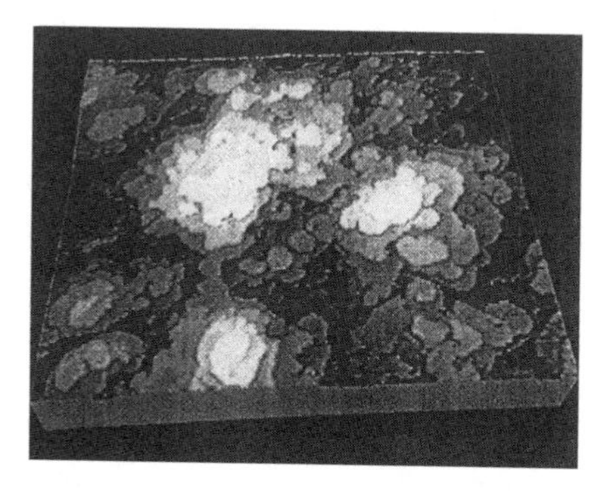

FIGURE 2.16 Scanning tunneling microscopy image of a $YBa_2Cu_3O_7$ film grown by pulsed-laser deposition.

faces, this is not the case. The wettability of CuO on the TiO_2 is very poor, leading to either the Y or BaO layer at the TiO_2 interface. In this case, the possible stacking sequences, such as BaO–CuO_2–Y–CuO_2–BaO, that yield a stable BaO surface also yield excess Cu–O on the surface. Figure 2.17 shows an AFM image of sub-unit-cell step heights for $GdBa_2Cu_3O_7$ nucleating on a TiO_2-terminated surface (75). Several studies suggest that this results in the nucleation of secondary-phase Cu-rich precipitates on the growing surface. Elimination of precipitate formation is possible either by starting with a SrO termination or by supplying a Cu-deficient flux during the first few monolayers of film growth.

Various defects have been observed in epitaxial films including twin boundaries, dislocations, and secondary phases. Some of these defects are observed in the bulk of the film, such as the double CuO chain layers related to the $YBa_2Cu_4O_8$ compound (76). Other secondary phases can appear as outgrowths on the film surface such as those seen in Figure 2.18. These outgrowths take the form of Y_2O_3, $CuYO_2$, and CuO impurity phases, as well as *a*-axis-oriented $YBa_2Cu_3O_7$ grains

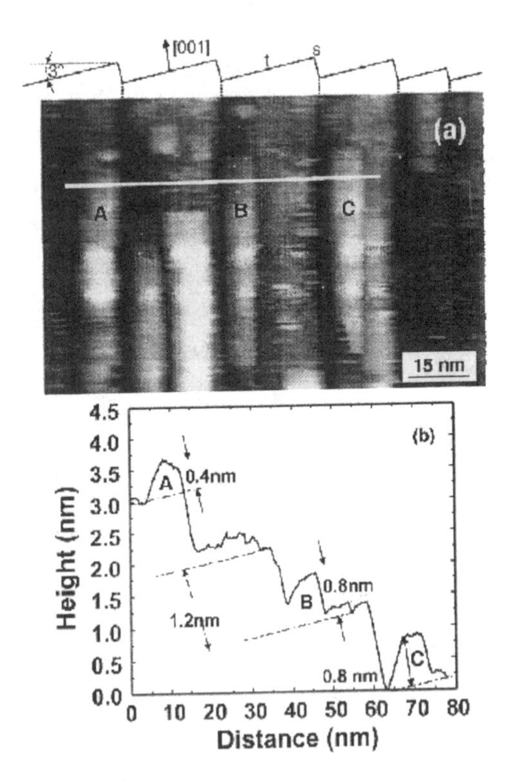

FIGURE 2.17 An AFM image of sub-unit-cell heights for $GdBa_2Cu_3O_7$ nucleating on a TiO_2-terminated $SrTiO_3$ surface. (From Ref. 75.)

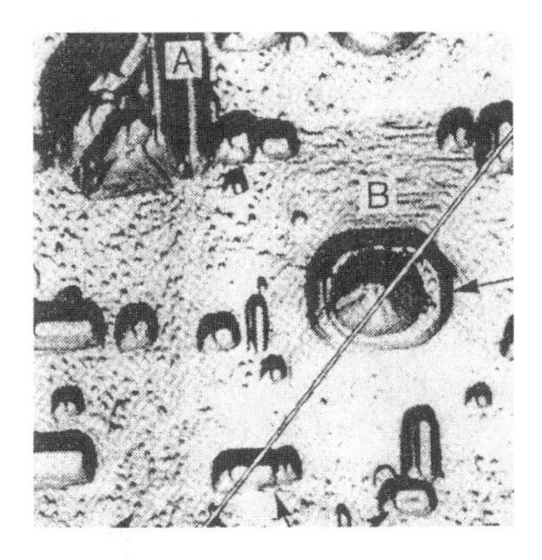

FIGURE 2.18 Scanning electron micrograph of a $YBa_2Cu_3O_7$ film showing secondary-phase outgrowths. Image size is ~ 2 μm × 2 μm. (From Ref. 77.)

(77). In some cases, they can be evident as impurity peaks in the x-ray diffraction patterns. Outgrowths formed on $YBa_2Cu_3O_7$ films are sensitive to the deposition conditions and surface terminations. They are obviously sensitive to the stoichiometry, with a high probability of nucleating when the composition deviates from the ideal $YBa_2Cu_3O_7$.

2.4.1.2 Lattice Mismatch and Epitaxy

Numerous studies have focused on the growth of $YBa_2Cu_3O_7$ on perovskites, with *c*-axis-oriented epitaxial films obtained on many of them. In nearly all cases, excellent cube-on-cube epitaxy is achieved with good superconducting properties. One of the better substrates for $YBa_2Cu_3O_7$ growth in terms of lattice match is $NdGaO_3$, with a lattice mismatch of only 0.2% at a typical growth temperature of 700°C (78,79). Comparisons have been made between the structural properties of ultrathin $YBa_2Cu_3O_7$ films grown on (100) $NdGaO_3$ and (100) $SrTiO_3$. The x-ray-diffraction rocking-curve width, which is a measure of crystalline perfection, was measured as a function of thickness for epitaxial $YBa_2Cu_3O_7$ deposited on these two substrates. For $YBa_2Cu_3O_7$ films grown on (100) $SrTiO_3$, the rocking-curve width increases by a factor of 3 when the $YBa_2Cu_3O_7$ film thickness exceeds 15 nm. This width increase reflects stress relief in the epitaxial film due to the 2% lattice mismatch. In contrast, films grown on $NdGaO_3$ show no increase in rocking-curve width with film thickness, resulting in a smoother morphology.

For microwave applications, a significant effort has focused on the growth of $YBa_2Cu_3O_7$ on MgO. The large 9% lattice mismatch between $YBa_2Cu_3O_7$ and MgO (a = 4.211 Å) leads to multiple in-plane orientations with either cube-on-cube $[100]_{film} \| [100]_{substrate}$ or 45° rotated $[110]_{film} \| [100]_{substrate}$ film orientations with respect to the substrate (80). These multiple in-plane orientations introduce large-angle grain boundaries and result in $YBa_2Cu_3O_7$ films with poor superconducting properties. Better results have been obtained by annealing the MgO substrate above 1000°C in an oxygen ambient to improve the crystallinity of the MgO (100) surface. c-Axis-oriented $YBa_2Cu_3O_7$ films with only one in-plane orientation and critical current densities greater than 1 MA/cm^2 at 77 K have been obtained on annealed MgO substrates (81). Excellent control of the in-plane texture for $YBa_2Cu_3O_7$ on (100) MgO can also be achieved with the use of a $SrTiO_3$ buffer layer (82,83). It has been shown that a relatively thin (~25 nm) $SrTiO_3$ buffer layer results in only one in-plane orientation. The $SrTiO_3$ film grows with the $[100]_{STO} \parallel [100]_{MgO}$ despite a large lattice mismatch. $YBa_2Cu_3O_7$ films can be reproducibly grown on $SrTiO_3$- buffered MgO with $T_c \sim 90$ K and $J_c(77\text{ K}) > 10^6$ A/cm^2. The adverse effects of the high dielectric constant for $SrTiO_3$ are minimal when used as a buffer layer, because the microwave losses are proportional to the volume of lossy material. A surface resistance of 260 μΩ at 77 K measured at 8.3 GHz has been reported for $YBa_2Cu_3O_7$ on $SrTiO_3$-buffered (100) MgO (84).

The chemical reactivity and large lattice mismatch of $YBa_2Cu_3O_7$ with *r*-plane sapphire requires the use of oxide buffer layers. Using pulsed-laser deposition, large-area $YBa_2Cu_3O_7$ films with critical current densities as high as 5×10^6 A/cm^2 at 77 K have been realized on 3-in.-diameter sapphire wafers with a CeO_2 buffer layer (85). Microwave loss measurements on similarly prepared films on CeO_2-buffered sapphire yielded $R_s(77\text{ K}) = 550$ μΩ at 9.5 GHz (86). The difference in the thermal expansion coefficients of $YBa_2Cu_3O_7$ and Al_2O_3 somewhat limits the film thickness that can be achieved without the appearance of cracks.

The epitaxial growth of $YBa_2Cu_3O_7$ on (100) YSZ has been extensively studied. The lattice mismatch of this substrate with $YBa_2Cu_3O_7$ results in multiple in-plane orientation possibilities. At high growth temperatures of ~ 760°C, the dominant in-plane orientation for *c*-axis-perpendicular $YBa_2Cu_3O_7$ is $[100]YBa_2Cu_3O_7 \parallel [100]_{YSZ}$, whereas at low temperatures, it is $[100]YBa_2Cu_3O_7 \parallel [110]YBa_2Cu_3O_7$ (87). YSZ also supports a third in-plane $YBa_2Cu_3O_7$ orientation in which the $YBa_2Cu_3O_7$ *a* axis makes a 9° angle with the YSZ ⟨100⟩ (88). In addition, an interfacial reaction of $YBa_2Cu_3O_7$ with YSZ occurs to form $BaZrO_3$. In order to prevent multiple orientations, a thin CeO_2 or Y_2O_3 layer on the (100) YSZ substrate surface eliminates all but the $[100]YBa_2Cu_3O_7 \parallel [110]YBa_2Cu_3O_7$ orientation (89). Similar results have also been obtained with monolayers of CuO or $BaZrO_3$ (88).

The growth of epitaxial $YBa_2Cu_3O_7$ on silicon presents the interesting possibility of integrating superconducting and semiconducting electronics. A buffer layer is required due to chemical interactions. High critical current densities have been obtained for structures employing an epitaxial YSZ buffer layer (90). For example, critical current densities greater than 2 MA/cm^2 at 77 K have been realized for 50-nm-thick $YBa_2Cu_3O_7$ films with a 50-nm-thick YSZ buffer layer. Superconducting films with high critical current densities have also been obtained with epitaxial MgO on Si(001) (91). A hydrogen termination of the Si surface is generally required in order to prevent oxidation of the Si prior to buffer-layer growth. MgO buffers are more effective than YSZ in preventing subsequent oxidation of the Si surface due to a smaller oxygen diffusion coefficient. In either case, the strain induced due to the large difference between thermal expansion coefficients of $YBa_2Cu_3O_7$ and Si significantly limits the thickness of the $YBa_2Cu_3O_7$ film.

The growth of epitaxial $YBa_2Cu_3O_7$ on metal substrates with rolling-induced biaxial texture, coupled with appropriate buffer-layer architectures, represents an interesting approach for producing long-length $YBa_2Cu_3O_7$-based superconducting tapes with a high J_c (92). Biaxially textured (001) Ni tapes, formed by recrystallization of cold-rolled pure Ni, have been used as the initial in-plane-aligned substrate material for subsequent $YBa_2Cu_3O_7$ film deposition. An epitaxial buffer layer is necessary in order to grow $YBa_2Cu_3O_7$ due to reactions between the superconductor and Ni. For example, the epitaxial growth of a (001)-oriented CeO_2/YSZ oxide buffer-layer architecture maintains the sharp crystallographic cube texture of the metal substrate while providing a barrier to chemical interaction of the Ni with the HTS film. Figure 2.19 shows the x-ray diffraction data for a YBCO/YSZ/CeO_2/Ni structure. Note that the epitaxial relationship is maintained throughout the structure. Figure 2.20 shows a cross-section transmission electron microscopic (TEM) image of the CeO_2/Ni interface revealing a NiO layer

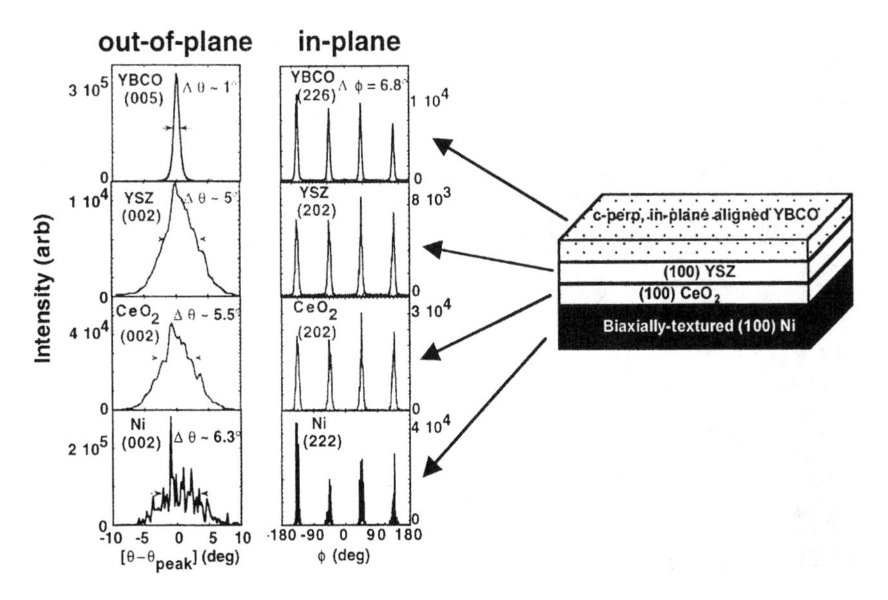

FIGURE 2.19 X-ray diffraction data for an epitaxial $YBa_2Cu_3O_7$/YSZ/CeO_2 multilayer on biaxially textured (001) Ni.

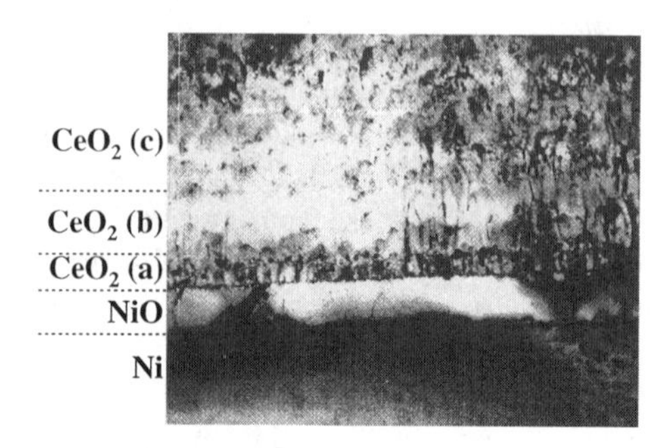

FIGURE 2.20 Cross-section TEM image of an epitaxial CeO_2/Ni structure, showing a NiO layer at the interface.

at the boundary. This layer presumably forms after the CeO_2 is nucleated on the (001) Ni surface. High-resolution TEM indicates that the NiO layer is epitaxial with respect to the Ni substrate. The epitaxial buffer layers compensate for poor lattice match and chemical incompatibility, although thermal expansion mismatch remains an issue. Subsequent growth of $YBa_2Cu_3O_7$ on this structure, referred to as a rolling-assisted, biaxially textured substrate (RABiTS™), results in *c*-axis-oriented in-plane-aligned films with superconducting critical current densities as high as 1 MA/cm^2 at 77 K. Figure 2.21 shows a cross-section TEM image of a

FIGURE 2.21 Cross-section TEM image of a completed $YBa_2Cu_3O_7$/YSZ/CeO_2/Ni structure, including a grain boundary.

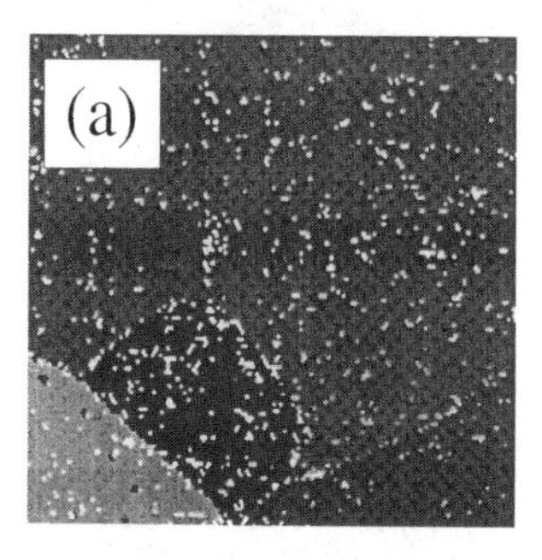

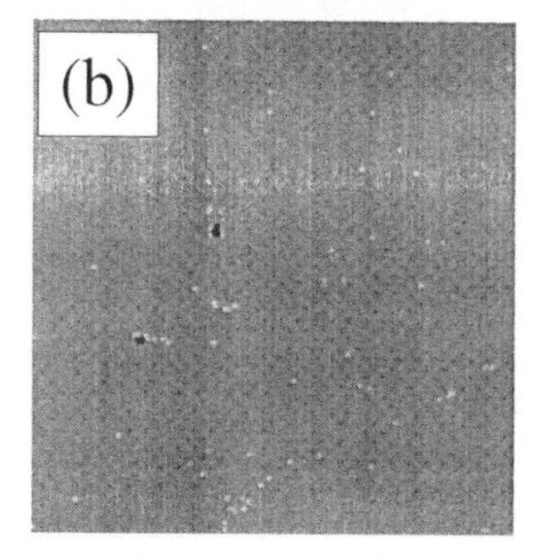

FIGURE 2.22 Electron backscattered pattern form a YSZ/CeO_2 biaxially textured Ni structure. The different shading corresponds to grain-to-grain misorientation of (a) 1° and (b) 5°.

grain boundary that extends through the Ni substrate into the oxide film. No NiO is observed in the boundary. Figure 2.22 shows an electron backscattered pattern of an epitaxial YSZ/CeO_2 multilayer on a biaxially textured substrate. The color shading is such that the grains with similar shading share a grain boundary with a grain misorientation angle of less than (a) 1° or (b) 5°. Recently it has been shown that buffer-layer architectures with single-buffer layers, such as YSZ, can be utilized.

In addition to the RABiTS concept described, two additional approaches have been identified whereby thick epitaxial $YBa_2Cu_3O_7$ films, with the high degree of in-plane and out-of-plane crystallographic texture necessary for achieving high J_c, can be obtained on long-length substrates for deposited conductors. In one case shown schematically in Figure 2.23, ion-beam-assisted deposition (IBAD) is used to achieve in-plane alignment of oxide buffer layers on polycrystalline metal tapes (93–95). In-plane texture on the order of 5°–10° can be realized with this technique. A second approach, known as inclined substrate deposition, involves directional impingement of the depositing flux on the substrate surface (96). Varying growth rates for different orientations of grains results in dominate crystal

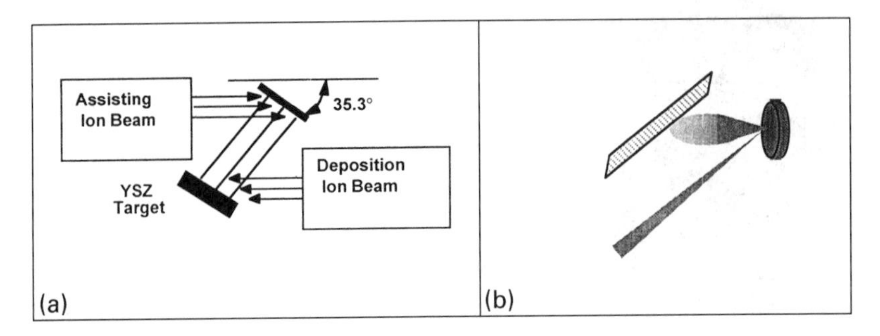

FIGURE 2.23 Schematic drawings for (a) ion-beam-assisted deposition and (b) inclined substrate deposition approaches to achieving biaxially textured films.

growth by specific orientations. Epitaxial $YBa_2Cu_3O_7$ films on these surfaces yield critical current densities approaching 1 MA/cm^2. The magnetic field dependence of J_c for films on either IBAD or RABiTS substrates are quite similar to that measured for the epitaxial $YBa_2Cu_3O_7$ film on $SrTiO_3$. In fact, the relative drop of J_c with applied external magnetic field is less for the $YBa_2Cu_3O_7$ film grown on IBAD or RABiTS tapes than for the film on (001) $SrTiO_3$, indicating the presence of additional flux pinning defects in the films. These pinning sites may be associated with growth-induced defect structures, low-angle grain-boundary pinning, or both. The performance of the epitaxial $YBa_2Cu_3O_7$ on either RABiTS or IBAD substrates is far superior, in terms of J_c, to other superconducting wire technologies, both for zero-field and high-field applications. Thus, the use of IBAD or RABiTS represents a possible means for producing long-length superconducting tapes for high-current, high-field applications at 77 K, particularly if high values of the "engineering" J_c, defined as the critical current per total conductor cross-sectional area (including substrate thickness), can be realized with thinner substrates and/or thicker $YBa_2Cu_3O_7$ films.

2.4.1.3 *a*-Axis-Oriented Epitaxy

The layered crystal structure in the HTS materials results in a significant anisotropy in the crystal growth rate. $YBa_2Cu_3O_7$ grows faster along the *a*-*b* plane than along the *c* axis. This anisotropic growth mechanism tends to favor epitaxial films in which the *c* axis is perpendicular to the substrate surface. For many applications, this orientation is sufficient. However, for some proposed device structures, thin films with either the *a* or *b* axis perpendicular to the substrate surface are preferred. For instance, *a*-axis-oriented $YBa_2Cu_3O_7$ films are potentially useful for sandwich-type junctions due to the longer coherence length along the *a* direction. One technique that has proven successful is to nucleate *a*-axis-oriented $YBa_2Cu_3O_7$ films at a reduced temperature for subsequent growth of additional *a*-axis-oriented material at elevated temperatures. Epitaxial $YBa_2Cu_3O_7$ films tend to grow *a*-axis oriented on (001) $SrTiO_3$ at growth temperatures that are approximately 100°C below that which is optimal for the growth of *c*-axis-oriented films with good superconducting properties. Figure 2.24 shows a cross-section TEM image of an *a*-axis $YBa_2Cu_3O_7$ film nucleated at low temperature and subsequently grown at high temperatures (97). Despite having inferior superconducting properties, the initial *a*-axis-oriented $YBa_2Cu_3O_7$ layer can serve as a template for the subsequent growth of additional *a*-axis-oriented $YBa_2Cu_3O_7$ at higher temperatures, resulting in *a*-axis-oriented films having a T_c greater than 90 K (97–99). Subsequent work has also shown that a template layer of $PrBa_2Cu_3O_7$ grown under similar conditions is also quite effective at nucleating *a*-axis-oriented $YBa_2Cu_3O_7$ films (100).

One limitation with *a*-axis-oriented epitaxial films grown as described above is that the orientation of the in-plane *c* axis is not uniquely determined. For *a*-axis-oriented $YBa_2Cu_3O_7$ films on cubic substrates, the *c* axis of individual grains have an equal probability of aligning with the substrate [100] or [010] axes, resulting in grain boundaries in which the *a-b* planes meet at 90°. This is seen in the plan-view TEM image of an *a*-axis film on $SrTiO_3$ (Fig. 2.25) (101,102). The grain boundaries significantly degrade the transport properties of the films. One solution is to use a noncubic single-crystal substrate, such as orthorhombic $NdGaO_3$ (a = 5.431 Å, b = 5.499 Å, c = 7.710 Å) or tetragonal $LaSrGaO_4$ (a = 3.843 Å, c = 12.681 Å), to break the in-plane fourfold symmetry and enable an unambiguous selection for the *c*-axis direction in the $YBa_2Cu_3O_7$ film (103,104).

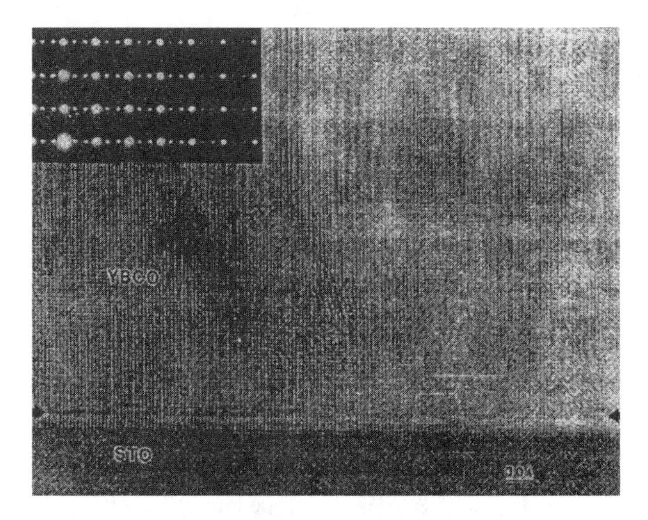

FIGURE 2.24 Cross-section TEM image of an *a*-axis-oriented $YBa_2Cu_3O_7$ film nucleated at low temperature and grown at high temperature. (From Ref. 97.)

FIGURE 2.25 Plan-view TEM image of *a*-axis-oriented $YBa_2Cu_3O_7$, showing the 90° grain boundaries.

Using (100) $LaSrGaO_4$, in-plane alignment of *a*-axis films with $J_c(77\ K) > 10^5$ A/cm^2 have been reported (102).

2.4.2 $NdBa_2Cu_3O_7$

Of the $REBa_2Cu_3O_7$ materials, the most studied, other than $YBa_2Cu_3O_7$, is the $NdBa_2Cu_3O_7$ system. $NdBa_2Cu_3O_7$ films possessing high crystallinity, surface stability, and relatively high T_c have been realized using several techniques, including pulsed-laser deposition and sputtering (104,105). However, control of the film properties is more difficult due to the possible antisite substitution of Nd for Ba. For $NdBa_2Cu_3O_7$ thin films produced by off-axis dc magnetron sputtering, the superconducting properties of the films was strongly dependent on target composition, substrate temperature, and oxygen partial pressure (105). As with many other rare-earth-based superconductors, $NdBa_2Cu_3O_y$ (NdBCO) is known to form a solid solution in which Nd^{3+} ions can substitute for Ba^{2+} ions because of their comparable radii, leading to the actual composition $Nd_{1+x}Ba_{2-x}Cu_3O_y$. The amount of Nd defects on the Ba sites (indicated by x) has a strong influence on the electronic properties of the material, and even in small amounts it significantly suppresses superconductivity in both bulk and thin films. Recent studies conducted on melt-textured and single-crystal materials obtained by different bulk synthesis techniques have demonstrated that the degree of Nd-for-Ba substitution is highly sensitive to processing conditions. In particular, bulk samples with nominal stoichiometric composition ($x = 0.0$) and T_c of 96 K are obtained by synthesis in a reduced oxygen atmosphere, with $P(O_2)$ ranging between 0.01 and 0.001 atm. These results suggest that the thermodynamic stability curve of NdBCO in the $P(O_2)$–T phase space is shifted toward reduced oxygen pressure values relative to the $YBa_2Cu_3O_{7-\delta}$ curve. Consequently, deposition conditions appropriate for the growth of NdBCO films by the in situ technique differ significantly from those successfully used for YBCO.

A systematic study of the growth of NdBCO films was performed, focusing on the correlation of T_c, J_c, and $\rho(T)$ with the temperature and oxygen partial pressure used during deposition for thin films grown by PLD (105,106). As a result, the line in the $1/T$-log[$P(O_2)$] plane was determined for optimal NdBCO thin-film synthesis. The NdBCO films were deposited on $LaAlO_3$ substrates by ablation of a stoichiometric target. The results show that the best NdBCO films are obtained at oxygen pressures in the range 0.2–1.2 m torr, depending on the substrate temperature. This is more than two orders of magnitude lower than the correspondent oxygen pressure appropriate for YBCO film growth. A plot of the optimal conditions [T_s, $P(O_2)$] in the log[$P(O_2)$]-$1/T$ phase space originates a line almost parallel to the corresponding Hammond and Bormann line for YBCO films, but shifted with respect to this by about two orders of magnitude, as shown in Figure 2.26. In the same diagram, we show the melting line for stoichiometric bulk $Nd_1Ba_2Cu_3O_y$ ($x = 0$). For comparison, we also show the melting line, the decomposition line, and the Hammond and Bormann line for YBCO. The latter refers to the CuO–Cu_2O–O_2 equilibrium and provides experimental conditions for optimal in situ YBCO film synthesis.

Additional work on deposition of NdBCO by pulsed-laser deposition showed that the use of single-crystal $NdBa_2Cu_3O_7$ as an ablation target is effective in producing superconducting films with enhanced smoothness as seen in Figure 2.27 (107). Particle density decreased by a factor of 10,000. $Nd_{1+x}Ba_{2-x}Cu_3O_{7-y}$ thin films have also been grown by metal–organic chemical vapor deposition (108). At oxygen partial pressures between 0.65 and 0.85 torr, better film properties are obtained for nonstoichiometric compositions. However, for an oxygen partial pressure of 0.01 torr, high-quality films were obtained with stoichiometric compositions. This dependence on oxygen pressure is related to the tendency for Nd–Ba substitution in this system. Even for the better films, T_c was only 50 K.

2.4.3 $Tl_2Ba_2Ca_{n-1}Cu_nO_{2n+4}$ and $TlBa_2Ca_{n-1}Cu_nO_{2n+3}$

The synthesis of Tl-based HTS thin films has attracted significant interest due in large part to the high superconducting transition temperatures ($\sim$ 125 K) that have

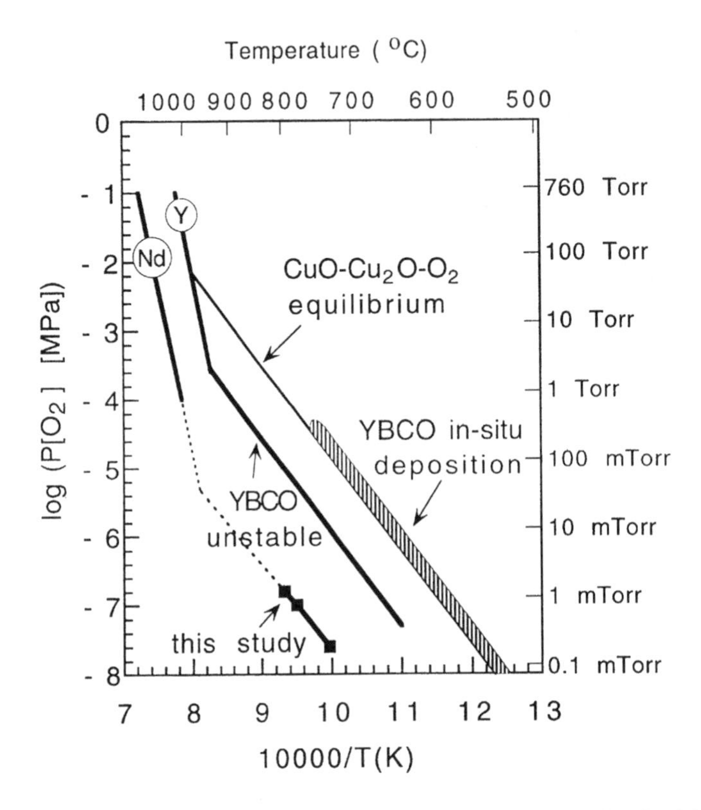

Figure 2.26 The $P(O_2)$–T phase diagram for $NdBa_2Cu_3O_7$ thin-film synthesis. (From Ref. 105.)

been achieved for these materials. Most of the application interest for epitaxial Tl-based HTS oxide films is in passive microwave devices. Within the HTS materials, the lowest measured values of surface resistance at high frequencies (10 GHz) have been for Tl-based films (109). The two related Tl-based cuprate series, $Tl_2Ba_2Ca_{n-1}Cu_nO_{2n+4}$ and $TlBa_2Ca_{n-1}Cu_nO_{2n+3}$, are shown schematically in Figure 2.28 (110). Most thallium-based HTS materials possess a tetragonal crystal structure and can be simply described as consisting of alternating single or double Tl–O layers and the perovskitelike $Ba_2Ca_{n-1}Cu_nO_{2n+1}$ layers. An exception to this is the $Tl_2Ba_2CuO_6$ compound, which exhibits orthorhombic and monoclinic symmetry. The unit cell for $Tl_2Ba_2Ca_{n-1}Cu_nO_{2n+4}$ contains a double Tl–O layer that is sandwiched between two Ba–O layers. The CuO_2 sheets are adjacent to the Ba–O layers. Bulk samples have been synthesized with $n = 1$–5, although thin-film work has been limited to $n \leq 3$. Extremely low surface resistance has been measured in these materials with $R_s \sim 0.024$ mΩ-cm at 5 K and 10 GHz for both

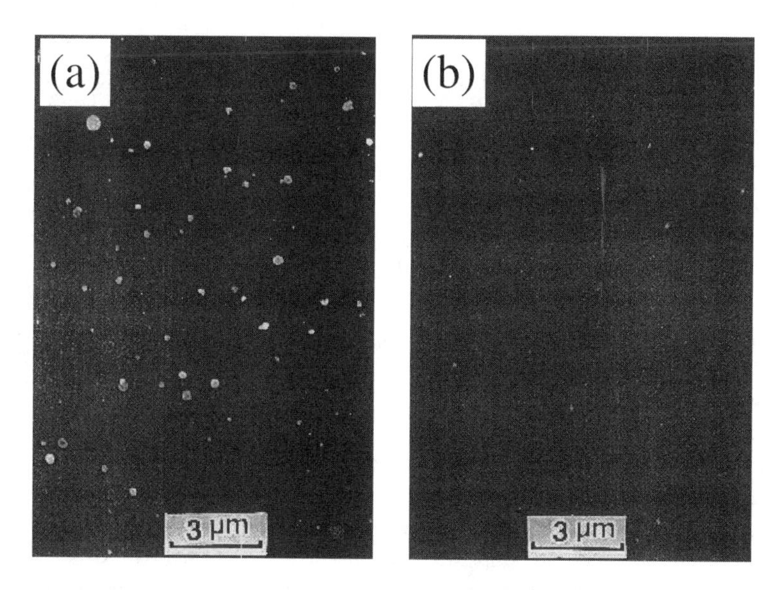

FIGURE 2.27 Scanning electron micrographs of $NdBa_2Cu_3O_7$ films grown by pulsed-laser deposition using (a) a ceramic and (b) a single-crystal ablation target.

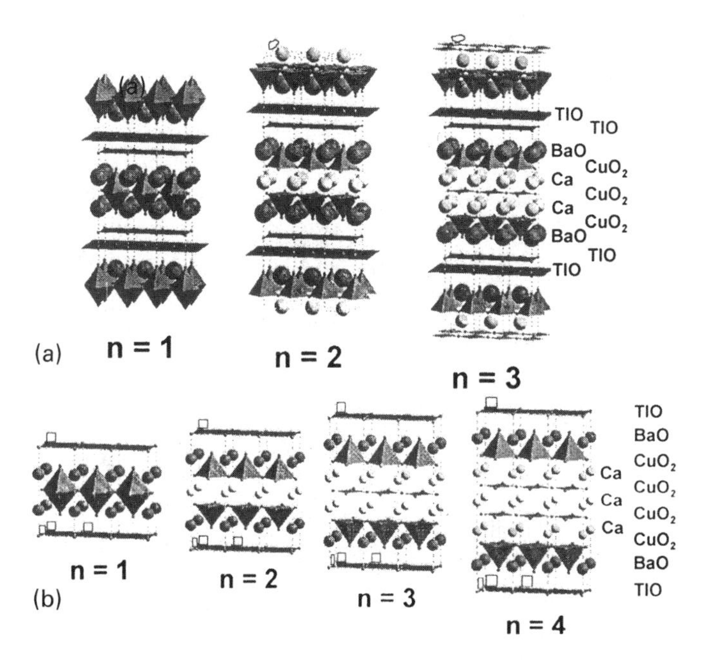

FIGURE 2.28 Schematic drawing of the (a) $Tl_2Ba_2Ca_{n-1}Cu_nO_{2n+4}$ and (b) $TlBa_2Ca_{n-1}Cu_nO_{2n+3}$ series of compounds.

$Tl_2Ba_2CaCu_2O_8$ and $Tl_2Ba_2Ca_2Cu_3O_{10}$ films (111). The second series, $TlBa_2Ca_{n-1}Cu_nO_{2n+3}$, differs in that each unit cell contains only a single Tl–O layer. These materials exhibit much stronger magnetic flux pinning than the double Tl–O layer compounds, presumably due to the reduced anisotropy when compared to $Tl_2Ba_2Ca_{n-1}Cu_nO_{2n+4}$. When optimally doped, the Tl-based HTS materials exhibit high superconducting transition temperatures. For optimally doped $Tl_2Ba_2Ca_{n-1}Cu_nO_{2n}0_{+4}$, $T_c = 90, 110$, and 125 K for $n = 1, 2$, and 3, respectively. For $TlBa_2Ca_{n-1}Cu_nO_{2n+3}$, T_c = 44, 105, and 124 K, respectively. Optimal doping and/or enhanced phase stability is achieved through control of the oxygen content and by chemical substitution, including rare-earth doping on the Ca site, Pb or Bi doping on the Tl site, or Sr doping on the Ba site.

The growth of epitaxial Tl-based HTS thin films is quite challenging due to the complex phase relationships of these materials and the high volatility of Tl oxides. The fabrication of thin films of the Tl-based HTS compounds involves a precarious balance between the high temperatures required to form the superconducting phases and the high volatility of Tl at these temperatures. The partial pressures of both oxygen and thallous oxide play key roles. Most TBCCO film synthesis approaches are based on a two-step process involving the deposition of the amorphous precursor, followed by ex situ annealing in the presence of thallous oxide vapor to form the desired crystalline phase. Initial efforts involving bulk materials suggested that the formation of Tl-Ba-Ca-Cu-O(TBCCO) compounds proceeds by the following sequences: Tl(2201)–Tl(2212)–Tl(2223)–Tl(1223)–Tl(1234)–Tl(1245) or Tl(2201)–Tl(2212)–Tl(2223)–Tl(1212). The toxicity of Tl oxide adds to the difficulty in synthesizing thin films. Substrates with attractive microwave properties have received the majority of the attention. For optimal film properties, the substrate of choice is usually $LaAlO_3$ due to a low lattice mismatch. MgO and Al_2O_3 have also been considered with some success (112,113). Due to the volatility of thallium oxide, most of the effort to grow epitaxial Tl-based HTS film has focused on ex situ annealing of precursor films. The precursor films can be deposited by a number of techniques, including laser ablation (114), thermal and electron beam evaporation (115), chemical vapor deposition (116), and sputtering (109). Epitaxial films of the various Tl-based HTS phases have been obtained both with and without Tl in the precursor films. Thallination and phase formation then takes place upon annealing in a furnace. Typically, the precursor film is placed in a closed crucible along with a bulk powder or pellet that has a composition similar to that of the desired phase (109). The bulk powder or pellet serves as the Tl source. For independent control of the sample annealing temperature, Tl-oxide partial pressure, and oxygen partial pressure, a two-zone furnace can be used in which the precursor film is placed in the high-temperature zone and thallium oxide in the low-temperature zone.

For $Tl_2Ba_2Ca_{n-1}Cu_nO_{2n+4}$, most of the effort in growing epitaxial films has focused on the $Tl_2Ba_2CaCu_2O_8$ (n = 2) phase due to its relatively high superconducting transition temperature and superior phase stability when compared to $Tl_2Ba_2Ca_2Cu_3O_{10}$. By postannealing sputter-deposited precursor films in the presence of $Tl_2Ba_2Ca_2Cu_3O_{10}$ and Tl_2O_3 powder, epitaxial films of $Tl_2Ba_2CaCu_2O_8$ films have been grown on $LaAlO_3$ with $T_c \sim 108$ K and $R_s(77\text{ K}) = 130$ μΩ measured at 10 GHz (109). Epitaxial growth of $Tl_2Ba_2CaCu_2O_8$ has also been achieved on both MgO and Al_2O_3, although the properties of the films are generally inferior to that obtained on $LaAlO_3$. Although physical deposition techniques are more commonly used to deposit the precursor film, MOCVD has also been utilized to deposit precursor BaCaCuO(F) films for subsequent annealing in the presence of Tl_2O vapor (116). $Tl_2Ba_2CaCu_2O_8$ films have been obtained with BaCaCuO(F) precursors with $T_c = 105$ K, $J_c(77\text{ K}) = 1.2 \times 10^5$ A/cm^2, and a surface resistance of 400 μΩ at 10 GHz and 5 K.

Formation of the $Tl_2Ba_2Ca_2Cu_3O_{10}$ ($n = 3$) phase is more challenging, requiring careful control of precursor stoichiometry and complex optimization of the annealing conditions (117). Phase evolution is rather complex and dependent on both thermodynamic and kinetic considerations. Some evidence suggests that the formation of the $n = 3$ member proceeds as the initial nucleation of the $n = 2$ members, followed by conversion to the higher-n phase (118). Although Tl(2212) can be formed by solid-state diffusion, evidence suggests that a liquid phase is required in order to transform Tl(2212) into Tl(2223). The partial pressure of Tl_2O is a key factor in the formation of specific phases. Epitaxial films of the $Tl_2Ba_2Ca_2Cu_3O_{10}$ compound that are nearly single phase have been synthesized on single-crystal (100) $LaAlO_3$ by annealing sputter-deposited BaCaCuO precursor films in the presence of Tl–Ba–Ca–Cu–O and Tl_2O powders (117). Films with $T_c \sim 120$ K and surface resistances measured at 10 GHz of 24 μΩ at 4.2 K, 180 μΩ at 77 K, and 360 μΩ at 100 K have been obtained. Several studies suggest that reducing the oxygen pressure promotes the formation of Tl(2223) at reduced temperatures. Because reducing the oxygen partial pressure promotes melting, a reduction in annealing temperature is apparently the result of a decrease in the temperature required to produce the liquid phase required for Tl(2223) formation.

Interest in the single Tl–O series, $TlBa_2Ca_{n-1}Cu_nO_{2n+3}$, primarily stems from the high-T_c and superior flux pinning properties when compared to $Tl_2Ba_2Ca_{n-1}Cu_nO_{2n+4}$. Epitaxial films of the single Tl–O layer $TlBa_2Ca_{n-1}Cu_nO_{2n+3}$ compounds can be synthesized by ex situ annealing of precursor films. The formation of the single versus the double Tl–O layer compounds is controlled to a large extent by the amount of Tl in the accompanying pellet or powder during annealing of the precursor. Using the ex situ precursor approach, $TlBa_2CaCu_2O_7$ thin films have been obtained with $T_c \sim 90$ K and $J_c(77\text{ K}) \sim 2 \times 10^5$ A/cm^2 (119). Related materials to the $TlBa_2Ca_{n-1}Cu_nO_{2n+3}$ system include the fractional cation substitution of Pb or Bi on the Tl site, and Sr on the Ba. The addition of Sr to the TBCCO compounds tends to stabilize the single Tl–O layered phases. It is interesting to note that $(Tl,Pb)Sr_2CaCu_2O_7$ and $TlBa_2CaCu_2O_7$ films show significantly different dependence on subsequent oxygen annealing. Whereas the superconducting transition temperature of $TlBa_2CaCu_2O_7$ changes when annealed in different oxygen pressures, the transition temperature of $(Tl,Pb)Sr_2CaCu_2O_7$ shows little change when annealed at 700°C in oxygen pressures ranging form 10^{-4} to 1 atm of oxygen (120,121). This apparent stability achieved with Pb doping may prove advantageous in device applications.

Some of the best results for the single Tl–O layer series of compounds have been realized for ex situ processed $(Tl,Bi)Sr_{1.6}Ba_{0.4}Ca_2Cu_3O_{9-d}$ films in which the precursor film includes Tl. Precursor $Tl_xBi_ySr_{1.6}Ba_{0.4}Ca_2Cu_3O_z$ films were deposited on (100) $LaAlO_3$ by single-target laser ablation and were sandwiched between two $(Tl,Bi)Sr_{1.6}Ba_{0.4}Ca_2Cu_3O_{9-d}$ pellets during the subsequent anneal to produce the superconducting $n = 3$ phase (122). The $(Tl,Bi)Sr_{1.6}Ba_{0.4}Ca_2Cu_3O_{9-d}$ epitaxial films exhibit a T_c as high as 111 K and a critical current density of 2 MA/cm^2 at 77 K. Epitaxial superconducting $Tl_{0.5}Pb_{0.5}Sr_{1.6}Ba_{0.4}Ca_2Cu_3O_9$ films have been synthesized by thermal spray followed by postannealing (123). This simple, low-cost process is performed by spraying a stoichiometric solution containing Tl, Pb, Ba, Sr, Ca, and Cu nitrates on substrates heated to 300–400°C. The precursor films are annealed in a sealed silver containment with unfired pellets of the same stoichiometry. Epitaxial films have been realized on $LaAlO_3$ with T_c = 106–109 K and J_c(77 K) = 1.1 MA/cm^2. Another interesting phase is the TlCu-1234 material, in which single-phase films have been prepared by solid-phase epitaxy (124). Films on $SrTiO_3$ show a T_c = 113 K, with J_c (77 K) = 1 MA/cm^2. The deposition of precursor films for Tl-based HTS epitaxy has also been achieved via electrodeposition (125). Nitrates have been used as the cation source materials. With this approach, $(Tl,Bi)_{0.9}Sr_{1.6}Ba_{0.4}Ca_2Cu_3Ag_{0.2}O_x$ films have been obtained with J_c (77 K) = 3×10^5 A/cm^2.

Unlike the double Tl–O members, the single Tl–O layer HTS compounds have been successfully grown by in situ techniques. Formation of the superconducting phase during the film deposition process is achieved by introducing a separate thallium oxide source during high-temperature film deposition. The properties of in situ Tl-based HTS films are not yet as good as those synthesized by ex situ postannealing. Typically, the best in situ films have been obtained with a partial substitution of Pb for Tl, Sr for Ba, and Y-doping on the Ca site. Epitaxial $(Tl,Pb)Sr_2CuO_5$ ($n = 1$) films have been grown in situ by sputtering in a thallium oxide ambient. Note that stoichiometric films of the $n = 1$ compounds $TlBa_2CuO_5$ and $(Tl,Pb)Sr_2CuO_5$ are overdoped and are not superconductors. The formal valence of copper must be reduced in order to realize superconductivity. This can be accomplished either by removal of oxygen or by substitution of a trivalent cation on the divalent Sr or Ba site. Superconducting transition temperatures as high as 44 K have been realized. Epitaxial $TlBa_2CaCu_2O_7$ ($n = 2$) thin films have been grown on $LaAlO_3$ and $NdGaO_3$ substrates by in situ off-axis magnetron sputtering in the presence of Tl_2O vapor (126). Although the T_c is relatively low in the as-grown epitaxial films, subsequent annealing in O_2 and Tl_2O vapor at 800°C produces a $T_c \sim 97$ K with little degradation of film morphology. (Tl,Pb)Sr_2 $Ca_{1-x}Y_xCu_2O_7$ thin films have also been grown in situ using off-axis magnetron sputtering in the presence of thallium oxide vapor (127). Film grown on $LaAlO_3$ and subjected to an in situ heat treatment exhibit T_c as high as 93 K for $x = 0.2$.

Some success has been realized in the in situ epitaxial growth of $TlBa_2Ca_2Cu_3O_9$ ($n = 3$) films (128). Laser ablation was used to deliver the Ba–Ca–Cu–O flux, whereas a thermal source provides the thallium oxide vapor. Smooth, epitaxial $TlBa_2Ca_2Cu_3O_9$ films were obtained on $LaAlO_3$ substrates with a superconducting transition temperatures near 100 K.

2.4.4 $HgBa_2Ca_{n-1}Cu_nO_{2n+2+\delta}$

The mercury-based cuprates currently possess the highest superconducting transition temperature, with $T_c \sim 135$ K, making these compounds obviously attractive for device applications. The $HgBa_2Ca_{n-1}Cu_nO_{2n+2+\delta}$ series is analogous to the $TlBa_2Ca_{n-1}Cu_nO_{2n+3}$ compounds with Hg replacing the single Tl–O layer. This series extends to $n = 4$, with the $n = 3$ member possessing the highest T_c. Efforts to synthesize epitaxial films have focused on the $HgBa_2CaCu_2O_{6+\delta}$ ($n = 2$) and $HgBa_2Ca_2Cu_3O_{8+\delta}$ ($n = 3$) phases (129–133). Difficulties arise in the synthesis of these materials due in part to the high volatility of Hg-based compounds, requiring significant effort in maintaining the Hg content during film processing. The toxic nature of the mercury oxides is also an issue that must be considered.

Success in obtaining epitaxial films of these two compounds has been limited to ex situ annealing of precursor films in a Hg atmosphere. As with the Tl-based cuprates, the ex situ synthesis of Hg-based HTS films can be accomplished both with or without Hg in the precursor. In one approach, a layer-by-layer mixing of HgO and $Ba_2CaCu_2O_x$ layers is realized by sequentially depositing the precursor film from two laser ablation targets (105). A cap layer was found to be useful in protecting the delicate precursor from adverse reaction with the ambient and in retaining the Hg in the film during the postannealing procedure. The precursor film is subsequently annealed in an evacuated quartz tube in the presence of bulk Hg-cuprate and $Ba_2CaCu_2O_x$. Using this approach, epitaxial $HgBa_2CaCu_2O_{6-d}$ films have been grown on (100) $SrTiO_3$ with $T_c \sim 124$ K and $J_c(5\text{ K}) \sim 10^7$ A/cm^2, as shown in Figure 2.29 (129–132). Epitaxial $HgBa_2CaCu_2O_{6-d}$ and $HgBa_2Ca_2Cu_3O_{8-d}$ films have also been obtained with Hg-free precursor films that are postannealed in the presence of Hg vapor (132,133). Using Hg-free precursors, $HgBa_2Ca_2Cu_3O_{8-d}$ films have been grown on (100) $LaAlO_3$ with $T_c = 128$ K and $J_c(77\text{ K}) = 1.4 \times 10^6$ A/cm^2. Some improvement in film properties has been observed when the temperature ramp for the annealing step is rapid, presumably by reducing film–substrate interactions and minimizing the formation

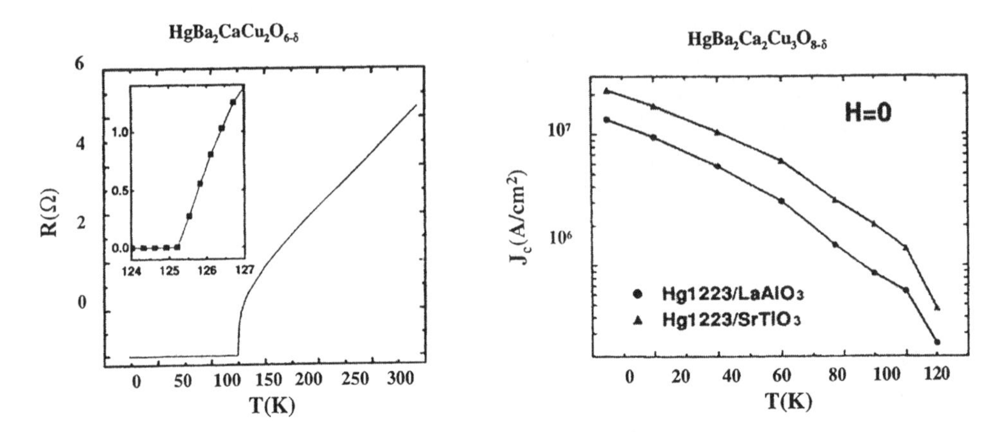

FIGURE 2.29 Resistivity and J_c for $HgBa_2CaCu_2O_6$ grown on (001) $SrTiO_3$. (From Ref. 132.)

of $CaHgO_2$ as an impurity phase (134). The deposition of precursor films by non-vacuum techniques has also been investigated for Hg-based HTS cuprate films. Using spray pyrolysis, 1–2-μm-thick epitaxial $HgBa_2Ca_2Cu_3O_y$ films with $T_c \sim$ 130 K and moderately high J_c have been grown on (100) $SrTiO_3$ substrates (135). Although the film surface morphology may not be suitable for multilayer device structures, the use of spray pyrolysis for epitaxial film growth is potentially useful for deposited HTS conductors. Re-doped Hg-1223 films have been synthesized by the two-step process of depositing a precursor and annealing a mercury vapor atmosphere (136). The precursor is a $HgO/Re_xBa_2Ca_2Cu_3O_y$ multilayer capped with a HgO protective layer. Re-doped Hg-1223 films exhibit a $T_c(R = 0)$ as high as 127.5 K with J_c (77 K) $= 1.5 \times 10^6$ A/cm^2, as seen in Figure 2.30. In addition, the Re doping increases the irreversibility field above 77 K due to enhanced coupling along the c axis.

Epitaxy of the Hg-based materials is difficult due to the volatile nature of Hg. Recently, a different approach involving cation exchange has been adopted to achieve epitaxial Hg-1212 (137). First, epitaxial Tl-1212 films are obtained and use as the precursor. Second, the Tl-1212 film is annealed in an evacuated quartz tube with Hg–Ba–Ca–Cu–O pellets. A gas/solid reaction leads to an exchange of the Tl with Hg, leading to Hg-1212 films possessing $T_c \sim$ 120–124 K and $J_c \sim 3.2 \times 10^6$ A/cm^2. The transport properties of these films are shown in Figure 2.31.

The in situ growth of Hg-based cuprate films is seriously hampered by the high vapor pressure of mercury. The only report of limited success is for the in situ sputter deposition of $HgBa_2CuO_4$ ($n = 1$) films (138). To prevent the loss of Hg from the film, the growth temperature was limited to 550°C. The result was an epitaxial film with the $HgBa_2CuO_4$ structure and a $T_c = 40$ K.

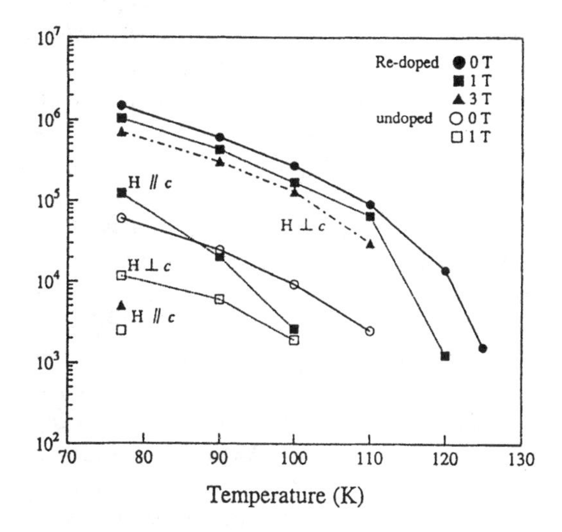

FIGURE 2.30 *J*yc(*T*) behavior for Re-doped Hg-1223 films. (From Ref. 136.)

2.4.5 $Bi_2Sr_2Ca_{n-1}Cu_nO_x$

The $Bi_2Sr_2Ca_{n-1}Cu_nO_{2n+4}$ system is represented by three related phases with $n = 1, 2$, or 3 in which each member contains n copper oxide planes per unit cell. The crystal structure of these materials is slightly orthorhombic due to an incommensurate modulation along the *b* direction. As with the other homologous HTS

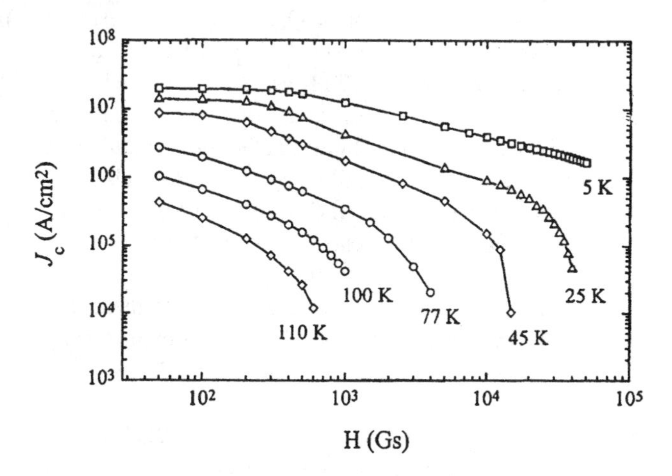

FIGURE 2.31 J_c behavior for $HgBa_2CaCu_2O_x$ films synthesized by cation exchange. (From Ref. 137.)

series, the value of T_c increases as n increases, with $T_c \sim 25, 90$, and 110 K for the $n = 1, 2$, and 3 phases, respectively. In all of the Bi-based HTS cuprates, carrier concentration and oxygen content are critical in optimizing T_c. For instance, bulk studies show that fully oxidized $Bi_2Sr_2CuO_6$ ($n = 1$) is overdoped, with the maximum T_c realized through a combination of trivalent doping on the Sr site and annealing in a reduced atmosphere (139). These materials are more anisotropic than $YBa_2Cu_3O_7$ in terms of their electronic and growth properties. This anisotropy leads to reduced flux pinning relative to $YBa_2Cu_3O_7$ that is particularly evident for $T > 40$ K. Most of the application interest in $Bi_2Sr_2Ca_{n-1}Cu_nO_{2n+4}$ films is directed toward junction devices.

In situ $Bi_2Sr_2Ca_{n-1}Cu_nO_{2n+4}$ films have been synthesized by a number of techniques, including pulsed-laser deposition (140), sputtering (141), molecular beam epitaxy (9), and metal–organic chemical vapor deposition (142). In general, epitaxial $Bi_2Sr_2Ca_{n-1}Cu_nO_{2n+4}$ films with good superconducting properties are more difficult to obtain than for $YBa_2Cu_3O_7$. The formation of a specific phase is highly sensitive to the growth conditions, including temperature, oxygen pressure, composition, and growth rate (140). Interest in the epitaxial growth of the $n = 1$ member has been limited due to its relatively low T_c, although some effort has been made. For example, the epitaxial growth of $Bi_2Sr_{2-x}La_xCuO_{6-d}$ has been studied using RF-magnetron sputtering (143). The highest value for T_c was 29 K. A significant effort has focused on the in situ growth of $Bi_2Sr_2CaCu_2O_8$ ($n = 2$) films. For instance, single-target PLD has been used to grow epitaxial films on (100) MgO with $T_c = 71$ K and $J_c(4.2\ K) = 5 \times 10^6\ A/cm^2$ (140). The observation that both $Bi_2Sr_2CuO_6$ and $Bi_2Sr_2CaCu_2O_8$ films could be grown from a nominally $Bi_2Sr_2CaCu_2O_8$ target simply by changing the growth temperature by 60°C illustrates the sensitivity of phase formation in the $Bi_2Sr_2Ca_{n-1}Cu_nO_{2n+4}$ system to growth conditions. In general, it is difficult to synthesize phase-pure $Bi_2Sr_2CaCu_2O_8$ ($n = 2$) films in situ with T_c near the optimal of ~ 90 K as reported in the bulk, as the oxygen content is difficult to optimize and intergrowths of the $n = 1$ and $n = 3$ members are difficult to avoid. By carefully controlling the oxidation conditions, dc sputtering has been used to obtain $Bi_2Sr_2CaCu_2O_8$ films with $T_c = 88$ K (141).

As with many of the HTS compounds, the $Bi_2Sr_2Ca_{n-1}Cu_nO_{2n+4}$ unit cell can be viewed as a layered stack of well-defined chemical units. For instance, the structure of $Bi_2Sr_2Ca_{n-1}Cu_nO_{2n+4}$ can be derived from a layering sequence of the $n = 1$ $Bi_2Sr_2CuO_6$ member with $n - 1$ layers of $CaCuO_2$. As such, these compounds are highly amenable to formation through layer-by-layer growth schemes at the atomic level. Layer-by-layer growth schemes using molecular beam epitaxy have proven most effective in controlling the in situ formation of $Bi_2Sr_2Ca_{n-1}Cu_nO_{2n+4}$ epitaxial films of a desired phase (9,14,144,145). With RHEED to monitor the film growth process in situ, it is possible to control the chemical reactions and layer formation in an atomic layer-by-layer approach to form not only the $n = 1$–3 phases, but also metastable phases for which $n \geq 4$ (14,145). Figure 2.32

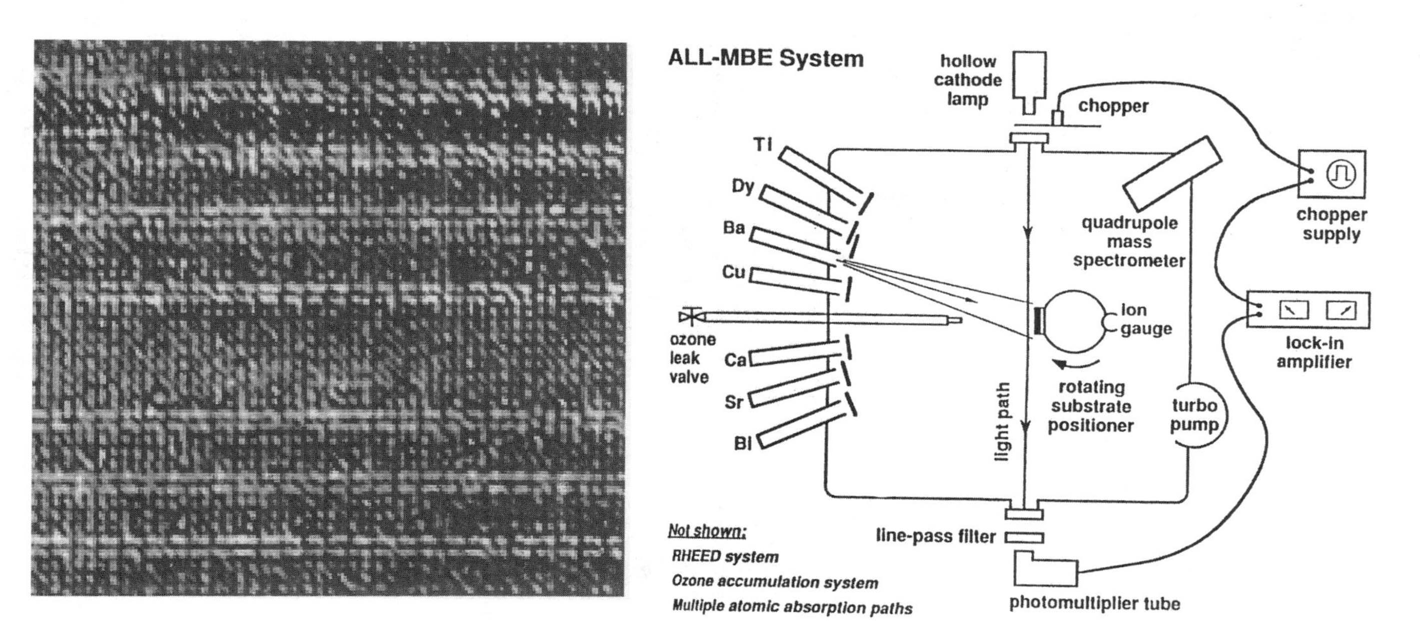

FIGURE 2.32 Cross-section TEM image of a Bi-2278 layer sandwiched in a Bi-2212 matrix using MBE. (From Ref. 145.)

shows a cross-section TEM image of a structure grown by MBE in which an $n = 8$ layer is sandwiched in an $n = 2$ matrix. Using the layer-by-layer approach, epitaxial $Bi_2Sr_2CaCu_2O_8$ films with $T_c \sim 81$ K have been realized by MBE (9).

With a superconducting transition temperature of 110 K, the synthesis of epitaxial $Bi_2Sr_2Ca_2Cu_3O_{10}$ has been the focus of numerous studies. Early efforts to grow $Bi_2Sr_2Ca_3Cu_3O_x$ films focused on postannealing at high temperatures to improve crystallinity and phase formation (146). As with bulk samples, the synthesis of the $n = 3$ phase is enhanced by the addition of Pb either homogeneously in the precursor or as multilayered $Bi_2Sr_2Ca_2Cu_3O_x$/PbO structures. Even then, phase-pure samples are difficult to realize and require detailed control of the annealing conditions. Recently, $Bi_2Sr_2Ca_3Cu_3O_x$ films with $T_c \sim 100$ K have been grown using in situ growth techniques. High-quality $Bi_2Sr_2Ca_3Cu_3O_x$ films with $T_c \sim 97$ K and $J_c(77\text{ K}) = 3.8 \times 10^5$ A/cm^2 have been grown in situ by MOCVD (142). RF sputtering has also been used to obtain as-grown superconducting $Bi_2Sr_2Ca_2Cu_3O_{10}$ films with $T_c \sim 102$ K (147). In this case, the films were synthesized by alternatively depositing $Bi_2Sr_2CuO_6$ and $CaCuO_2$ blocks.

Epitaxial $Bi_2Sr_2Ca_{n-1}Cu_nO_{2n+4}$ films have been grown on a number of oxide substrates, with most of the effort focusing on perovskites or MgO. AFM and scanning electron microscopy (SEM) studies of Bi–Sr–Ca–Cu–O films grown on various substrates reveal a two-dimensional growth mode for films on substrates with a small ($<1°$) miscut and a step flow-growth mode for larger miscut angles (148). One interesting case is the growth of untwinned $Bi_2Sr_2CaCu_2O_8$ films on orthorhombic Nd : $YAlO_3$ (149). Due to a rectangular symmetry in the crystal structure of Nd : $YAlO_3$, *c*-axis-oriented $Bi_2Sr_2Ca_{n-1}Cu_nO_{2n+4}$ films grow untwinned with a uniquely defined in-plane *b*-axis direction. Similar results have also been obtained for $Bi_2Sr_2Ca_{n-1}Cu_nO_{2n+4}$ films grown on (100) $SrTiO_3$ that are miscut by $\sim 4°$ (9). In this case, the surface steps on the substrate sufficiently break the symmetry of the surface, leading to complete in-plane alignment.

The highly anisotropic nature of the film-growth process for the $Bi_2Sr_2Ca_{n-1}Cu_nO_{2n+4}$ system is evidenced from detailed microstructural analysis of epitaxial films. Cross-section TEM images reveal a strong tendency to grow laterally as flat, uninterrupted planes. This is evident in the tendency for $Bi_2Sr_2Ca_{n-1}Cu_nO_{2n+4}$ films to grow over substrate surface defects instead of incorporating stacking faults at the substrate–film interface (150). Unfortunately, this highly anisotropic growth nature of $Bi_2Sr_2Ca_{n-1}Cu_nO_{2n+4}$ also leads to significant meandering of grain boundaries for epitaxial films on bicrystalline substrates (151).

2.4.6 $La_{2-x}Sr_xCuO_4$

$La_{2-x}Sr_xCuO_4$ has one of the simplest crystallographic structures of the HTS materials, with only a single copper-oxide plane per unit cell. The crystal structure of $La_{2-x}Sr_xCuO_4$ is nearly tetragonal with $a = 3.78$ Å and $c = 13.2$ Å. Stoichiometric La_2CuO_4 is an insulator that can be hole doped with the substitution of a divalent alkaline earth, such as Sr, for some of the La. Bulk samples of $La_{1.85}Sr_{0.15}CuO_x$ are superconducting with T_c as high as 40 K. As with all of the HTS materials, the superconducting transition temperature of $La_{2-x}Sr_xCuO_4$ is very sensitive to oxygen content. With T_c substantially lower that that for many other HTS phases, $La_{2-x}Sr_xCuO_4$ has drawn limited attention in terms of applications. As a HTS oxide possessing a single CuO_2 layer per unit cell, $La_{2-x}Sr_xCuO_4$ provides an excellent opportunity to understand the role of interplane coupling in achieving high values for T_c. For this reason, a significant effort has focused on the epitaxial growth of $La_{2-x}Sr_xCuO_4$ films as a model HTS material.

Early work showed that $La_{1.85}Sr_{0.15}CuO_x$, like $YBa_2Cu_3O_7$, tends to grow by a two-dimensional-like growth mode based on oscillations in the RHEED intensity for films deposited on (001) $SrTiO_3$ (67). It is generally difficult to synthesize epitaxial $La_{1.85}Sr_{0.15}CuO_4$ thin films with values of the T_c near that achieved in bulk (152). $La_{1.85}Sr_{0.15}CuO_4$ films grown in situ using molecular oxygen typically have a maximum T_c of only 30 K, even after postannealing in 1 atm oxygen. High-oxygen-pressure annealing (153) has proven useful in increasing the T_c in epitaxial films to near the bulk value. Both the normal state and superconducting properties of $La_{1.85}Sr_{0.15}CuO_4$ films were examined as a function of lattice strain and oxygen content (154). Experiments with both single-layer and multilayer $La_{1.85}Sr_{0.15}CuO_4$-based structures indicate a profound dependence of T_c on both film thickness and film/substrate lattice mismatch, suggesting that strain has a significant influence on T_c (155,156). This is consistent with bulk studies in which a large, positive pressure effect on T_c has been observed. Epitaxial $La_{1.85}Sr_{0.15}CuO_4$ films, with an in-plane lattice parameter of $a = 3.777$ Å, are in tension when grown on (100) $SrTiO_3$ ($a = 3.905$ Å) and exhibit a suppressed T_c. Films deposited on substrates with a much smaller lattice mismatch exhibit a T_c close to that of bulk material. In fact, the epitaxial growth of $La_{1.85}Sr_{0.15}CuO_4$ on (100) $LaSrAlO_4$, with a lattice constant ($a = 3.756$ Å) that is slightly smaller than the film, places the $La_{1.85}Sr_{0.15}CuO_4$ film in the compression in-plane with an expansion of the c axis (157). In this case, the pressure effect increases T_c to 44 K, which is higher than that observed in bulk samples.

Due to the relatively low value of T_c for this material, only a limited amount of consideration has been given for device applications. Some of this effort has focused on investigating the properties of bicrystal grain boundaries based on $La_{1.85}Sr_{0.15}CuO_{4-\delta}$ films (158). One intriguing aspect of this system is the ability to induce metallic and superconducting behaviors in insulating La_2CuO_4 films by ozone or electrochemical oxidation (159,160). Superconducting La_2CuO_{4+d} thin films have been prepared simply by cooling in ozone gas, as seen in Figure 2.33. Films with T_c (onset) $\sim$ 52 K and $T_c(R = 0) \sim$ 48 K were realized on $LaSrAlO_4$ substrates. The transition temperature for films on $LaSrAlO_4$ was higher that that observed for films on $SrTiO_3$, indicating, again, that strain caused by lattice mismatch affects the T_c.

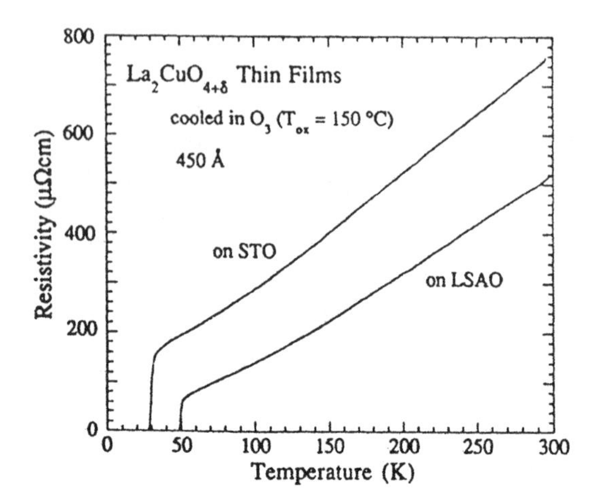

FIGURE 2.33 Resistivity behavior for La_2CuO_4 films exposed to ozone. (From Ref. 160.)

2.4.7 (Nd,Ce)CuO_4

$Nd_{1.85}Ce_{0.15}CuO_{4-\delta}$ differs from nearly all other HTS cuprates in that superconductivity is induced by electron doping. The crystal structure of $Nd_{1.85}Ce_{0.15}CuO_{4-\delta}$ is related to $(La,Sr)_2CuO_4$ except that the Nd_2O_2 layer has a fluorite structure instead of rock salt. As such, the placement of the oxygen atoms in the Nd_2O_2 layer are well removed from the apical Cu positions, resulting in Cu atoms that are only fourfold coordinated. With optimized Ce and O content, T_c is as high as 24 K. Several subtleties remain to be resolved in this system, including evidence that conduction may involve both holes and electron (161).

Epitaxial $Nd_{1.85}Ce_{0.15}CuO_{4-\delta}$ films have been obtained by in situ growth, with most efforts focusing on pulsed-laser deposition. The synthesis of superconducting films requires precise control over stoichiometry, with a very narrow window of Ce composition required in order to obtain the maximum T_c. For this material, optimal superconducting properties also require the removal of oxygen from the crystal lattice of as-grown films typically by annealing in vacuum. The is especially true for films prepared in an oxygen ambient. Unfortunately, the optimal oxygen content appears to be near the phase decomposition limit. For films that are reduced by a vacuum anneal, this often results in a significant degradation of the film surface. This is somewhat remedied by synthesizing the films at an oxygen pressure just below the CuO/Cu_2O phase stability line and cooling in reduced oxygen partial pressure (162). However, the best results have been obtained for $Nd_{1.85}Ce_{0.15}CuO_{4-\delta}$ films synthesized by PLD with N_2O as the oxidant during growth, as seen in Figure 2.34 (163). Epitaxial films that are grown in a N_2O background are superconducting as deposited, with only a short vacuum anneal necessary to produce $T_c \sim 20$ K. As with $(La,Sr)_2CuO_4$, the T_c of relatively thin (< 1000 Å) $Nd_{1.85}Ce_{0.15}CuO_{4-\delta}$ films is influenced by strain due to film/substrate lattice mismatch.

Despite having a superconducting transition temperature of only 24 K, the electron-doped superconductor $Nd_{1.85}Ce_{0.15}CuO_{4-\delta}$ may prove useful for device applications. This HTS compound has the largest in-plane coherence length ($\xi_{ab} \sim 80$ Å) of the cuprates, which is potentially advantageous in junction device applications.

2.4.8 Oxycarbonate Cuprates

A limited amount of effort has focused on the growth of the oxycarbonate cuprates (164–166). The structure of the Ba–Ca–Cu oxycarbonate superconductors consists of $CaCuO_2$ infinite-layer blocks separated by charge carrier blocks containing CO_3 groups, either sandwiched, alone, or in combination with CuO_x groups between the Ba layers. The compounds are given by the formula $(C_xCu_{1-x})_mBa_{m+1}Ca_{n-1}Cu_nO_{2n+m+2}$ with $n = 1, 2, 3, 4, \ldots$. Epitaxial growth has been achieved for a few of the oxycarbonate structures. The range of suitable growth parameters for obtaining optimized film properties is limited. In general, film growth proceeds in an atmosphere of a CO_2/O_2 mixture. Films typically possess intergrowths of the various phases. Films with T_c(onset) $= 115$ K and $T_c(R{=}0) = 78$ K have been reported for films with the $n{=}4$ member as the majority phase. Epitaxial cuprate oxycarbonate, $Sr_2CuO_2(CO_3)$, has been synthesized by metal–organic chemical vapor deposition. The conditions for growth are

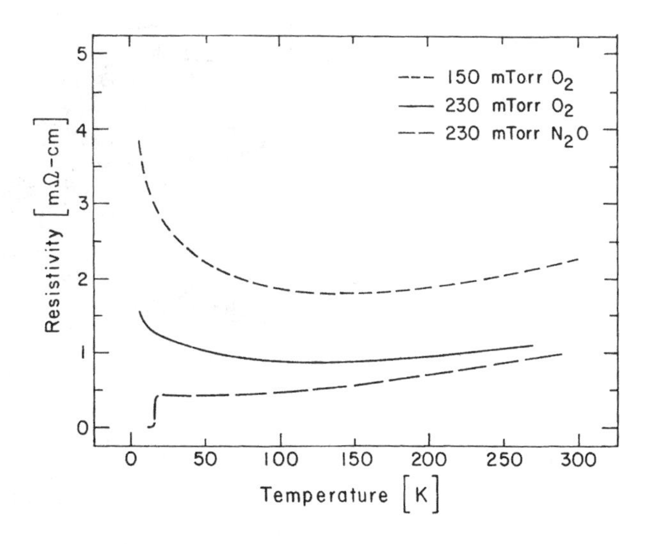

FIGURE 2.34 Resistivity behavior for $Nd_{1.85}Ce_{0.15}CuO_2$ film grown in either N_2O or O_2 ambient. (From Ref. 163.)

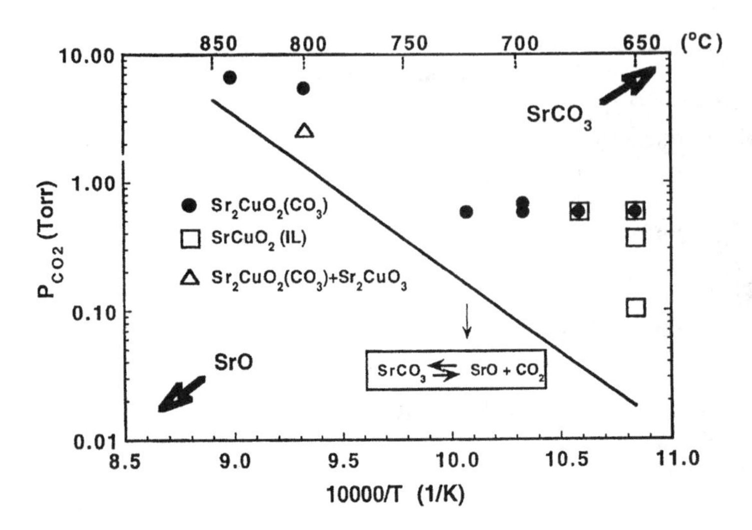

FIGURE 2.35 Phase stability for oxycarbonate films. (From Ref 165.)

primarily determined by the $SrCO_3$ decomposition line, as seen in Figure 2.35. Doping is achieved via the partial substitution of BO_3 for CO_3. Superconducting films with T_c(onset) = 34 K and $T_c(R=0)$ = 20 K have been obtained.

2.4.9 "Infinite-Layer" (Ca,Sr)CuO_2

The parent compound of the copper oxide superconductors, $Ca_{1-x}Sr_xCuO_2$, has a relatively simple, layered structure consisting of CuO_2 planes separated by planes

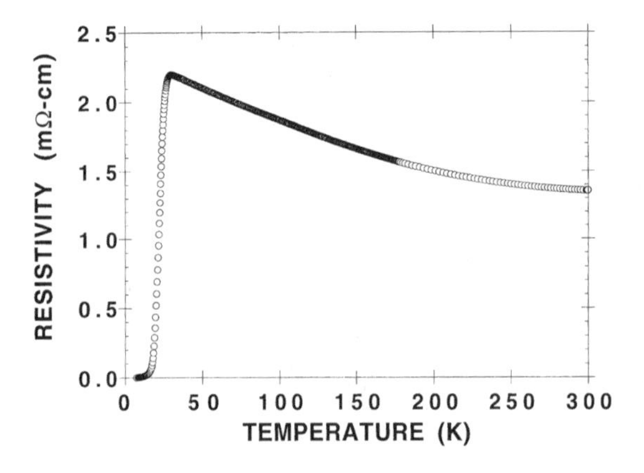

FIGURE 2.36 Resistivity for a $Sr_{0.9}Nd_{0.1}CuO_2$ film deposited by pulsed-laser deposition.

of alkaline earth elements (167). It has been shown that this material can be made to superconduct by electron doping (trivalent substitution on the alkaline earth site) (168,169). However, bulk synthesis of $Ca_{1-x}Sr_xCuO_2$ with the tetragonal, infinite-layer structure, in either the undoped insulating phase or the doped superconducting phase is difficult. At ambient pressure, only $Ca_{1-x}Sr_xCuO_2$ with $x \sim 0.14$ can be synthesized in bulk, with other values of x possible only with high-pressure and high-temperature bulk processing techniques. However, experiments show that this metastable compound can be epitaxially stabilized at less than atmospheric pressure by utilizing thin-film-growth techniques (170–172). In particular, infinite-layer $Ca_{1-x}Sr_xCuO_2$ thin films with near-single-crystal-like character have been grown by pulsed-laser deposition, using both codeposition and layer-by-layer growth schemes target for the entire range of composition $0.15 \leq x \leq 1.0$. X-ray diffractometry indicates that these $Ca_{1-x}Sr_xCuO_2$ thin films are essentially single crystals with extremely narrow diffraction peaks, complete in-plane crystalline alignment with the (100) $SrTiO_3$ substrate, and virtually no impurity phases present. Superconductivity is observed in $Sr_{1-y}Nd_yCuO_2$ films with T_c(onset) $\sim$ 28 K for y=0.10, as illustrated in Figure 2.36 (173). A Nd solubility limit of y=0.10 is observed with the appearance of a new phase with $c \sim 0.37$ nm for y>0.10. Superlattice structures, consisting of $SrCuO_2$ and $(Sr,Ca)CuO_2$ layers in the tetragonal, "infinite-layer" crystal structure, have been grown by pulsed-laser deposition (174). Superlattice chemical modulation is observed for structures with $SrCuO_2$ and $(Sr,Ca)CuO_2$ layers as thin as a single unit cell ($\sim$3.4 Å). X-ray-diffraction (XRD) intensity oscillations due to the finite thickness of the film shown in Figure 2.37 indicate that these films are extremely flat with a thickness variation of only $\sim$ 20 Å over a length scale of several thousand angstroms.

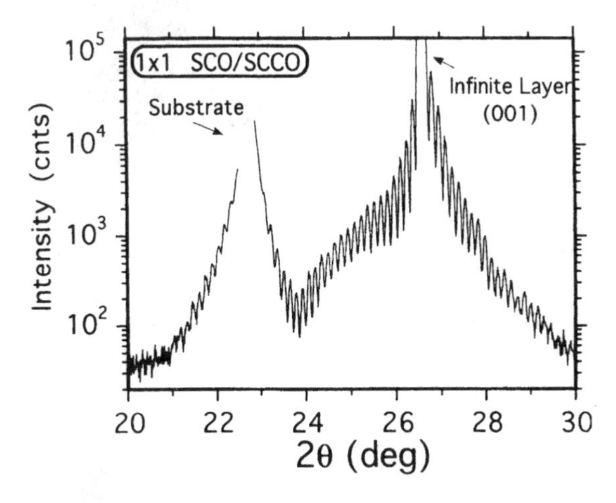

FIGURE 2.37 X-ray-diffraction intensity oscillations for a $SrCuO_2$ film indicating a thickness variation of only 2 nm.

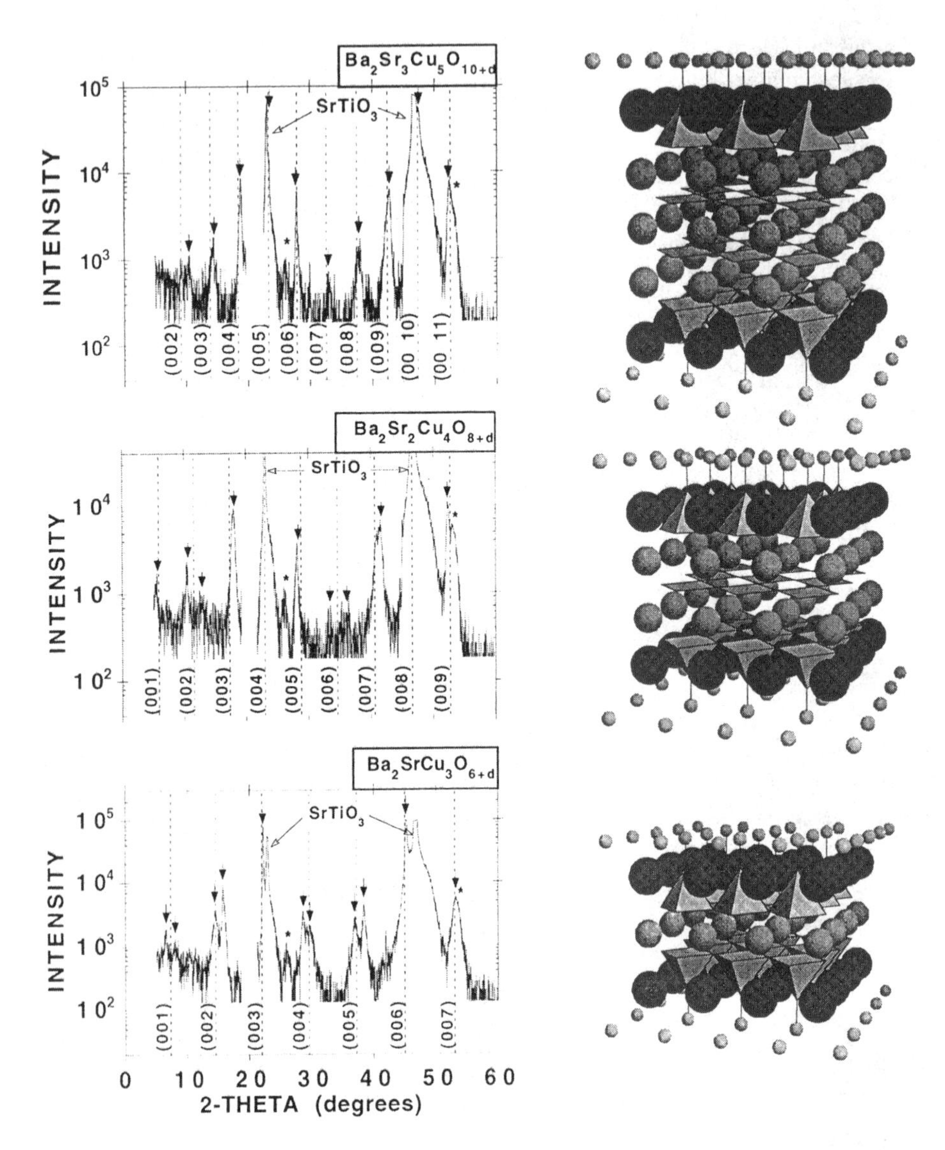

FIGURE 2.38 X-ray diffraction data for three artificially layered $SrCuO_2$/$BaCuO_2$ superlattice structures.

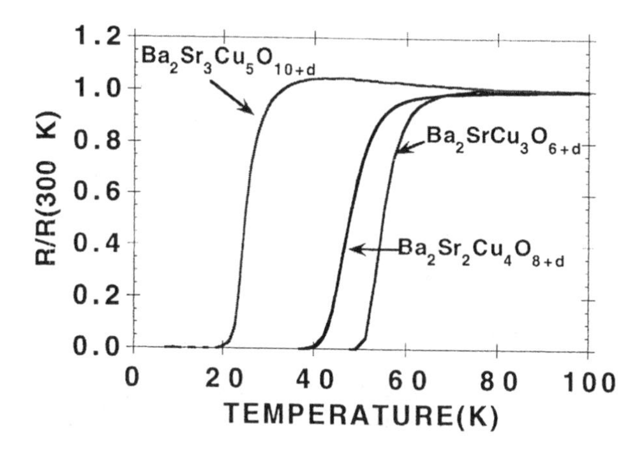

FIGURE 2.39 Superconducting transitions for $SrCuO_2/BaCuO_2$ superlattice structures.

These results show that unit-cell control of $(Sr,Ca)CuO_2$ growth is possible using conventional pulsed-laser deposition. Pulsed-laser deposition and epitaxial stabilization have been effectively used to engineer artificially layered thin-film materials. Novel cuprate compounds have been synthesized using the constraint of epitaxy to stabilize $(Ca,Sr)CuO_2/(Ba,Ca,Sr)CuO_2$ superconducting superlattices in the infinite-layer structure (175). Figure 2.38 shows the XRD data for three artificially layered structures. With the incorporation of $BaCuO_2$ layers, superlattice structures have been synthesized which superconduct at temperatures as high as 70 K, as seen in Figure 2.39. The dc transport measurements indicate that $(Ca,Sr)CuO_2/BaCuO_2$ superlattices are two-dimensional superconductors with the superconducting transition primarily associated with the $BaCuO_2$ layers.

2.5 SUMMARY

Significant progress has been made in the development of HTS thin films and devices. High-quality epitaxial films have been synthesized for many of the HTS phases. Several techniques have proven very successful in obtaining epitaxial HTS films, with an emerging understanding of how the growth of these complex oxide materials proceeds. Despite the significant progress, significant issues remain unresolved. A detailed understanding of how various defect structures in the films affect material properties is currently underdeveloped. The epitaxial growth of films with fully optimized, bulklike properties has yet to be realized for many HTS materials. In addition, a viable approach for the reproducible fabrication of smooth films suitable for multilayer device applications needs to be identified. Progress in resolving these on other issues will define the future directions in HTS film research, both in terms of fundamental understanding and technology development.

ACKNOWLEDGMENT

This research was cosponsored by Oak Ridge National Laboratory, managed by Lockheed Martin Energy Research Corp., for the U.S. Department of Energy Office of Energy Research and the Office of Energy Efficiency and Renewable Energy, under contract DE-AC05-96OR22464.

REFERENCES

1. JG Bednorz, KA Müller. Z Phys B 64:189–193, 1986.
2. C Park, RL Snyder. J Am Ceram Soc 78:3171–3194, 1995.
3. ZF Ren, JH Wang, DJ Miller. Appl Phys Lett 71:1706–1708, 1997.
4. DK Christen, JR Thompson, HR Kerchner, BC Sales, BC Chakoumakos, L Civale, AD Marwick, F Holtzberg. In: H Kwok, D Shaw, M Naughton, eds. Proceedings of Superconductivity and its Applications. New York: American Institute of Physics, 1992, pp 24–36.
5. NF Heinig, RD Redwing, IF Tsu, A Gurevich, JE Nordman, SE Babcock, DC Larbalestier. Appl Phys Lett 69:577–579, 1996.
6. D Dimos, P Chaudhari, J Mannhart. Phys Rev B 41:4038–4049, 1990.
7. P Berberich, B Utz, W Prusseit, H Kinder. Physica C 219:497–504, 1994.
8. K Shinohara, V Matijasevic, P Rosenthal, AF Marshall, RH Hammond, MR Beasley. Physica C 185–189:2119–2120, 1991.
9. JN Eckstein, I Bozovic, DG Schlom, JS Harris Jr. Appl Phys Lett 57:1049–1051, 1990.
10. JP Locquet, A Cantana, E Mächler, C Gerber, JG Bednorz. Appl Phys Lett 64:372–374, 1994.
11. A Inam, MS Hedge, XD Wu, T Venkatesan, P England, PF Miceli, EW Chase, CC Chang, JM Tarascon, JB Wachtman. Appl Phys Lett 53:908–910, 1988.
12. CB Eom, JZ Sun, K Yamamoto, AF Marshall, KE Luther, TH Geballe, SS Laderman. Appl Phys Lett 55:595–597, 1989.
13. JM Triscone, Ø Fischer, O Brunner, L Antognazza, AD Kent, MG Karkut. Phys Rev Lett 64:804–807, 1990.
14. M Kanai, T Kawai, S Kawai. Jpn J Appl Phys 31:L331–L333, 1992.
15. M Leskelä, H Mölsä, L Niinistö. Supercond Sci Technol 6:627–656, 1993.
16. C Klemenz, HJ Scheel. Physica C 265:126–134, 1996.
17. ME Klausmeier-Brown, JN Eckstein, I Bozovic, GF Virshup. Appl Phys Lett 60:657–659, 1992.
18. SJ Benerofe, CH Ahn, MM Wang, KE Kihlstrom, KB Do, SB Arnason, MM Feyer, TH Geballe, MR Beasley, RH Hammond. J Vac Sci Technol B 12:1217–1220, 1994.
19. T Shimizu, H Nonaka, S Hosokawa, S Ichimura, K Arai. Physica C 185–189: 2003–2004, 1991.
20. A Sawa, H Obara, S Kosaka. Appl Phys Lett 64:649–651, 1994.
21. JP Locquet, E Mächler. J Vac Sci Technol A 10:3100, 1992.
22. T Siegrist, DA Mixon, E Coleman, TH Tiefel. Appl Phys Lett 60:2489–2490, 1992.
23. I Bozovic, JN Eckstein, GF Virshup, A Chaiken, M Wall, R Howell, M Fluss. J Supercond 7:187–195, 1994.

24. I Bozovic, JN Eckstein. MRS Bull 20:32–38, 1995.
25. JP Locquet, A Cantana, E Machler, C Gerber, JG Bednorz. Appl Phys Lett 64: 372–372, 1994.
26. JT Cheung, H Sankur. CRC Crit Rev Solid State Mater Sci 15:63–109, 1988.
27. XD Wu, RE Muenchausen, S Foltyn, RC Estler, RC Dye, C Flamme, NS Nogar, AR Garcia, J Martin, J Tesmer. Appl Phys Lett 56:1481–1483, 1990.
28. EV Pechen, AV Varlashkin, SI Krasnosvobodtsev, B Brunner, KF Renk. Appl Phys Lett 66:2292–2294, 1995.
29. B Holzapfel, B Roas, L Schultz, P Bauer, G Saemann-Ischenko. Appl Phys Lett 61:3178–3180, 1992.
30. Z Trajanovic, S Choopun, RP Sharma, T Venkatesan. Appl Phys Lett 70: 3461–3463, 1997.
31. T Schauer, L Weber, J Hafner, O Kus, EV Pechen, AV Variashkin, T Kaiser, KF Rank. Supercond Sci Technol 11:270, 1998.
32. GJHM Rijnders, G Koster, DHA Blank, H Rogalla. Appl Phys Lett 70:1888–1890, 1997.
33. U Kruger, R Kutzner, R Wordenweber. IEEE Trans Appl Supercond AS-3:1687, 1993.
34. XX Xi, T Venkatesan, Q Li, XD Wu, A Inam, CC Chang, R Ramesh, DM Hwang, TS Ravi, A Findikoglu, D Hemmick, S Etemad, JA Martinez, E Wilkens. IEEE Trans Magnetics M-27:982, 1991.
35. DJ Lichtenwalner, CN Soble II, RR Woolcott Jr., O Auciello, AI Kingon. J Appl Phys 70:6952, 1991.
36. JM Phillips. J Appl Phys 79:1829–1848, 1996.
37. R Sum, HP Lang, HJ Guntherodt. Physica C 242:174–182, 1995.
38. M Kawasaki, K Takahashi, T Maeda, R Tsuchiya, M Shinohara, O Ishiyama, T Yonezawa, M Yoshimoto, H Koinuma. Science 266:1540–1542, 1994.
39. FS Galasso. Perovskites and High T_c Superconductors. New York: Gordon and Breach, 1990.
40. G Koster, BL Kropman, GJHM Rijnders, DHA Blank, H Rogalla. Appl Phys Lett 74:3729–3731, 1999.
41. T Hikita, T Hanada, M Kudo, M Kawai. J Vac Sci Technol A11:2649–2654, 1993.
42. J Gallop. Supercond Sci Technol 10:A120–A141, 1997.
43. B Dam, JM Huijbregtse, FC Klaassen, RCF van der Geest, G Doornbos, JH Rector, AM Testa, S Freisem, JC Martinez, B Stauble-Pumpin, R Griessen. Nature 399:439, 1999.
44. MP Siegal, JM Phillips, AF Hebard, RB Van Dover, RC Farrow, TH Tiefel, JH Marshal. J Appl Phys 70:4982–4988, 1991.
45. R Feenstra, DK Christen, JD Budai, SJ Pennycook, DP Norton, DH Lowndes, CE Klabunde, MD Galloway. In: L Correra, ed. Proceedings, High Temperature Superconducting Thin Films, ICAM 91. Amsterdam: Elsevier Science, 1992, pp 331–342.
46. R Feenstra, TB Lindemer, JD Budai, MD Galloway. J Appl Phys 69:6569, 1991.
47. A Mogro-Campero, LG Turner. J Supercond 6:37–41, 1993.
48. RH Hammond, V Matijasevic, R Bormann. In: RD McConnell, R Noufi, eds. Science and Technology of Thin Films 2. New York: Plenum, 1990, pp 395–401.

49. JA Smith, MJ Cima, N Sonnenberg. IEEE Trans Appl Supercond AS-9:1531–1534, 1999.
50. T Terashima, K Shimura, Y Bando, Y Matsuda, A Fujiyama, S Komiyama. Phys Rev Lett 67:1362–1365, 1991.
51. DK Christen, CE Klabunde, R Feenstra, DH Lowndes, D Norton, HR Kerchner, JR Thompson, ST Sekula, JD Budai, LA Boatner, J Narayan, R Singh. Mater Res Soc Symp Proc 169:883–886, 1990.
52. ED Specht, CJ Sparks, AG Dhere, J Brynestad, OB Cavin, DM Kroeger, HA Oye. Phys Rev B 37:7426, 1988.
53. V Olsan, M Jelinek. Physica C 207:391–396, 1993.
54. M Lorenz, H Hochmuth, D Natusch, H Borner, T Tharigen, DG Patrikarakos, J Frey, K Kreher, S Senz, G Kastner, D Hesse, M Steins, W Schmitz. IEEE Trans Superconductors S-7:1240–1243, 1997.
55. DW Face, C Wilker, JJ Kingston, ZY Shen, FM Pellicone, RJ Small, SP McKenna, S Sun, PJ Martin. IEEE Trans Superconductors S-7:1283–1286, 1997.
56. RA Rao, Q Gan, CB Eom, Y Suzuki, AA McDaniel, JWP Hsu. Appl Phys Lett 69:3911–3913, 1996.
57. O Nakamura, EE Fullerton, J Guimpel, IK Schuller. Appl Phys Lett 60:120–122, 1992.
58. DH Lowndes, DP Norton, JD Budai. Phys Rev Lett 65:1160–1163, 1990.
59. DP Norton, DH Lowndes, SJ Pennycook, JD Budai. Phys Rev Lett 67:1358–1361, 1991.
60. Z Lu, JK Truman, ME Johansson, D Zhang, CF Shih, GC Liang. Appl Phys Lett 67:712–714, 1995.
61. Q Zhong, PC Chou, QL Li, GS Taraldsen, A Ignatiev. Physica C 246:288–296, 1995.
62. A Ignatiev, Q Zhong, PC Chou, X Zhang, JR Liu, WK Chu. Appl Phys Lett 70:1474–1476, 1997.
63. HJ Scheel, C Klemenz, FK Reinhart, HP Lang, HJ Güntherodt. Appl Phys Lett 65:901–903, 1994.
64. Y Yamada, Y Niiori, I Hirabayashi, S Tanaka. Physica C 278:180, 1997.
65. M Yoshida, T Nakamoto, T Kitamura, OB Hyun, I Hirabayashi, S Tanaka, A Tsuzuki, Y Sugawara, Y Ikuhara. Appl Phys Lett 65:1714–1716, 1994.
66. H Von Känel, N Onda, L Miglio. In: FC Matacotta, G Ottaviani, eds. Science and Technology of Thin Films. Singapore: World Scientific, 1995, pp 29–56.
67. T Terashima, Y Bando, K Iijima, K Yamamoto, K Hirata, K Hayashi, K Kamigaki, H Terauchi. Phys Rev Lett 65:2684–2687, 1990.
68. CC Tsuei, T Frey, CC Chi, T Shaw, DT Shaw, MK Wu. In: H Kwok, D Shaw, M Naughton, eds. Proceedings, Superconductivity and its Applications, New York: American Institute of Physics, 1992, pp 12–23.
69. XY Zheng, DH Lowndes, S Zhu, JD Budai, RJ Warmack. Phys Rev B 45: 7584–7587, 1992.
70. KH Wu, RC Wang, SP Chen, HC Lin, JY Juang, TM Uen, YS Gou. Appl Phys Lett 69:421–423, 1996.
71. M Hawley, ID Raistrick, JG Berry, RJ Houlton. Science 251:1587–1589, 1991.

72. C Gerber, D Anselmetti, JG Bednorz, J Mannhart, DG Schlom. Nature 350:279–280, 1991.
73. DP Norton, DH Lowndes, XY Zheng, S Zhu, RJ Warmack. Phys Rev B 44: 9760–9763, 1991.
74. B Dam, NJ Koeman, JH Rector, B Stauble, U Poppe, R Griessen. Physica C 261:1–11, 1996.
75. T Haage, J Zegenhagen, HU Habermeier, M Cardona. Phys Rev Lett 80:4225–4228, 1998.
76. Y Zhu. In: D Shi, ed. High-Temperature Superconducting Materials Science and Engineering. Oxford: Pergamon, 1995, pp 199–258.
77. A Catana, JG Bednorz, C Gerber, J Mannhart, DG Schlom. Appl Phys Lett 63:553–555, 1993.
78. M Mukaida, S Miyazawa, M Sasaura, K Kuroda. Jpn J Appl Phys 29:L936–L939, 1990.
79. CC Chin, H Takahashi, T Morishita, T Sugimoto. J Mater Res 8:951–956, 1993.
80. SK Streiffer, BM Lairson, CB Eom, BM Clemens, JC Bravman, TH Geballe. Phys Rev B 43:13,007–13,018, 1991.
81. T Minamikawa, T Suzuki, Y Yonezawa, K Segawa, A Morimoto, T Shimizu. Jpn J Appl Phys 34:4038–4042, 1995.
82. V Boffa, T Petrisor, L Ciontea, U Gambardella, S Barbanera. Physica C 260: 111–116, 1996.
83. C Prouteau, JF Hamet, B Mercey, M Hervieu, B Raveau, D Robbes, L Coudrier, G Ben. Physica C 248:108–118, 1995.
84. Rao MR. 1996. Appl Phys Lett 69:1957–1959
85. M Lorenz, H Hochmuth, D Natusch, H Borner, G Lippold, K Kreher, W Schmitz. Appl Phys Lett 68:3332–3334, 1996.
86. AG Zaitsev, R Wördenweber, G Ockenfuß, R Kutzner, T Königs, C Zuccaro, N Klein. IEEE Trans Appl Supercond AS-7:1482–1485, 1997.
87. JA Alarco, G Brorsson, ZG Ivanov, PA Nilsson, E Olsson, M Lofgren. Appl Phys Lett 61:723–725, 1992.
88. DK Fork, SM Garrison, M Hawley, TH Geballe. J Mater Res 7:1641–1651, 1992.
89. GL Skofronick, AH Carim, SR Foltyn, RE Muenchausen. J Mater Res 8:2785–2798, 1993.
90. DK Fork, DB Fenner, RW Barton, JM Phillips, GAN Connell, JB Boyce, TH Geballe. Appl Phys Lett 57:1161–1163, 1990.
91. DK Fork, FA Ponce, JC Tramontana, TH Geballe. Appl Phys Lett 58:2294–2296, 1991.
92. DP Norton, A Goyal, JD Budai, DK Christen, DM Kroeger, ED Specht, Q He, B Saffian, M Paranthaman, CE Klabunde, DF Lee, BC Sales, FA List. Science 274:755–757, 1996.
93. Y Iijima, N Tanabe, O Kohno, Y Ikeno. Appl Phys Lett 60:769, 1992.
94. XD Wu, SR Foltyn, PN Arendt, WR Blumenthal, IH Campbell, JD Cotton, JY Coulter, WL Hults, MP Maley, HF Safar, JL Smith. Appl Phys Lett 67:2367–2399, 1995.
95. CP Wang, KB Do, MR Beasley, TH Geballe, RH Hammond. Appl Phys Lett 71:2955–2957, 1997.

96. K Hasegawa, N Yoshida, K Fujino, H Mukai, K Hayashi, K Sato. Proceedings of the 9th International Symposium on Superconductors, Sapporo, 1996.
97. JG Wen, S Mahajan, W Ito, T Morishita, N Koshizuka. Appl Phys Lett 64: 3334–3336, 1994.
98. H Takahashi, T Hase, H Izumi, K Ohata, T Morishita, S Tanaka. Physica C 179:291–294, 1991.
99. U Jeschke, R Schneider, G Ulmer, G Linker. Physica C 243:243–251, 1995.
100. CB Eom, AF Marshall, SS Laderman, RD Jacowitz, TH Geballe. Science 249:1549–1552, 1990.
101. GY Sung, JD Suh. Appl Phys Lett 67:1145–1147, 1995.
102. Z Trajanovic, I Takeuchi, PA Warburton, CJ Lobb, T Venkatesan. Physica C 265:79–88, 1996.
103. KH Young, JZ Sun. Appl Phys Lett 59:2448–2450, 1991.
104. M Badaye, JG Wen, K Fukushima, N Koshizuka, T Morishita, T Nishimura, Y Kido. Supercond Sci Technol 10:825, 1997.
105. Y Hakuraku, Z Mori, S Koba, N Yokoyama, T Doi, T Inoue. Supercond Sci Technol 12:481, 1999.
106. C Cantoni, DP Norton, DM Kroeger, M Paranthaman, DK Christen, D Verebelyi, R Feenstra, DF Lee, ED Specht, V Boffa, S Pace. Appl Phys Lett 74:96–98, 1999.
107. Y Li, X Yao, K Tanabe. Physica C 304:239, 1998.
108. Y Kumagai, Y Yoshida, M Iwata, M Hasegawa, Y Sugawara, T Hirayama, Y Ikuhara, I Hirabayashi, Y Takai. Physica C 304:35, 1998.
109. WL Holstein, LA Parisi, C Wilker, RB Flippen. Appl Phys Lett 60:2014–2016, 1992.
110. AP Bramley, JD O'Conner, CRM Grovenor. Supercond Sci Technol 12:R57–R74, 1999.
111. WL Holstein, LA Parisi, C Wilker, RB Flippen. IEEE Trans Appl Supercond AS-3:1197–1200, 1993.
112. AP Bramley, BJ Glassey, CRM Grovenor, MJ Goringe, JD O'Conner, AP Jenkins, KS Kale, KL Jim, D Dew-Hughes, DJ Edwards. IEEE Trans Appl Supercond AS-7:1249–1252, 1997.
113. M Nemoto, S Yoshikawa, K Shimaoka, N Niki, I Yoshida, Y Yoshisato. IEEE Trans Superconductors S-7:1895–1898, 1997.
114. RB Hammond, GV Negrete, LC Bourne, DD Strother, AH Cardona, MM Eddy. Appl Phys Lett 57:825–827, 1990.
115. DS Ginley, JF Kwak, RP Hellmer, RJ Baughman, EL Venturini, B Morosin. Appl Phys Lett 53:406–408, 1988.
116. BJ Hinds, RJ McNeely, DB Studebaker, TJ Marks, TP Hogan, JL Schindler, CR Kannewurf, XF Zhang, DJ Miller. J Mater Res 12:1214–1236, 1997.
117. WL Holstein, LA Parisi. J Mater Res 11:1349–1366, 1996.
118. MP Siegal, DL Overmyer, EL Venturini, PP Newcomer, R Dunn, F Dominguez, RR Padilla, SS Sokolowski. IEEE Trans Appl Supercond AS-7:1881–1186, 1997.
119. MP Seigal, EL Venturini, PP Newcomer, B Morosin, DL Overmyer, F Dominguez, R Dunn. Appl Phys Lett 67:3966–3968, 1995.
120. DJ Kountz, PL Gai, C Wilker, WL Holstein, FM Pellicone, RJ Brainard. IEEE Trans Appl Supercond AS-3:1222, 1993.

121. KE Meyers. In: I Bozovic, D. Pavuna, eds. Oxide Superconducting Physics and Non-Engineering II. Bellingham, WA: SPIE, 1996, pp 160–171.
122. ZF Ren, CA Wang, JH Wang. Appl Phys Lett 65:237–239, 1994.
123. WD Li, Z Wang, JY Lao, ZF Ren, JH Wang, M Paranthaman, DT Verebelyi, DK Christen. Supercond Sci Technol 12:L1, 1999.
124. NA Khan, Y Sekitaa, F Tateai, T Kojima, K Ishida, N Terada, H Ihara. Physica C 320:39, 1999.
125. RN Bhattacharya, RD Blaugher, ZF Ren, W Li, JH Wang, M Paranthaman, DT Verebelyi, DK Christen. Physica C 304:55, 1998.
126. DW Face, JP Nestlerode. Appl Phys Lett 61:1838–1840, 1992.
127. KE Myers, DW Face, DJ Kountz, JP Nestlerode. Appl Phys Lett 65:490–492, 1994.
128. N Reschauer, U Spreitzer, W Brozio, A Piehler, R Berger, G Saemann-Ischenko. Appl Phys Lett 68:1000–1002, 1996.
129. CC Tsuei, A Gupta, G Trafas, D Mitzi. Science 263:1259–1261, 1994.
130. A Gupta, JZ Sun, CC Tsuei. Science 265:1075–1077, 1994.
131. L Krusin-Elbaum, CC Tsuei, A Gupta. Nature 373:679–681, 1995.
132. SH Yun, JZ Wu. Appl Phys Lett 68:862–864, 1996.
133. JZ Wu, SH Yun, A Gapud, BW Kang, WN Kang, SC Tidrow, TP Monahan, XT Cui, WK Chu. Physica C 277:219–224, 1997.
134. SH Yun, JZ Wu, SC Tidrow, DW Eckart. Appl Phys Lett 68:2565–2567, 1996.
135. Y Moriwaki, T Sugano, C Gasser, A Fukuoka, K Nakanishi, S Adachi, K Tanabe. Appl Phys Lett 69:3423–3425, 1996.
136. Y Moriwaki, T Sugano, A Tsukamoto, C Gasser, K Nakanishi, S Adachi, K Tanabe. Physica C 303:65, 1998.
137. SL Yan, YY Xie, JZ Wu, T Aytug, AA Gapud, BW Kang, L Fang, M He, SC Tidrow, KW Kirchner, JR Liu, WK Chu. Appl Phys Lett 73:2989, 1998.
138. K Mizuno, H Adachi, K Setsune. J Low Temp Phys 105:1571–1576, 1996.
139. BC Sales, BC Chakoumakos. Phys Rev B 43:12,994–13,000, 1991.
140. S Zhu, DH Lowndes, BC Chakoumakos, JD Budai, DK Christen, XY Zheng, E Jones, B Warmack. Appl Phys Lett 63:409–411, 1993.
141. J Auge, U Rüdiger, H Frank, HG Roskos, G Güntherodt, H Kurz. Appl Phys Lett 64:378–380, 1994.
142. K Endo, H Yamasaki, S Misawa, S Yoshida, K Kajimura. Nature 355:327–328, 1992.
143. YZ Zhang, L Li, DG Yang, BR Zhao, H Chen, C Dong, HJ Tao, HT Yang, SL Jia, B Yin, JW Li, ZX Zhao. Physica C 295:75–79, 1998.
144. JN Eckstein, I Bozovic, M Klausmeier-Brown, G Virshup. Thin Solid Films 216:8–13, 1992.
145. JN Eckstein, I Bozovic, GF Virshup. MRS Bull 19:44–50, 1994.
146. A Sajjadi, IW Boyd. Appl Phys Lett 63:3373–3375, 1993.
147. K Ohbayashi, T Ohtsuki, H Matsushita, H Nishiwaki, Y Takai, H Hayakawa. Appl Phys Lett 64:369–371, 1994.
148. T Sugimoto, N Kubota, Y Shiohara, S Tanaka. Appl Phys Lett 63:2697–2699, 1993.
149. I Tsukada, K Uchinokura. J Appl Phys 78:364–371, 1995.
150. A Chaiken, MA Wall, RH Howell, I Bozovic, JN Eckstein, GF Virshup. J Mater Res 11:1609–1615, 1996.

151. B Kabius, JW Seo, T Amrein, U Dahne, A Scholen, M Siegel, K Urban, L Schultz. Physica C 231:123–130, 1994.
152. J Talvacchio, MG Forrester, JR Gavaler, TT Braggins. In: R. D. McConnell, R. Noufi, eds. Science and Technology of Thin Film Superconductors-2. New York: Plenum, pp 57–66.
153. IE Trofimov, LA Johnson, KV Ramanujachary, S Guha, MG Harrison, M Greenblatt, Z Cieplak, P Linderfeld. Appl Phys Lett 65:2481–2483, 1994.
154. W Si, HC Li, XX Xi. Appl Phys Lett 74:2839, 1999.
155. M Suzuki Phys Rev B 39:2312, 1989.
156. MZ Cieplak, M Berkowski, S Guha, E Cheng, AS Vagelos, DJ Rabinowitz, B Wu, IE Trofimov, P Lindenfeld. Appl Phys Lett 65:3383–3385, 1994.
157. H Sato, M Naito. Physica C 274:221–226, 1997.
158. A Beck, OM Froehlich, D Koelle, R Gross, H Sato, M Naito. Appl Phys Lett 68:3341–3343, 1996.
159. JP Locquet, C Gerber, A Cretton, Y Jaccard, E Williams, E Machler. Appl Phys A 57:211, 1993.
160. H Sato, M Naito, H Yamamoto. Physica C 280:178–186, 1997.
161. SN Mao, W Jiang, XX Xi, Q Li, JL Peng, RL Greene, T Venkatesan, DP Beesabathina, L Salamanca-Riba, XD Wu. Appl Phys Lett 66:2137–2139, 1995.
162. WT Lin, GJ Chen. J Mater Res 10:2422–2427, 1995.
163. A Kussmaul, JS Moodera, PM Tedrow, A Gupta. Appl Phys Lett 61:2715–2717, 1992.
164. G Calestani, A Migliori, U.Spreitzer, S Hauser, M Fuchs, H Barowski, T Schauer, W Assmann, KJ Range, A Varlashkin, O Waldmann, P Muller, KF Renk. Physica C 312:225, 1999.
165. KW Chang, BW Wessels, W Qian, VP Dravid, JL Schindler, CR Kannewurf, DB Studebaker, TJ Marks, R Feenstra. Physica C 303:11, 1998.
166. Y Miyazaki, H Yamane, T Hirai, T Kajitani. Jpn J Appl Phys 35:L1053, 1996.
167. T Siegrist, SM Zahurak, DW Murphy, RS Roth. Nature 334:231, 1998.
168. MG Smith, A Manthiram, J Zhou, JB Goodenough, JJ Markert. Nature 351:549, 1991.
169. G Er, Y Miyamoto, F Kanamaru, S Kikkawa. Physica C 181:206, 1991.
170. M Kanai, T Kawai, S Kawai. Appl Phys Lett 58:771, 1991.
171. M Yoshimoto, H Nagata, J Gong, H Ohkubo, H Koinuma. Physica C 185–189:2085, 1991.
172. DP Norton, BC Chakoumakos, JD Budai, DH Lowndes. Appl Phys Lett 62:1679, 1993.
173. DP Norton, BC Chakoumakos, JD Budai, EC Jones, DK Christen, DH Lowndes. Bull Electrotech Lab 58:69, 1994.
174. DP Norton, JD Budai, DH Lowndes, BC Chakoumakos. Appl Phys Lett 65:2869, 1994.
175. DP Norton, BC Chakoumakous, JD Budai. Proc SPIE 2697:295, 1996.

3

High-Temperature Superconducting Multilayer Ramp-Edge Junctions

Q. X. Jia
Los Alamos National Laboratory, Los Alamos, New Mexico, U.S.A.

3.1 INTRODUCTION

There has recently been tremendous progress in the development of high-temperature superconducting Josephson junctions and superconducting quantum interference devices (SQUIDs) fabricated from $YBa_2Cu_3O_7$ (YBCO) thin films. Josephson junctions have potential applications to high-speed low-power digital logic, whereas SQUIDs are the most sensitive sensors to the magnetic field. A Josephson junction is composed of two superconductors separated by a barrier. Cooper pairs can tunnel or diffuse through the barrier and current flows through it with no voltage appearing across the junction (1,2).

Josephson junctions and their related devices have been fabricated using conventional low-temperature superconductors where the main building block of the device is based on a configuration of superconductor/insulator/superconductor (SIS). In contrast to low-temperature superconducting junctions, many junction structures have been investigated for high-temperature superconductors in order to fabricate reproducible and controllable junctions. Since the first report on the fabrication of a natural grain-boundary Josephson junction using YBCO (3), there have been many efforts in the fabrication of Josephson junctions and

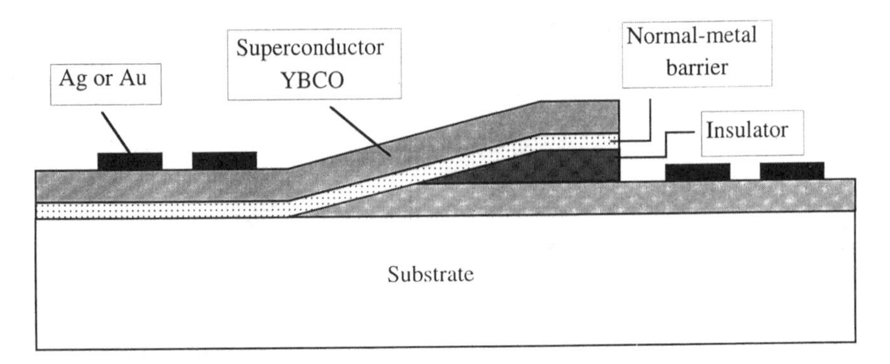

FIGURE 3.1 Cross-sectional diagram of a ramp-edge superconductor/normal-metal/superconductor (SNS) junction. The active area of the device is located on the ramp, where the normal metal is sandwiched between the top and bottom superconductor electrodes.

SQUIDs based on different device constructions. Bicrystal grain boundary (4), biepitaxial grain boundary (5), step-edge grain boundary (6), step-edge superconductor/normal-metal/superconductor (SNS) (7), ramp-edge SNS (8), locally damaged superconducting line by ion beams (9), and interface-engineered junctions (10) are the most commonly investigated configurations.

The ramp-edge SNS is one of the very attractive device structures used to fabricate high-temperature superconducting junctions because it provides many advantages. Shown in Figure 3.1 is a generic structure of a ramp-edge SNS junction. The device is composed of top and bottom superconductor electrodes isolated in the overlapping region by an insulating layer. The active area of the device is located on the ramp, where the N-layer is sandwiched between the top and bottom superconductor electrodes. This ramp-edge SNS structure uses *c*-axis-oriented superconducting films, where the Josephson current flows along the *a-b* planes of the electrodes. The most important feature of this SNS scheme compared to that of the grain boundary is that the device can be put anywhere on a chip without affecting other devices. This feature allows flexibility in device design and substrate choice, which makes it possible to fabricate more complicated circuitry. Because the junction performance depends on the N-layer thickness and resistivity in the ideal case, one can control the physical properties of the N-layer to fabricate Josephson junctions for specific applications. This chapter describes the processes and materials issues to fabricate ramp-edge SNS junctions and SQUIDs. It also discusses the recent progress in the fabrication of high-temperature superconducting ramp-edge SNS junctions and SQUIDs.

3.2 PROCESS AND FABRICATION OF MULTILAYER RAMP-EDGE SNS JUNCTIONS

To fabricate the multilayer ramp-edge SNS junction shown in Figure 3.1, one has to go through several processing steps, such as multilayer thin-film deposition, patterning, metallization, packaging, and so on. The most commonly used technique to deposit high-temperature superconducting electrodes and the normal-metal layer is pulsed-laser deposition (PLD), although high-temperature superconducting thin films have been deposited by other techniques. The substrates commonly used are highly polished single-crystal $LaAlO_3$, MgO, $NdGaO_3$, $SrTiO_3$, $(LaAlO_3)_{0.3}$–$(Sr_2AlTaO_6)_{0.7}$, and yttria-stabilized zirconia (YSZ). The most commonly used insulating materials to isolate the superconducting electrodes are $PrBa_2Cu_3O_7$ (PBCO), $LaAlO_3$, $NdGaO_3$, $SrTiO_3$, MgO, and CeO_2. The insulating layer should be highly resistive at operating temperatures. It should also be thermally stable at a deposition temperature as high as 800°C and not be poisonous to superconductors or N-layer materials.

The following presents an example of processing steps in the fabrication of a multilayer ramp-edge SNS junction by using YBCO as electrodes and PBCO as the N layer. The bottom YBCO (250 $\pm$ 20 nm) electrode and the insulating CeO_2 layer (300 $\pm$ 20 nm) are deposited first on the $LaAlO_3$ substrate. The first photolithographic mask is used to define the location and geometry of the bottom YBCO electrode. Ion milling with 200–250-eV Ar ions is used to etch the CeO_2/YBCO and to form the active ramp edge of the device. The angle between the edge and the substrate surface is controlled in the range of 15° $\pm$ 3° (11).

The N-layer PBCO and the top YBCO electrode are deposited after stripping off the photoresist (PR) and cleaning the edge surface. The second mask is used to define the geometry of the active area of the device. Ion milling is used again to remove unnecessary material and to expose the bottom YBCO for electric contact.

The third mask is a lift-off mask that is used to define the location and geometry of contact pads. The contact electrodes can be either Ag or Au. The finished chip is annealed at 400–500°C in oxygen for a desired period of time before packaging for electrical measurements. In total, three masks and two ion-milling steps are used to finish the device fabrication. Figure 3.2 shows the processing sequence to fabricate ramp-edge SNS junctions. Figure 3.3 shows a top view of a finished SNS junction near the active area of the device. Top and bottom YBCO electrodes could be clearly distinguished from this photograph. The contrast of the electrodes shown in Figure 3.3 is due to different surface finishes.

It is very important to have a shallow angle ($< 30°$) between the plane of the exposed ramp edge and the substrate plane. This is essential because a high-quality epitaxial *c*-axis-oriented N layer cannot be grown on a steep edge. A shallow

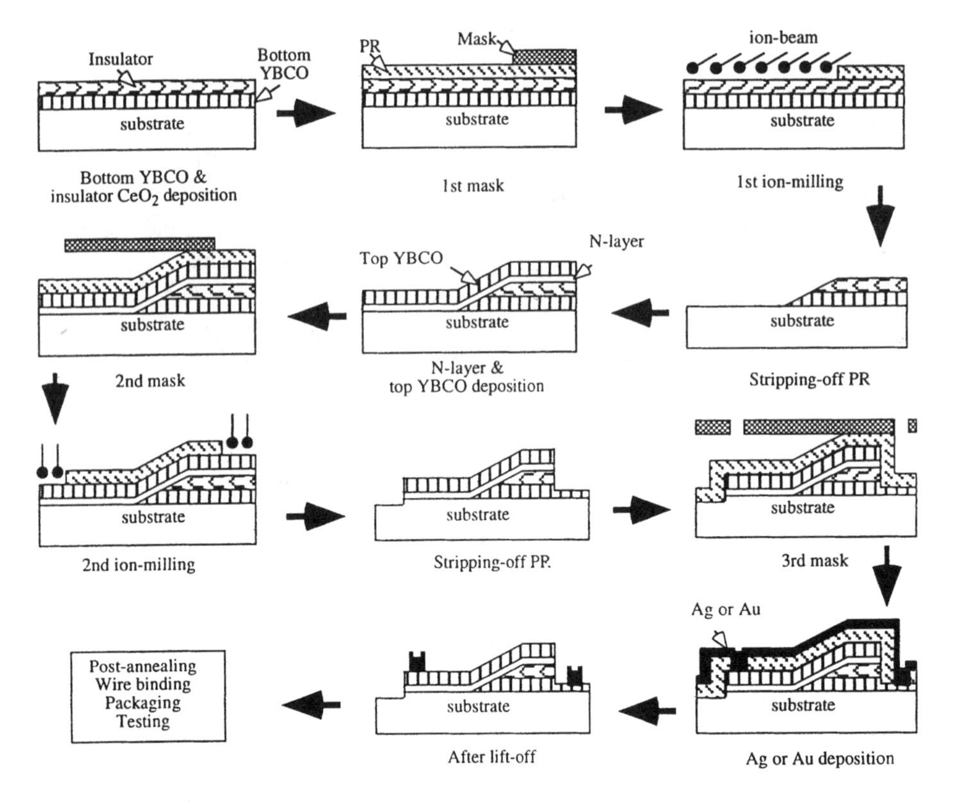

FIGURE 3.2 Processing sequences in the fabrication of high-temperature superconductor multilayer ramp-edge SNS Josephson junctions, where PR represents photoresist.

angle also makes it easier for the N layer to cover the ramp region with high uniformity and homogeneity. The degree of the angle can be controlled by baking the PR before the first ion milling. For example, the angle between the substrate surface and the edge is above 70° with no bake of PR before ion-milling, but around 15° with a bake of PR at 170°C for 1 min (11).

As an alternative to ion milling for the formation of the ramp edge, a microshadow mask can be used to form a very shallow angle and damage-free edge. This can be done in a completely in situ process (12,13). The in situ growth and the avoidance of any treatment of the interfaces result in junctions with a high I_cR_n product and in the possibility of using a rather thick N-layer barrier.

The ion-milling energy and the beam orientation have been found to be crucial to success in the fabrication of high-performance junctions. A series of samples have been fabricated by varying the ion-milling energy (200, 300, 600, and

900 eV) while fixing the orientation of the ion beam at 60° from normal into the edge of the PR. The experimental results have shown that the higher ion beam energies lead to less control of the device performance. The device yields are also lower when using a high ion-milling energy (14). It has been reported that the deposition of PBCO on an ion beam (with energy of 600 eV) etched and annealed YBCO surface produces an additional layer of cubic and cation-disordered YBCO or PBCO in a few nanometers thickness. The annealing treatment of the damaged surface at 800°C in oxygen atmosphere does not restore the perfect lattice structure of the film (15). However, it should be noted that the degree of the damage to the crystal structure and of the suppression of superconducting properties of YBCO due to ion irradiation depends on the energy and fluence of the ions used (16). In general, ion beam energy of around 250 eV or lower should be used in order to minimize the surface damage. Substrate cooling during ion milling is also important to preserve electrical properties of superconductor electrodes and the mechanical properties of the PR. Cooling the substrate stage with chilled water should be sufficient. Liquid nitrogen has also been used to cool the substrate stage during ion milling (17).

Ion-milling damage to the ramp edge can be partially removed by Br–ethanol etching after the ion-milling process (18). It has shown that the interfaces produced involving Br–ethanol etching are essentially of the same structural quality as those produced by the microshadow mask technique, which leads to

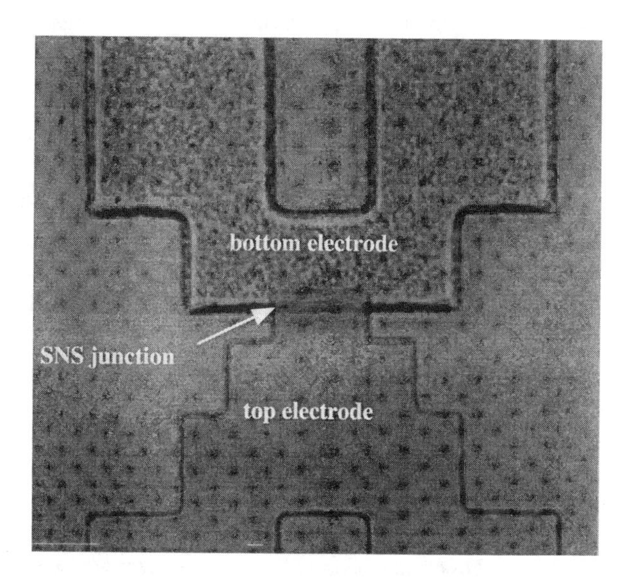

FIGURE 3.3 Top view of a finished SNS junction, where the ramp-edge and top/bottom superconductor electrodes are clearly evident.

abrupt and coherent interfaces (15). However, it should be noted that the angle of the edge could be significantly increased by increasing the time of the chemical etching or in combination with ion beam etching (19). In addition, for using low ion-milling energy, annealing the sample in oxygen can be also performed prior to the N-layer deposition to reduce the surface damage.

3.3 SUPERCONDUCTING ELECTRODES FOR MULTILAYER RAMP-EDGE SNS JUNCTIONS

High-quality epitaxial superconducting electrodes are essential for high-performance devices because these films show less critical current fluctuation and vortex hopping. For ramp-edge SNS junctions, the superconducting electrode needs to be oriented with the *c* axis normal to the substrate surface. The typical thickness of the superconductor electrode is in the range 200–300 nm. The general requirements of physical and superconducting properties of the electrodes are as follows: The orientation mosaic from both the out of plane and in plane should be as small as possible; the surface should be as smooth as possible; and the zero-resistance temperature and the critical current density at 77 K should be as high as possible.

The bottom superconducting electrode plays an important role in determining the properties of the ramp-edge SNS junctions. Thus far in high-temperature superconductor ramp-edge SNS junction development, YBCO is the most widely investigated electrode material. Recently, $DyBa_2Cu_3O_7$ (20), $GdBa_2Cu_3O_7$ (21), Ag-doped YBCO (22), $Y_{1-x}Ca_xBa_{2-y}La_yCu_3O_7$ (23), $YBa_{1.95}La_{0.05}Cu_3O_7$ (24), and $NdBa_2Cu_3O_7$ (25) have been investigated as superconductor electrodes. It has been shown that both the base-electrode material and deposition technique can have a strong effect on SNS device resistance (25).

The choice of $GdBa_2Cu_3O_7$ instead of YBCO as the superconductor electrode is mainly due to the close lattice match (either *a* and *b* or *c*) between $GdBa_2Cu_3O_7$ and Pr-doped YBCO (21). The $GdBa_2Cu_3O_7$ also tends to give a higher zero-resistance temperature than YBCO. Ag-doped YBCO is used as electrodes because it provides superior environmental stability compared to pure YBCO (26). Ramp-edge SNS junctions fabricated from Ag-doped YBCO superconducting electrodes exhibit little sign of degradation in air (27). Importantly, the controllability and reproducibility of the processing is improved substantially when using Ag-doped YBCO for the electrode (28).

The high corrosion resistance of $Y_{1-x}Ca_xBa_{2-y}La_yCu_3O_7$ makes it attractive as an electrode in ramp-edge SNS junctions. It has been found that a cosubstitution of Ca^{2+} for Y^{3+} and La^{3+} for Ba^{2+} in $Y_{1-x}Ca_xBa_{2-y}La_yCu_3O_7$ can compensate the Cu valence and maintain the transition temperature above 80 K. For this system, an orthorhombic to tetragonal transition is found to occur at $y \sim 0.4$ (29). Experimental results have shown that $Y_{0.6}Ca_{0.4}Ba_{1.6}La_{0.4}Cu_3O_7$ possesses high corrosion resistance in water as well as enhanced processability (30).

$YBa_{2-x}La_xCu_3O_7$ with $x = 0.025$–0.05 is chosen as the electrode because a small amount of La doping can help suppress *a*-axis grain formation (24). The use of $NdBa_2Cu_3O_7$ is probably due to the fact that the newly optimized $NdBa_2Cu_3O_7$ films are superior to YBCO thin films, with respect the transition temperature, crystallinity, surface stability and smoothness, and oxygenation properties (31).

3.4 NORMAL-METAL BARRIERS FOR MULTILAYER RAMP-EDGE SNS JUNCTIONS

Many N-layer materials have been investigated to fabricate ramp-edge SNS junctions. Conductive oxides and doped YBCO are the main choice for N-layer materials because of their favorable electrical and structural properties. The thickness of the N layer depends on the material used and the requirements of the specific designs. It is necessary to consider the following factors when choosing N-layer material. It should be lattice and thermal expansion matched with the superconductor electrode; there should be negligible chemical reactions with the electrode; the growth conditions should be compatible with the stability of the superconductor electrode; and the thin N layer should be smooth and pinhole free. The electrical properties of the N-layer material should also be considered in order to tune the device performance for specific applications. Table 1 outlines the N-layer materials reported in the literature for the fabrication of ramp-edge SNS junctions (32–46). For completeness, Table 3.1 also outlines the superconductor electrodes

TABLE 3.1 Different N-Layer Materials Used in Edge-Geometry SNS Junctions

Electrodes	N-Layer barrier material	Ref.
YBCO	$PrBa_2Cu_3O_{7-\delta}$	8,19,32,33
Ag-doped YBCO		22
YBCO	$Y_{0.3}Pr_{0.7}Ba_2Cu_3O_{7-\delta}$	34
YBCO	$Y_{0.6}Pr_{0.4}Ba_2Cu_3O_{7-\delta}$	35
$GdBa_2Cu_3O_7$		21
YBCO	Nb: SrTiO	36
YBCO	$YBa_2Cu_3O_x$	14,37
YBCO	$CaRuO_3$, $SrRuO_3$	38, 39
YBCO	$Y_{0.7}Ca_{0.3}Ba_2Cu_3O_{7-\delta}$	11,40
YBCO	$YBa_2Cu_{2.79}Co_{0.21}O_{7-\delta}$	40, 41
$YBa_{1.95}La_{0.05}Cu_3O_7$		24
YBCO	$La_{0.5}Sr_{0.5}CoO_3$, $La_{1.4}Sr_{0.6}CuO_4$	40
YBCO	$PrBa_2Cu_{3-x}Ga_xO_7$ ($x = 0.15, 0.3$)	42
$DyBa_2Cu_3O_7$	($x = 0.1, 0.4$)	20
YBCO	$Y_{1-x}Pr_xBa_2Cu_3O_7$ (radient)	43
YBCO	$NdBa_2Cu_3O_{7-x}$	44
YBCO	Ga-doped YBCO	45
YBCO	Indium–tin–oxide	46

used to fabricate these devices. Currently, the most commonly used N-layer materials for ramp-edge SNS junctions are Co-doped YBCO and PBCO.

It should be noted that the N layer may not always determine the junction properties in ramp-edge SNS configuration due to the interface and the inhomogeneity of the N layer. Depending on the N-layer materials and fabrication processes, the resistance of junctions can be controlled by the interface instead of the N-layer barrier. For example, the majority of the junction resistance comes from the interface between YBCO and the barrier with conductive oxides such as $CaRuO_3$ and $SrRuO_3$ as N-layer material. It is speculated that the stress due to the thermal expansion mismatch between YBCO and $CaRuO_3$ may give rise to oxygen disorder in the vicinity of the interface and thereby increase the interface resistance (40). Resistance of junctions can be only determined by the physical properties of the N layer if the interface resistance is negligibly small compared with the N-layer resistance. For example, the use of Co-doped YBCO as a barrier for ramp-edge SNS junctions seems quite promising based on the published results. No significant interface resistance between YBCO and the barrier has been observed (40). In this case, the temperature dependence of the critical current of the junction can be well described by the conventional proximity effect (47). It should be noted also that the inhomogeneity of the N layer can lead to pinholes or microshorts in the barrier. The rough ramp-edge morphology of YBCO may even induce the nucleation of secondary phases which have completely different electrical characteristics, thus changing the behavior of the junctions (48). In this case, the junction current and resistance may not be controlled by the N-layer thickness. Therefore, it is important to carefully control the morphology of ramp-edge.

3.5 OTHER MULTILAYER RAMP-EDGE JUNCTIONS

Recently, interface-engineered ramp-edge Josephson junctions have been fabricated by modification of the edge surface prior to counterelectrode deposition. These devices appear to be uniform and reproducible. A detailed description of the processing procedures can be found in Ref. 10. It is well known that the crystal structure and chemical composition strongly influence electrical properties of YBCO materials. The idea in this scheme is to create a few-nanometer-thick surface layer of YBCO on the junction edge by altering the structure or chemistry of the existing YBCO to form an effective barrier.

It should be noted that the nature of the barrier based on this technique is still unclear. It is speculated that the barrier material created in this way is near some sort of metal–insulator transition. It is also argued that a normal conducting barrier is formed by the depression in the transition temperature of the YBCO due to the induced particle damage from this process (49).

3.6 CHARACTERISTICS OF MULTILAYER RAMP-EDGE SNS JUNCTIONS

The current–voltage (I–V) characteristic of an ideal superconductor/insulator/superconductor Josephson junction is described by a resistively and capacitively shunted junction model for which the device shows a hysteretic I–V curve. For the high-temperature superconducting ramp-edge SNS junction, on the other hand, the capacitance of the device is negligibly small. In this case, the I–V characteristic can be described by a resistively shunted junction (RSJ) model:

$$V = I_c R_n \left[\left(\frac{I}{I_c} \right)^2 - 1 \right]^{1/2} \qquad \text{for } I > I_c \tag{1}$$

where the device shows a nonhysteretic I–V curve as shown in Figure 3.4. In Eq. (1), the I_c is the critical current of the junction and the R_n is the junction resistance determined from the slope of the dashed line shown in Figure 3.4.

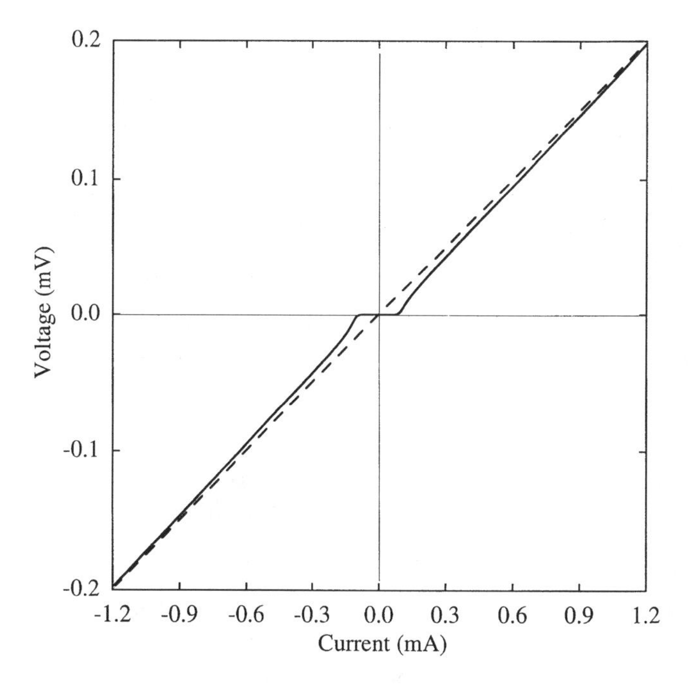

FIGURE 3.4 Current versus voltage characteristic of a ramp-edge SNS junction. The current versus voltage characteristic can be described by a RSJ model. The dashed line in the figure is used to determine the junction resistance.

For an ideal SNS junction, the conventional theory of the proximity effect should apply. The proximity theory predicts that

$$I_c \sim I_{c0} \exp\left(\frac{-L_n}{\xi_n}\right) \tag{2}$$

where ξ_n is the coherence length of the normal-metal barrier and L_n is the effective barrier thickness ($L_n >> \xi_n$). I_{c0} is somewhat temperature and superconducting electrode dependent. The coherence length in the N layer is given by

$$\frac{1}{\xi_n^2} = \frac{1}{\xi_{nc}^2} + \frac{1}{\xi_{nd}^2} \tag{3}$$

where $\xi_{nc} = \hbar v_F/2\pi KT$ and $\xi_{nd} = (\xi_{nc} l/3)^{1/2}$ are the clean-limit ($l >> \xi_n$) and dirt-limit ($l << \xi_n$) coherence lengths in the N layer. Here, $\hbar$ is $h/2\pi$, v_F is the Fermi velocity in the material, and l is the carrier mean free path. An excellent review article on the theoretical understanding of SNS junctions is Ref. 50. It should be noted that the effective N-layer thickness L_n is proportional to the N-layer thickness (t) with the relationship $t/\sin\theta$, where θ is the angle between the substrate surface and the ramp edge. In a very shallow-angle ramp-edge junction, the L_n can be much larger than t.

The presence of well-defined Shapiro steps in the I–V curve under microwave irradiation is widely used as a verification of the Josephson effect. The ramp-edge SNS junctions fabricated using variety of N-layer materials and high-temperature superconducting electrodes also show clear Shapiro steps under microwave irradiation with frequencies in the gigahertz range. Figure 5 shows a typical I–V curve under microwave irradiation for a ramp-edge SNS junction fabricated using Ag-doped YBCO as the electrode and PBCO as an N-layer barrier (51). The measured voltage step height agrees well with the theoretical calculation based on the Josephson relation of $V_n = nhf/2e$ ($n = \pm1, \pm2, \ldots$), where f is the frequency of the applied microwaves and the other symbols have their usual meaning.

The actual device performance, on the other hand, is affected by many factors. The interface between S/N or N/S has been recognized as the most important controlling factor in determining the performance of SNS junctions (40). The difficulty in controlling the interface comes from several intrinsic and extrinsic sources. The interface between S/N or N/S can be degraded due to the anisotropic nature of high-temperature superconductor materials, mismatch in the lattice and thermal expansion coefficient between the superconductor and N-layer, chemical incompatibility between the superconductor electrode and N-layer, growth of the multilayer thin film on a ramp edge instead of on a flat surface, damage to the superconductor bottom electrode from the ion beam used to pattern the film, or the unavoidable grain-boundaries intrinsic to the oxides.

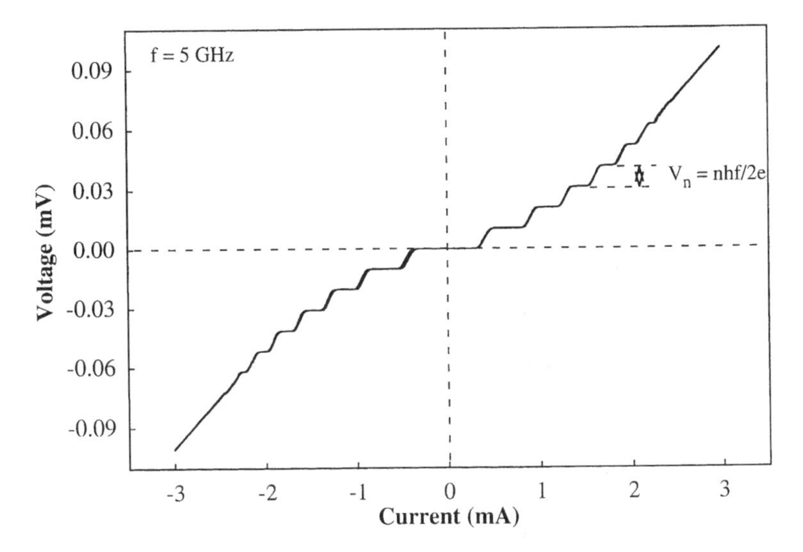

FIGURE 3.5 Current versus voltage characteristic of a ramp-edge SNS junction under microwave irradiation. (From Ref. 51.)

3.7 APPLICATIONS OF MULTILAYER RAMP-EDGE SNS JUNCTIONS AND SQUIDS

One of the very attractive applications of Josephson junctions is in single-flux-quantum (SFQ) digital circuit which can perform logic operation at extremely high speed while dissipating very low power. Ramp-edge SNS configuration is the most studied structure for this purpose due to its many advantages mentioned in Section 5.1. However, the simple SNS configuration shown in Figure 1 has limitations in high-speed operation due to the parasitic inductance and capacitance. To keep the circuit inductance small, a superconducting ground plane is necessary. Recently, multilayer high-temperature superconducting ramp-edge SNS junctions integrated with superconducting ground plane have been demonstrated (52,53). Figure 6 shows the schematic cross section of a YBCO/Co-doped YBCO/YBCO SNS ramp-edge junction integrated with an epitaxial YBCO ground plane. The multilayer structure includes six epitaxial layers and contact vias to the ground plane and base electrode. *I–V* data at 65 K for a chip with junction parameters suitable for SFQ logic are shown in Figure 7. These nineteen 4-μm-wide junctions with a 5-nm Co-doped YBCO barrier exhibit an average resistance of 0.97 Ω ($1 - \sigma = 6\%$). The average I_c is 327 μA ($1 - \sigma = 13\%$) and the average I_cR_n product is 315 μV ($1 - \sigma = 9\%$) (25).

Ramp-edge SNS junction is also a very promising technology for the fabrication of SQUIDs (22,27,28,32,43,54,55). This technology provides an alterna-

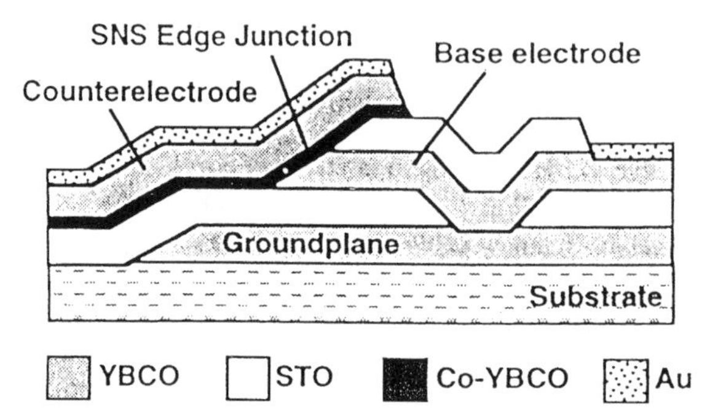

FIGURE 3.6 Schematic cross section of an YBCO/Co-doped YBCO/YBCO ramp-edge SNS junction integrated with an epitaxial YBCO ground plane. (From Ref. 52.)

tive to bicrystals in terms of substrate cost, flexibility, stability, and integration. Most significant external magnetic fields up to about 1 G do not influence the noise of the SQUIDs. Faley et al. speculated that the junctions are intrinsically shielded from external magnetic fields by the Meissner effect in the top electrode (55,56).

The SQUID based on ramp-edge SNS junction shows well-defined voltage modulation which is as good as that exhibited by the SQUIDs fabricated by any

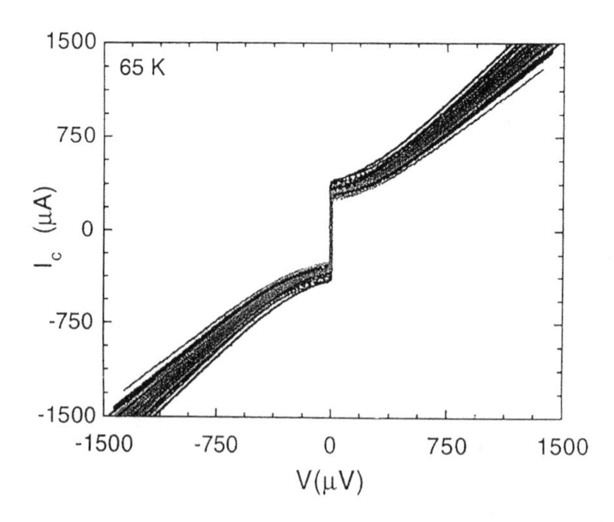

FIGURE 3.7 Current versus voltage characteristics at 65 K for nineteen 4-μm-wide junctions on a single chip. (From Ref. 25.)

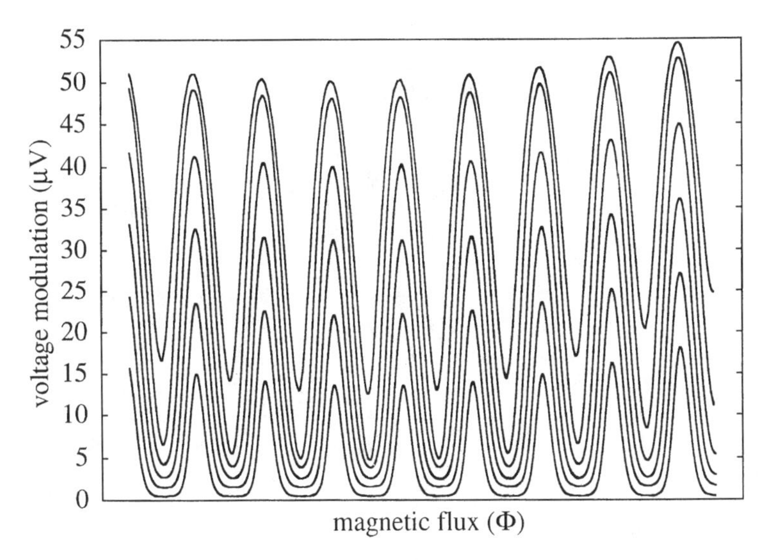

FIGURE 3.8 Voltage modulation versus magnetic flux (Φ) characteristic of a dc SQUID operated at 75 K with different bias currents. (From Ref. 22.)

other technologies. Figure 3.8 shows the voltage modulation versus magnetic flux (Φ) with different bias currents at 75 K from a SQUID fabricated using Ag-doped YBCO as the electrode and PBCO as a N-layer barrier (22). The curves are perfectly periodic and show no hysteresis while sweeping Φ back and forth. Bare SQUIDs with a flux noise of $5 \times 10^{-6}\Phi_0$ $\text{Hz}^{-1/2}$ ($\Phi_0 = 2.07 \times 10^{-15}$ Wb) at 1 kHz and liquid-nitrogen temperature have been demonstrated using ramp-edge SNS device geometry (55–57). Recently, directly coupled SQUID magnetometers based on a ramp-edge SNS technology have been successfully demonstrated also (28).

With the low-noise magnetometers based on ramp-edge SNS junctions, mobile applications of SQUID magnetometry become possible. One example is the demonstration of using an integrated YBCO magnetometer mounted in a handheld cryostat with a content of 100 cm^3 of liquid nitrogen for biomagnetic measurements such as for recording human heart traces (58). In another case, directly coupled SQUID magnetometers based on ramp-edge SNS junctions have been used as a receiver for low-frequency radio waves (59). The primary application of such a receiver is to communicate in underground areas where the overburden results in significant losses at the usual radio frequencies. The most significant advantage of SQUIDs for this application is that they allow the compact construction of three-axis receivers that are necessary to overcome a dominant source of vibrational or motional noise.

REFERENCES

1. BD Josephson. Possible new effects in superconductive tunneling. Phys Lett 1:251–253, 1962.
2. PW Anderson, JM Rowell. Probable observation of the Josephson superconducting tunneling effect. Phys Rev Lett 10:230–232, 1963.
3. RH Koch, CP Umbach, GJ Clark, P Chaudhari, RB Laibowitz. Quantum interference devices made from superconducting oxide thin films. Appl Phys Lett 51:200–202, 1987.
4. D Dimos, P Chaudhari, J Mannhart, FK LeGoues. Orientation dependence of grain boundary critical currents in $YBa_2Cu_3O_{7-\delta}$ bicrystals. Phys Rev Lett 61:219–222, 1988.
5. K Char, MS Colclough, SM Garrison, N Newman, G Zaharchuk. Bi-epitaxial grain-boundary junctions in $YBa_2Cu_3O_7$. Appl Phys Lett 59:733–735, 1991.
6. RW Simon, JB Bulman, JF Burch, SB Coons, KP Daly, WD Dozier, R Hu, AE Lee, JA Luine, CE Platt, SM Schwarzbek, MS Wire, MJ Zani. Engineered HTS microbridges. IEEE Trans Magnetics 27:3209–3014, 1991.
7. MS DiIorio, S Yoshizumi, KY Yang, J Yang, M Maung. Practical high-T_c Josephson junctions and dc SQUIDs operating above 85 K. Appl Phys Lett 58:2552–2554, 1991.
8. J Gao, WAM Aarnink, GJ Gerritsma, H Rogalla. Controlled preparation of all high-T_c SNS-type edge junctions and dc SQUIDs. Physica C 171:126–130, 1990.
9. MJ Zani, JA Luine, RW Simon, RA Davidheiser. Focused ion-beam high-T_c superconductor dc SQUIDs. Appl Phys Lett 59:234–236, 1991.
10. BH Moeckly, K Char. Properties of interface-engineered high-T_c Josephson junctions. Appl Phys Lett 71:2526–2528, 1997.
11. QX Jia, DW Reagor, SR Foltyn, M Hawley, C Mombourquette, XD Wu. Superconducting $YBa_2Cu_3O_{7-x}$ based edge junctions with $Y_{0.7}Ca_{0.3}Ba_2Cu_3O_{7-x}$ barriers. Physica C 228:160–164, 1994.
12. G Koren, E Aharoni, E Polturak, D Cohen. Properties of all YBCO Josephson edge junctions prepared by *in situ* laser ablation deposition. Appl Phys Lett 58:634–636, 1991.
13. MD Strikovski, F Kahlmann, J Schubert, W Zander, V Glyantsev, G Ockenfuss, CL Jia. Fabrication of YBCO thin-film flux transformers using a novel microshadow mask technique for *in situ* patterning. Appl Phys Lett 66:3521–3523, 1995.
14. D Reagor, R Houlton, K Springer, M Hawley, QX Jia, C Mombourquette, F Garzon, XD Wu. Development of high temperature superconducting Josephson junctions and quantum interference devices using low deposition temperature $YBa_2Cu_3O_{7-x}$ barriers. Appl Phys Lett 66:2280–2282, 1995.
15. CL Jia, MI Faley, U Poppe, K Urban. The effect of chemical and ion-beam etching on the atomic structure of interfaces in $YBa_2Cu_3O_7$/$PrBa_2Cu_3O_7$ Josepshon junctions. Appl Phys Lett 67:3635–3637, 1995.
16. Y Li, S Linzen, F Machalett, F Schmidl, P Seidel. Recovery of superconductivity and recrystallization of ion-damaged $YBa_2Cu_3O_{7-x}$ films after thermal annealing treatment. Physica C 243:294–302, 1995.
17. H Schneidewind, F Schmidl, S Linzen, P Seidel. The possibilities and limitations of ion-beam etching of $YBa_2Cu_3O_{7-x}$ thin-films and microbridges. Physica C 250:191–201, 1995.

18. RP Vasquez, BD Hunt, MC Foote. Non-aqueous chemical etch for YBCO. Appl Phys Lett 53:2692–2694, 1988.
19. MI Faley, U Poppe, H Soltner, CL Jia, M Siegel, K Urban. Josephson junctions, interconnects, and crossovers on chemically etched edges of $YBa_2Cu_3O_{7-x}$. Appl Phys Lett 63:2138–2140, 1993.
20. MAJ Verhoeven, GJ Gerritsma, H Rogalla, AA Golubov. Ramp type HTS Josephson-junctions with PrBaCuGaO barriers. IEEE Trans Superconductors 5:2095–2098, 1995.
21. QX Jia, XD Wu, D Reagor, SR Foltyn, RJ Houlton, P Tiwari, C Mombourquette, IH Campbell, F Garzon, DE Peterson. Superconductor $GdBa_2Cu_3O_{7-\delta}$ edge junctions with lattice matched $Y_{0.6}Pr_{0.4}Ba_2Cu_3O_{7-\delta}$ barriers. J Appl Phys 78:2871–2873, 1995.
22. QX Jia, XD Wu, D Reagor, SR Foltyn, C Mombourquette, DE Peterson. Edge-geometry dc SNS SQUIDs using Ag-doped $YBa_2Cu_3O_{7-x}$ electrodes. Electron Lett 32:499–501, 1996.
23. JP Zhou, RK Lo, JT McDevitt, J Talvacchio, MG Forrester, BD Hunt, QX Jia, D Reagor. Development of a reliable materials base for superconducting electronics. J Mater Res 12:2958–2975, 1997.
24. BD Hunt, MG Forrester, J Talvacchio, RM Young, JD McCambridge. High-T_c SNS edge junctions with integrated $YBa_2Cu_3O_x$ groundplanes. IEEE Trans Appl Supercond 7:2936–2939, 1997.
25. BD Hunt, MG Forrester, J Talvacchio, RM Young. High-resistance HTS SNS edge junctions. IEEE Trans Appl Superconduct 5:365–371, 1998.
26. D Kumar, PR Apte, R Pinto, M Sharon, LC Gupta. Aging effects and passivation studies on undoped and Ag-doped $YBa_2Cu_3O_x$ laser ablated thin films. J Electrochem Soc 141:1611–1616, 1994.
27. QX Jia, D Reagor, C Mombourquette, Y Fan, J Decker, P D'Alessandris. Stability of dc superconducting quantum interference devices fabricated using ramp-edge superconductor/normal-metal/superconductor technology. Appl Phys Lett 71:1721–1723, 1997.
28. QX Jia, Y Fan, C Mombourquette, DW Reagor. Directly coupled direct current superconducting quantum interference device magnetometers based on ramp-edge Ag:$YBa_2Cu_3O_{7-x}$/$PrBa_2Cu_3O_{7-x}$/Ag: $YBa_2Cu_3O_{7-x}$ junctions. Appl Phys Lett 72:3068–3070, 1998.
29. A Manthiram, JB Goodenough. Factors influencing T_c in 123 copper-oxide superconductors. Phyica C 159:760–768, 1989.
30. JP Zhou, SM Savoy, RK Lo, J Zhao, M Arendt, YT Zhu, JT McDevitt. Improved corrosion-resistance of cation substituted $YBa_2Cu_3O_{7-\delta}$. Appl Phys Lett 66:2900–2902, 1995.
31. M Badaye, JG Wen, K Fukushima, N Koshizuka, T Morishita, T Nishimura, Y Kido. Superior properties of $NdBa_2Cu_3O_y$ over $YBa_2Cu_3O_x$ thin films. Supercond Sci Technol 10:825–830, 1997.
32. M Schilling, D Reimer, U Merkt. $YBa_2Cu_3O_7$ direct current superconducting quantum interference devices with artificial $PrBa_2Cu_3O_7$ barriers above 77 K. Appl Phys Lett 64:2584–2586, 1994.
33. JB Barner, BD Hunt, MC Foote, WT Pike, RP Vasquez. $YBa_2Cu_3O_{7-\delta}$-based, edge-geometry SNS Josephson junctions with low resistivity $PrBa_2Cu_3O_{7-\delta}$ barriers. Physica C 207:381–390, 1993.

34. C Stolzel, M Siegel, G Adrian, C Krimmer, J Sollner, W Wilkens, G Schulz, H Adrian. Transport properties of $YBa_2Cu_3O_{7-\delta}/Y_{0.3}Pr_{0.7}Ba_2Cu_3O_{7-\delta}/YBa_2Cu_3O_{7-\delta}$ Josephson junctions. Appl Phys Lett 63:2970–2972, 1993.
35. E Polturak, G Koren, D Cohen, A Aharoni, Proximity effect in $YBa_2Cu_3O_7/Y_{0.6}Pr_{0.4}Ba_2Cu_3O_{7-\delta}/YBa_2Cu_3O_7$ edge junction. Phys Rev Lett 67:3038–3041, 1991.
36. DK Chin, T Van Duzer. Novel all-high T_c epitaxial Josephson junction. Appl Phys Lett 58:753–755, 1991.
37. BD Hunt, MC Foote, LJ Bajuk. All high-T_c edge-geometry weak links utilizing Y–Ba–Cu–O barrier layers. Appl Phys Lett 59:982–984, 1991.
38. K Char, MS Colclough, TH Geballe, KE Myers. High T_c superconductor–normal–superconductor Josephson junctions using $CaRuO_3$ as the metallic barrier. Appl Phys Lett 62:196–198, 1993.
39. L Antognazza, K Char, TH Geballe, LLH King, AW Sleight. Josephson coupling of $YBa_2Cu_3O_{7-x}$ through a ferrpmagnetic barrier $SrRuO_3$. Appl Phys Lett 63: 1005–1007, 1993.
40. K Char, L Antognazza, TH Geballe. Study of interface resistances in epitaxial $YBa_2Cu_3O_{7-x}$/barrier/ $YBa_2Cu_3O_{7-x}$ junctions. Appl Phys Lett 63:2420–2422, 1993.
41. K Char, L Antognazza, TH Geballe. Properties of $YBa_2Cu_3O_{7-x}/YBa_{2.79}Co_{0.21}Cu_3O_{7-x}/YBa_2Cu_3O_{7-x}$ edge junctions. Appl Phys Lett 65:904–906, 1994.
42. MI Faley, U Poppe, CL Jia, K Urban, Y Xu. Proximity effect in edge type junctions with $PrBa_2Cu_3O_7$ barriers prepared by Br-ethanol etching. IEEE Trans Appl Supercond 5:2091–2094, 1995.
43. QX Jia, XD Wu, D Reagor, SR Foltyn, C Mombourquette, P Tiwari, IH Campbell, RJ Houlton, DE Peterson. High-temperature superconductor Josephson-junctions with a gradient Pr-doped $Y_{1-x}Pr_xBa_2Cu_3O_{7-\delta}$ ($x = 0.1, 0.3, 0.5$) as barriers. Appl Phys Lett 65:2866–2868, 1994.
44. B Oh, YH Choi, SH Moon, HT Kim, BC Min. $YBa_2Cu_3O_{7-\delta}/NdBa_2Cu_3O_{7-\delta}/YBa_2Cu_3O_{7-\delta}$ edge junctions and SQUIDs. Appl Phys Lett 69:2288–2290, 1996.
45. IH Song, EL Lee, BM Kim, I Song, G Park. High-T_c edge junctions with a Ga-doped $YBa_2Cu_3O_{7-\delta}$ barrier and interface resistances. Appl Phys Lett 74:2053–2055, 1999.
46. T Hato, H Aso, Y Ishimaru, A Yoshida, N Yokoyama. Ramp-edge Josephson junction with indium–tin–oxide barrier. Jpn J Appl Phys 38:L123–L125, 1999.
47. AW Kleinsasser, KA Delin. Demonstration of the proximity effect in $YBa_2Cu_3O_{7-x}$ edge junctions. Appl Phys Lett 66:102–104, 1995.
48. E Olsson, K Char. Origin of nonuniform properties of $YBa_2Cu_3O_{7-x}/CaRuO_3/YBa_2Cu_3O_{7-x}$ Josephson edge junctions. Appl Phys Lett 64:1292–1294, 1994.
49. SS Tinchev, S Alexandrova. Comment on: Properties of interface-engineered high T_c Josephson junctions. Appl Phys Lett 73:1745–1746, 1998.
50. KA Delin, AW Kleinsasser. Stationary properties of high-critical-temperature proximity effect Josephson junctions. Supercond Sci Technol 9:227–269, 1996.
51. QX Jia, Y Fan, C Mombourquette, D Reagor. Development of ramp-edge SNS junctions and SQUIDs. Mater Sci Eng B 56:95–99, 1998.
52. BD Hunt, MG Forrester, J Talvacchio, JD McCambridge, RM Young. High-T_c superconductor/normal-metal/superconductor edge junctions and SQUIDs with integrated groundplanes. Appl Phys Lett 68:3805–3807, 1996.

53. WH Mallison, SJ Berkowitz, AS Hirahara, MJ Neal, K Char. A multilayer $YBa_2Cu_3O_x$ Josephson junction process for digital circuit applications. Appl Phys Lett 68:3808–3810, 1996.
54. R Scharnweber, M Schilling. Integrated $YBa_2Cu_3O_7$ magnetometer with flux transformer and multiloop pick-up coil. Appl Phys Lett 69:1303–1305, 1996.
55. MI Faley, U Poppe, K Urban, H Hilgenkamp, H Hemmes, W Aarnink, J Flokstra, H Rogalla. Noise properties of direct-current SQUIDs with quasiplanar $YBa_2Cu_3O_{7-x}$ Josephson junctions. Appl Phys Lett 67:2087–2089, 1995.
56. MI Faley, U Poppe, CL Jia, U Dahne, Yu Goncharov, N Klein, K Urban, VN Glyantsev, G Kunkel, M Siegel. Application of Josephson edge type junctions with a $PrBa_2Cu_3O_7$ barrier prepared with Br-ethanol etching or cleaning. IEEE Trans Appl Supercond 5:2608–2611, 1995.
57. QX Jia, D Reagor, XD Wu, C Mombourquette, SR Foltyn. Characterization of ramp edge-geometry Ag: $YBa_2Cu_3O_{7-x}$/$PrBa_2Cu_3O_{7-x}$/Ag: $YBa_2Cu_3O_{7-x}$ junctions and dc SQUIDs. IEEE Trans Appl Supercond 7:3005–3008, 1997.
58. M Schilling, S Key, R Scharnweber. Biomagnetic measurements with an integrated $YBa_2Cu_3O_7$ magnetometer in a hand-held cryostat. Appl Phys Lett 69:2749–2751, 1996.
59. D Reagor, Y Fan, C Mombourquette, QX Jia, L Stolarczyk. A high-temperature superconducting receiver for low-frequency radio waves. IEEE Trans Appl Supercond 7:3845–3849, 1997.

4

Step-Edge Josephson Junctions

F. Lombardi
Chalmers Institute of Technology and Göteborg University,
Göteborg, Sweden

A. Ya. Tzalenchuk
National Physical Laboratory, Middlesex, England

4.1 INTRODUCTION

In the framework of high-critical-temperature Josephson Junctions (HTS-JJs) the development of grain-boundary (GB) junctions has represented an important innovation. The structures involving "extrinsic" interfaces are well established for metallic low-temperature superconductors. A similar approach in HTS multilayer technology remains very difficult for both physical (the short coherence length) as well as chemical (surface instability) reasons, although significant progress has recently been achieved in the fabrication of the ramp-type multilayer JJs. These difficulties have motivated the fabrication methods of HTS-JJs to deeper exploit the unique combination of structure and properties of the high-critical-temperature superconductors. Soon after the discovery of the HTS superconductors, it was realized that at least some of the grain boundaries in a polycrystalline material behave as weak links for the superconducting current. The IBM group (1) first managed to separate a single grain boundary and proved that it worked as a Josephson junction. Later, the same group found a method to artificially create individual grain boundaries in otherwise single-crystalline thin films—the bicrystal technology (2). The intrinsic Josephson effect in the c-axis direction (3) or the use of a controlled nucleation of grains with different crystallographic orientation [the biepitaxial technique (4)] are other examples of valid alternative methods to realize Josephson structures. In these methods, "intrinsic" interfaces and/or barrier layers determine the Josephson junction properties. The fabrication of devices, which are useful in complex circuits, requires, however, the optimization and a precise control of these intrinsic interfaces/barriers.

Among the possibilities for producing grain-boundary junctions, step-edge junctions (SEJs) represent a step up in technological complexity, compared to bicrystal junctions, but, at the same time, bring the topological freedom necessary for the design and the integration on small and large scales. The idea and the first demonstration are due to Daly et al. (5). SEJs are obtained by the epitaxial growth of a high-T_c (transition temperature) film on a step etched in a substrate prior to the film deposition. The preparation of well-defined microstructurally reproducible steps is the key point in the step-edge junction technology.

The step pattern is defined by either photolithography or electron beam (e-beam) lithography. The step is then produced in the substrate by ion milling. Because etching rates of the common substrate materials are slow and ion etching is very directional, the microstructure of the step depends greatly on the mask properties and especially on its profile. Reflown photoresist masks are commonly used to produce shallow steps. Hard materials, such as carbon (diamondlike or amorphous) and chromium, often in combination with e-beam lithography are used to produce straight steep steps. The step angle and morphology directly affect the film growth on the substrate and the structural and transport properties of the GBs that are subsequently formed.

In this chapter, we will give an overview of the current state of art of $YBa_2Cu_3O_{7-\delta}$ (YBCO) step-edge Josephson junctions. First, we will make some general remarks about YBCO growth on differently oriented substrates. It will be followed by a detailed description of the structural properties of the GBs, obtained on the most commonly substrates used for HTS film growth, in correlation with the step-edge profile. A description of the principal fabrication techniques used to form a step in a substrate will then be given. The transport properties of the GBs will be widely discussed in the framework of the well-established theory of the Josephson effect and of the up-to-date understanding of the HTS phenomenology. Then, we will characterize the dc superconducting quantum interference devices (SQUIDs), which represent the most successful application of the SEJs. Throughout the chapter, the performances of the SEJs will be also discussed in comparison with other HTS Josephson junction technologies.

4.2 YBCO GROWTH ON EXACT AND VICINAL CUT (100)/(110) SUBSTRATES

In this section, we will briefly summarize some aspects of YBCO growth on differently oriented substrates. What is discussed in Sections 4.2.1 and 4.2.2 is meant to give a general understanding of the mechanisms which lead to the nucleation of grain boundaries on stepped substrates. A direct correlation between the YBCO growth habits and the microstructure observed is, in fact, needed to clarify the technological key points necessary for the development of reproducible engineered grain-boundary Josephson devices.

4.2.1 General Remarks About YBCO Growth on Exact Substrates

For a good epitaxial growth of HTS superconductors, the choice of the substrate is quite crucial, as it must be compatible with the superconductor material both structurally and chemically. The types of substrate most commonly used for YBCO growth, are perovskites, such as $LaA1O_3$, $SrTiO_3$, and $NdGaO_3$, but also ZrO_2, MgO, and sapphire buffered with thin films of CeO or MgO. In Table 4.1, the values of the lattice parameters for these compounds and for the YBCO are summarized.

On perovskite substrates, the similarity of the crystal structure and the low lattice mismatch allow epitaxial growth of the YBCO, leading to films with excellent superconducting properties. The crystal cut of the substrate affects the orientation of the films. For example, on (100) $SrTiO_3$ and $LaA1O_3$, depending on the deposition conditions, (100) or (001) YBCO orientation is achieved; (110) or (103) growth is, instead, obtained on (110)-oriented perovskites. As a general rule, low substrate temperature, high oxygen pressure, and high deposition rates during film deposition favor in-plane alignment of the YBCO *c* axis (6–8). Opposite conditions are required for optimal growth of (001) and (103) YBCO.

On (110) perovskite surfaces, the fourfold rotational symmetry inherent to the (001) surface is broken. In other words, two orthogonal in-plane directions (e.g., the [100] and $[1\bar{1}0]$) are no longer equivalent. For a (110) growth, the lattice matching is obtained by aligning the *c* axis and the $[1\bar{1}0]$ YBCO directions respectively with the [001] and $[1\bar{1}0]$ in-plane directions of the substrate

TABLE 4.1 Lattice Parameters for the YBCO and for Some Compounds

Material	Lattice parameter (nm)
YBCO	$a = 0.382$, $b = 0.389$, $c = 0.1169$
$LaAlO_3$	0.379
$SrTiO_3$	0.391
$NdGaO_3$	0.386
MgO	4.21
Al_2O_3	$a = b = 0.476$, $c = 0.13$

($[001]_{YBCO} \parallel [001]_{sub}$ and $[1\bar{1}0]_{YBCO} \parallel [1\bar{1}0]_{sub}$). The *a* and *b* YBCO axes will therefore be out of the plane of the substrate, forming an angle of approximately 45° with respect to the normal **n** to the substrate surface (see Fig. 4.1a). For a (103) growth (Fig. 4.1b), the *b* axis and the $[10\bar{3}]$. YBCO direction are aligned respectively with the [001] and $[1\bar{1}0]$, directions of the substrate. In this case, the *c* and *a* YBCO axes are out of the substrate plane at an angle of about 45° with respect to **n.** Owing to the twofold axial symmetry of the substrate surface, a ($\bar{1}03$) YBCO growth is also possible (see Fig. 5.1b). The (103) and ($\bar{1}03$) domains differ by a 90° rotation around the *b* axis (i.e., the *a* and *c* axes are exchanged). The coalescence and growth of {103} nuclei leads to the formation of triangular grains delimited by the *a-b* plane at 45° with the substrate normal. This can be interpreted in the following way (9).

Figure 4.2a shows the presence of both the (103) and ($\bar{1}03$) domains. They terminate on one side by a basal plane face, which is smooth but slow growing due to the layer-by-layer growth mode of the YBCO and the lack of favorable sites for the nucleation of new layers. The other side, which is very rough, should grow relatively fast due to the abundance of steps and kinks provided by the grain morphology. On the basis of this hypothesis, the (103) face expands faster on the left (L) side of one grain than it does on the ($\bar{1}03$) face on the right (R) side of another grain (see Fig. 4.2a). At the meeting point, they form symmetrical 90° tilt grain

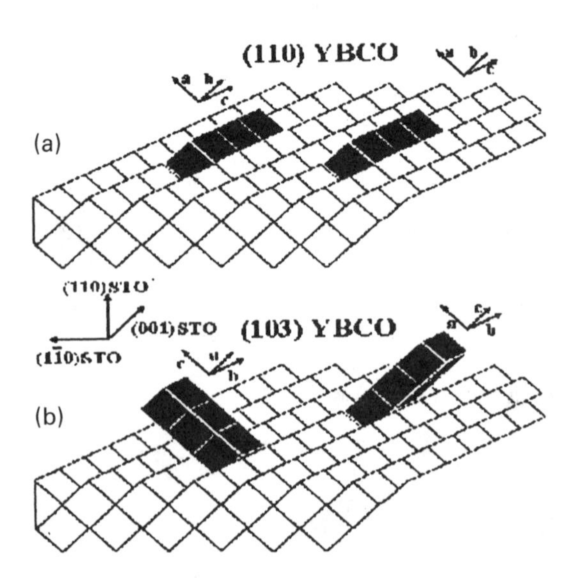

FIGURE 4.1 Sketch of the possible epitaxial relations between the YBCO cell and a (110) $SrTiO_3$ substrate: (a) (110) YBCO growth; (b) (103) YBCO growth. The two domains (103) and ($\bar{1}03$) are tilted 90° with respect to each other.

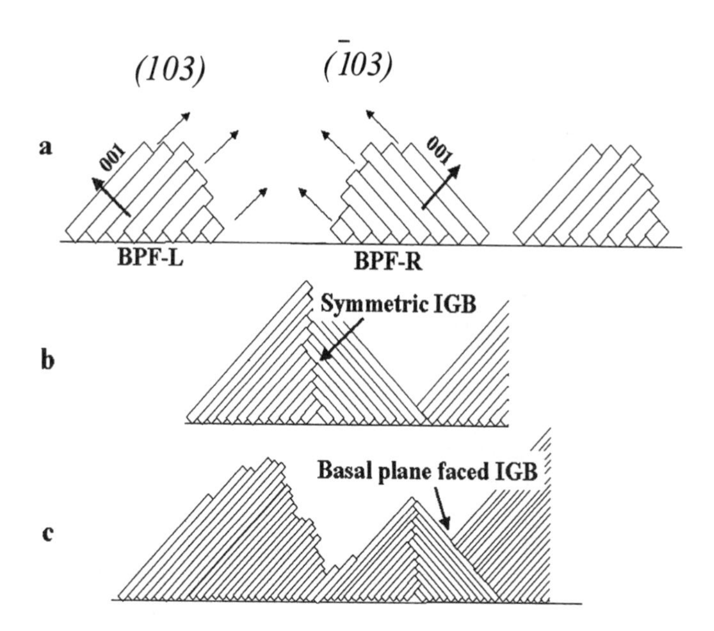

FIGURE 4.2 Schematic representation of the nucleation and domain coalescence of a (103) YBCO thin film: (a) (103) and ($\bar{1}$03) nuclei characterized by a rough side and a side bound by a basal plane (BPF). The coalescence of grains leads to the formation of a (b) 90° symmetric intrinsic GB or a (c) basal-plane-faced intrinsic GB.

boundaries (SGB) (Fig. 4.2b). If the two meeting grains have roughly the same height, they will form a triangular grain, with both sides terminating by basal plane face. In the following stages of the film growth, the triangular grain may be covered and embedded by larger nuclei expanding either in the [$1\bar{1}0$] or in the [$\bar{1}10$] direction. This leads to the formation of 90° tilt boundaries of the basal-plane-faced (BPF) type (see Fig. 4.2c). These features account for the characteristic morphology of (103) YBCO films (10) determined by the formation of intrinsic grain boundaries.

On poorly matched substrates such as the MgO for example, the growth habits of the YBCO are quite different. On (100) and (110) substrates, the growth is almost *c* axis in a wide range of values of deposition parameters. Few reports on *a*-axis growth on MgO are present in the literature (11). Moreover, a (103) orientation has only been obtained on $SrTiO_3$-buffered (110) MgO (12). This behavior is a consequence of the large mismatch between the lattice parameters of the YBCO and the MgO. The substrate lattice cannot be a template for the growth, so the orientation which minimizes the free energy at the interface with the MgO is dominant.

4.2.2 YBCO Growth on Vicinal Cut Substrates

On perovskite (100) substrates, the alignment between the YBCO lattice parameters and the crystallographic axis of the substrate is kept, even when the in-plane symmetry of the surface substrate is broken by the introduction of a vicinal cut (13). Figure 4.3a shows the YBCO growth on a perovskite substrate with a small vicinal angle α in the (100) [or (010)] direction. On the atomic scale, the substrate surface is not flat. The ratio between the height h and the width w of the steps is defined by the angle α ($\mathrm{tg}\,\alpha = h/w$). It is clear that the normal $\mathbf{n}$ to the horizontal surface of the atomically defined steps is still the (001) crystallographic direction of the substrate, whereas the normal $\mathbf{n}'$ to the macroscopic substrate surface is rotated by an angle α with respect to $\mathbf{n}$. The growth mode, characterized by the c axis of the YBCO parallel to the direction locally defined by $\mathbf{n}$ is energetically favorable because of the small mismatch between the lattice parameters of the YBCO and that of the substrate. On poorly matched substrates such as the MgO, the YBCO will, instead, preferably grow, aligning the c direction with the normal $\mathbf{n}'$ to the macroscopic substrate surface (14) (see Fig. 4.3b).

From a microscopic point of view, the presence of a step with an angle α and with the edges aligned with one of the two in-plane directions is equivalent to the introduction of a (001) surface, in the substrate, with a vicinal cut α. For small values of α, in a perovskite substrate, the direction of the YBCO c axis on the step surface will therefore remain parallel to the c-axis orientation on the top and the bottom flat parts of the step (see Fig. 4.3a). At this point, a question naturally arises: What is the maximum value of α compatible with this kind of growth? Or alternatively, what is the minimum value of α which allows rotation of the YBCO c axis on the step surface and the formation of GBs at the edges of the step? For α approaching 45°, the step surface will have twofold symmetry. On such a kind of

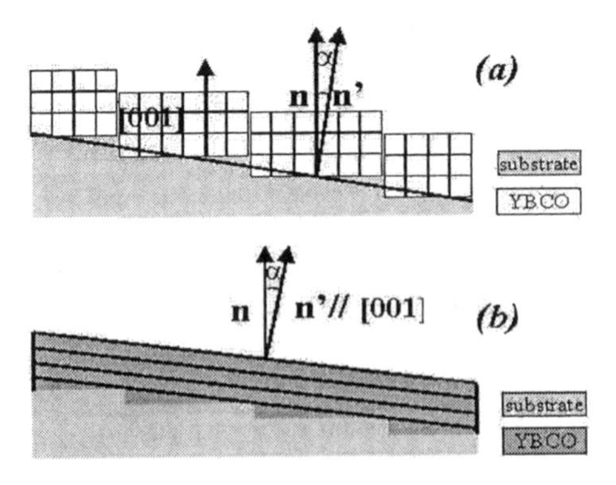

FIGURE 4.3 YBCO growth on a vicinal cut (a) and perovskite substrate (b) on a MgO substrate.

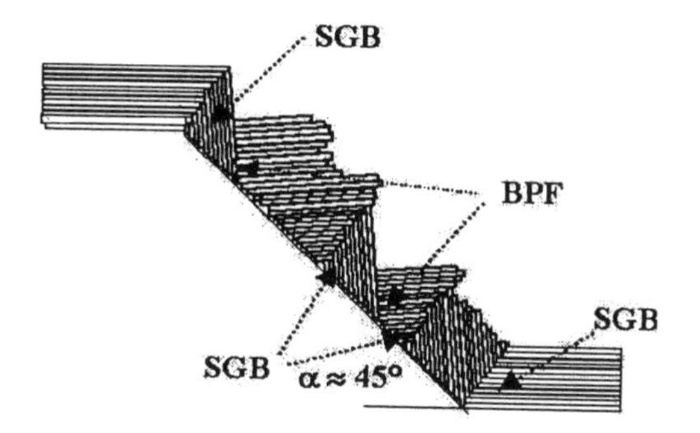

FIGURE 4.4 Grain-boundary formation on a step with α approaching 45°.

surface, as discussed in the previous paragraph, depending on the growth conditions, two different orientations are possible: the (110) and the (103). From these considerations, we can, therefore, guess that at the temperatures typical for a *c*-axis growth on a (001) substrates ($T \approx 730$–$800°C$), on steps with $\alpha \approx 45°$, we will have a *c*-axis growth on the upper and bottom flat parts and a preferential (103) growth on the step surfaces. For a step angle $\alpha > 45°$, we would expect a situation similar to that illustrated in Figure 4.4 with two 90° tilt symmetric grain boundaries at the defined edges of the step and a combination of 90° symmetric and basal-plane-faced tilt GBs on the step surface. A more detailed correlation among the morphology of the step, the step angle, and the exact structural properties of the GBs will be discussed in the following sections.

4.2.2.1 Do These Grain Boundaries Behave as Weak Links?

Electrical measurements performed on steps with different angles fabricated on a $SrTiO_3$ substrate give some indications about the transport properties of the YBCO film grown on the step. Figure 4.5 shows the dependence of the critical current density J_{cm} of a microbridge across the step, normalized by J_{cs} of a stripline defined on the flat part of the substrate, as a function of the step angle (15). In this case, the thickness of the YBCO film is less than the step height. It is clear that, up to an angle of 10°, no evident degradation of the superconducting properties is observed. In the range of values $10° < \alpha < 40°$, J_{cm} is reduced by almost one order of magnitude compared to J_{cs}. This reduction is related both to the presence of defects between *c*-axis grains nucleated on two adjacent microscopic steps (16) (like antiphase boundaries) and to the strong out-of-plane anisotropy of the YBCO superconductor. Indeed, the current flows partially along the *c*-axis direction throughout the step, where the critical current density is almost one order of magnitude lower than the corresponding value in the *a-b* plane. For $\alpha \approx 65°$, J_{cm} is reduced by two orders of magnitude compared with J_{cs} and this is a sign for a weak-link-like behavior (17).

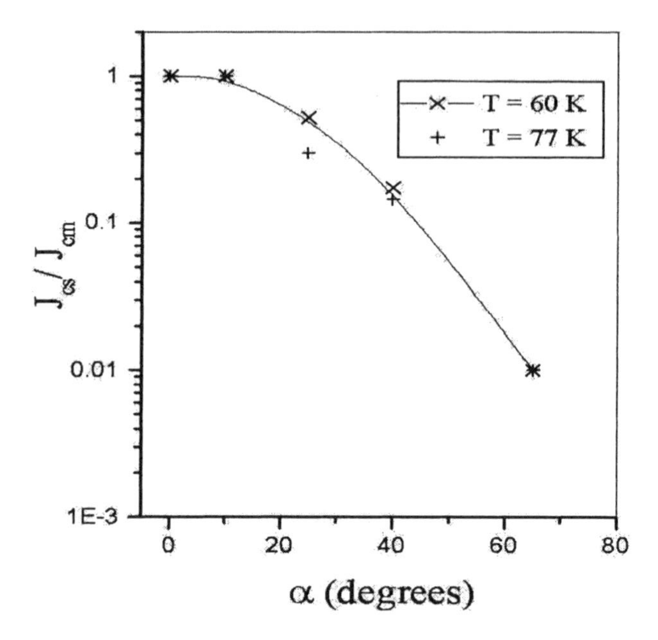

FIGURE 4.5 Critical current density J_{cs} across the step normalized to the J_{cm} of a stipline, as a function of the step angle α for two different temperatures. (After Ref. 15.)

On poorly matched substrates, such as MgO, the situation is quite different. As discussed earlier, the YBCO *c* axis on the step should be aligned with the normal **n** of the macroscopic surface. In this case, the step angle α determines the misalignment between the *c* axes on the flat substrate and on the step surface. If the step edges are well defined (not rounded), even a small step angle may, therefore, introduce [100]/[010] tilt grain boundaries into the structure. These kinds of GBs have been extensively explored by Dimos et al. (18) using bicrystal junctions. From their transport measurements, also in the presence of an external magnetic field, there is clear evidence that [100]/[010] tilt GBs, with an angle as small as 10°, act as Josephson weak links. An interesting features of SEJs on MgO substrates is, therefore, represented by the possibility to explore [100]/[010] tilt GBs in a wide angular range.

4.3 MICROSTRUCTURE OF EPITAXIAL YBCO FILMS ON STEP-EDGE PEROVSKITE SUBSTRATES

The description of the step surface in terms of atomically defined steps has given a qualitative understanding about the possibility of forming GBs during the

YBCO deposition. In order to study the microstructure of YBCO films across a step, structural investigation on the atomic scale is required. In what follows, a description of the GBs structure is given on the basis of high-resolution transmission electron microscopy (HRTEM) analysis performed on samples fabricated by pulsed laser deposition.

The microstructure of the YBCO film is found to vary with the steepness of the step characterized by the angle α. Figure 4.6 shows the YBCO growth on two different steps fabricated in $SrTiO_3$ substrates (19). No grain boundaries are observed for a mild slope, $\alpha \approx 40°$ (Fig. 4.6a). The *c* axis does not change the orientation across the step, remaining perpendicular to the substrate surface. When α is about 45° (Fig. 4.6b) the microstructure is just like that observed in a (103) growth on a (110) surface (for comparison, see Fig. 4.2c). Multiple GBs consisting of a complex combination of 90° symmetric and basal-plane-faced (BPF) GBs are formed along the step and at the edges. This behavior is, therefore, in agreement with the arguments presented in the previous section for YBCO growth on well-matched substrates.

Figure 4.7 shows the YBCO film growth on a step with $\alpha \approx 58°$ in $SrTiO_3$ (19). Two similar GBs can be clearly distinguished. They are almost of the symmetrical type, although the presence of small facets and misfit dislocations is also observed. This is typical for symmetric grain boundaries (20). They are located near the top and the bottom edges of the step. A combination of symmetrical GBs and BPF grain boundaries are also visible near the interface between the film and the step surface.

When the thickness of the film on the step exceeds 30 nm, these domains are shunted by a larger unidirectional domain. This behavior may be interpreted by

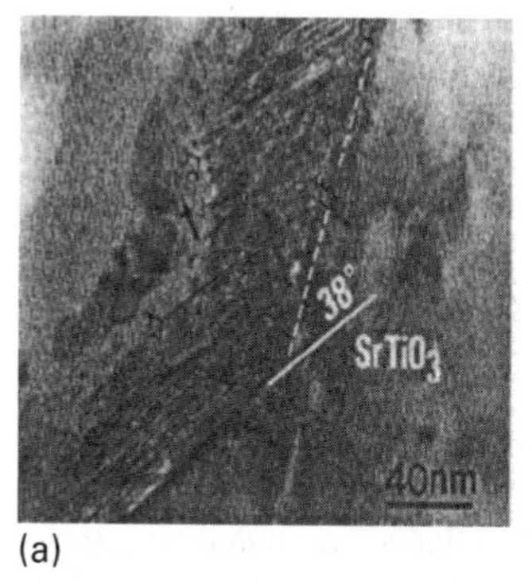

(a)

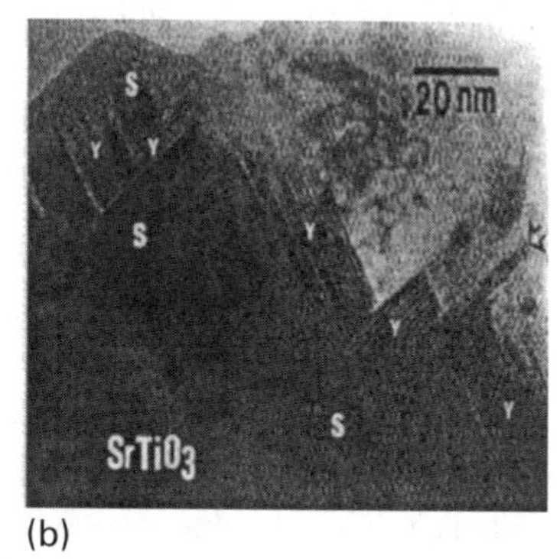

(b)

FIGURE 4.6 High-resolution TEM pictures showing the microstructure of the YBCO film (a) on a 38° step and (b) on a 45° step for ratios $t/h \approx 1/3$ and 1 respectively between the film thickness and the step height. (After Ref. 19.)

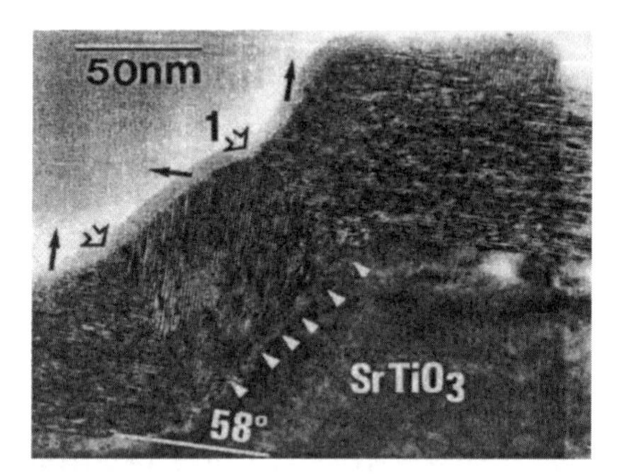

FIGURE 4.7 Lattice image of the YBCO film grown on a 58° step. Open arrows indicate the two grain boundaries, and the triangles point to the 90° domains at the interface with the step surface. (After Ref. 19.)

considering the surface of a step with an angle α exceeding 45° as the surface of the (110) substrate with a vicinal angle $\beta \approx \alpha - 45°$ in the $[1\bar{1}0]$ direction. On such a kind of surface, one of the two growth modes (103) or $(\bar{1}03)$ can be selected (12). In this way, the (103) YBCO film presents a single domain. At the early stage of growth, however, local variations of the step angle may induce nucleation of both domains. By increasing the film thickness on the step, the triangular grain formed by the coalescence of (103) and $(\bar{1}03)$ domains are covered and embedded by larger nuclei of the dominant orientation (see Fig. 4.2 for comparison).

The YBCO microstructure substantially changes as the angle α is increased. Figure 4.8 shows a low-resolution TEM image of a YBCO grown on a steep step ($\alpha \approx 80°$) in a $LaAlO_3$ substrate (21). Two well-defined 90° GBs separate the YBCO *c*-axis film on the substrate from the film grown on the step, which, in this case, presents a single domain. The film on the step flank is then often referred to as the YBCO *a* axis. From Figure 4.8, it is evident that the upper YBCO *c* axis has overgrown the film on the step. Furthermore, in contrast with shallow steps, the following is found:

1. The *a-b* plane termination on the step are not exposed to the environment (see for comparison, Fig. 4.7).
2. The *a*-axis YBCO film thickness is much reduced compared with the *c*-axis component on the top and bottom parts of the step (about one-third of the nominal YBCO film thickness).
3. The top and bottom grain boundaries are very dissimilar.

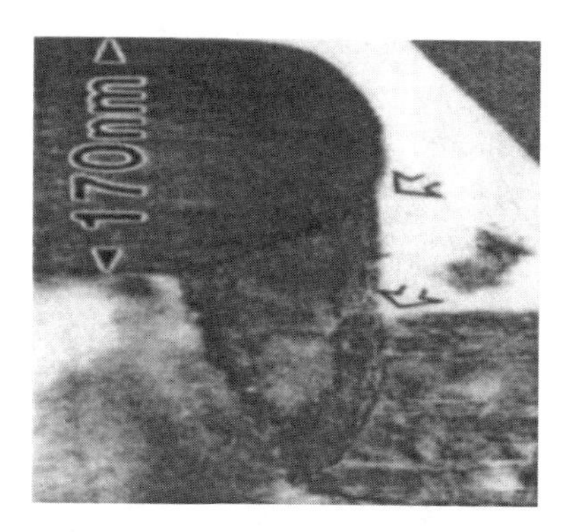

FIGURE 4.8 YBCO microstructure across a steep $LaAlO_3$ step. Arrows indicate the two grain boundaries. The film thickness reduction on the step flank is evident. (After Ref. 21.)

In Figure 4.9 is shown the microstructure of the top GB, for two steps obtained with different fabrication procedures. In Figure 4.9a, the step has been obtained using an amorphous carbon mask and e-beam lithography. In Figure 4.9b, a Nb mask and ordinary photolithography have been used to pattern the geometry of the step. In both cases, the YBCO presents a single domain on the step flank. The microstructure shown is typical for the data presented in the literature. From Figure 4.9a (22), it is evident that the boundary plane varies with the distance from the substrate. Two parts can be distinguished. A BPF-type grain boundary starts at the interface between the film and the substrate, extended by 10–20 nm. The boundary plane deviates toward an angle of almost 45° with respect to the substrate normal as the thickness is increased. The grain boundary, which nucleates

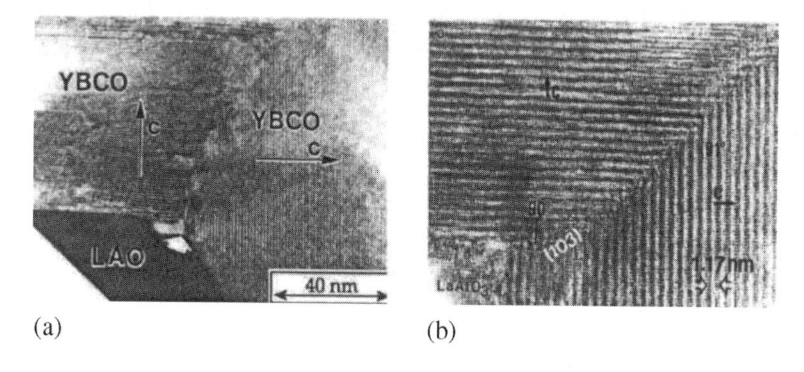

FIGURE 4.9 Microstructure of the grain-boundary nucleate at the top edge of a $LaAlO_3$ substrate. (a) from Ref. 22; (b) from Ref. 21.)

on top of the BPF part, consists of a random alternating sequence of segments of (100)(001) or (010)(001) boundaries (still of the BPF type) and segments of (013)(013) or (103)(103) boundaries [of the symmetrical(s) type]. The length of each facet does not exceed a few unit cells. In Figure 4.9b, the top GB consists, instead, essentially of symmetric segments. The general trend, however, is that the top GB is a combination of BPF and S facets, the detailed geometry of which is influenced by the thin-film evolution during the growth.

The growth dynamics is determined by many factors. Both the anisotropy of the YBCO growth rate, which is intrinsically higher in the *a-b* plane than in the *c*-axis direction, and the growth conditions (i.e., the deposition method, temperature, pressure, substrate, etc.), can affect the grain-boundary structure at different stages of the growth evolution. A possible explanation for the BPF grain-boundary presence at the interface between the substrate and the film (see Fig. 4.9a) is, in fact, that the *c*-axis film nucleates prior to the *a*-axis part and does not expand beyond the step corner, because of the presence of a ledge barrier. The *a*-axis-oriented particle nucleated on the step, however, quickly expands vertically, reaching the top corner and the upper part of the *c*-axis film with the subsequent formation of a BPF grain boundary. Moreover, the presence of 90° facets of the grain-boundary plane of Figure 4.9a can be explained in a similar way.

The late nucleation of *a*-axis grain may also be related to the high directionality of the plasma plume in the plasma laser deposition (PLD). The growth rate turns out to be dependent on the incident angle of the plume with the substrate surface, resulting in thicker films on surfaces normal to the plume axis. Under usual deposition conditions, the plume is perpendicular to the substrate, so the nucleation probability on the step is reduced. This can also justify the presence of the overgrowth of the *c*-axis orientation with respect to the *a*-axis part in Figure 4.8.

Figure 4.10 shows the HRTEM image of a typical bottom grain boundary. It is quite irregular and has the tendency to become vertical (i.e., to evolve into an orientation parallel to the normal to the substrate). It consists mostly of BPF grain boundaries with a small percentage of symmetric facets near the interface with the step. This behavior is essentially a consequence of the higher growth rate in the *a-b* plane compared with the *c* direction.

The step often meanders around the predefined line, and the nucleation rate of the YBCO across the step is affected by the microscopic orientation of the meandering line. Depending on the technique used to define the step pattern, the meandering profile may present faceting. When this is the case, the facets are randomly oriented and, in general, are not aligned with the (100) and (010) in-plane orientations of the substrate.

Figure 4.11a schematically represents an intentionally wavy patterned step-edge profile used by Gustafsson et al. (22,23,24) to simulate the step faceting. They have studied the nucleation and growth of YBCO on wavy steps and compared it with the corresponding growth on straight steps. In Figure 4.11b is shown

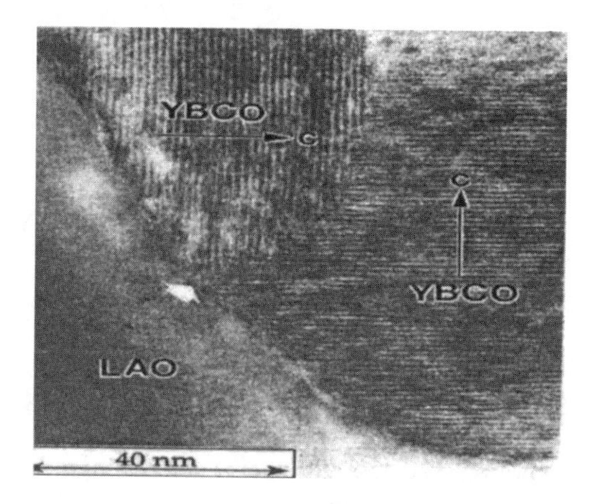

FIGURE 4.10 High-temperature TEM image of the GB formed at the bottom edge of a high-angle $LaAlO_3$ step. (From Ref. 22.)

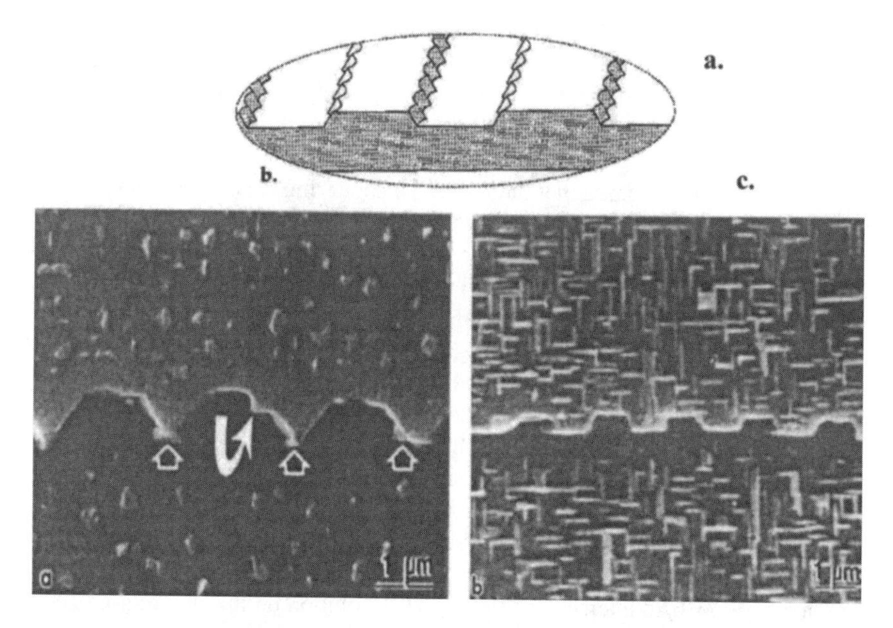

FIGURE 4.11 (a) Sketch of the wavy step-edge profile patterned in a $LaAlO_3$ substrate. SEM micrograph after the deposition of (b) a 50-nm-thick YBCO film and (c) a 200-nm-thick YBCO film. In (b), the open arrows indicate the *a*-axis grains which nucleates at the apex of the step. (from Ref. 23.)

a scanning electron microscopy (SEM) picture of a thin YBCO film grown on a wavy step with an average angle of 60°–65°. The film is *c*-axis oriented, except in the step region, where the YBCO grows *a* axis (referring to the usual terminology). As it is clear from the picture, the nucleation of *a*-axis grains varies along the step profile. They primarily nucleate at the apex and the bottom of the step, where the deviations from the (100) and (010) in-plane directions of the substrate are small. As the YBCO thickness is increased (Fig. 4.11c), the *a*-axis grains, because of their typical elongated shape, become a support for the *c*-axis film on the horizontal surface. In the step region next to the apex, the *c*-axis film grows over the step until it meets the *a*-axis grains. This determines a strong variation of the original step profile, which now shows a flatter morphology (Fig. 4.11c). The above considerations are confirmed by the TEM analysis, which shows a variable thickness of *a*-axis particles according to the position along the wavy edge at which they nucleate. Grain boundaries similar to those shown in Figure 4.9a, but with a much more irregular microstructure, have been detected only near the apex and the bottom of the wavy edges. CuO particles nucleate on the flank surface in the vicinity of these regions, where the profile is no longer parallel to the substrate (100) and (010) directions. In an extreme case, the combined effect of poor matching between the YBCO lattice parameter and the wavy edge of the step and the lower nucleation probability on the flank surface leads to the absence of an YBCO film on this part of the step profile. In some of these regions, as discussed earlier, the film continuity is established through the overgrowth of the *c*-axis film at the edges which join the *a*-axis grains protruding from the apex edge.

It is worth mentioning that the use of the sputtering technique (both on-axis cylindrical magnetron and off-axis planar magnetron) to grow YBCO films on stepped substrates gives rise to a significantly different film nucleation and growth habits at the step region. This deposition technique has been very successfully employed for the growth of high-quality YBCO films on bare substrates and, in general, for the multilayer technology. A comparative study on YBCO growth on steep $LaAlO_3$ steps using both laser ablation and the sputtering technique has been made by Gustaffson et al. (25). Their results demonstrate that the sputtering technique is hardly applicable for the SEJ technology. Compared to YBCO laser-ablated films, no regular grain boundaries are observed at both edges of the step. The boundary plane often deviates from a 45° orientation, evolving toward 90° and leading to the formation of facets mostly of the basal-plane-faced type. Moreover, a significant amount of secondary phases frequently nucleate on top of the first layers of YBCO, interrupting the growth evolution on the step region. As a consequence, the effective grain-boundary area can be substantially reduced. These features can be related to the lower deposition rate of sputtering compared to the laser ablation technique as well as to the intrinsically different deposition regime (diffusive in one case, mostly directional in the other).

4.4 STEP-EDGE FABRICATION

Step-edge fabrication is a very tricky process, even though the degree of complexity is not that high. All the research groups dealing with this technology have in mind clearly that once the procedure has been established, it has to be strictly respected in order to obtain reproducible results. Even very small variations in the fabrication procedure, which at first glance may look insignificant, may cause dramatic changes in the performances of the junctions. It is, therefore, senseless to give a detailed recipe. Instead, we will address different aspects of the fabrication procedure which influence the final step profile at different stages of the process.

The quality of the step edge is strongly affected by the profile of the protecting mask. The step pattern is usually defined by ion milling after either photolithography or electron beam lithography. Wet etching of the steps has also been used especially for MgO substrates (26,27). However, this method is complicated by selectiveness and anisotropy and has not become widespread. The resulting step angle is determined by the ratio between the ion-milling rate of the substrate and the protecting mask and by the edge angle of the mask itself. As a general rule for fabricating a high-angle step, a hard protecting mask with sharp profiles is needed. The harder the mask, however, the less crucial is the second requirement. The ion-milling rates for a few practically important materials and substrates are summarized in Table 4.2. For relatively soft masks, such as Nb or photoresist, a steep angle step ($\alpha \approx 60°$) can be obtained only by defining the mask edges close to 90°.

An additional problem when using soft materials as masks for ion milling is the formation of an amorphous layer of redeposited material near the upper edge

TABLE 4.2 Ion-Etching Rates of Various Materials for Argon Ions of 500 eV with Normal Incidence on the Substrate and Ion Current Density of 1 mA/cm^2

Material	Etching rate (nm/min)
$LaAlO_3$	18
$SrTiO_3$	17
MgO	19
Nb	40
Photoresist S-1813	50
α-C	5

of the step. A much higher ion-milling rate of the protecting mask compared to the substrate causes shrinking of the mask edges and the formation of a "bump" at the top edge of the step. The growth of such layer, usually a few nanometers thick, can be detected by SEM or atomic force microscopic (AFM) inspection of the substrates. The material is usually redeposited in an amorphous form. The formation of such a layer may strongly affect the uniformity of the grain boundary in the film, and more in general, its nucleation. In some cases, it was observed that the YBCO film deposited on the stepped substrate is not continuous, as the "bump" inhibits the mobility of atoms in the vicinity of the top edge (28). This drawback can be overcome by using a special ion-milling procedure as reported in Ref. 29. The ion beam is to be parallel to the mask edge plane at an angle $\gamma = 45°$ with respect to the normal to the substrate (instead of the usual condition where the ion-milling plate rotates in order to obtain a better uniformity of the step). An alternate milling at $\gamma \pm 45°$ can avoid the effect of redeposition. In this configuration, the amorphous material etches faster than the substrate. The resulting steps show a sharp and uniform profile.

The use of a hard mask automatically helps to avoid the problem of redeposition. To suppress any redeposition whatsoever, we have been using etching at a small, 4°–6°, angle to the substrate normal and rotation of the substrate holder. High-quality steps in terms of uniformity and steep profile have been obtained by Sun et al. (30) using diamondlike carbon. In this form, however, the carbon is difficult to deposit, so this technique is rather complicated to set up. Amorphous carbon (α-C), instead, can be easily deposited by using e-beam evaporation or, alternatively, a high-power plasma decomposition of methane. The ion-milling rate of α-C is slightly higher than that of the diamondlike form. Very good results in terms of a step-edge profile have been obtained by using an e-beam-defined α-C mask (31). The use of e-beam lithography, in place of ordinary photolitography, strongly improves the straightness of the step. In the previous section we have shown that this point is rather crucial: the grain-boundary microstructure is strongly affected by the meandering of the step edge.

Figure 4.12 schematically summarized the most important steps of the fabrication procedure used by the authors:

1. An α-C film, 100 nm thick, is deposited on the substrate by e-beam evaporation and in situ covered by a 50-nm-thick Au film.
2. The step pattern is then defined by e-beam lithography.
3. The pattern is transferred from the resist to the gold:
   (a) by ion milling and from the gold to the α-carbon
   (b) by oxygen reactive ion etching (RIE) (low power)
4. The step pattern is then obtained by an ion-milling etching of the sample.
5. The residual α-C is removed by oxygen RIE.

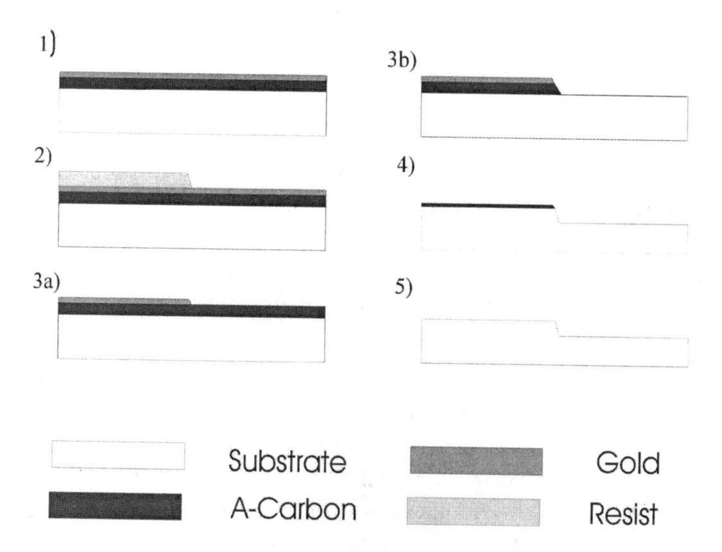

FIGURE 4.12 Schematic representation of the procedure used to fabricate a step in a substrate using α-C mask and e-beam lithography.

4.5 TRANSPORT PROPERTIES OF STEP-EDGE JUNCTIONS

4.5.1 Common Properties

Results of transport measurements performed on step-edge junctions are consistent with a weak-link-like behavior of the grain boundaries. The majority of step-edge junctions show the following common properties:

- Electromagnetically small junctions (32,33) ($w/\lambda_j < 4$, where w is the geometrical width and λ_j is the Josephson penetration depth) show a current voltage (I–V) characteristic typical of a resistively and capacitively shunted junction (RCSJ) model (34,35).
- As the ratio w/λ is increasing, the I–V curves exhibit an increasing amount of excess current, consistently with the transition to a long junction regime.
- The junctions are generally overdamped, corresponding to a value of the McCumber parameter $\beta_c = 2\pi J_c \rho_N^2/\Phi_0$ less than 1. At low temperatures, however, some junctions are underdamped ($\beta_c > 1$) with a hysteretic I–V characteristic.
- The critical parameters such as the critical current density J_c and the normal resistance per unit area ρ_N show a larger spread than bicrystal junctions. One can only roughly estimate $J_c(4.2\text{ K}) \approx 10^4$–$10^5$ A/cm^2 and $J_c(77\text{ K}) \approx 10^3$–$10^4$ A/cm^2. The value of ρ_N is of the order of 10^{-7}–10^{-9} $\Omega \cdot \text{cm}^2$ and is almost independent of the temperature.

- The characteristic voltage $V_c = J_c\rho_N$ lies in the range 1 mV $< V_c(T = 4.2) <$ 5 mV and $V_c(T = 77) <$ 0.5 mV and is much smaller than the gap Δ/e.
- V_c generally scales with J_c
- The working temperature can be as high as 80 K. Low-J_c junctions (36) (J_c less than 10^4 A/cm^2 at T = 4.2K), however, do not usually work at 77 K.
- J_c is generally spatially inhomogeneous, as evidenced by complicated magnetic fingerprints. The use of advanced technologies for the step definition can, however, noticeably improve the uniformity of the junctions.

Most of the above-listed properties are in common with other kinds of HTS-JJs. In particular, the scaling behavior of the I_cR_N product is also observed in bicrystal (37) and biepitaxial (38) GB junctions. Furthermore, this scaling law has also been found in ramp-edge and planar-type junction with artificial barriers (39). This is in clear contrast with a low-T_c superconductor–insulator–superconductor (S–I–S) Josephson junction, where V_c is independent of J_c and is almost equal to Δ/e. It is likely, therefore, that the same physical mechanism is responsible for the scaling behavior observed in different types of HTS-JJ. Different approaches have been proposed to describe these particular features. Gross et al. (40,41) for example, considered a tunneling-type transport through a large density of localized states. The quasiparticle current is dominated by the resonant tunneling, whereas for Cooper pairs, the transport only occurs through direct tunneling, because of the strong on-site Coulomb repulsion. This model accounts for the $V_c \infty (J_c)^p$ with $p \approx 0.5$ scaling and for the temperature-independent ρ_N.

4.5.2 SEJs on Perovskite Substrates Depending on the Step Profile

The I–V characteristics of step-edge junctions reflect the different microstructure of the YBCO across the steps of different angles. The following behavior can be observed:

1. On shallow steps, in the absence of GBs, the I–V curves are of the flux–flow type up to T_c (42). The critical current density is an order of magnitude less than in the absence of the step. It is not affected by a weak, up to 50 G, magnetic field and no Shapiro steps are observed.
2. As the step angle approaches 45°, a multidomain microstructure with several 90° $a(b)$ axis tilted GBs result in I–V characteristics (See Fig. 4.13) with the RSJ shape at low bias currents. Multiple singularities or "kinks" were observed at higher biases (42,43). The "kink" voltage positions can be shifted back and forth by applying a weak magnetic field. This suggests that the features observed in the I–V curves correspond to

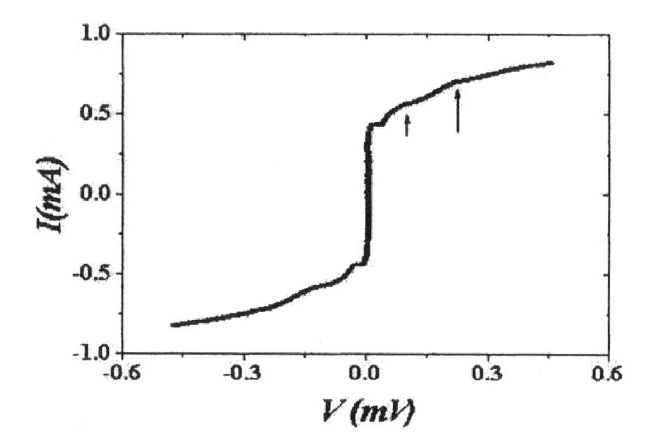

FIGURE 4.13 *I–V* characteristic of a step-edge junction with a step angle approaching 45°. The arrows indicate the presence of weak links in series.

the weak links connected in series with the weakest one, which defines the critical current of the junction. This statement is further confirmed by the fact that the R_N value increases for biases above each "kink."

3. On steps with steeper angles, the transport properties are determined by only two effective GBs formed at the top and bottom edges of the step.
   (a) For $50° < \alpha < 60°$, the GBs are quite similar, almost of the symmetric type (Fig. 4.7). The presence of the second junction is seldom detected in the *I–V* characteristic, due to the very close critical currents of the two GBs.
   (b) At higher angles, the top and the bottom GBs are, instead, very dissimilar. In this case, the *I–V* characteristics often show a rather distinct "kink" at finite biases. This circumstance, however, strongly depends on the microstructure of the two GBs; there are, in fact, reports where no features related to the second junction were detected (see Sec. 4.6).
4. On intentionally fabricated wavy steps, with an angle $\alpha \approx 65°$, the *I–V* characteristics of the junctions are no longer RSJ-like (44). The critical current I_c shows a weak dependence on the magnetic field; its suppression is, in fact, less than 10%. This value does not improve by increasing the temperature.

The transport properties of the junctions can be modified by varying the step height h and the film thickness t. Referring to steep steps ($\alpha \geq 65°$), the ratio t/h is quite crucial. RSJ behavior is observed for t less than h. For $t/h > 1$, both the top and bottom GB can be shunted by YBCO overgrowth. This may happen, for example, when the bottom GB nucleates not exactly at the edge, but somewhere on the flank part of the step (see Fig. 4.10). In this case, a shorting path may be created by the conjunction between the *c*-axis film overgrown on the top part of the step (compare with Fig. 4.8) and the *c*-axis film grown on the lower flat part of the step and extending up to the bottom GB. The Josephson behavior has been observed for t/h in the range $0.4 < t/h < 0.9$ (31,45). A much smaller value of the t/h ratio may, however, result in a discontinuity of the YBCO film across the step related to a much thinner YBCO film thickness on the flank part (about one-third of the nominal film thickness). Therefore, it is possible to trim the values of J_c and ρ_N by decreasing the YBCO film thickness over a fixed step.

For less steep steps ($\alpha \approx 60°$), an RSJ behavior has recently been observed for t/h in the range 0.6–0.8 (46). In this interval, a constant value of J_c and a specific resistance ρ_N linearly increasing by reducing t/h were observed. Moreover, both I_c and R_N scale with the junction width. For higher values ($t/h > 0.8$), the usual flux–flow-like *I–V* characteristics were measured.

The majority of devices employing the step-edge junctions have been realized on steep steps. The different structure of the top and bottom GBs leads, in fact, to quite different values of the two critical currents and deviations from the RSJ behavior by the undesirable features ("kinks") in the *I–V* characteristic appear at the voltages much higher than the I_cR_n product of the junction (42,47). This warrants a successful use of SEJs in applications such as SQUIDs. What, instead, appears to be rather limiting for the development of devices involving a great number of junctions [e.g., discrete flux flow devices, rapid single flux quantum (RSFQ) logic] is the spread in the critical parameters. The best results (47,48), in fact, address to values not better than 20–30%.

4.5.3 Josephson Phenomenology

Most of the phenomenology typical of the Josephson effect has been observed in step-edge junctions. Here, we will illustrate this statement on an example of a low-J_c junction at a step defined by e-beam lithography and ion etching through a carbon mask. The *I–V* characteristic in Figure 4.14 shows a clear hysteretic RSJ behavior, with the presence of a small amount of excess current. The magnetic field dependence of the maximum Josephson current I_c is shown in Figure 4.15 for the same junction. The diffraction pattern is quite regular, corresponding to an almost uniform current density distribution inside the junction. Such a behavior, however, is rarely observed in step-edge junctions. More in general, in fact, the SEJ shows quite a irregular pattern compared with bicrystal junctions (49). This effect is related to the waviness of the meandering line which leads to different YBCO growth along the step edge and to an intrinsic filamentary nature of SEJs. However, the use of advanced technologies can improve the uniformity of the junctions.

Although the comprehensive theory describing superconductivity in cuprates remains to be created, there is, by now, a growing number of experimen-

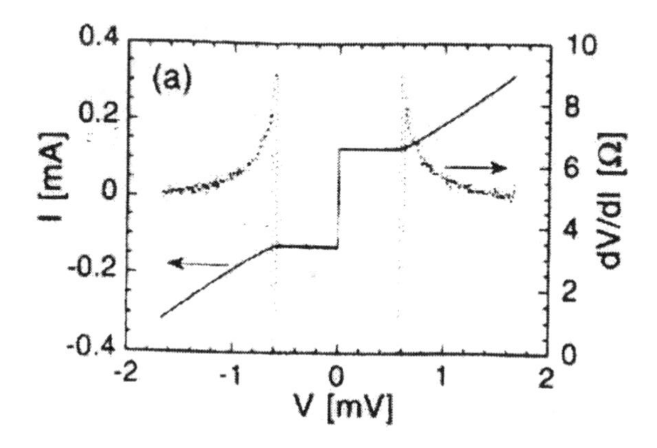

FIGURE 4.14 *I–V* characteristic (solid line) and differential resistance the bias voltage (dotted line) for a 16-μm-wide SEJ at $T = 4.2$ K. (After Ref. 35.)

tal facts, which point toward the dominant *d*-wave symmetry of the superconducting wave function (50). Therefore, it is important to analyze how this symmetry may influence the transport through the step-edge junctions.

It does have a serious impact on the (001) tilt grain-boundary junctions, such as bicrystaline or biepitaxial. In particular, Hilgenkamp et al. (51) have shown that the amount of π facets, with a negative critical current, increases by increasing the misorientation angle ϑ of the bicrystal. This leads to an inhomogeneous spatial de-

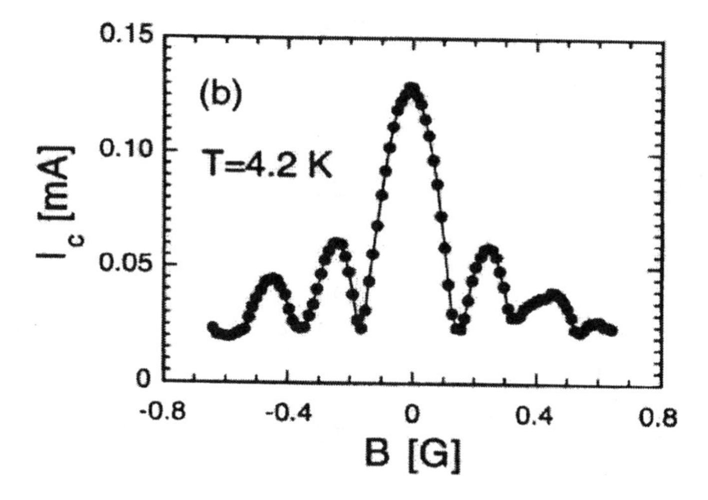

FIGURE 4.15 Magnetic field dependence of the critical current for the same SEJ junction as in Figure 6.14 at $T = 4.2$ K. (After Ref. 35.)

pendence of J_c and, therefore, to deviations from a Fraunhofer-like $I_c(B)$ dependence. The effect become very prominent for $\vartheta \approx 45°$. Indeed, for asymmetric 45° (001) tilt boundaries, obtained both with the bicrystal (51) and biepitaxial technique (52), a non-Fraunhofer $I_c(B)$ dependence is observed with symmetric maxima values corresponding to $B \neq 0$. This peculiar behavior has been attributed to the sequence of facets with 0 and π phase shift along the boundary line—the so-called π loops. If the junctions in a π loop are identical (53) or if the inductance of the π loop is sufficiently large (54), there might be a spontaneously generation of unquantized magnetic flux. If magnetic flux occurs spontaneously in the boundary, the current phase of the grain-boundary Josephson junction will also deviate from an ideal sinusoidal behavior. This circumstance has been demonstrated for asymmetric 45° (001) tilt bicrystal junctions (55). Moreover, the un-

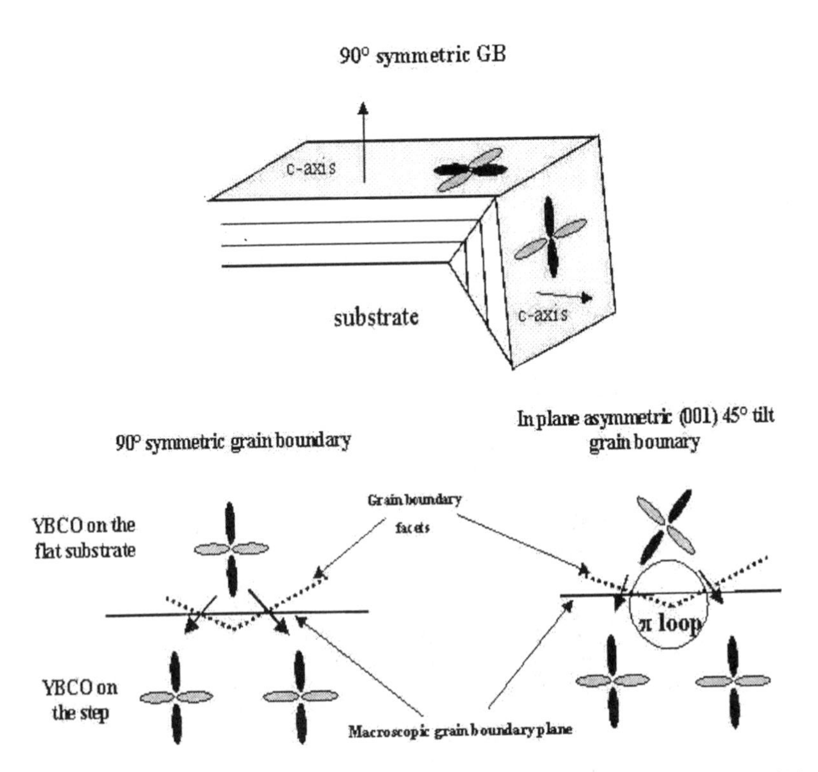

FIGURE 4.16 The *d*-wave component expected in 90° symmetric (S) grain boundaries, compared with the one expected in asymmetric in plane 45° (001) tilt bicrystal GB, with the intrinsic faceting taken into account. For step junctions, the order-parameter orientation does not produce additional π shift along the junction, in contrast with 45° (001) tilt bicrystal junctions.

quantized flux generation, also measured in 45° (001) tilt grain boundaries (56), may lead to a noise increase in the junction, particularly low $1/f$ noise. The situation is strikingly different in the case of 90° symmetric step-edge Josephson junctions (SEJJs) illustrated in Figure 4.16. In the absence of a (001) tilt component between the YBCO grains grown on the flat surface and grains on the step flank, the Josephson tunneling across the junction cannot, of course, produce any π shift. The maximum of the critical current I_c corresponds always to $B = 0$ and the current–phase relation has the usual sinusoidal form [measured by Il'ichev et al. (57)]. The absence of π loops also helps to avoid additional $1/f$ noise.

4.5.4 Electrodynamics of the Step-Edge Junction

Various mechanisms have been proposed to account for the transport properties of grain boundaries in high critical temperature superconductors. Among them, we have the mechanism based on the structural properties of the grain boundaries and the related bending of the electronic band structure and the mechanism based on deviations of the ideal stoichiometry at the boundary interface—oxygen deficiency or oxygen disorder—effects arising from the unconventional order-parameter symmetry and others based on a direct suppression of the pair potential at the interface (58). Several of them may, however, contribute simultaneously. In this subsection, we will address some aspects of the transport properties of step-edge junctions, related to the phenomenology of the Josephson effect, that require an insulating nature of grain-boundary region with a nonvanishing junction capacitance.

In SEJs, in analogy with low-T_c Josephson tunnel junctions, Fiske and Eack resonances have been observed. Figure 4.17a shows the I–V characteristic at different values of the external magnetic field (59), where the presence of two steps at voltages $V_1 = 490\ \mu\text{V}$ and $V_2 = 970\ \mu\text{V}$ ($\approx 2V_1$) respectively is evident. In Figure 4.17b, the same curves are shown after the subtraction of the background current V/R_n. The steps at voltages V_1 and V_2 do not move by changing the magnetic field; moreover, the I–V characteristics after the quasiparticle current subtraction show a Lorentzian-like shape, typical of a resonant mode. Therefore, these current singularities have been interpreted as Fiske steps, due to the interaction between the ac Josephson effect and the electromagnetic modes of the junction seen as a resonant cavity. They, in fact, appear at voltages $V_n = \Phi_0 f_n$, where $f_n = nc/2w$ are the frequencies of the normal modes of the cavity, c represents the phase velocity of the electromagnetic wave in the cavity, and w is the width of the junction. Figures 4.17c and 4.17d show the magnetic field dependence of the Josephson current and of first Fiske step. The experimental data are reasonably well fitted by the theoretical curves also presented for comparison. From the voltage position of the Fiske steps, one can derive c and, further, the ratio t/ε_r using the expression

$$c = c_0 \sqrt{\frac{t_d}{\varepsilon_r d}}$$

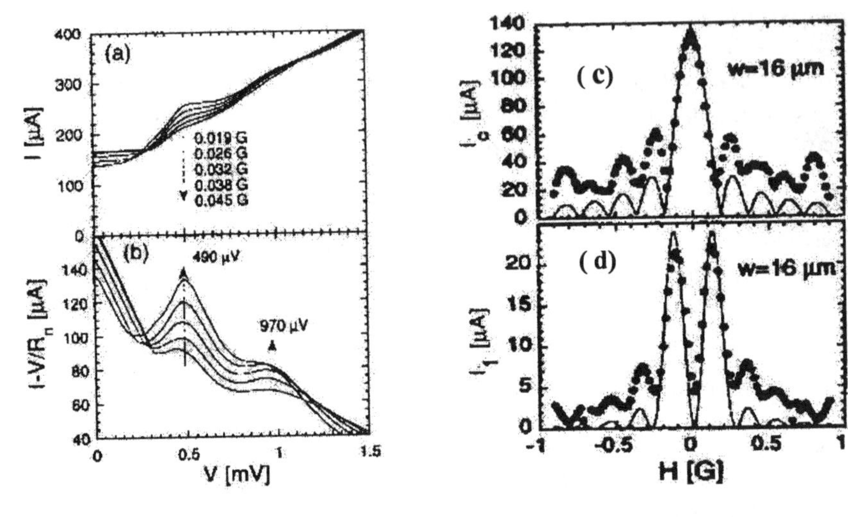

FIGURE 4.17 *I–V* curve of a 32-μm-wide SEJ at $T = 4.2$ K at different values of the applied magnetic field *B*. In (a), the largest value of the Josephson current corresponds to the lowest *B*. In (b), the quasiparticle background current has been subtracted. The arrows indicate the Fiske step position. The magnetic field dependence of the critical current I_c for a SEJ 16 μm wide is illustrated in (c), whereas that of the additional dc current I_1 due to the Fiske resonance is illustrated in (d). In (c) and (d), the theoretical dependence is also reported for comparison. (After Ref. 59.)

where t is the thickness of the dielectric barrier of the GB, ε_r is the relative dielectric constant, c_0 is the light velocity in vacuum, and $d \approx 2\lambda_L$, with λ_L being the London penetration depth.

Typical values for c/c_0 lie in the interval 0.03–0.05, in agreement with data reported for bicrystal junctions (60) and depend on the junction width. The corresponding values of the capacitance per unit area derived from the expression $C = \varepsilon_0\varepsilon_r/t$ are in the range 10–30 fF/μm^2.

In very long step-edge Josephson junctions ($w \gg \lambda_j$, λ_j being the Josephson penetration depth), another type of singularity, referred to as the Eck step, has been observed in the *I–V* curve. In the presence of an external magnetic field, when a voltage *V* is applied to the junction (ac Josephson effect), the Josephson current density is spatially and temporally modulated. When the phase velocity associated with the Josephson current density distribution matches the phase velocity of the electromagnetic field in the junction, a current step, with a resonant shape, appears in the *I–V* characteristic (61). The voltage position of the step depends linearly on the magnetic field according to the relation $V = dcFB$, where the flux focusing coefficient *F* has been also considered (62).

Figure 4.18 shows the *I–V* characteristic of a long SEJ ($w/\lambda_j = 13$ at 4.2 K) at different values of the magnetic field *B*. A field-dependent resonantlike current step is evident in the *I–V* characteristic. The structure is made more prominent by subtracting the background current (Fig. 4.18b). Moreover, the step voltage V_m increases linearly by increasing the external magnetic field. By calculating *F* from the I_c versus *B* dependence ($F = W\Phi_0/\Delta B$, where ΔB is the experimental periodicity of the magnetic pattern) from the linear fit of $V_m(B)$, one can estimate the light velocity *c* in the junction and, from it, all related junction characteristic parameters. For this particular junction, the values of *C* derived by the detection of an Eck step are very close to those obtained by the Fiske steps position (calculated in the previous paragraph).

We now discuss a recent observation of the so-called displaced linear slope (DLS)—a phenomenon quite rarely detected and not fully understood even in low vertical temperature junctions. In the *I–V* curve, one of the authors has observed the appearance of linear branches (63) under quasiparticle injection in YBCO electrodes. The first observations of such phenomenon are related to the early studies on the extended Josephon junction when the linear branches were called

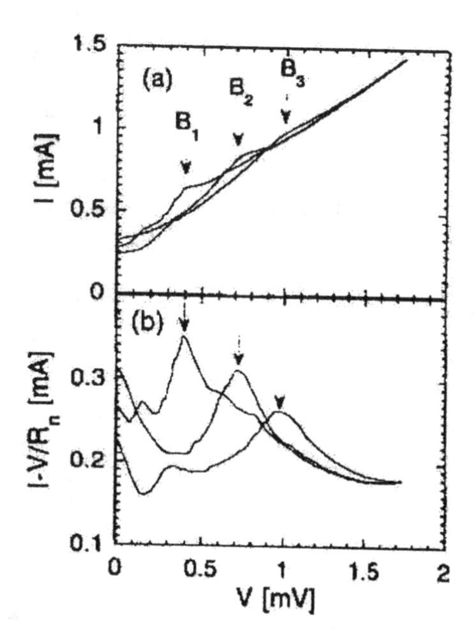

FIGURE 4.18 *I–V* curve of a 60-μm-wide SEJ with nonuniform bias at different values of the external magnetic field $B_1 = 0.12$ G, $B_2 = 0.24$ G, $B_3 = 0.37$ G. (b) The normal background current has been subtracted. $T = 4.2$ K. (After Ref. 35.)

displaced linear slope (DLS). The explanation of such features has been given in terms of *flux–flow* in the sine–Gordon Hamiltonian (64,65). Within the HTS family of Josephson junctions (JJs) the DLS has been observed also in $Bi_2Sr_2CaCu_2O_y$ mesas (66,67) consisting of intrinsic (naturally stacked) Josephson junctions. The observation of DLS in a single step-edge YBCO junction therefore represents an important issue to clarify some aspects of fluxon dynamics. The layout of the investigated sample is sketched in Figure 4.19.

The structure consists of a step-edge junction and a control line made up by two Au/YBCO junctions situated at the sides of the SEJ. The injection (or control) current I_{cr} flows across the series of two Au/YBCO interfaces and across the YBCO film, in the region next to the grain boundary. The device can, therefore, operate in a four-terminal configuration. Figure 4.20 shows the dependence of I_c on I_{cr} at $T = 4.2$ K. I_c decreases monotonically by increasing I_{cr}, differently than the dependence on the external magnetic field, which, instead, is Fraunhofer-like (63).

The dependence of I_c on I_{cr} can, therefore, be attributed to the combined action of the magnetic field and nonequilibrium. Due to the misalignment of the control line related to the lithographic process and also to the intrinsic nonplanar geometry of the step-edge junctions, the injection current density is not symmetric with respect to the grain boundary. Then, the current I_{cr} generates a nonvanishing magnetic field in the junction plane. Moreover, I_{cr} induces local nonequilibrium in the YBCO electrodes due to quasiparticle injection through the Au/YBCO interfaces (68).

For values of I_{cr} larger than about 3 mA at 4.2 K, the *I–V* characteristic clearly exhibits a DLS, as shown in Figure 4.21. The inset shows the same structures after subtracting the background current $I = V/R_N$. The shape is very differ-

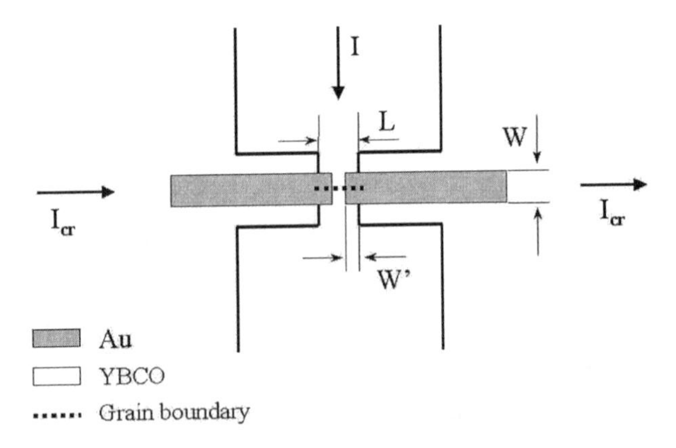

FIGURE 4.19 Schematic representation of the device. $W \times W'$ defines the injection area.

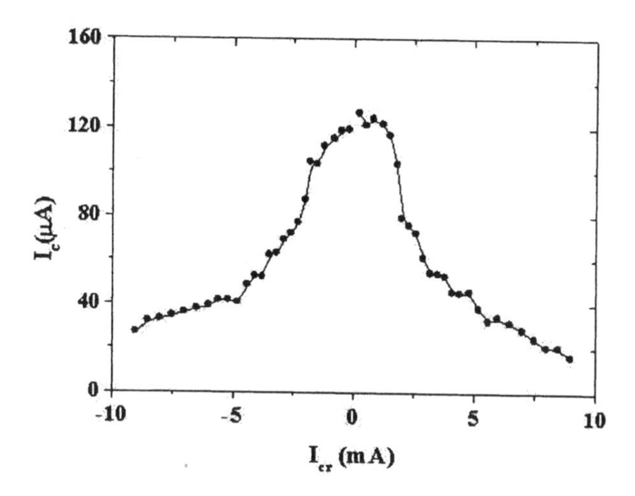

FIGURE 4.20 Dependence of the maximum Josephson current on the control current I_{cr} at $T = 4.2$ K for a SEJ with $L/\lambda_j \approx 4$, L being the width of the junction.

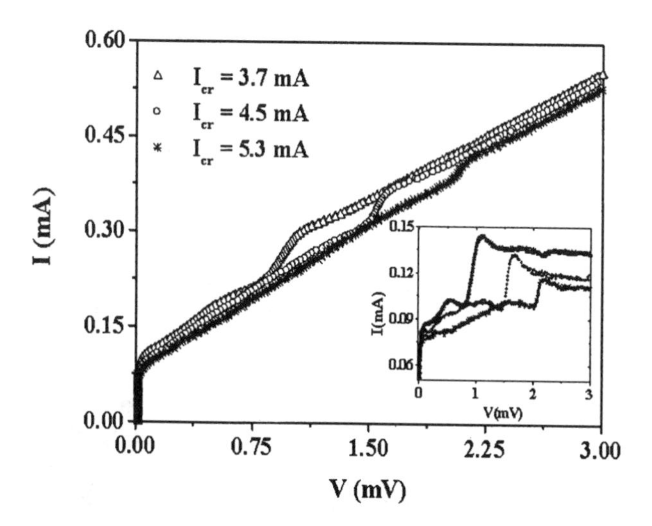

FIGURE 4.21 I–V characteristic ($L/\lambda_j \approx 4$) at $T = 4.2$ K at different values of the control current I_{cr}. The inset shows the same curves after subtracting the normal current V/R_N.

ent from that exhibited by resonant structures like the Fiske and Eck steps (compare Figs. 4.17 and 4.18) and cannot be fitted by a Lorentzian.

Let us define V_M as the voltage at which the current has increased to 50% of the total current step I_M. At a given value of I_{cr}, a single step appears in one branch of the I–V characteristic (i.e., at $V_M > 0$). Then, a symmetrical structure is observed on the opposite branch (i.e., at $V_M < 0$) by reversing the direction of I_{cr}. The observation of this strong asymmetry in the I–V characteristic leads one to conclude that the linear branch is originated by the motion of flux quanta. The direction of the motion is selected by the most favorable combination of the Lorentz forces generated by injection and bias current.

The described features and the independence of the dynamical resistance of DLS of Figure 4.21 on the value of the control current indicate that one is dealing with a classical manifestation of the phenomenon of flux–flow. A relevant parameter is, in this context, the damping coefficient α:

$$\alpha = \frac{1}{I_c R_N} \left(\frac{J_c \Phi_0}{2\pi C} \right)^{1/2}$$

(here, J_c is the critical current density, Φ_0 is the flux quantum, and C is the intrinsic capacitance of the junction per unit area). Estimating the McCumber parameter β_c from the hysteresis of the I–V characteristic and using the relation $\alpha = (1/\beta_c)^{1/2}$, we get $\alpha = 0.7$. Typical values of α for experiments on the DLS in low vertical temperature junctions (69,70) are of the order of 0.1 or less. Higher values are nevertheless possible, especially for moderate values of the ratio L/λ_j, as in pioneering investigations of DLS (71) for which $\alpha \approx 0.5$ was determined.

It is worth noting that in this experiment the application of the external magnetic field did not produce any DLS, but only resonant structures like Fiske steps. In the device considered, the effect of the control current is twofold. The Cooper pair-breaking (due to nonequilibrium) locally reduces the capability of the junction to screen the magnetic field, so that fluxons are more easily injected in the junction. Moreover, they experience a stronger Lorentz force due to the combined action of the field of bias and control current, which leads to DLS.

The above physical phenomenon offers wide and stable margins for possible applications in the field of the generation of electromagnetic radiation. This argument is confirmed by the dependence of V_M on the injection current I_{cr}. As already observed for the DLS (69), V_M increases proportionally to I_{cr}, with minor deviations at low V_M values; the 2-mV tunability range of V_M leads to a bandwidth of $\pm$500 GHz around 1 THz. The available dc power of the oscillations, when the junction is dc biased on the DLS at a voltage, for example, $V = 1.5$ mV and a current $I = 0.3$ mA, is $P = 0.45$ μW: Under conditions of perfect matching however, the available power will not be more than half of this amount.

An estimate of the linewidth of the radiation emitted from the junction can be made following classical dynamical resistance arguments (32). The expression for the linewidth of a Josephson oscillator at a temperature T is given by

$$\Delta\nu = \frac{4\pi k_B T\, R_D^2}{\Phi_0^2\ \ R_S}$$

where R_D is the dynamical resistance at the bias point, R_S is the static resistance corresponding to the same bias point, and k_B is the Boltzmann constant. For the linear branch shown in Figure 4.21, we get $\Delta\nu \approx 170$ MHz at $T = 4.2$ K, taking $R_D = 2.7\ \Omega$ and $R_S = 7\Omega$.

The stability of the observed phenomenon over a wide frequency range and the ease of its control makes these classes of device a potentially interesting component for submillimeter-wave HTS electronics.

4.5.5 SEJs on MgO Substrate

In this subsection, we will briefly discuss some general properties of SEJs fabricated on MgO substrates. Good junctions (72) and low-noise SQUIDs (73) have been demonstrated. Applications have however been hindered by the lack of systematic and fundamental studies. The direct competitors, the SEJs fabricated on perovskite substrates, appear technologically more reliable, as a clear correlation between transport and structural properties of the GBs has been established.

As discussed in Section 4.2, the YBCO grows with the c axis aligned with the normal to the macroscopic surface of the vicinal cut MgO substrates. Growth of a YBCO film on a step with a moderate angle α is schematically shown in Figure 4.22. This growth leads to the formation of two grain boundaries of the [100] or [010] tilt type, with an angle almost equal to the step slope. This has been confirmed by various TEM studies on samples with different step angles. For a shallow step ($\alpha \approx 20°$), however, the slope generally changes smoothly at the lower edge, and as a consequence, there is no GB formation at the bottom of the step

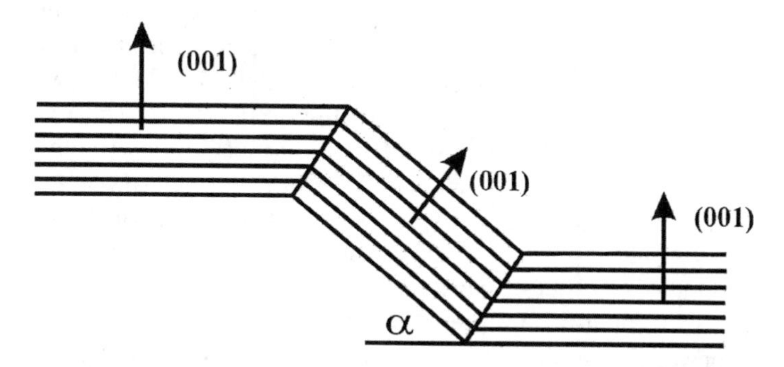

FIGURE 4.22 Typical YBCO growth on a step in a MgO substrate for a moderate value of the angle α.

(74). For steeper steps instead, more complicated structures have been reported at the bottom edge. In particular, the presence of an additional GB has been observed (75,76), sometimes in correlation with a deep trench structure of the step itself.

The transport properties of step-edge junctions on MgO substrates are, therefore, not characterized by 90° [100]-tilted GBs. Instead, it is possible to continuously modify the GB structure by changing the step angle in a rather wide range. In analogy with bicrystal junctions, the J_c has a tendency to decrease as the step angle increases (in the range 0°–45°) (77); however, the spread in data presented in the literature does not allow one to extract an analytical angular dependence. An RSJ-like behavior has been observed in the range of values 0.3–1.1 of the ratio t/h (77). This interval is wider than that for step-edge junctions on perovskite substrates, probably because of the relatively low value of the MgO step angle (generally less than 45°). In this case, the YBCO film thickness is not reduced on the step and no overgrowth is expected on the top flat part of the step. Moreover, these circumstances also explain the independence of J_c on the ratio t/h, once the value of the step height has been fixed (77).

Finally, in analogy with (100) *c*-axis tilt Josephson junctions obtained by a biepitaxial technique (78), the dependence of the Josephson current on the magnetic field is Fraunhofer-like (72). Moreover, Fiske resonances have been observed also in some junctions (79).

4.6 TRANSPORT PROPERTIES OF 90° [100] TILT GBS

Once a reliable technology for the substrate preparation is established, the step-edge junctions have the layout flexibility necessary to develop a great variety of devices based on the Josephson effect. The quality factor of the SEJ, defined by the product $J_c\rho_N$ is, in fact, comparable with the best values obtained with bicrystal junctions. What, instead, is rather crucial for applications is the large spread in the critical parameters J_c and ρ_N. The reason for the lack of reproducibility of the transport properties is obviously related to the microstructure of the YBCO film across the step. As widely discussed previously, the GB structure depends on the step profile, the local presence of step defects, and the kind of substrate. On perovskite substrates, step angles of the order of 60° are strictly required to obtain two well-defined grain boundaries at the edges of the step. However, even when this is the case, each GB is a complex combination of 90° *a*-(*b*)-axis tilted GBs of the symmetrical and BPF tilted types, whose relative proportions are affected by factors not strictly controllable. As a direct consequence, it becomes difficult also to distinguish which kind of GB is effectively responsible for the transport properties in SEJs or, alternatively, if it is the presence of defects in the grain-boundary planes that is responsible for the Josephson behavior of step-edge junctions (42). The weak-link properties of grain boundaries obtained by a 90° tilt of the YBCO *c* axis are, in fact, still an open question. In (103) films, for example, microbridges patterned along the $[1\bar{1}0]$ direction, where SGBs and BPF GBs are formed, do not show any weak-link behavior.

A useful method recently developed by one of the authors and co-workers (80) allows one to distinguish the superconducting transport properties of the two GBs formed at the top and the bottom edges of the step. This is achieved by using a tilted Ar ion-beam etching and the shadowing effect of the step in the process of defining the electrodes and the microbridge (81). The fabrication procedure allows one to study the transport properties of the isolated top GB and a comparison with the properties of the step-edge junction as a whole. Such an analysis is important for a complete characterization of the junctions with which one is dealing. At the same time, by studying SEJs with two very dissimilar GBs at the top and the bottom edges of the step, it is possible to obtain more basic information about the nature of transport in 90° [100] ([010]) tilt GBs.

Microbridges are defined by using an Ar-ion milling oriented at 60° with respect to the substrate normal. The shadowing effect of the step allows one to obtain a continuous YBCO stripe along the step which enables an electrical contact to the middle and the bottom parts of the SEJ (see Fig. 4.23).

By a proper choice of contact pads, it is possible to perform four-point measurements on the whole junction, the YBCO stripe, and the top GB. The devices investigated were fabricated by an α-C mask patterned by e-beam lithography. The critical temperature of the step-edge junction, T_c^{SEJ}, ranged between 70 and 80 K, whereas that of the thin stripes was between 60 and 85 K. For samples with T_c^{stripe}, less than T^{SEJ}, the pad's configuration makes it possible to measure the critical current of the bottom GB (I_c^{botGB}) in the interval $T_c^{\mathrm{stripe}} < T < T_c^{\mathrm{SEJ}}$. Referring to Figure 4.23, when the YBCO stripe is normal, I_c^{botGB} is determined by

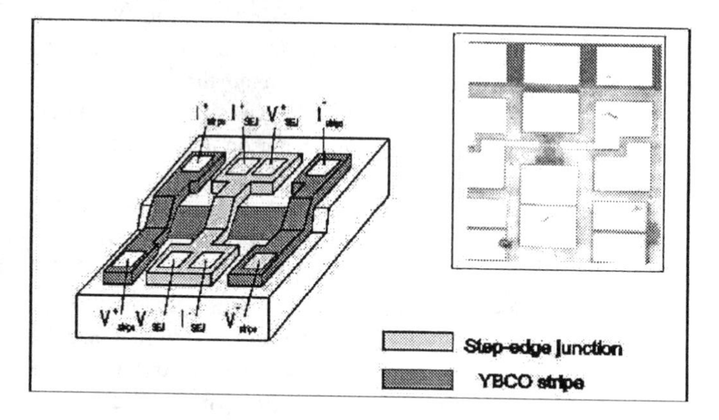

FIGURE 4.23 A schematic view of a step-edge junction with the presence of a YBCO stripe along the step. The voltage and current leads are also represented. The inset shows a micrograph of a 4-μm junction with contact pads.

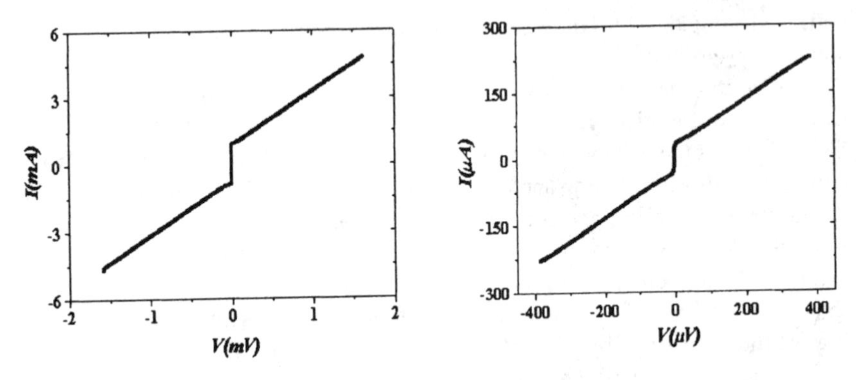

FIGURE 4.24 Typical *I–V* characteristics of the top GB of (a) an 8-μm-wide SEJ at 4.2 K and (b) an 8-μm-wide SEJ at $T = 77$ K. The differential resistance versus voltage does not show any features at finite voltage in both cases.

sending current through the whole junction and measuring the voltage difference between the pads V^{-}_{SEJ} and V^{+}_{stripe} or V^{-}_{stripe}.

Typical current–voltage characteristics of the top GB is shown in Figure 4.24 at two temperatures: $T = 4.2$ K and at $T = 77$ K. They shows a shape typical of the RSJ model, although an excess current is present.

Furthermore, there is no evidence of the presence of a second junction in series with the top GB. Within the measurement accuracy, in fact, the *I–V* characteristic of the whole SEJ is identical to the that of the top GB. This has been verified for several junctions on different chips in the range of temperature between 4.2 K and the critical temperature T_c of the specific junction. After removal of the YBCO stripes, by additional ion milling the *I–V* characteristics of the SEJs remained unchanged. Therefore, these data allow one to correlate the transport properties of these SEJs with the occurrence of only one effective weak link, namely the top GB. However, a clear understanding of the Josephson nature of the junctions needs an atomic-level inspection of the microstructure of these specific GBs. HRTEM analysis performed on the step-edge junctions showed that the GBs formed at the top and at the bottom edges of the step were very dissimilar (see Fig. 4.25). The top GB (Fig. 4.25a) consists of two parts: a 10–20-nm-thick BPF GB and an almost regular S GB of 130–140 nm on top of it. The bottom GB consists (Fig. 4.25b), instead, essentially of a BPF GB. Comparing the different sizes of BPF GBs in the top and bottom GBs, it is possible to attribute the Josephson current to the S GB part and a shunting, more strongly coupled part to the BPF GB. This would result in a weak-link behavior of the top GB and in a stronger superconductive link in the bottom part, as it, indeed, was confirmed by the transport measurements.

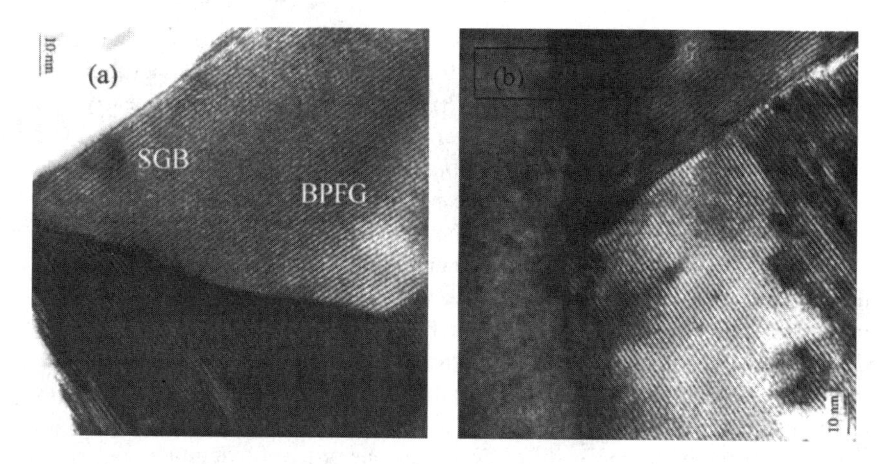

FIGURE 4.25 Microstructure of the grain boundaries (a) at the top edge and (b) at the bottom edge of a typical step used in this experiment. The geometry of the two GBs can vary from one sample to another. A general trend, however, is that the top GB consists essentially of a symmetric GBs, whereas in the bottom GB, the basal plane component is dominant.

Figure 4.26 shows a typical dependence of critical current of the top GB, I_c^{Top}, as a function of the external magnetic field B at $T = 77\mathrm{K}$. At this temperature, the junction is in the short limit $w/\lambda_j \approx 2$.

For most of the measured junctions, in the short limit, I_c^{Top} oscillated with the applied field, with a modulation depth ranging between 50% and 80%. In some

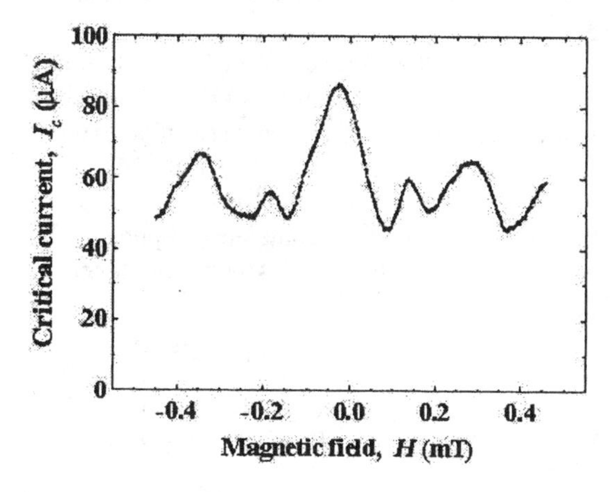

FIGURE 4.26 Magnetic field dependence of the critical current I_c for the top GB of an 8-µm-wide SEJ at 77 K.

cases, I_c^{Top} could be completely suppressed by the external magnetic field. Referring to the microscopic structure of the top GB (Fig. 4.25), one can attribute the maximum modulation depth of the I_c^{Top} (B) pattern to the S GB and the nearly constant background current to the BPF GB. From the data of Figure 4.26, both the S GB and the BPF GB contribute approximately 50% each to I_c^{Top}. With respect to the different dimensions of the GBs, one can roughly calculate the critical current density of the BPF GB and the S GB. They are summarized in Table 4.3.

$J_c^{\mathrm{S\ GB}}$ can vary from one device to another (typical variations are of the order of 50%). The value of J_c is the lowest for the BPF-type GB. There is a direct proportion between the percentage of the BPFGB detected by HRTEM in the top GB and the total critical current of the latter. One can, therefore, argue that the observed spread is related to the difference in the BPF GB percentage present in the top grain boundary. Furthermore, at any temperature, it is likely to expect that the J_c of the top and bottom GBs differ by at least one order of magnitude.

The critical current density of the top GB of a 8-μm-wide SEJ normalized to the corresponding value at $T = 10$ K is shown as a function of normalized temperature, T/T_c, in Figure 4.27. This dependence is typical of an (superconductor normal metal superconductor) SNS structure (82). The ratio $L/\xi_N(T_c) = 4$ between the weak-link length L and the correlation length in the normal region ξ_N at $T = T_c$ fits the experimental data reasonably well in the whole temperature range. Experimental data on BPF GB obtained by Lew et al. (83) are also shown for comparison. They found that the $I_c(T)$ dependence of a pure BPF GB can be equally well fitted by a flux creep model or a Josephson junction model. It is seen from Figure 4.27 that within the scatter in the data, a pure BPF GB (open squares) performs qualitatively in the same way as the top junction of our SEJ (solid circles). This indicates that both the S GB and the BPF GB have the same critical current versus temperature dependence.

Because of the high values of $J_c^{\mathrm{BPF\ GB}}$, the presence of the bottom GB may be detected by reducing the width of the junctions. In a recent work of Tzalenchuk et al. (84), submicron SEJs have been fabricated with a procedure similar to those considered in this section. They shows two GBs in series with an order of magnitude different J_c. In Figure 4.28, the I–V characteristic of a 0.4-μm-wide SEJ is reported. The critical current of the first GB is very low (completely suppressed by thermal fluctuation). In fact, the I–V characteristic presents only a nonlinearity

TABLE 4.3 Estimated Value of the Critical Current for S and BPF Grain Boundaries

	J_c (A/cm^2), $T = 4.2$ K	J_c (A/cm^2), $T = 77$ K
S GB	3×10^4	10^3
BPF GB	5×10^5	3×10^4

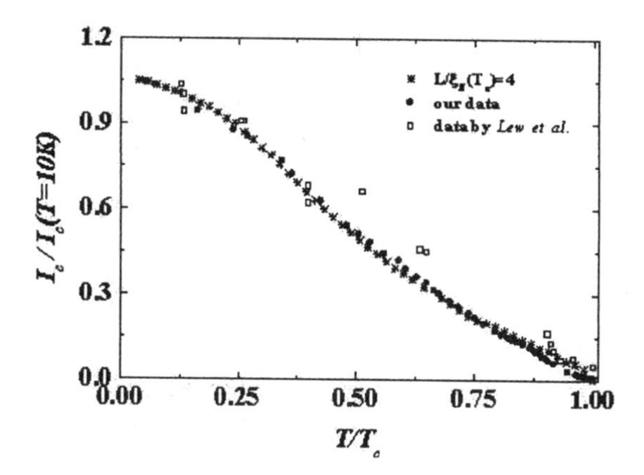

FIGURE 4.27 The critical current I_c normalized to the corresponding value at $T = 10K$ (solid circle) of the top GB of a 4-μm-wide SEJ is shown as a function of the normalized temperature T/T_C. For comparison, data derived from the theory asterisk for the ratio $L/\xi_N(T_c) = 4$ and data by Lew et al. (83) (square) on pure BFP GB are shown.

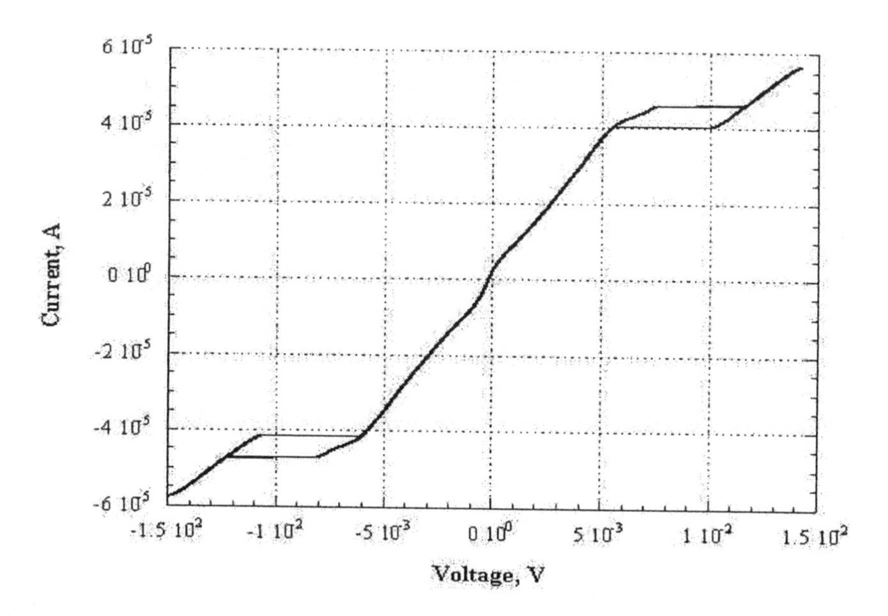

FIGURE 4.28 *I–V* characteristic of a 0.4-μm-wide SEJ. The presence of a highly hysteretic second junction in series is evident.

around zero voltage. For the second GB, I_c is much higher; the corresponding value of critical current density is of the order of 10^5 A/cm^2. Therefore, it is possible to correlate the low-J_c and high-J_c junctions respectively with the top and bottom GBs. Moreover, by assuming that the bottom GB consists essentially of the BPF component, the value $J_c^{\mathrm{BPF\ GB}} \approx 10^5$ A/cm^2 calculated from Figure 4.26 is only slightly reduced with respect to the value reported in Table 4.3. This may be related to the submicron dimensions of the junction. The critical current of the bottom GBs showed little dependence on the external magnetic fields up to 50 G. Because of the small size of the junction, in fact, a much higher field ($\approx$ 0.1 T) might be required.

Other submicron SEJ junctions studied by the same authors revealed a number of interesting features which can be understood in terms of the evolution of BPF GB formed at the top edge of the step. The $R(T)$ dependence was of the semiconductor type and the T_c was reduced compared to the plain microbridges of the same width. Nevertheless, at 4.2 K, the critical current density of this type of junctions was high, of the order of 5×10^5 A/cm^2, with a weak dependence on the magnetic field and the excess current comparable to the Josephson current (see Fig. 4.29). Current steps were observed in the I–V curves in all magnetic fields up to 5 T with the amplitude decreasing as the magnetic field increases. The voltage positions of the steps roughly scaled with the critical current and all steps behave

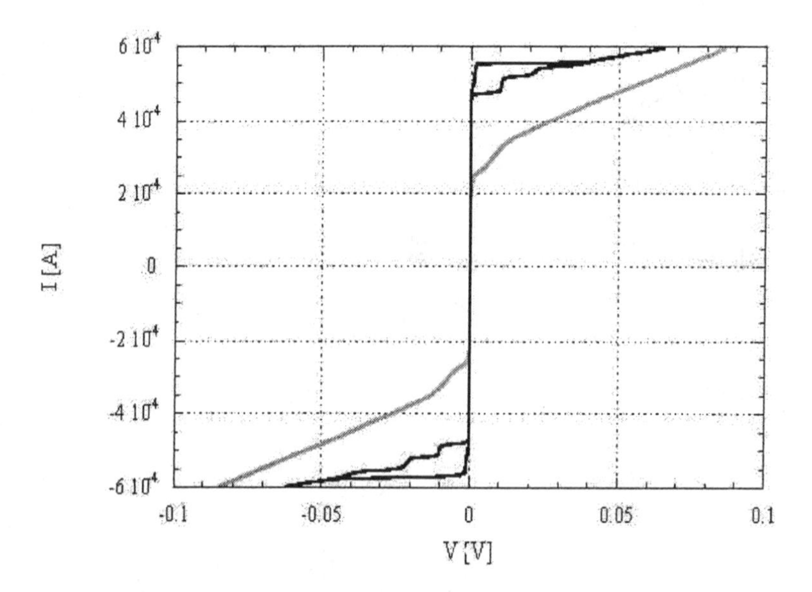

FIGURE 4.29 I–V characteristics at 4.2 K of a 0.8-μm-junction having both high critical current and high excess current. The black plot is the I–V characteristics at zero magnetic field; the gray plot is the same at 5 T.

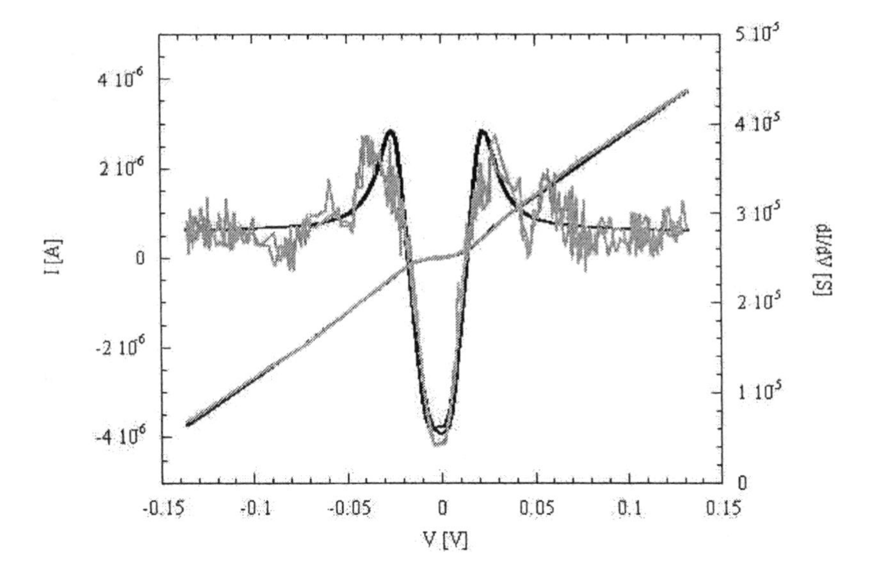

FIGURE 4.30 A 0.6-μm-wide tunnel junction measured at 4.2 K. The gray plot shows the *I–V* characteristic and the dynamic conductance calculated from the *I–V* curve. The black plots depict the best fit to the model of Dynes et al. (92) with the following parameters: $\Delta_1 + \Delta_2 = 21.2$ meV, $\Gamma = 5.2$ meV, $G = 2.8 \times 10^{-5}$ S.

in the same way as the magnetic field is varied. All these observations and computer simulations rule out Fiske resonant modes, phase-slip process (32), and a multiple Andreev reflection (85–87). A plausible explanation is that the steps were due to intrinsic tunneling in between the YBCO layers, as is commonly observed in BSCCO and TBCCO single-crystal stacks (88,89) and also in oxygen-depleted YBCO (90). Remarkably, a similar behavior was also observed in $Tl_2Ba_2CaCu_2O_8$ high-angle ($\alpha = 58°$) step-edge Josephson junctions grown on $LaAlO_3$ substrates (91). In the latter system, the intrinsic Josephson effect is much more pronounced as a consequence of a higher degree of anisotropy of the Tl-2212 compound compared to $YBa_2Cu_3O_x$.

The junctions were rather unstable to repeated thermal cycling and eventually deteriorated to quasiparticle tunneling (Fig. 4.30). As a matter of fact, in a number of experiments, the transition was observed in situ upon cooling and could temporarily be reversed by reversing the temperature. The authors concluded that the continuity of the junction was broken and a break-type junction was formed. The *I–V* characteristics in the quasiparticle tunneling regime were fitted to a theoretical model developed by Dynes et al. (92). The model accounts for a broadening of the characteristics by introducing a complex parameter Δ-$i\Gamma$ instead of Δ

into the BCS expression for the density of states. $\Gamma = h/\tau$ describes broadening due to inelastic scattering of quasiparticles with a finite lifetime τ. The fitting parameters were the gap Δ, broadening Γ, and conductance G. Note that the $\Delta_1 + \Delta_2$ value found from the fit is a factor of 2 smaller than the value obtained in the break-junction experiments (93,94). This is in agreement with tunneling from a axis to the *c*-axis direction compared to the in-plane tunneling in the break-junction experiments. The broadening appears to be rather big suggesting a high rate of inelastic scattering of quasiparticles on the break-junction surface defects.

The data presented in this section give some indications about the different nature of the transport in 90° [100] tilt grain boundaries. S GBs do act as a Josephson weak link. The coupling in BPF GB appears to be stronger. This is in fairly good agreement with the data reported in Ref. 83, where almost pure BPF GB, did not show any weak-link behavior. Their junctions, in the presence of an external magnetic field, behaved in the same way as *a*-axis films, patterned in the [001] direction, where BPF grain boundaries are formed because of the exchange between the *b* and *c* axes (95). The critical current could, in fact, be noticeably suppressed only by a magnetic field of the order of 1 T, even at a temperature close to the short-junction regime. It is worth mentioning, however, that there are reports (96) in which a BPF GB formed along the boundary between an *a*-axis-oriented YBCO grain and *c*-axis YBCO grains showed a Josephson weak-link behavior. This further confirmed that experiments are needed to obtain a clearer picture about the transport in 90° tilt GBs.

The weak-link behavior of symmetric grain boundaries do contrast with the observations of many authors on (103) YBCO films (97). The absence of a Josephson behavior of microbridges patterned along the $[1\bar{1}0]$ direction represents a controversial question. In some recent reports, it was argued that a helical current flow through 90° [100] twist boundaries (98) can explain the transport in (103) films. It is, in fact, well established that these GBs, present along the [100] direction, are strongly coupled (99). Figure 4.31 shows schematically *a-b* plane configurations for two joint (103) and $(\bar{1}03)$ grains.

The current flow in the $[1\bar{1}0]$ direction would occur through a helical path involving 90° twist boundaries. In this way, the current flow is always in the *a-b* plane. Therefore, there is no direct involvement of 90° [100] tilt boundaries and no weak-link-like behavior is expected.

This model would also provide an explanation for the anisotropy in the normal and superconducting states observed in (103) films. The *a-b*-plane current paths for an average current flow in the [001] direction are simply longer and narrower than for [001] transport. The corresponding value of the resistivity is expected to be higher. The transport along the substrate [001] direction occurs throughout the volume of the film, whereas the current flow along the $[1\bar{1}0]$ direction is confined within a reduced area, defined by the mean distance between twist boundaries in the [001] direction and the thickness of the film. This would result in that J_c in the $[1\bar{1}0]$ direction is reduced compared to the $J_c \| [001]$.

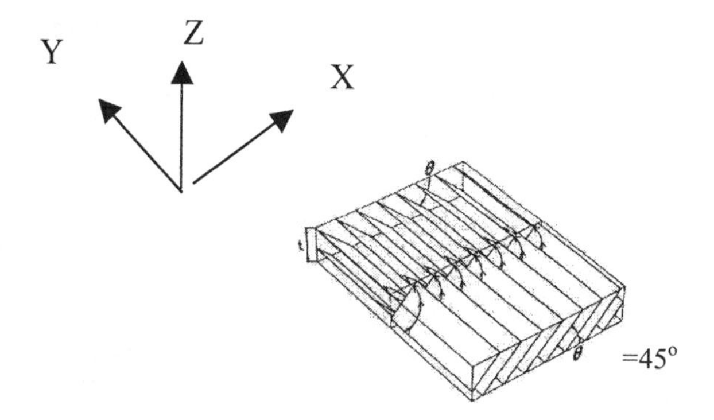

FIGURE 4.31 Two tilted (103) and ($\bar{1}$03) grains of equal thickness are joined at a junction (90° [100] twist) which lies in the *X–Z* plane. The electric field is applied in the *X* direction.

4.7 SQUID DEVICES WITH STEP-EDGE JOSEPHSON JUNCTIONS

A general discussion on the systems based on superconducting quantum interference devices (SQUIDs) made of HTS materials can be found in Chapter 8. Another excellent review on SQUIDs was recently prepared by Koelle et al. (100). It is probably fair to say that the majority of SQUIDs up to date have been made using either bicrystal or step-edge Josephson junctions. Here, we shall emphasize only some junction-specific aspects of the step-edge SQUIDs in comparison to the bicrystal ones.

1. *Availability.* A step-edge Josephson junction can be made on almost any substrate, whereas the choice of bicrystals is limited and, in addition, they are rather expensive. The SEJJ SQUIDs were successfully fabricated on MgO, $SrTiO_3$, $LaAlO_3$, and $NdGaO_3$ substrates.
2. *Design flexibility.* As already mentioned in Section 4.1, integrated superconducting circuits based on Josephson junctions demand elements that can be placed rather freely on a chip. Sensitive SQUID magnetometers and gradiometers, in addition, require large-area antennas, which should be able to withstand large induced currents. Weak links in the antennas should be avoided, which unavoidably leads to a larger linewidth in the case of bicrystals. Being an ideal model system, bicrystal junctions obviously do not fit these criteria of flexibility, whereas the SEJJs can be made locally anywhere on the chip in almost any shape. Significantly, especially for submicron circuits, the superconducting and contact layers can be easily aligned with respect to the step accord-

ing to the alignment marks produced in the substrate along with the step. In contrast, alignment to the bicrystal grain boundary represents problems and cannot reproducibly be made better than about 1 μm. For example, Dörrer et al. (101) have investigated planar galvanometer SQUID gradiometers with bicrystal and SE junctions. They have shown that in gradiometers for operation in a highly disturbed environment, the use of SEJJ is advantageous compared to bicrystals. The parallel gradiometer structure leads to a large permanent current in the antenna if the gradiometer is placed in a homogeneous field. In the case of bicrystal substrates, this current is able to exceed the critical current of the grain boundary through the antenna. For using this type of gradiometers in a real environment, a larger linewidth across the grain boundary and/or a larger film thickness is necessary. In the case of step-edge junctions, this problem does not exist. Reached gradient sensitivity values in the white-noise region are 0.46 pT/(cm/Hz$^{1/2}$) in the case of bicrystal junctions and 0.69 pT/(cm/Hz$^{1/2}$) for step-edge junctions.

3. *Reproducibility.* Bicrystal junctions and SQUIDs are still unbeaten. Despite a significant progress in fabrication methods, the SEJJs still suffer from a rather large spread in parameters. As already mentioned, a 30% spread in the parameters of the junctions prepared on the same chip have been reported (see also Ref. 102). Fortunately, it is less critical for SQUIDs with only a few Josephson junctions than for more sophisticated superconducting circuits (e.g., rapid single-flux quantum logics, RSFQ).
4. *Noise.* In general, the noise levels of the step-edge SQUIDs are only slightly above those for the bicrystal devices. Two types of low-frequency noise are commonly observed in the step-edge SQUIDs: random telegraph noise, giving a Lorentzian-type spectrum and the $1/f$ noise. The $1/f$ noise is dominated by junction critical current fluctuations and can be suppressed almost completely by current reverse biasing, whereas the the Lorentzian part remains unaffected. Also, it was found out that the $1/f$ noise level is reproducible after repeated thermal cycling, but the Lorentzian noise attains an arbitrary level independent of the cooling field. The Lorentzian noise originates at least partly from a single long-range hopping of a trapped Abrikosov flux vortex between two pinning sites, and the $1/f$ noise arises from many uncorrelated short-range telegraph noise processes (103). An additional source of the Lorentzian noise was discovered in Ref. 104. The authors found out that, for their junctions, the Lorentzian component of the noise oscillates in the magnetic field with the period corresponding to Φ_0 in the junction. Each individual SEJ was treated as a multijunction interferometer which forms a two-level fluctuator when the junction is placed inside a SQUID loop.

5. *Structure–properties relation.* It was shown (102) that the SQUID parameters do not change in a wide range of step height-to-film thickness ratios, $0.3 < t/h < 0.6$. For smaller values of t/h, the amplitude of the voltage modulation $V(\Phi)$ decreases, whereas the critical current density j_c and characteristic voltage I_cR_n are not affected. The parameters I_c and R_n of the SEJJ are connected through the relation $I_cR_n \sim I_c^q$ or $I_c \sim R_n^{1/(q-1)}$. Dillmann et al. (105) made use of the scaling law to tune the parameters of the SQUIDs for optimum I_cR_n and $\beta_L = 2LI_c/\Phi_0$. This is done by trimming the oxygen content in the junction by oxygenating/reducing annealing. A significant improvement in the step microstructure and SQUID properties has been made by Francke et al. (28). By using the alternating angle of ion incidence during the ion milling, they managed to avoid redeposition of the substrate material in the vicinity of the step. It resulted in a better yield and excellent properties of the SQUIDs. For the t/h ratios increasing from 0.6 to 0.8, the critical current density was found constant at about 8×10^3 A/cm^2, whereas the specific resistance decreased linearly from more than $20 \times 10^{-9}\ \Omega\ \text{cm}^2$ down to $5 \times 10^{-9}\ \Omega\ \text{cm}^2$, the former value being close to the optimum found by Dillmann et al. (105).
6. *Applications.* Step-edge Josephson junction SQUIDs were successfully used in systems for nondestructive evaluation (106–108), magnetocardiography (109–111), simple digital circuits (112,113), and scanning SQUID microscopy (84,114). An important issue for most of the real-life SQUID applications is the possibility of operating the system in the unshielded environment. Apart from the design considerations and introduction of strong pinning centers, the use of submicrometer-wide junctions help to reduce the influence of the external field. Glyantsev et al. (115) studied the stability of both dc and radio-frequency (rf) SQUIDs to the external magnetic fields. In particular, they have fabricated SEJJ-based devices with the junction width down to 0.3 μm. They have shown that the transfer function of a washer-type SQUID with the smallest junction width was not depressed by external magnetic fields up to 3 G. The same stability was demonstrated for direct-coupled rf and dc SQUIDs with 1-μm-wide junctions.

   Scanning SQUID magnetometry with high spatial resolution dictates a compromise between the SQUID sensitivity and spatial resolution. The resolution close to the size of the SQUID loop of 10 μm was achieved in Ref. 84 using the 0.5-μm-wide step-edge Josephson junctions.

4.8 CONCLUSION

To conclude this chapter, we can say that the properties of step-edge Josephson junctions are by now well understood and correlated to the YBCO growth modes and microstructure at the step region. The parameters of SQUIDs based on SEJJs are not inferior to the bicrystal ones. In the majority of up-to-date real-life applications, however, the SEJJ SQUIDs have an advantage of flexibility and availability. Despite a significant progress in fabrication methods, the lack of reproducibility still hinders their use in more complicated systems. It remains to be seen if this can be improved.

ACKNOWLEDGMENTS

All of the co-authors of our articles on HTS Josephson junctions have contributed to this review. In particular, we would like to mention Zdravko Ivanov, Francesco Tafuri, Fabio Miletto, and Umberto Scotti di Uccio. We are grateful to the authors of the cited articles and their publishers for permission to reproduce some of their results. Most of our own work was done at Chalmers University of Technology, utilizing the facilities of the Swedish Nanometer Laboratory. We are indebted to Tord Claeson for constant support and encouragement and we wish to thank Bengt Nilsson for advising us on many aspects of technology.

REFERENCES

1. P Chaudhari, RH Koch, RB Laibowitz, TR McGuire, RJ Gambino. Phys Rev Lett 58:2684, 1987.
2. P Chaudhari, J Mannhart, D Dimos, CC Tsuei, J Chi, MM Oprysko, M Scheuermann. Phys Rev Lett 60:1653, 1988.
3. R Kleiner, F Steinmeyer, G Kunkeland, P Muller. Phys Rev Lett 68:2394, 1992.
4. K Char, MS Colclough, SM Garrison, N Newman, G Zaharchuk. Appl Phys Lett 59:733, 1991.
5. KP Daly, WD Dozier, JF Burch, SB Coons, R Hu, CE Platt, RW Simon. Appl Phys Lett 58:543, 1990.
6. A Cassinese, A Di Chiara, FM Granozio, S Saiello, USD Uccio. J Mater Res 10:11, 1995.
7. S Polders, R Auer, G Linker, R Smithey, R Schneider. Physica C 247:309, 1995.
8. FM Granozio, USD Uccio. J Alloys Comp 251:332, 1997.
9. F Tafuri, FM Granozio, F Carillo, AD Chiara, K Verbist, GV Tendeloo. Phys Rev B 59:11,523, 1999.
10. C Eom, AF Marshall, Y Suzuki, TH Geballe, B Boyer, RF Pease, RBV Doover, J Philips. Phys Rev B 46:11,902, 1992.
11. U Jeschke, R Schneider, G Ulmer, G Linker. Physica C 243:243, 1995.
12. USD Uccio, F Lombardi, FM Granozio. Physica C 323:51, 1999.
13. F Wellhofer, P Woodall, DJ Norris, S Johnson, D Vassiloyannis, M Aindow, M Slaski, CM Muirhead. Appl Surface Sci 127:525, 1998.
14. SK Streiffer, BM Lairson, JC Bravman. Appl Phys Lett 57:2501, 1990.
15. G Friedl, B Roas, M Rmheld, L Schultz, W Jutzi. Appl Phys Lett 59:2751, 1990.
16. T Haage, JQ Li, B Leibold, M Cardona, J Zegenhagen, HU Habermeier, A Forkl, Gh Jooss, R Warthmann, H Kronmuller. Solid State Commun 99:553, 1996.

17. DK Lathrop, SE Russek, BH Mockley, D Chamberlain, L Peseson, RA Burhrman, DH Shin, J Silcox. IEEE Trans Magnetics M-27:3203, 1991.
18. D Dimos, P Chaudhari, J Mannhart. Phys Rev B 41:4038, 1991.
19. CL Jia, B Kabius, K Urban, K Hermann, GJ Cui, J Schubert, W Zander, AI Braginski, C Heiden. Physica C 175:545, 1991.
20. AF Marshall, CB Eom. Physica C 207:239, 1993.
21. CL Jia, B Kabius, K Urban, K Hermann, J Schubert, W Zander, AI Braginski. Physica C 196:211, 1992.
22. M Gustafsson, E Olsson, HR Yi, D Winkler, T Claeson. Appl Phys Lett 70:2903, 1997.
23. M Gustafsson, E Olsson, HR Yi, D Winkler, T Claeson. J Alloys Comp 251:19, 1997.
24. M Gustafsson, E Olsson, HR Yi, D Winkler, T Claeson. Physica C 282:623, 1997.
25. M Gustafsson, E Olsson, HR Yi, M Vaupel, R Woerdenweber. Appl Supercond Proc EUCAS 158:241, 1997.
26. J Gohng, EH Lee, IH Song, J Sok, SJ Park, JW Lee, CY Dosquet. IEEE Trans Appl Supercond As-7:3694, 1997.
27. C Park, J Hong, I Song, E Lee, C Moon, S Song, J Lee. Physica C 234:85, 1994.
28. C Francke, A Kramer, M Offner, R Strade, L Mex, J Muller. Appl Supercond 6:735, 1998.
29. C Francke, M Offner, A Kramer, L Mex, J Muller. Supercond Sci Technol 11:1311, 1998.
30. JZ Sun, WJ Gallagher, AC Callegari, V Foglietti, RH Koch. Appl Phys Lett 1561, 1993.
31. HR Yi, ZG Ivanov, D Winkler, YM Zhang, H Olin, P Larsson, T Claeson. Appl Phys Lett 65:1177, 1994.
32. A Barone, G Paternò. Physics and Applications of the Josephson Effect. New York: John Wiley & Sons, 1981.
33. KK Likharev. Dynamics of Josephson Junctions and Circuits. Gordon & Breach, Philadelphia, 1986.
34. K Hermann, Y Zhang, HM Muck, J Schubert, W Zander, AI Braginski. Supercond Sci Technol 4:583, 1991.
35. HR Yi, M Gustafsson, D Winkler, E Olsson, T Claeson. J Appl Phys 79:9213, 1996.
36. H Kohlstedt, J Schubert, K Hermann, M Siegel, CA Copetti, W Zander, AI Braginski. Supercond Sci Technol 4:583, 1991.
37. R Gross, P Chaudhari, M Kawasaki, A Gupta. Phys Rev B 42:10,735, 1990.
38. K Char, MS Colclough, SM Garrison, N Newman, G Zaharchuk. Appl Phys Lett 59:733, 1991; 59:2177, 1991.
39. R Gross, L Alff, A Beck, OM Froehlich, D Koelle, A Marx. IEEE Trans Appl Supercond As-7:2929, 1997 and references therein.
40. R Gross, B Mayer. Physica C 180:235, 1991.
41. R Gross. DH Blank, ed. Proceedings of the 2nd Workshop on HTS Applications and New Materials. University of Twente. 1995, Vol 8.
42. K Hermann, G Kunkel, M Siegel, J Schubert, W Zander, A Braginski, CL Jia, B Kabious, K Urbano. J Appl Phys 78:1131, 1995.
43. F Lombardi, ZG Ivanov, P Komissinski, GM Fisher, P Larsson, T Claeson. Appl Supercond 6:437, 1998.

44. HR Yi, D Winkler, ZG Ivanov, T Claeson. IEEE Trans Appl Supercond As-5:2778, 1995.
45. M Vaupel, G Ockenfuss, R Wordenweber. Appl Phys Lett 68:3623, 1996.
46. C Francke, M Offer, A Kramer, L Mex, J Muller. Supercond Sci Technol 11:1311, 1998.
47. W Reuter, M Siegel, K Herrmann, J Schubert, W Zander, A Braginski, P Muller. Appl Phys Lett 62:2280, 1993.
48. J Luine, J Bulman, J Burch, K Daly, A Lee, C Pettiette-Hall, S Schwarzbech, D Miller. Appl Phys Lett 61:1128, 1991.
49. YM Zhang, D Winkler, G Brorsson, and T Claeson. IEEE Trans Appl Supercond As-5:2200, 1995.
50. DJ Van Harlingen. Rev Mod Phys 67:515, 1995.
51. H Hilgenkamp, J Mannhart, B Mayer. Phys Rev B 53:14,586, 1996.
52. CA Coppetti, F Ruders, B Oelze, C Buchal, B Kabius, JW Seo. Physica C 253:63, 1995.
53. CA Coppetti, F Ruders, B Oelze, C Buchal, B Kabius, JW Seo. Physica C 253:63, 1995.
54. CC Tsuei, JR Kirtley, CC Chi, LSY Jahnes, A Gupta, JZ Sun, MB Ketchen. Phys Rev 73:593, 1994.
55. E Il'ichev, V Zakosarenko, RPJ IJsselsteijn, V Schultze, HG Meyer, HE Hoenig, H Hilgenkamp, J Mannhart. Phys Rev Lett 81:894, 1998.
56. J Mannhart, H Hilgenkamp, B Mayer, C Gerber, JR Kirtley, KA Moler, M Sigrist. Phys Rev Lett 77:2782, 1996.
57. E Il'ichev, V Zakosarenko, V Schultze, HG Meyer, HE Hoenig, VN Glyantsev. Appl Phys Lett 72:731, 1998.
58. H Hilgenkamp, J Mannhart. Appl Phys Lett 73:265, 1998.
59. HR Yi, D Winkler, T Claeson. Appl Phys Lett 66:1677, 1995.
60. D Winkler, YM Zhang, PÅ Nilsson, EA Stepantsov, T Claeson. Phys Rev B 72:1260, 1994.
61. RE Eck, DJ Scalapino, BN Taylor. Phys Rev Lett 13:15, 1964.
62. PA Rosenthal, MR Beasley, K Char, MS Colclough, G Zaharchuk. Appl Phys Lett 99:3482, 1991.
63. F Lombardi, USD Uccio, Z Ivanov, T Claeson, M Cirillo. Appl Phys Lett 76:2591, 2000.
64. AC Scott, WJ Johnson. Appl Phys Lett 14:316, 1969.
65. JR Waldram, AB Pippard, J Clark. Phil Trans R Soc London Ser A 268:265, 1970.
66. JU Lee, P Guptasarma, D Hornbaker, A El-kortas, D Hinks, KE Gray. Appl Phys Lett 71:1412, 1997.
67. G Hechtfisher, R Kleiner, K Schlenga, W Walkenhorst, P Muller. Phys Rev B 55:14,638, 1997.
68. F Lombardi, Z Ivanov, T Claeson, USD Uccio. IEEE Trans Appl Supercond As-9:3652, 1999.
69. AV Ustinov, H Kohlstedt, P Henne. Phys Rev Lett 77:3617, 1996.
70. M Cirillo, V Merlo, N Gronbech-Jensen. IEEE Trans Appl Supercond AS-9:4137, 1999.
71. A Barone. J Appl Phys 42:2747, 1971.

72. JA Edwards, JS Satchell, NG Chew, RG Humphreys, MN Keen, OD Dosser. Appl Phys Lett 60:2433, 1992.
73. J Ramos, M Seitz, GM Daalmans, D Uhl, Z Ivanov, T Claeson. Physica C 220:50, 1994.
74. T Mitzuka, K Yamaguchi, S Yoshikawa, K Hayashi, M Konishi, Y Enomoto. Physica C 218:229, 1993.
75. S Kuriki, T Kamiyama, D Suzuki, M Matsuda. IEEE Trans Appl Supercond AS-3:2461, 1993.
76. CP Foley, GJ Sloggett, KH Muller, S Lam, N Savvides, A Katsaros, DN Matthewes IEEE Trans Appl Supercond AS-5:2805, 1995.
77. CP Foley, S Lam, B Sankrithyan, Y Wilson, JC Macfarlane, L Hao. IEEE Trans Appl Supercond AS-7:3185, 1997.
78. F Tafuri, F Miletto, F Carillo, F Lombardi, U Scotti, K Verbiest, O Lebedev, GV Tendeloo. Physica C 326:63, 1999.
79. YJ Feng, YQ Shen, J Myging, NF Pedersen, PH Wu. Physica C 282:2459, 1997.
80. F Lombardi, ZG Ivanov, GM Fischer, E Olsson, T Claeson. Appl Phys Lett 72:249, 1998.
81. M Grove, R Dittmann, M Bode, M Siegel, AI Braginski. Appl Phys Lett 69:696, 1996.
82. KK Likharev. Rev Mod Phys 51:101, 1979.
83. DJ Lew, Y Suzuki, AF Marshall, TH Geballe, MR Beasley. Appl Phys Lett 65:1584, 1994.
84. AY Tzalenchuk, ZG Ivanov, S Pehrson, T Claeson, A Lohmus. IEEE Trans Appl Supercond AS-9:4115, 1999.
85. AW Kleinsasser, RE Miller, WH Mallison, GB Arnold. Phys Rev Lett 72:1738, 1994.
86. YG Ponomarev, NB Brandt, CS Khi, SV Tchesnokov, EB Tsokur, AV Yarygin, KT Yusupov, BA Aminov, MA Hein, G Muller, H Piel, D Wehler, VZ Kresin, K Rosner, K Winzer, T Wolf. Phys Rev B 52:1352, 1995.
87. U Zimmermann, D Dikin, S Abens, K Keck, T Wolf. Physica C 235:1901, 1994.
88. R Kleiner, P Müller. Phys Rev B 49:1327, 1994.
89. A Yurgens, D Winkler, NV Zavaritsky, T Claeson. Phys Rev B 53:R8887, 1996.
90. M Rapp, A Murk, R Semerad, W Prusseit, Phys Rev Lett 77:928, 1996.
91. YF Chen, ZG Ivanov, LG Johansson, RI Kojouharov, IM Angelov, E Olsson, VA Roddatis, EA Stepantsov, AY Tzalenchuk, AL Vasiliev, T Claeson. IEEE Trans Appl Supercond AS-7:2498, 1997.
92. RC Dynes, JP Garno, GB Hertel, TP Orlando. Phys Rev Lett 53:2437, 1984.
93. YG Ponomarev, NB Brandt, CS Khi, SV Tchesnokov, EB Tsokur, AV Yarygin, KT Yusupuv, BA Aminv, MA Hein, G Muller, H Piel, D Wehler, VZ Kresin, K Rosner, K Winzer, T Wolf. Phys Rev B 52:1352, 1995.
94. U Zimmermann, S Abens, D Dikin, K Keck, T Wolf. Physica C 235:1901, 1994.
95. Y Gao, G Bai, DJ Lam, KL Merkle. Physica C 173:487, 1991.
96. Y Ishimaru, J Wen, N Koshizuka, Y Enomoto. Phys Rev B 55:11,851, 1997.
97. S Poelders, R Auer, G Linker, R Smithey, R Schneider. Physica C 247:309, 1995.
98. RP Campion, JR Fletcher, PJ King, SM Morley, A Polimeni, RG Ormson. Supercond Sci Technol 11:730, 1998.

99. CB Eom, AF Marshall, Y Suzuki, B Boyer, RFW Pease, TH Geballe. Nature 353:544, 1991.
100. D Koelle, R Kleiner, F Ludwig, E Dantsker, J Clarke. Rev Mod Phys 71:631, 1999.
101. L Dorrer, S Wunderlich, F Schmidl, H Schneidewind, U Hubner, P Seidel. Appl Supercond 6:349, 1998.
102. Y Shen, Z Sun, R Kromann, T Holst, P Vase, T Freltoft. IEEE Trans Appl Supercond AS-5:2505, 1995.
103. MJ Ferrari, M Johnson, FC Wellstood, JJ Kingston, TJ Shaw, J Clarke. J Low Temp Phys 94:1, 1994.
104. V Glyantsev, M Siegel, J Schubert, W Zander, A Braginski, Supercond Sci Technol 7:253, 1994.
105. F Dillmann, VN Glyantsev, M Siegel. Appl Phys Lett 69:1948, 1996.
106. H Itozaki, S Tanaka, H Toyoda, T Hirano, Y Haruta, M Nomura, T Saijou, H Kado. Supercond Sci Technol 9:A38, 1996.
107. F Schmidl, S Wunderlich, L Dorrer, H Specht, S Linzen, H Schneidewind, P Seidel. IEEE Trans Appl Supercond AS-7:2756, 1997.
108. S Wunderlich, F Schmidl, H Specht, L Dorrer, H Schneidewind, U Hubner, P Seidel. Supercond Sci Technol 11:315, 1998.
109. R Weidel, S Brabetz, F Schmidl, F Klemm, S Wunderlich, P Seidel. Supercond Sci Technol 10:95, 1997.
110. S Krey, KO Subke, D Reimer, M Schilling, R Scharnweber, B David. Appl Supercond 5:213, 1997.
111. P Seidel, R Weidl, S Brabetz, F Schmidl, H Nowak, U Leder. Appl Supercond 6:309, 1998.
112. JD McCambridge, MG Forrester, DL Miller, BD Hunt, JX Pryzbysz, J Talvacchio, RM Young. IEEE Trans Appl Supercond AS-7:3622, 1997.
113. T Umezawa, T Fujita, Y Higashino. Jpn J Appl Phys 2:L981, 1996.
114. AY Tzalenchuk, ZG Ivanov, SV Dubonos, T Claeson. Appl Supercond Proc EUCAS 1999, IOP Conf. Series 167, 581–584, 2000.
115. VN Glyantsev, Y Tavrin, W Zander, J Schubert, M Siegel. Supercond Sci Technol 9:A105, 1996.

5

Conductance Noise in High-Temperature Superconductors

László Béla Kish
Texas A&M University, College Station, Texas, U.S.A.

5.1 INTRODUCTION

High-T_c superconductor (HTS) materials have the potential to revolutionize low-noise electronics, because superconductor electronics can be realized at relatively high temperatures. These temperatures might soon be achievable by solid-state cooling elements at the commercial level if both the HTS and the cooling element development continues at the current level. However, it should be kept in mind that this task might not compromise the requirements of low-noise. The real condition of success of HTS materials will always be the potentially low level of noise, because their high-speed semiconductor and nanoelectronics are strong competitors at convenient working temperatures. Due to the different physics of superconductors, their potentially achievable noise properties seem to be unbeatable. However, as far as HTS devices are concerned, although their noise properties are good, the same properties are achievable by semiconductor circuits also, but the circuits need a carefully designed circuitry. Therefore, understanding the source of noise and reducing the noise level in the HTS materials has a high priority in research.

Although the details of the mechanism of noise generation in HTS materials is not fully understood, it has been proven that percolation effects are the key to understanding the characteristic behavior of the noise in the conductor–superconductor temperature regime. The percolation models are superior also in the sense that they can predict the behavior of normalized noise in a very wide range, with variations up to nine orders of magnitude. Such a success of a model to predict behavior can rarely be found in condensed matters physics.

In the present survey, we give a basic overview (based on Refs. 1–3) of low-frequency ($f < 10^5$ Hz) conductance noise in the conductor–superconductor transition region of HTS materials. We are concerned only about the "essence," namely understanding the origin and mechanism of dominant and generally occurring noise effects. For readers who are interested in learning more about the base of other, more sample- and material-specific noise effects and their theories, we recommend Refs. 1–3 and the references therein. The original concept has been enriched by the inclusion of the *biased percolation effect;* see Section 5.4.4, which explains the lack of (universal) scaling when the control parameter is not the temperature but the current or magnetic field. Those who are interested in learning some fundamental limits of the present approaches and some relevant unsolved problems, we recommend Ref. 4. Since the publication of the percolation picture (1–4), the field has become very active; however, up to now, no comparable breakthrough has appeared. Some interesting experimental and theoretical additions can be found in Refs. 5–18. It can also be of interest to review Refs. 19–27. Noise results relevant for applications, such as bolometers, can be found in Refs. 28–33.

Finally, some concern to those readers who go beyond reading the present chapter. As the aim of this chapter is to show a coherent, and reliable frame of thinking which can be the base of further studies, I will not deal with and will not take any responsibility for all the materials described in Refs. 5–33. Moreover, in certain cases, I have some strong reservations about the reliability of some of the published data and theories, see the above relevant comments about the number and nature of mistakes in this field. However, as the evolution of science has been manifested by disputes, when some inspiring thoughts were presented in an article, I decided to include it even if I could not always fully trust its content.

5.2 BASIC TERMS

Conductance noise in a normal conductor material (34) means a fluctuation of the resistance which can be described as a stationary, random, stochastic process. Conductance noise of a superconductor material, either in the conductor–superconductor transition temperature region or in the superconductor state, is a nontrivial issue, as the material is non-Ohmic and the energy dissipation can contain components due to vortex motion. Therefore, in order to avoid any misunderstanding, in this section, we clarify what we call conductance noise and relate the defined quantities to well-known quantities of classical noise research.

For the measurement of the conductance noise of the superconducting material, it is assumed that a four-terminal measurement method is used (see Fig. 5.1) or an equivalent arrangement to avoid contact noise. Otherwise, contact noise could dominate the noise due to the low resistivity of HTS samples when being close to the superconducting state. For simplicity and due to the low resistivity, here we neglect thermal noise, however, the thermal noise voltage of the voltage contacts can be a problem. The measured resistance fluctuation is defined as

$$\Delta R(t) = \frac{\Delta U(t)}{I} \tag{1}$$

where $\Delta U(t)$ is the measured voltage noise and I is the dc current through the current contacts. Therefore, the power density spectra of the resistance noise is related to the power density spectrum of $\Delta U(t)$ by

$$S_R(f) = \frac{S_U(f)}{I^2} \tag{2}$$

Both the measured noise and the resistance $R(T)$ of the sample strongly depend on the temperature and the dependence varies between samples made by different technologies or made of different materials. The normalized resistance noise spec-

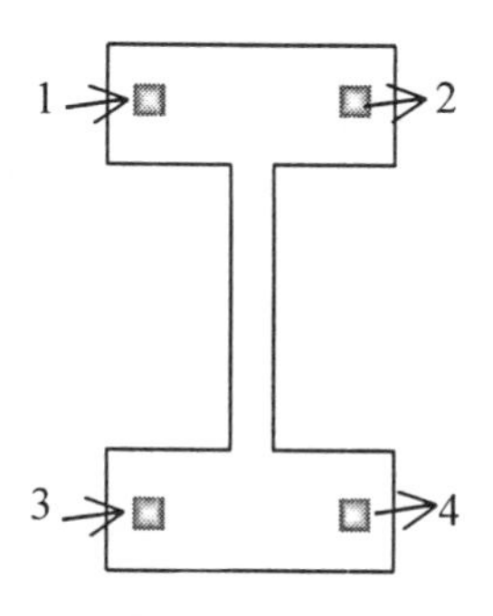

FIGURE 5.1 Four-terminal sample arrangement. The current is fed by a low-noise current generator via contacts 1–3 (current contacts), and the voltage and voltage noise are measured between 2–4 (voltage contacts), presumably by a device which does not have dc input current. Thus the resistance fluctuations of contacts 2 and 4 cannot be seen because of the lack of dc current, and the resistance fluctuations of contacts 1 and 3 cannot be seen because of the current generator driving.

trum is defined as

$$C(T) = \frac{S_R(f, T)}{R^2(T)} \tag{3}$$

History has shown that to find a coherent, conclusive, and reproducible behavior of the noise of various HTS samples and materials, the temperature has to be used as a control parameter (hidden variable) and the $C(T)$ versus $R(T)$ curve should be analyzed. In this way, the various temperature dependencies are put out of the picture and the $C[R(T)]$ or $C(R)$ curve with a hidden temperature variable supplies information about the spatial distribution of the microscopic current density distribution in the sample. This sort of plot made it possible to identify percolation in HTS films already at an early stage of HTS technology in 1989 (35).

5.3 TEMPERATURE DEPENDENCE OF THE MEASURED NOISE

5.3.1 General Temperature Dependence of the Normalized Noise

We study the most characteristic temperature-dependent behavior of the noise at a fixed frequency. According to thorough investigations, in a significant fraction of samples two fundamental temperature regimes exist; see Figure 5.2.

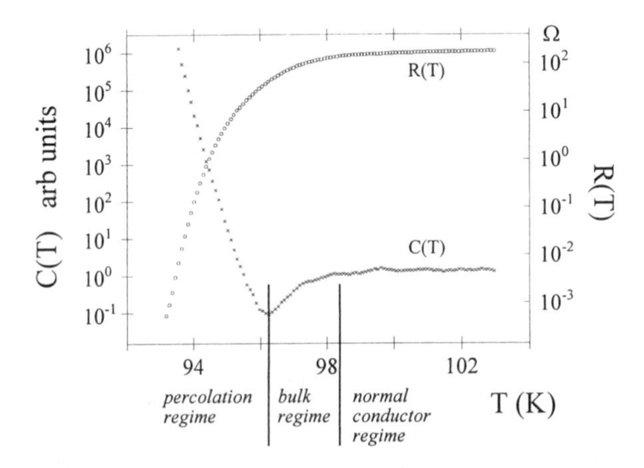

FIGURE 5.2 Qualitative temperature dependence of the normalized noise and the resistance. In the bulk regime, the normalized noise is not increasing, and sometimes it is even decreasing with decreasing temperature. In the percolation regime, the normalized noise is radically increased while the temperature is decreased.

5.3.1.1 "Bulk" Temperature Regime

At the high-temperature part of the conductorn–superconductor transition region, when the resistance has significantly decreased (the onset of superconductivity has started), $C(R)$ can be constant, or sometimes even decreasing, with decreasing R (i.e., with decreasing T). In this regime, the microscopic current density is spatially homogeneous in the sample (as in a bulk conductor) or, at least, the distribution is independent of the temperature. *Note:* Although a significant part of HTS samples show this behavior especially materials of lower quality, the bulk regime is often missing from the $C(R)$ curve of the films (2).

5.3.1.2 "Percolation" Temperature Regime

$C(R)$ increases many orders of magnitude with decreasing R (i.e., with decreasing T). This regime is reported in almost all articles in the literature. In this regime, the microscopic current density is spatially random in the sample and the distribution randomly changes when the temperature is varied. The distribution of current density has the properties (percolation) of conductor–superconductor random composites.

5.3.2 Scaling of *C* with *R*

This is a very frequently occurring behavior (see Figs. 5.3–5.5) which can often be quantitatively explained (1–4). When, at fixed dc measuring current, the nor-

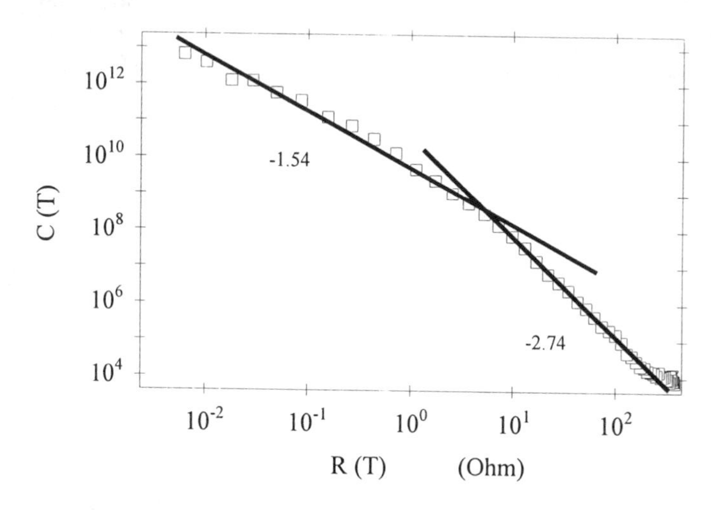

FIGURE 5.3 Scaling with universal exponents through nine orders of magnitudes of the normalized noise. The sample (4) is an Y-based HTS film, representing low-noise and good technology (laser ablation technique with in situ annealing).

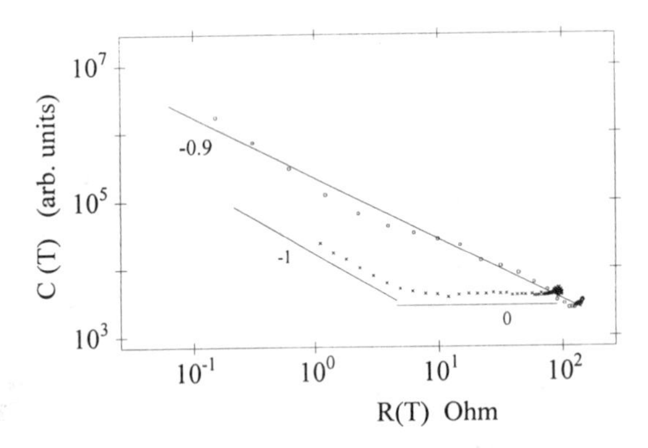

FIGURE 5.4 Universal scaling in Y-based superconductor films (1) representing an older technology (evaporation with ex situ annealing).

malized noise and the resistance R is controlled by varying the temperature, C can be approximated as a power function of $R(T)$:

$$C \propto R^x \tag{4}$$

where the exponent x can have various values in different temperature ranges and in different samples. The occurring x values are usually close to the following values: -2.74, -1.54, -1, 0, and $+2$. These values are predicted by a simple theory (1). Note that the existence of relation (4) remains hidden if only the $R(T)$ and $S_R(T)$ curves are plotted.

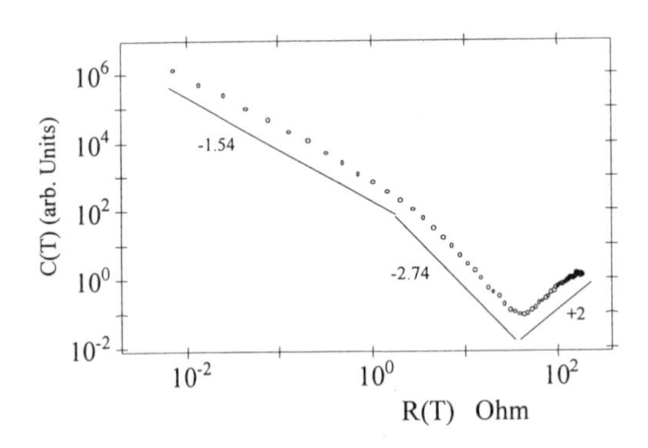

FIGURE 5.5 Universal scaling in a Y-based superconductor film (1,2) representing a middle-class technology (evaporation with in situ annealing).

5.4 SIMPLE THEORETICAL PICTURE

5.4.1 Framework of the Model: Two-Stage Transition Picture

For simplicity, we call normal conducting charge carriers "electrons" and superconducting ones "Cooper pairs." The model is based on a physical picture of the conductor–superconductor transition (1), which is a slight modification of the two-stage transition picture of HTS materials. In sufficiently homogeneous HTS materials, at the high-temperature part of the transition there are no superconducting grains present yet. The conductance increases with decreasing temperature due to the increasing number and lifetime of Cooper pairs. The current density distribution is homogeneous, so the name "bulk regime" is used here. In the low-temperature part of the transition region, there are superconductor grains of random sizes at random locations, because the Cooper pairs are very "fragile" due to their extremely short coherence length, which means that even an atomic scale disorder can prohibit superconductivity in small subvolumes. That implies a random distribution of current density and naturally leads to percolation effects in this regime. The neighboring grains can form superconducting islands via Josephson coupling, and the lower the temperature, the larger the mean linear size (percolation length) of these islands. When the percolation length reaches the thickness of the film, a three-dimensional (3D)/pro-dimensional (2D) crossover occurs. At the effective T_c, where the macroscopic superconductivity sets in (then the system is at the "percolation threshold"), there is at least one large island between the electrodes.

Note: If the T_c in the microscopic subvolumes of the material is strongly inhomogeneous ($\Delta T_c > \Delta T_{\mathrm{tr}}$, where ΔT_{tr} is the width of the transition region in the subvolumes and ΔT_c is the root mean square spatial fluctuation of T_c), the bulk region does not exist. Percolation occurs in the whole transition region. This behavior can be observed at high-tech HTS materials with a very narrow transition region.

5.4.2 Effects at the High-Temperature End of the Transition Region (Bulk Region). Number of Fluctuations of Electrons and Cooper Pairs. Mobility Fluctuations of Electrons

In the present model, it is assumed that, in the bulk region, the noise basically originates from the electrons, as in normal conductors. The knowledge of low-frequency noise in conductors (36) implies that the microscopic origin of the noise is rather independent of the temperature in the few Kelvin range around 100 K, which is the typical width and location of the conductor–superconductor transition region. This condition makes the explanation of the generality of the scaling behavior easier.

5.4.2.1 Number of Fluctuations of Electrons and Cooper Pairs

It is assumed that the density n of free electrons fluctuates due to trapping. When the proportion of the Cooper pairs is small, at a given temperature the density of Cooper pairs is proportional to the density of electrons. In that way, the fluctuation of the electron density causes a correlated fluctuation of the Cooper-pair density and the normalized fluctuations $\Delta n/n$ of electrons and Cooper pairs will be equal. It can be easily shown (1) that this effect leads to a normalized conductance noise which is independent of the temperature, no matter which class of carriers dominates the dc current. Thus,

$$C(R) \propto R^0 \tag{5}$$

An example can be seen in Figure 5.4.

5.4.2.2 Mobility Fluctuations of Electrons

It is assumed that the mobility μ of free electrons is the only fluctuating quantity. As the mobility of electrons (unlike their density) is not coupled to the Cooper pairs, the resulting system can be modeled as two parallel conductors: one of them (the electronic) is noisy and its conductance is independent of the temperature, whereas the other one (the Cooper pairs) is noise-free and strongly temperature dependent. It can be easily shown (1) that this effect leads to a normalized conductance noise which satisfies the following relation:

$$C(R) \propto R^2 \tag{6}$$

An example can be seen in Figure 5.5.

5.4.3 Effects in the Percolation Regime. Classical Percolation Noise and *p*-Noise

Percolation effects in random resistor networks have been intensively studied during the last two decades (see Ref. 3 and references therein). The study of HTS noise enriched the field of percolation by the appearance of p-noise, which is a new type of percolation noise (see Sec. 5.4.1.2).

5.4.3.1 Classical Percolation Noise

The relevant classical model is a random resistor network, where some resistors at randomly located places are short-circuited. The resistors represent the normal conducting materials, whereas the short circuits represent the superconducting grains, so the lower the temperature, the larger the number of short circuits. *The noise originates from the normal conducting material.* The resistors are noisy and their resistance fluctuations are uncorrelated and independent of the temperature. In such a system the normalized noise satisfies the following relation:

$$C(R) \propto R^x \tag{7}$$

with x close to -1 and its value depending on the geometrical dimension of the sample. Examples are presented in Figure 5.4, at the low-resistance ends of the curves.

5.4.3.2 Novel Percolation Noise Effect: *p*-Noise (1,2)

Assume that we have the same random resistor network as above, except that the resistors are noise-free and some of the short circuits are "noisy"; that is, they randomly switch "on" and "off" in time, controlled by independent random processes $w(t)$. These switching elements represent unstable superconductivity in grains or intergrin junctions. Such switching may occur because of defect motion, electron trapping, or flux motion at the unstable elements. The large number of switching elements causes a *fluctuation of the volume fraction of superconducting material,* which is called p in earlier works on percolation ($0 < p < 1$). We call this fluctuation "p-noise." In order to calculate the behavior of the resultant normalized noise C, it is assumed that in a given temperature range, the number and dynamics of switching elements do not change. For that temperature range, simple calculations yield (1,2)

$$C(R) \propto R^x \tag{7}$$

with x values given as -2.74 in three dimensions, -1.54 in two dimensions, and 0 in one dimension. Figures 5.3 and 5.5 are examples for the 2D and 3D behavior and the 3D/2D dimensional crossover, respectively, described earlier. The difference between classical percolation noise exponents and p-noise exponents is remarkable.

5.4.3.3 Summary of the General Model

Under the applied assumptions, the normalized noise can be approximated as $C(R) \propto R^x$, where the x exponent can vary as the conditions determine (Fig. 5.6). Apart from p-noise, the microscopic origin of the noise is the noise of the normal conducting charge carriers. If the normal conductor phase is noise-free, then only p-noise would exist in terms of this model.

Figures 5.7 and 5.8 show a rough prediction based on this simple theoretical picture. A number fluctuation Δn of charge carriers and a p-noise Δp are assumed. For simplicity, these microscopic noise sources (Δn and Δp) are assumed to have a constant strength in the whole temperature range. An exponentially decaying resistance $R(T)$ is assumed. The 3D/2D crossover of percolation is assumed to be abrupt, which makes the relevant peak on the noise curve sharper than real peaks of that kind.

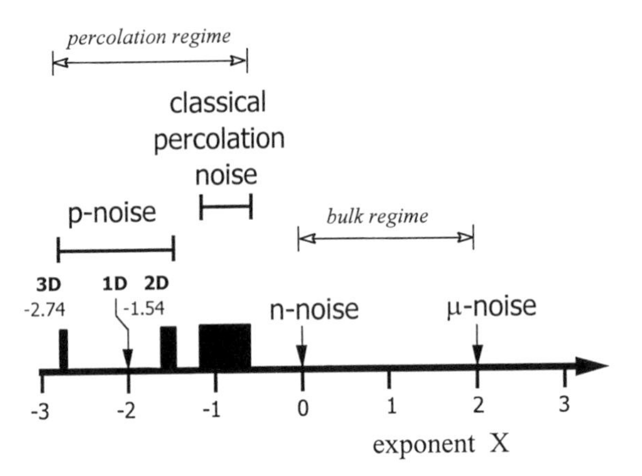

FIGURE 5.6 Summary of the power exponents by which the $C(R)$ function can most frequently be approximated.

5.4.4 When the Temperature Is Kept Constant and the Current or Magnetic Field Is Varied: Biased Percolation

Until now, we assumed that the temperature was the control parameter which causes the change of R and C. Another interesting questions is what happens when

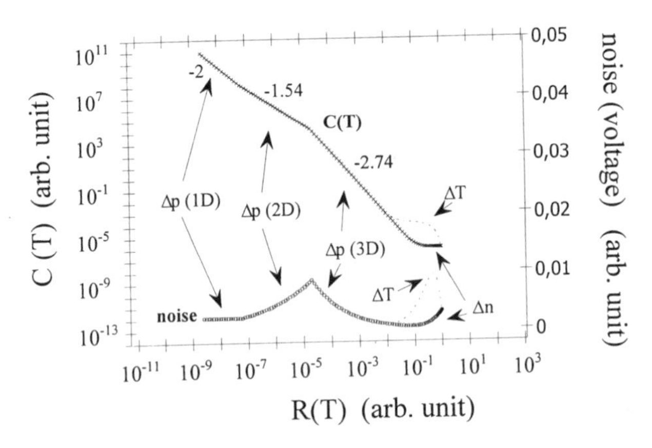

FIGURE 5.7 Illustration of the scaling of normalized noise versus resistance and the behavior of measured noise voltage in terms of the simple model. The dashed hairline shows the change due to temperature fluctuations (not inherent in the general model).

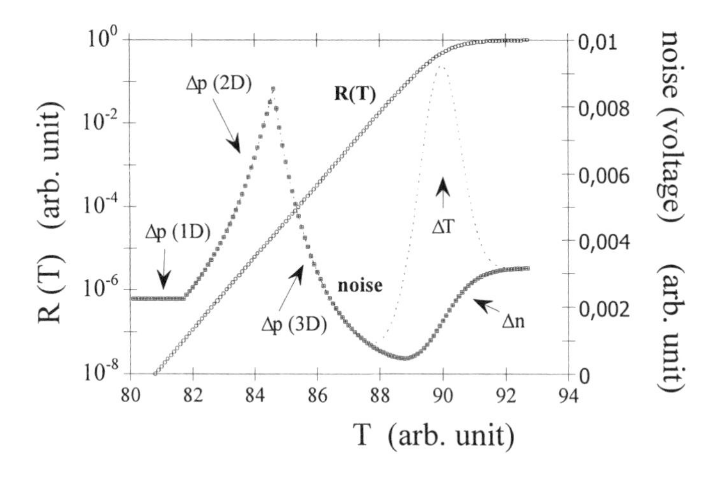

FIGURE 5.8 A possible temperature-dependent behavior is shown, as an illustration of the simple model. A number fluctuation Δn of charge carriers and a p-noise Δp are assumed. For simplicity, these microscopic noise sources (Δn and Δp) are assumed to have a constant strength in the whole temperature range. An exponentially decaying resistance $R(T)$ is assumed for the sake of simplicity. The 3D/2D crossover of percolation is assumed to be abrupt, which makes the relevant peak on the noise curve sharper than real peaks of that kind.

the temperature is kept constant and the current density or the magnetic field is the control parameter (4); see Figure 5.9. Surprisingly, in these cases, scaling between R and C often cannot be found or the scaling exponent is nonuniversal (i.e., sample and temperature dependent). This fact was a mystery until 1995, when inspired by this problem, Kiss introduced the biased percolation concept, which, with other co-workers, was later applied to explain degradation and abrupt failure of electronic devices (37,38).

The picture is very simple. Let us assume that in a superconductor with homogeneous current density, an isolated island of normal conducting phase suddenly occurs, due to the increase of external current. As the electrical field in the superconductor is zero, no current will flow through this island. As a consequence, the local current density around the island will increase. More precisely, the increase will happen around the edge of the cross section perpendicular to the direction of the original current density, and to the contrary, a decrease is observed around that region of the island surface, which is, at most, far from this edge (projection of the center of that cross section to the island surface). The situation is very similar to the case of a real island surrounded by a laminar flow of water: The flow around the corresponding edges of the island will be higher than without the island.

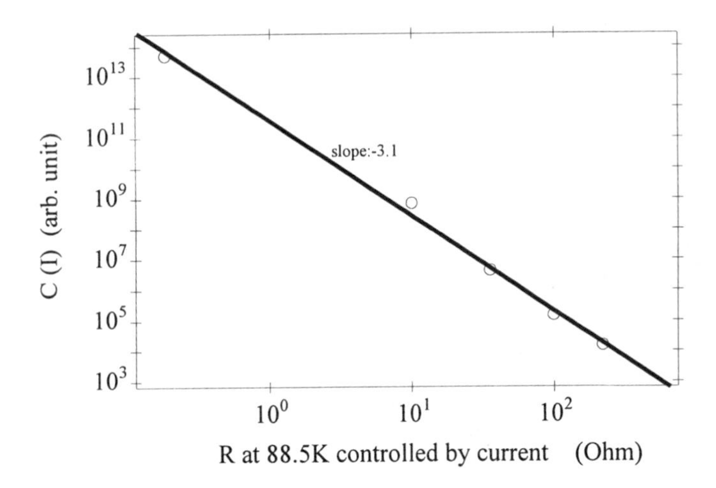

FIGURE 5.9 Nonuniversal scaling (4) in the same sample as shown in Figure 5.3; when at fixed temperature, the control parameter is the current. Usually, the scaling exponent depends on the temperature.

The increased local current density implies a higher probability of the occurrence of a superconductor–conductor transition at this edge. Therefore, this effect manifests itself by causing growth of the conductor island in a perpendicular direction against the electrical field. These effects has been observed by computer simulations (37,38) of vacancy generation due to high current density in metal films and, recently, in real experiments on nanocrystalline gold films. The simulation of percolation noise effects (37,38) shows that due to the strongly anisotropic nature of biased percolation, the effects do not show universal scaling exponents, and the scaling is only approximate or its existence is doubtful. As a similar argumentation can be made for magnetic fields by invoking screening effects, we can conclude that the nonuniversal behavior is a biased percolation effect.

5.5 ON NOISE DUE TO TEMPERATURE FLUCTUATIONS

The normalized temperature derivative

$$\gamma = R^{-1}(T)\frac{dR(T)}{dT} \tag{7a}$$

describes the strength of coupling of temperature variations to resistance variations and this is the relevant quantity to describe the behavior of the normalized noise $C(T)$. In certain temperature regions, the γ of HTSC materials can be orders of magnitude greater than that of metals and semiconductors. As a consequence, temperature noise caused by heating power fluctuations can significantly contribute to the conductance noise (2,39–41). Heating power fluctuations naturally occur due to the noise of the thermometer (especially silicon diode thermometers), mechanical noise (helium-flow temperature regulators), or the noise of the temperature control amplifiers which can originate from the method of regulation (switching regulators). The frequency range of interest (1 Hz to 10 kHz) implies that the very short-term stability is the crucial property.

For a good HTSC sample, to avoid this kind of artificial noise *in the most sensitive temperature region,* the temperature noise typically has to be smaller than 10^{-8} K/Hz$^{-1/2}$. Such stability typically needs a stable ac bridge with a metal wire thermometer, a tuned PID regulator, and a low-noise dc heater amplifier. A large (several 100 s) thermal time constant of the sample holder is beneficial. An example that this is shown on Figure 5.10.

For practical applications, the effect of temperature fluctuations can easily be avoided by using the device at that temperature region where γ is low. There is one exception—the bolometer, which has to be used where γ is large.

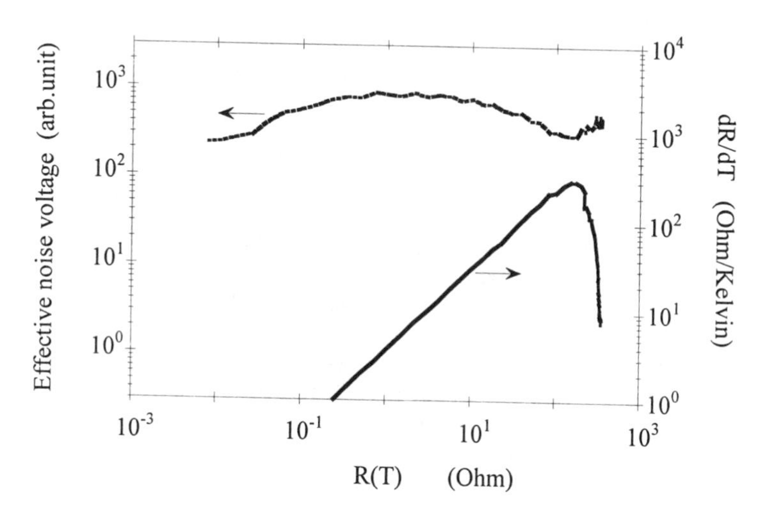

FIGURE 5.10 An example in which, due to the careful temperature control design, the temperature fluctuations do not show up at any part of the conductor–superconductor transition regime. The sample is the same, which is shown in Figures 5.3 and 5.4 (4). The noise voltage amplitude and the temperature derivative of the resistance have opposite temperature dependence in the whole regime.

REFERENCES

1. LB Kiss, T Larsson, P Svedlindh, L Lundgren, H Ohlsen, M Ottoson, J Hudner, L Stolt. Physica C 207:318–332, 1993.
2. LB Kiss, P Svedlindh. Noise in high-T_c superconductors, IEEE Trans Electr Dev 41:2112–2122, 1994.
3. LB Kiss, P Svedlindh. New noise exponents in conductor–superconductor and conductor–insulator random composites. Phys Rev Lett 71:2817, 1993.
4. LB Kiss, P Svedlindh, LKJ. Vandamme, CM Muirhead, Z Ivanov, T Claeson. Current controlled percolation exponents in the noise of high-temperature superconductor thin films. In: ChR Doering, LB Kiss, MF Shlesinger, eds. Unsolved Problems of Noise. London: World Scientific, 1997, pp 306–311.
5. M Baziljevich, AV Bobyl, H Bratsberg, et al. Fractal structure near the percolation threshold for $YBa_2Cu_3O_7$ epitaxial films. J Phys IV 6:259–264, 1996.
6. M Celasco, R Eggenhoffner, E Gnecco, et al. Noise dependence on magnetic field in granular bulk high-T_c superconductors. Phys Rev B 58:6633–6638, 1998.
7. A Taoufik, S Senoussi, A Tirbiyine. Characteristic noise anisotropy of the normal conductor–superconductor transition in $YBa_2Cu_3O_{7-\delta}$. Ann Chim Sci Mater 24:227–232, 1999.
8. L Cattaneo, M Celasco, A Masoero, et al. Current noise in HTC polycrystalline superconductors—A comparison between experiments and different types of percolation models. Physica C 267:127–146, 1996.
9. I Puica, V Popescu, P Mazzetti, et al. Current-influenced voltage noise in bulk polycrystalline high-T_c superconductors. Physica C 290:303–310, 1997.
10. E Granato, D Dominguez. Current–voltage characteristics of diluted Josephson-junction arrays: Scaling behavior at current and percolation threshold. Phys Rev B 56:14,671–14,676, 1997.
11. DG McDonald, RJ Phelan, LR Vale, et al. Noise from YBCO films: Size and substrate dependence. IEEE Trans Appl Supercond AS-7:3091–3095, 1997.
12. AV Bobyl, ME Gaevski, SF Karmanenko, et al. Magneto-depending noise of a single latent weak link in $YBa_2Cu_3O_{7-x}$ film. Physica C 266:33–43, 1996.
13. LB Kiss, U Klein, J Smithyman, et al. Experimental study of flux noise in YBCO/PBCO superlattices. Inst Phys Conf Ser 148:1027–1030, 1995.
14. LB Kiss, U. Klein, CM Muirhead, et al. Diffusive fluctuations, long-time and short-time cross-correlations in the motion of vortice-pancakes in different layers of YBCO/PBCO superlattices. Solid State Commun 101:51–56, 1997.
15. G Jung, B Savo, A Vecchione A, et al. Intrinsic high-T_c Josephson junctions in random-telegraph-noise fluctuators Phys Rev B 53:90–93, 1996.
16. B Savo B, C Coccorese. Vortex-induced voltage instabilities in a superconducting BSCCO thin film. IEEE Trans Appl Supercond-AS-9:2336–2339, 1999.
17. G Jung, B Savo. Josephson mechanism in random telegraph voltage noise in high-T_c superconductors. Appl Supercond 6:391–397, 1998.
18. G Jung, B Savo. Elementary and macroscopic two-level fluctuations in high-T_c superconductors. J Appl Phys 80:2939–2948, 1996.
19. SK Arora, R Kumar, R Singh, et al. Electronic transport and $1/f$ noise studies in 250 MeV Ag_{107} ion irradiated $La_{0.75}Ca_{0.25}MnO_3$ thin films. J Appl Phys 86:4452–4457, 1999.

20. SK Arora, R Kumar, D Kanjilal, et al. 1/*f* Noise properties of a $La_{1-x}Ca_xMnO_3$ thin film. Solid State Commun 108:959–963, 1998.
21. YP Chen, GL Larkins, CM Van Vliet, et al. Current noise of contacts to $YBa_2Cu_3O_7$ high-T_c superconducting platelets through the transition region IEEE Trans Appl Supercond AS-9:3429–3431, 1999.
22. AI Abou-Aly, MT Korayem, NG Gomaa, et al. Synthesis and study of the ceramic high-T_c superconductor $Hg_{1-x}Tl_xBa_2Ca_{1.8}Y_{0.2}Cu_3O_{8+\delta}$ ($x = 0.3, 0.5, 0.7, 0.9$ and 1). Supercond Sci Technol 12:147–152, 1999.
23. A Peled, RE Johanson, Y Zloof, et al. 1/*f* Noise in bismuth ruthenate based thick-film resistors. IEEE Trans Compon Pack ACP-20:355–360, 1997.
24. M Prester. Current transfer and initial dissipation in high-T_c superconductors. Supercond Sci Technol 11:333–357, 1998.
25. V Sandu, S Popa, E Cimpoiasu. Fluctuation conductivity in Li-doped $YBa_2Cu_3O_{7-\delta}$. J Supercond 9:487–492, 1996
26. L Liu, ER Nowak, HM Jaeger, et al. High-angle grain-boundary junctions in $YBa_2Cu_3O_7$—Normal-state resistance and 1/*f* noise. Phys Rev B 51:16,164–16,167, 1995.
27. K Frikach, A Taoufik, S Senoussi, et al. Vortex lattice in the superconductor–conductor transition region and the voltage noise in a highly textured thin film of $YBa_2Cu_3O_{7-\delta}$. Physica C 282:1977–1978, 1997.
28. FS Galasso, DB Fenner, L Lynds, et al. An assessment of synthesis effects on the high-T_c superconducting transition: With application to thin-film bolometer devices. Appl Supercond 4:119–133, 1996.
29. IA Khrebtov, AD Tkachenko. Studying the noise of high-temperature superconductor bolometers on silicon membranes. J Opt Technol 66:1064–1067, 1999.
30. M Fardmanesh, A Rothwarf, KJ Scoles. Noise characteristics and detectivity of $YBa_2Cu_3O_7$ superconducting bolometers: Bias current, frequency, and temperature dependence. J Appl Phys 79:2006–2011, 1996.
31. IA Khrebtov, AD Tkachenko. High-temperature superconductor bolometers for the IR region. J Opt Technol 66:735–741, 1999.
32. IA Khrebtov, VN Leonov, AD Tkachenko, et al. Noise of high-T_c superconducting films and bolometers. J Phys IV 8:293–296, 1998.
33. H Neff, IA Khrebtov, AD Tkachenko, et al. Noise, bolometric performance and aging of thin high-T_c superconducting films on silicon membranes. Thin Solid Films 324:230–238, 1998.
34. MB Weissman. Low-frequency noise as a tool to study disordered materials. Annu Rev Mater Sci 26:395–429, 1996.
35. LB Kiss, P Svedlindh, L Lundgren, J Hudner, H Ohlsen, L Stolt. Spontaneous conductivity fluctuations in Y–Ba–Cu–O thin films: Scaling of fluctuations, experimental evidence of percolation at the superconducting transition. Solid State Commun 75:747, 1990.
36. ChR Doering, LB Kiss, MF Shlesinger, eds. Unsolved Problems of Noise. London: World Scientific, 1997.
37. C Pennetta, Z Gingl, LB Kiss, L Reggiani. A percolative simulation of dielectric-like breakdown. Microelectr Reliab 38:249–253, 1998.
38. Z Gingl, C Pennetta, LB Kiss, L Reggiani. Biased percolation and abrupt failure of electronic devices. Semicond Sci Technol 11:1770–1775, 1996.

39. KH Han, MK Joo, SHS Salk, HJ Shin. Phys Rev B 46:11,835, 1992.
40. S Jiang, P Hallemeier, Ch Suria, JM Phillips. In: PH Handel, AL Chung, eds. Noise in Physical Systems and $1/f$ Fluctuations. AIP Conf Proc No. 285. New York: American Institute of Physics, 1993, p 119.
41. RD Black, LG Turner, A Mogro-Campero, TC McGee, AL Robinson. Appl Phys Lett 55:2233, 1989.

6

Noise in High-Temperature Superconductor Josephson Junctions

J.C. Macfarlane, L. Hao,* and C.M. Pegrum
University of Strathclyde, Glasgow, Scotland

6.1 INTRODUCTION

6.1.1 Sources of Electronic Noise

Noise is an important problem in science and engineering because it degrades the accuracy of any measurement and the quality of electronically processed signals. To understand and minimize these effects, one must measure this noise simply and accurately. Perhaps the two most commonly encountered types of noise are thermal noise and shot noise. Thermal noise arises from the random velocity fluctuations of the charge carriers (electrons and/or holes) in a resistive material. The mechanism is sometimes said to be Brownian motion of the charge carriers due to the thermal energy in the material. Thermal noise is present when the resistive element is in thermal equilibrium with its surroundings, and it is often referred to as Johnson noise (or Nyquist noise) in recognition of two early investigators of this phenomenon (1,2). Thermal noise is usually represented by the equation

$$S_v(f) = 4kTR \quad (\mathrm{V^2/Hz}) \tag{1}$$

* Current affiliation: Centre for Basic Metrology, National Physical Laboratory, Teddington, England

where k is Boltzmann's constant (1.38×10^{-23} J/K), R is the resistance of the conductor, T is the absolute temperature, and S_v is the voltage noise power spectral density.

Shot noise occurs when the current flows across a barrier. It was first discussed by Schottky (3). It is often found in solid-state devices when a current passes a potential barrier such as the depletion layer of a *p–n* junction. The stream of charge carriers fluctuates randomly about a mean level. The fluctuations (i.e., the shot noise) are due to the random, discrete nature of the tunneling process. The shot noise has a constant spectral density of

$$S_i(f) = 2eI_{DC} \quad (\mathrm{A^2/Hz}) \tag{2}$$

where e is the electronic charge (1.6×10^{-19} C) and I_{DC} is the average current. In both of the above cases, the noise spectral density is independent of frequency. In many devices, however, there is additional noise which varies with frequency as $|f|^{-\alpha}$, where α usually lies between 0.8 and 1.2. This is commonly known as $1/f$ noise or flicker noise or excess noise. For homogeneous materials, the $1/f$ noise can be represented by the empirical formula (4)

$$\frac{S_i(f)}{(\bar{I})^2} = \frac{S_v(f)}{(\bar{V})^2} = \frac{\alpha_H}{f\bar{N}} \tag{3}$$

where $\bar{N}$ is the total number of charge carriers in the specimen. $\bar{I}$ is the mean current, and $\bar{V}$ is the mean voltage. α_H is Hooge's parameter (4) and is $\sim 2 \times 10^{-3}$ in metal film. The value of Hooge's parameter varies very widely in semiconductor and junction devices.

The fourth type of noise is sometimes found in transistors and other devices. It is called burst noise or random telegraph noise. It consists typically of random pulses of variable length and equal height. The above-described noise processes are all intrinsic noise; that is, they are caused by processes inside the devices. External noise due to interference from electrical or magnetic disturbances are a separate topic. Appropriate steps must be taken to eliminate external effects from the measurements.

6.1.2 Noise in Superconducting Devices

It is logical to consider first the subject of noise in liquid-helium-temperature Josephson junction and superconducting quantum interference devices (SQUIDs). This work forms a foundation for later studies of high-temperature devices.

6.1.2.1 The Josephson Effects

The Josephson effects (5) are well known, and only those details necessary for an understanding of the noise mechanisms (described below) will be included here.

When two superconducting electrodes are weakly linked (see Fig. 6.1a), each can be described by a single quantum state. Let Ψ_L and Ψ_R be the macroscopic wave function, where $|\Psi|^2$ is the Cooper-pair density ρ. The phase φ of the wave function is constant over each electrode, but can change across the weak link. If the phase remains between $\pm\pi/2$, the current flow ($\pm I_c$) generates no voltage drop; this is the dc Josephson effect with I_c (critical current).

If the current across the weak link exceeds I_c, then a voltage appears and the phase φ slips at a rate that is proportional to the voltage:

$$\frac{d\varphi}{dt} = \frac{2e}{\hbar} V \tag{4}$$

or

$$f_J = \frac{2e}{h} V$$

This is the ac Josephson effect; the frequency $2e/h = 484$ THz/V.

The I–V characteristic for the RSJ (resistively shunted junction) model is given by the equation in limit of low capacitance for $I > I_c$:

$$V = I_c R_n \left[\left(\frac{I}{I_c} \right)^2 - 1 \right]^{1/2} \tag{5}$$

where I_c is the critical current and R_n is the normal resistance of the junction at a high-voltage bias (Fig. 6.1b). The critical voltage, $V_c = I_c R_n$, is a figure of merit for the device.

The weak link which enables the Josephson effects to occur has been produced in several ways. Two of the more common types are the SNS (superconducting–normal–superconducting) (Fig. 6.2a), and the SIS (superconductor–insulator–superconductor) tunnel junction (Fig. 6.2b). Typical features of the I–V characteristics are indicated. The SNS and SIS types of structure have been realized in high-temperature superconducting (HTS) materials, but not, so far, in the tunnel barrier. The junctions commonly rely on grain boundaries which are formed during growth of the YBCO film by suitable modification of the substrate at predetermined positions. The grain boundary junction (GBJ) is considered to be of the SNS type, with intrinsic normal conducting paths.

The unique features of the Josephson effects are well illustrated by the application of a magnetic field to the junction. These phenomena have a strong bearing on the noise properties of junctions and will be referred to again in Section 6.3.4). If a magnetic field is applied to the SIS structure (perpendicular to paper) the value of I_c is modulated in the Fraunhofer diffraction pattern:

$$I_c(B) = I_c(0) \left| \frac{\sin[\pi(\Phi_B/\Phi_0)]}{\pi(\Phi_B/\Phi_0)} \right| \tag{6}$$

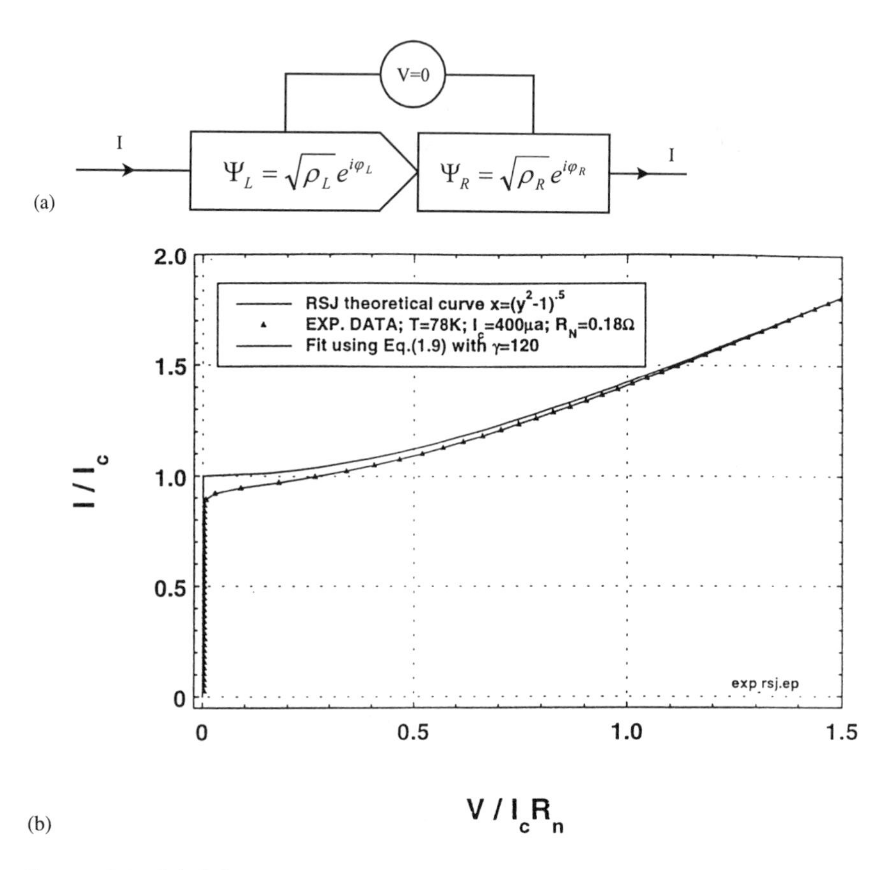

FIGURE 6.1 (a) Schematic diagram of a Josephson junction. (b) Theoretical noise-free resistively shunted junction (RSJ) current–voltage characteristic and experimental points fitted to noise-rounded characteristic [Eq. (9)] with $\gamma = 120$.

where $I_c(0) = J_0WL$ (W = junction width, L = junction length, t = film thickness), $\Phi_B = BLt$ (magnetic flux applied), and $\Phi_0 = 2 \times 10^{-15}$ Wb (magnetic flux quantum).

A uniform and small SIS junction, free of self-shielding effects (i.e., having a width $w < 4\lambda_J$, where λ_J is the Josephson penetration depth), exhibits a Fraunhofer-like pattern for magnetic field dependence of the dc critical current, $I_c(B)$. Other striking effects are observed when high-frequency electromagnetic fields are applied. The Josephson junction is a highly nonlinear device and external electromagnetic fields (frequency $= f_x$) can "mix" with the internal Josephson oscillation. Microwave irradiation induces a high-frequency current, I_{rf}, in the junction, producing interference with the Josephson oscillation. When

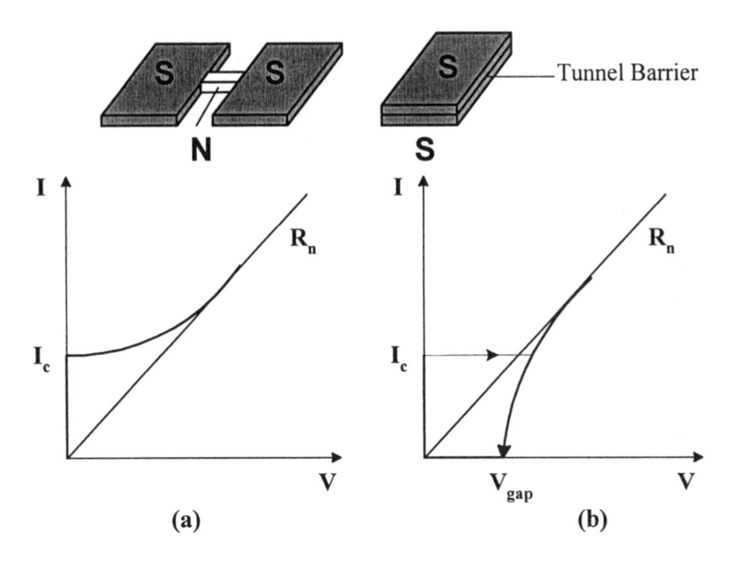

FIGURE 6.2 Schematic and I–V characteristics of (a) SNS and (b) SIS tunnel junctions.

the internal frequency f_J is equal to an integral multiple of f_x, zero-frequency components appear as constant–voltage steps (Shapiro steps) in the I–V curve (6):

$$V_n = n\frac{h}{2e}f_x \tag{7}$$

This effect is now used in many countries to define the standard of voltage (e.g. Refs. 7 and 8). A uniform and small SIS junction exhibits a predictable dependence of Shapiro step heights $\Delta I(n)$ on the intensity of the applied radio-frequency (rf) current I_{rf}, where n is the step number. Measurements of $I_c(B)$ and $\Delta I(n)$ versus I_{rf} for the first few steps (n = 0, 1, 2) are unambiguous tests for Josephson junction behavior and uniformity.

6.1.2.2 Thermal Noise

The effects of intrinsic thermal fluctuations on Josephson junction characteristics were studied by Ambegaokar and Halperin (9). They considered the equations for the resistively shunted junction (RSJ) (or a Josephson junction in parallel with a large external resistance and capacitor) and added a term for thermal fluctuation:

$$\begin{aligned} \frac{d\varphi}{dt} &= \frac{2eV}{\hbar} \\ C\frac{dV}{dt} &= I - I_c(T)\sin\varphi - \frac{V}{R} + \overline{L}(t) \end{aligned} \tag{8}$$

where φ is the difference in the phases of the order parameter on opposite sides of the junction, V is the potential difference, C is the capacitance of the junction, $I_c(T)$ is the maximum Josephson current at temperature T in the absence of noise, R is the resistance of the junction, and $L(t)$ is a fluctuation noise current.
They showed that the problem is entirely equivalent to the Brownian motion of a particle of mass M in the potential U, where $M = (\hbar/2e)^2C$, $U = -1/2\gamma T(I_0\varphi + \cos\varphi)$, $\gamma = \hbar I_c(T)/eT$, and $I_0 = I/I_c(T)$. This analysis allowed them to simulate the effect of thermal noise on the current and voltage characteristics.

Ambegaokar and Halperin (9) gave the following analytical expression for the noise-rounded RSJ I–V curve:

$$\nu = 2(1 - x^2)^{1/2} \exp\{-\gamma[(1 - x^2)^{1/2} + x \sin^{-1}x]\} \sinh\left(\frac{\pi\gamma x}{2}\right) \tag{9}$$

where $x \ll 1$, $\gamma > 1$, $\nu = V/I_cR_n$, and $x = I/I_c(T)$; $I_c(T)$ is the maximum value of critical current at temperature T and

$$\gamma = \frac{hI_c(T)}{2\pi ekT} \tag{10}$$

The noise-rounded curve is fitted to experimental data in Figure 6.1b, for comparison with the RSJ noise-free characteristic.

An expression similar to Eq. (9) has been used (10–12) to describe noise rounding of Shapiro steps. In this case, $I_c(T)$ is replaced by $I_s(T)$, the maximum step current amplitude. {Note that Likharev (12) and others used the symbol γ, which transforms to the Ambegaokar–Halperin form [Eq. (10)] by $\gamma \equiv (2\gamma)^{-1}$.} Kautz et al. (10) have discussed and experimentally demonstrated the range of validity of Eq. (9) when applied to Shapiro steps.

The linewidth Γ of the thermally broadened Josephson radiation is given by (13)

$$\Gamma = \frac{4\pi kTR_d^2}{R_n\Phi_0^2} \tag{11}$$

$$\Rightarrow \Gamma = 40.64 \text{ MHz}/\Omega\text{ K for } R_d \sim R_n$$

Here, T is the temperature, R_d is the dynamic resistance of the junction at the relevant operating point, R_n is the normal tunneling resistance at $I \gg I_c$, $\Phi_0 = 2.06 \times 10^{-15}$ Wb, and $k = 1.38 \times 10^{-23}$ J/K. The above analysis is valid for white thermal noise. Likharev (12) stated that the effect of low-frequency excess noise on the linewidth of the Josephson oscillation can be described by the quantity Γ_1, which is obtained by integrating the voltage noise spectral density from a low-frequency f_0 up to the characteristic frequency ($f_c = I_cR_n/\Phi_0$) of the junction. In the case of $1/f$ noise, this can be evaluated for the nth step as

$$\Gamma_1 = n\left(\frac{4\pi e}{h}\right)\left\{S_v(1\text{Hz})\, 2\ln\left(\frac{f_c}{f_0}\right)\right\}^{1/2} \tag{12}$$

where the lower limit of integration, arbitrarily taken here as 1 Hz, is set by the response time (bandwidth) of the measurement system.

6.1.2.3 Voltage Spectral Density

Likharev and Semenov (14) have calculated the voltage noise spectral density of a current-biased RSJ with zero capacitance when the current noise of the shunt is in the classical limit $\hbar\omega_J \ll k_BT$, where $\omega_J = 2eV/\hbar$ is the Josephson (angular) frequency at the average voltage V. The spectral density of the voltage noise is given by

$$S_v(0) = \frac{4k_BTR_d^2}{R}\left[1 + \frac{1}{2}\left(\frac{I_0}{I}\right)^2\right] \tag{13}$$

Here, R and R_d are the normal tunneling resistance and the dynamic resistance, respectively. The first term represents Nyquist noise in the dynamic resistance of the junction. The second term is due to thermal noise at high-frequency "mixed down" by the nonlinear junction to appear at the measurement frequency. It is assumed that the I–V characteristic is close to the ideal RSJ characteristic [Eq. (5)].

Voss (15) made numerical calculations of the noise in the thermally rounded regime with finite values of capacitance. In terms of a mechanical analog, he described the motion of a particle moving between a series of wells separated by potential barriers. Where the particle is overdamped ($\beta_c \ll 1$), thermal activation produces jumps between wells. Each jump produces a voltage pulse. These pulses produce a phase-slip shot noise $S_v(f) = 8\pi^2 r$, where r is the average rate of the jumps. This extended the work of Ambegaokar and Halperin (9) and calculated the noise voltage as function of the bias voltage. It has to be noted that Voss did not calculate noise curves for the condition $0.10 < \beta_c < 1$, which is the range of interest for GBJs, because, in this range, the Josephson oscillation is highly nonsinusoidal, making it difficult to analyze the solutions.

The existence of thermal noise has a useful application at low temperatures in Josephson noise thermometry which can be used to set up an absolute scale of temperature. In this application, Soulen and Giffard (16) carried out a very careful measurement of the thermal noise in the junction and compared the result with Eq. (13). Their principal aim was to develop an absolute noise thermometer based on the linewidth of Josephson oscillation. Their measurement technique is shown in Figure 6.3. A dc current I greater than the critical current of the Josephson junction J is applied to the junction and the resistor, R, in parallel, so that the junction is held at a steady voltage V. By the ac Josephson effect, a radio-frequency oscillation occurs in the junction with frequency $f = (2e/h)V$. A stable external radio-frequency (local oscillator) voltage V_{rf} is mixed with the Josephson frequency in the junction. Thermal noise in the resistor causes random frequency modulation of the signal. The signal frequency is measured by repetitive counting of the frequency of the output $V(t)$ over a fixed time interval τ, and the result of many measurements can be expressed as a mean frequency with variance

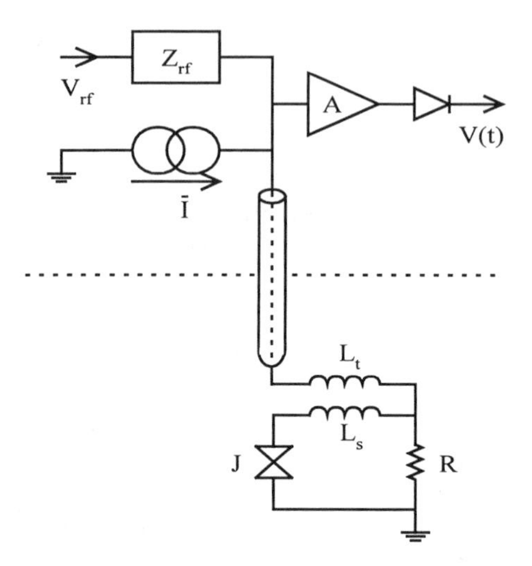

FIGURE 6.3 Diagram of circuit used for Josephson linewidth noise thermometry.

$$\sigma_f^2 = \left(\frac{1}{\tau}\right)\left(\frac{2e}{h}\right)^2 (2kTR) \tag{14}$$

from which the temperature can be derived. Thermal noise in the junction has to be considered and can be accurately calculated (16) using the Likharev–Semenov equation [Eq. (13)]. This work was done at liquid-helium temperatures. Applicability of the Likharev–Semenov equation to high-T_c Josephson junctions was demonstrated by Hao et al. (6).

6.1.2.4 Quantum Noise

Koch et al. (17) extended the calculation of thermal noise in a Josephson junction to the quantum limit $\hbar\omega_J >> k_B T$, where zero-point energy fluctuations in the shunt become significant. The equation of motion for the junction is

$$\frac{\hbar c}{2e}\varphi + \frac{\hbar}{2eR}\varphi + I_0 \sin\varphi = I + I_N \tag{15}$$

where φ is the phase difference across the junction and the noise current $I_N(t)$ has a spectral density

$$S_I(v) = \frac{2hv}{R}\coth\left(\frac{hv}{2k_BT}\right) = \frac{4hv}{R}\left(\frac{1}{\exp(hv/k_BT) - 1} + \frac{1}{2}\right) \tag{16}$$

in the limit $0 < \beta_c \equiv 2\pi I_0 R^2 C/\Phi_0 \ll 1$ ($\Phi_0 \equiv h/2e$).

Equation (16) reduces to Eq. (13) in the limit $\hbar\omega_J \ll k_B$T and in the extreme quantum limit $\hbar\omega_J \gg k_B T$. Equation (16) reduces to

$$\frac{S_v(0)}{R_d^2} = \frac{2eV}{R}\left(\frac{I_0}{I}\right)^2 \tag{17}$$

when the observed noise is generated solely by zero-point fluctuation in the shunt resistor. The same authors (18) reported the experimental results of the quantum noise in the RSJ, which were in good agreement with theoretical predictions.

6.1.2.5 1/*f* Noise

The origin of $1/f$ noise in Josephson junctions has been a topic of much recent research. Clarke and Hawkins (19) suggested that the noise results from temperature fluctuations in the shunted Josephson junctions. Koch (20) reported measurements of the spatial correlation length of the fluctuations causing the $1/f$ noise in Josephson junctions. The measured length is less than about 1 μm, much smaller than the prediction of the thermal fluctuation model. Koch proposed a model in which the observed noise results from a thermal activation process with a range of activation energies, similar to that introduced by Dutta and Horn for metal films (21). In the process of tunneling through the barrier, an electron becomes trapped on a defect in the barrier and is subsequently released. While the trap is occupied, there is a local change in the height of the tunnel barrier and, hence, in the critical current density of that region. As a result, the presence of a single trap causes the critical current of the junction to switch randomly back and forth between values, producing a random telegraph signal. If the mean time between pulses is τ, the spectral density of this process is a Lorentzian,

$$S(f) \propto \frac{\tau}{1 + (2\pi f \tau)^2} \tag{18}$$

namely white at low frequencies and falling off as $1/f^2$ at frequencies above $1/2\pi\tau$. In many cases, the trapping process is thermal activated and τ is of the form

$$\tau = \tau_0 \exp\left(\frac{E}{k_B T}\right) \tag{19}$$

where τ_0 is a constant and E is the barrier height.

In general, there may be several traps in the junction, each with its own characteristic time τ_i. One can superimpose the trapping processes, assuming them to be statistically independent, to obtain a spectral density where $D(E)$ is the distribution of activation energies. The term

$$S(f) \propto \int dE\, D(E)\left[\frac{\tau_0 \exp(E/k_B T)}{1 + (2\pi f \tau_0)^2 \exp(2E/k_B T)}\right] \tag{20}$$

in square brackets is a strongly peaked function of E, centered at $\widetilde{E} \equiv k_B T \ln(1/2\pi f \tau_0)$, with a width $\sim k_B T$. Thus, at a given temperature, only traps with energies within a range $k_B T$ of $\widetilde{E}$ contribute significantly to the noise. If one now assumes that $D(E)$ is broad with respect to $k_B T$, one can take $D(E)$ outside the integral and carry out the integral to obtain

$$S(f, T) \propto \frac{k_B T}{f} D(E) \tag{21}$$

In fact, one obtains a $1/f$-like spectrum from just a few traps.

The magnitude of the $1/f$ noise in a low-temperature superconducting (LTS) Josephson junction depends strongly on the quality of the junction as measured by the current leakage at a voltage below $(\Delta_1 + \Delta_2)/e$ where Δ_1 and Δ_2 are the energy gaps of two superconductors. Traps in the barrier enable electrons to tunnel in this voltage range, a process producing both leakage current and $1/f$ noise.

Koch et al. (22) measured the spectral density of $1/f$ voltage noise in a current-biased resistively shunted Josephson tunnel junction and dc SQUIDs. They found that the $1/f$ voltage noise in low-T_c SQUIDs was not due to critical current fluctuations, but apparently arose from an unidentified source of flux noise. Clarke (23) proposed that this source was due to the motion of flux lines trapped in the superconducting films. This mechanism manifests itself as a flux noise; for all practical purposes, the noise source behaves as if an external flux noise scales as V_Φ^2 and, in particular, vanishes at $\Phi = (n \pm 1/2)\,\Phi_0$ when $V_\Phi = 0$. By contrast, critical current noise is still present when $V_\Phi = 0$, although its magnitude does depend on the applied flux. It is now accepted that the level of $1/f$ flux noise depends strongly on the microstructure and quality of the superconducting thin films.

6.2 EXPERIMENTAL MEASUREMENTS

6.2.1 Noise Measurement System

The measurement of intrinsic noise requires careful design of the measurement system and attention must be paid to several key issues. An example of a system successfully used over several years is described next.

It is known that thermal noise power in an ideal resistor at 77 K is given by the Johnson–Nyquist equation [see Eq. (1)]. Thus, a superconducting tunnel junction having a typical normal resistance at 77 K of $\sim 1\ \Omega$ generates Nyquist noise of only 0.021 $nV/Hz^{1/2}$. The noise measurement system was required to measure this fundamental thermal noise limit. It was also required to measure unknown levels of noise due to flux motion, trapping of charge carriers, critical current fluctuations and resistance fluctuations, and so forth, and their frequency dependence.

The noise performance of commercially available "low-noise" amplifiers was not adequate for all of the above requirements. The system was therefore designed to measure noise by three complementary techniques:

1. Very high-sensitivity (<0.04 nV/Hz$^{1/2}$) in a restricted frequency band around 60 kHz.
2. Moderate sensitivity (0.8 nV/Hz$^{1/2}$) by spectrum analyzer or fast Fourier transform, in a broad band, 0.01 Hz to 10 kHz.
3. Measurements in real time of junction voltage to record random telegraph signals and other slowly varying noise.

The same measurement system was also required to provide conventional data such as junction current–voltage characteristics as a function of temperature and magnetic field.

6.2.2 Low-Noise Techniques

To obtain reliable data on noise, sources of external interference must be removed or minimized, and intrinsic system noise (e.g., in pre-amplifiers) must be measured.

6.2.2.1 Thermal emfs

For dc or low-frequency ac measurements, a major source of noise is thermal electromagnetic forces (emfs) due to temperature gradients or contacts between dissimilar metals. These can be minimized by using the same metal (usually copper) in continuous wires from the terminals of the sample device all the way to the preamplifier input. This technique is used for the dc I–V curves and for the spectrum analyzer measurements of wide-band ($1/f$) noise. The level of remaining thermal emfs was found to be <1 μV and was stable during measurements. High-frequency measurements ($\sim$60 kHz) are practically unaffected by thermal emfs, but they are usually limited by the levels of Johnson noise at the input to the preamplifier.

6.2.2.2 Impedance Matching

By improving the impedance match between the sample (<10 Ω) and the preamplifier EG&G 5004 ($\sim$10 kΩ), the effective system base noise can be reduced by one or two orders of magnitude. A toroidal transformer (1 : 5.5 turns ratio), to which the junction is connected for noise measurement, operates at the sample temperature (see Fig. 6.4). An additional toroidal transformer (1 : 5.5 ratio) at room temperature further improves the impedance match to the preamplifier. The low-temperature transformer was made by winding 0.2-mm-diameter copper wire on a toroidal core of sintered iron powder. The primary and secondary turns were

100 and 550, respectively. A capacitor selected for use at low-temperature was connected in series with the primary, forming a tuned circuit with resonant frequency about 60 kHz. This further enhanced the impedance match. A transformer of turns ratio n increases the impedance by n^2, so the overall impedance ratio from the primary of the low-temperature transformer to the secondary of the room-temperature transformer was nominally 1 : 1000. The calibration technique automatically takes into account the actual transformer ratio.

6.2.2.3 Elimination of rf and Other Interferences

The probe assembly was enclosed in the metal cryostat which provided electrostatic screening. All leads outside the cryostat had coaxial screening, and they entered the cryostat through three stages of low-pass *R–C* filters to keep out radio-frequency interference. Magnetic shielding was provided by a high-permeability mu-metal cylinder which was placed around the lower part of the cryostat tail. Interference at 50 Hz frequency due to ground-loop effects was reduced by connecting the cryostat and coaxial screens to ground at one point only. The dc potential leads were isolated from the ground. The noise voltage signal leads at the preamplifier input were isolated from the ground. Current leads had a common ground return. This arrangement minimized ground-loop interference.

6.2.3 System Layout

The principal features of the noise measurement system will be separately described in the following subsections.

6.2.3.1 Cryostat Probe

The superconducting devices were mounted together with a diode temperature sensor on a copper plate at the lower end of the probe. Coaxial current leads entered the probe body via three-stage low-pass filters to minimize rf interference. The dc voltage leads passed through similar filters and terminated in copper screw terminals to minimize thermal emfs. Noise measurement leads were connected to the samples via capacitor and transformer coupling designed for operation in the 40–80-kHz band.

6.2.3.2 High-Sensitivity Components at Fixed Frequency

The electrical layout of the high-sensitivity measurement system at fixed bandwidth is shown in Figure 6.4. The secondary of the low-temperature toroidal transformer is connected through the room-temperature input transformer to an EG&G 5004 preamplifier. The amplified noise signal is passed through a lock-in acting as a band-pass amplifier to the noise detector, which squares and integrates the noise waveform. A low-pass filter further reduces ac components. The dc output,

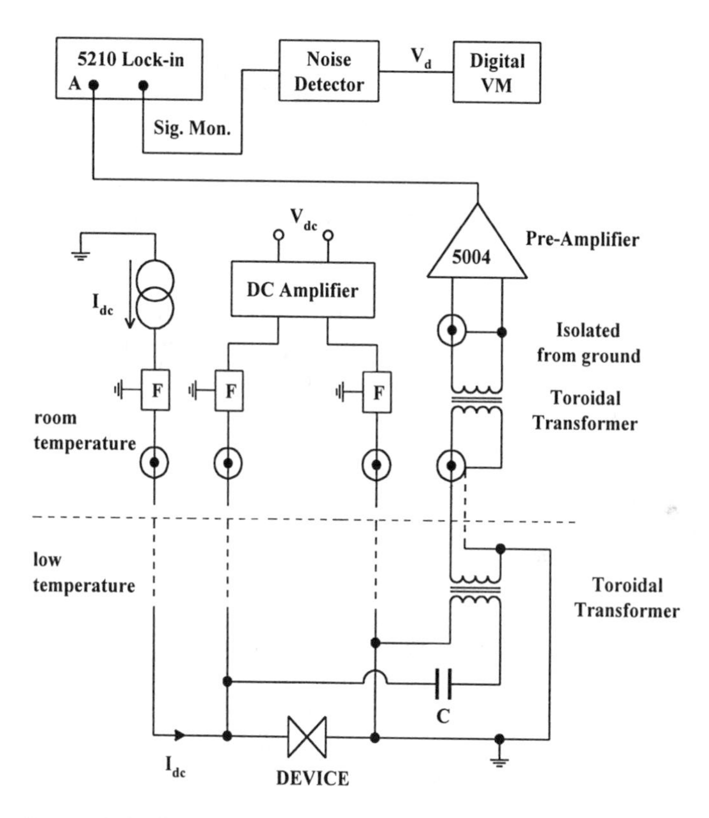

FIGURE 6.4 Circuit diagram of a noise measurement system.

which is proportional to the square of noise voltage, is finally measured by a digital voltmeter.

6.2.3.3 Broadband Components

The system also can be used with lesser sensitivity to measure the noise of samples in a broad band (frequency between 0.01 Hz to 10 kHz). For these measurements the output dc voltage leads from the sample were connected to a custom-built dc amplifier (Fig. 6.4), which has a dc sensitivity of about 200 nV with no significant shift of zero during short-term measurements. The HP 3561 A Dynamic spectrum signal analyzer following the dc amplifier was used to record the low-frequency components of the sample noise in the bandwidth below 1 kHz. The sensitivity is limited by the $\sim$1-nV/Hz$^{1/2}$ white voltage noise of the amplifier.

6.2.3.4 Cryostat Design and Temperature Measurement

So that the electrical noise characteristics of the superconducting devices can be studied over a wide range of temperature, a continuous-flow helium cryostat (Oxford Instruments Type CF1200) is used. Liquid helium is drawn from a storage dewar through a flexible transfer tube at a controllable rate. The helium flow rate is normally in the range 0.5–1.0 L (liquid) per hour. The cryostat itself provides a high degree of electrical shielding. Magnetic shielding is provided by an external high-permeability can surrounding the cryostat tail.

Precise temperature control is achieved by means of a Lakeshore 330 temperature controller. Diode temperature sensors A and B are located in the cryostat base and on the sample stage, respectively. The sample temperature is controlled to ± 0.01 K in the range 4–90 K, by means of feedback current to a heater located in the cryostat base.

6.2.4 Method of Operation

The above-described system measures noise in three independent but complementary ways. High-sensitivity spot measurements of the voltage noise are made at 60 kHz; broadband measurements of noise spectral density are made in the frequency band 0.01 Hz to 10 kHz; and continuous voltage or noise measurements are made in the time domain. Low-temperature transformers and tuned circuits give improved matching of the low-impedance junction device to a low-noise preamplifier. This technique enables a very low system noise to be achieved within a defined frequency band. The system is calibrated with reference to the Nyquist noise of wire-wound resistors, so that measured noise can be defined in terms of absolute noise power spectral density. The resolution of the system is of order 0.04 $\mathrm{nV/Hz^{1/2}}$ in the frequency band 40–80 kHz. In addition to the 60-kHz spot measurement, broadband measurements of noise power spectral density are recorded at frequencies between 0.01 Hz to 10 kHz by means of a HP 3561 Dynamic signal analyzer and custom-built preamplifier having a noise floor of 0.8 $\mathrm{nV/Hz^{1/2}}$. The noise voltage of a device is usually first recorded as a function of bias current. In this way, bias conditions giving rise to high levels of excess noise are quickly identified. The frequency dependence of the noise can then be recorded at chosen, fixed values of bias current.

Coils attached to the sample stage allow the noise to be recorded while varying magnetic fields are applied to the sample. The sample temperature is recorded and controlled with 0.01 K precision, so that repeated I–V and noise curves may be rapidly obtained at different temperatures.

Devices are measured by a four-terminal method with computer control of current supply, digital voltmeter, and data acquisition. A schematic diagram of the experimental circuit is shown in figure 6.4. Noise curves are first recorded with high-sensitivity ($\sim$0.04 $\mathrm{nV/Hz^{1/2}}$) by means of a tuned-transformer technique at a fixed frequency of 60 kHz. This technique enables quite subtle details of noise behavior to be monitored. The frequency spectrum of the noise (0.01 Hz to 10 kHz) at chosen bias-current settings is then determined by a fast Fourier transform method. Real-time recordings of the junction voltage are also made when large low-frequency noise intensities (random telegraph signals) are present. Details of the noise detection system have been given elsewhere (6).

6.2.5 Absolute Noise Calibration and Noise Models

A readily accessible reference standard is required against which noise data can be compared in a reproducible way. In the first place, the measurement system is calibrated using the Johnson noise generated by standard wire-wound resistors at known temperatures. Then, each junction device itself provides a calculable source of thermal noise based on Eq. (13).

The junction parameters R, R_d, and I_0 as measured on the junction are inserted in Eq. (13) to calculate a curve of $(S_v)_{\text{L-S}}$ versus I at the relevant temperature, with which the actual measured noise data can be directly compared. If the measured data exceed the calculated curve, then we define the difference as "excess noise." Usually the excess noise will have a frequency dependence $\sim 1/f$, whereas Eq. (13) is independent of frequency.

6.3 EXPERIMENTAL RESULTS AND DISCUSSION

6.3.1 Preparation and Properties of the Junctions

The preparation of typical bicrystal YBCO junctions is briefly outlined. Y–Ba–Cu–O films are deposited by pulsed laser ablation on MgO bicrystal substrates (24). Junctions are formed in microbridges 5 or 10 μm wide across the grain boundary by conventional photolithography and ion beam etching. The samples are installed in a magnetically shielded and electrostatically screened continuous-flow helium cryostat. Current and voltage leads into the cryostat are filtered and it has been verified that external apparatus, such as computer monitors and certain types of programmable current source, produces negligible levels of rf interference at the junction. Sample temperature is stabilized to 0.01 K and junction characteristics are recorded under computer control, as more fully described elsewhere (6,25).

6.3.2 Excess Noise in the Josephson Junction

Excess noise is conventionally expressed in terms of critical current fluctuations

$$\left|\frac{\delta I_c}{I_c}\right| = \frac{S_v^{1/2}}{I_c R_d} = (\check{S}_i)^{1/2} \quad (\text{at } I \sim I_c)\ (\text{Hz}^{-1/2}) \tag{22}$$

and resistance fluctuations

$$\left|\frac{\delta R_n}{R_n}\right| = \frac{S_v^{1/2}}{I_b R_n} = (\check{S}_R)^{1/2} \quad (\text{at } I >> I_c) \quad (\text{Hz}^{-1/2}) \tag{23}$$

Miklich et al. (26) showed for biepitaxial junctions with RSJ characteristics that the measured excess voltage noise power spectral density $S_v(f)$ could be expressed as

$$S_v(f) = \check{S}_i(f)(V - IR_d)^2 + \check{S}_R(f)V^2 + k(V - IR_d)V\,\check{S}_{iR}(f) \tag{24}$$

where

$$\check{S}_i(f) = \left|\frac{\delta I_c}{I_c}\right|^2$$

$$\check{S}_R(f) = \left|\frac{\delta R_n}{R_n}\right|^2$$

$$\check{S}_{iR}(f) = \left|\frac{\delta I_c}{I_c}\right|\left|\frac{\delta R_n}{R_n}\right|$$

The above normalized quantities by definition have units Hz^{-1} and

$$\text{k} = 2\,\langle\cos[\Delta\phi(\text{t})]\rangle$$

where k is a cross-correlation coefficient. It equals two times the time average of the cosine of the phase difference of the I_c and R_n fluctuations. $\check{S}_{iR}(f)$ is the cross-spectral density of the fluctuations.

$\check{S}_i(f)$ and $\check{S}_R(f)$ are first estimated from the experimental data at $I \sim I_c$ and $I >> I_c$, respectively, and then adjusted slightly to obtain the best fit between the calculated curve and the experimental points.

With $SrTio_3$ bicrystal junctions at $f \sim 100$ Hz, Kawasaki et al. (27) found that $S_v(I \sim I_0)$ is proportional to $I_0^{2.6}$. The same authors found that, in many cases, the ratio $p = |\,\delta I_c/I_c\,|/|\,\delta R_n/R_n\,|$ was close to 2.5, in agreement with the "intrinsically shunted junction" model described by Gross and Mayer (28).

Typical results for a step-edge junction showing nearly ideal RSJ characteristics are illustrated in Figures 6.5 and 6.6. The V–I curve is differentiated to yield the calculated noise curve (a) from Eq. (13). The measured noise curve (e) at higher bias currents has a linear current-dependent term which can be ascribed to resistance fluctuations of amplitude $\delta R/R = 1.1 \times 10^{-7}$. Comparing the peak values of curves (b) and (e), the excess noise can be evaluated in terms of critical current fluctuations as $\delta I_c/I_c = (S_v^{1/2} I_c R_d) = 3.3 \times 10^{-7}$ (R_d is the dynamic resistance of the device at the operating current). The noise power is strongly affected by magnetic fields, as shown in Figure 6.7. Following exposure to magnetic fields, the curves often develop an asymmetry due to flux trapping. The broadband measurements of $1/f$ noise are found to be consistent with the high-sensitivity spot measurements at 60 kHz, as illustrated in Figure 6.8.

6.3.3 Possible Sources of Low-Frequency Noise

Gross and colleagues (28,29) proposed that the low-frequency noise is caused by fluctuations in the tunneling currents due to the random trapping and subsequent release of charge carriers at localized sites in the grain boundary. The spectral density of the noise, which is attributed to critical current fluctuations δI_c and resistance fluctuations δR, often has a $1/f$ frequency dependence. Comparisons between different junctions (critical current I_c, resistance R) and for varying experimental conditions are facilitated by introducing the normalized spectral densities, defined as $S_I = [\delta I_c/I_c]^2$ and $S_R = [\delta R/R]^2$, which are found to be independent of temperature (30). Typically, the excess noise at frequency f scales with junction resistance R as $S_I \sim a^2/f$, where $a^2 \sim 10^{-8}R/\Omega$, and $S_R \sim 0.25S_I$, for temperatures between 20 and 77 K. Other works (e.g., Refs. 29 and 31) have indicated that where the junctions are "large" (i.e., wider than four times the Josephson penetration depth), excess noise may be associated with the movement of Josephson vortices across the junction. Although junctions frequently fall into this size

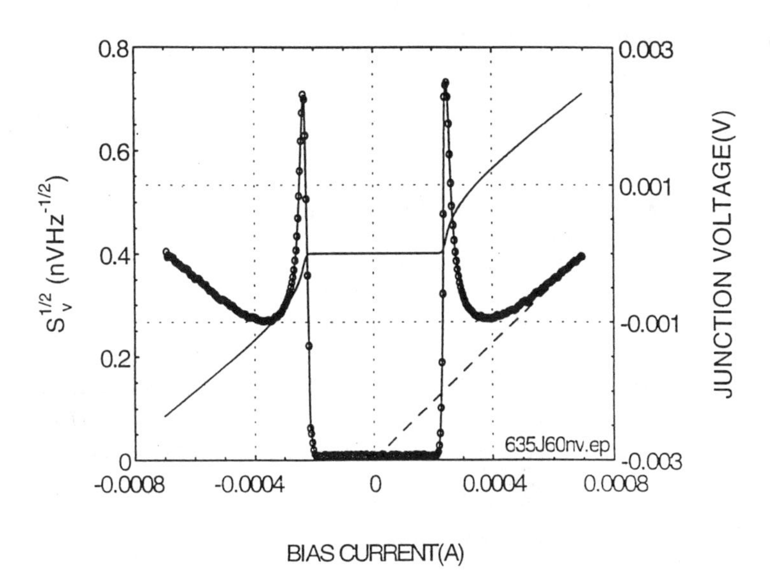

FIGURE 6.5 Bias current dependence of junction voltage (solid line, right axis) and voltage noise (experimental points, nV/Hz$^{0.5}$) (left axis) for a step-edge junction at $T = 39.58$ K.

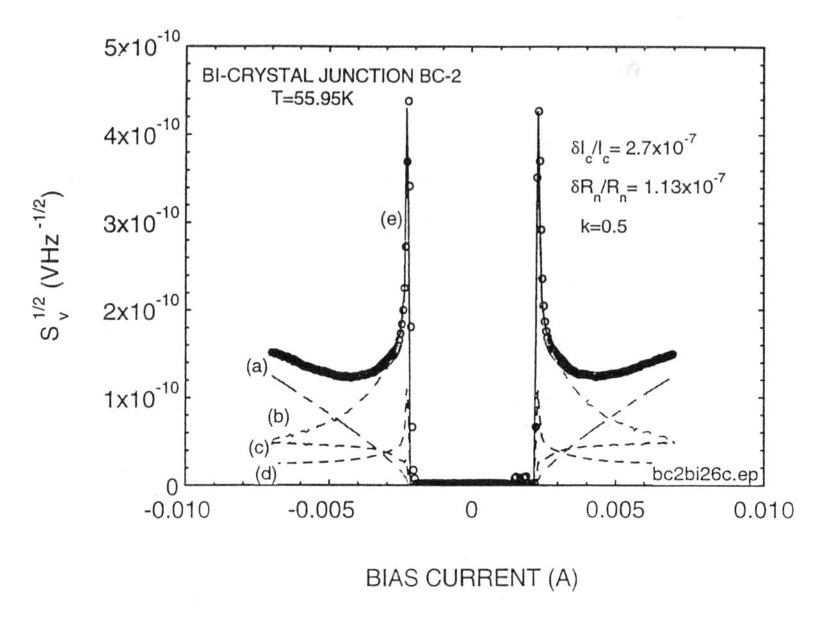

FIGURE 6.6 Calculated (lines) and experimental measurements (points) of voltage noise for a bicrystal junction ($T = 55.95$ K).

regime at sufficiently low temperatures, their general noise behavior is consistent with many of the expected properties of the trapping model.

6.3.4 Effect of a Magnetic Field on Junction Noise

The effect of an applied magnetic field on the noise in grain-boundary junctions has been studied in a systematic way using the above-described techniques (32). [Other works (29,30,33) using a SQUID as a noise detector, which is itself sensitive to external fields, excluded the application of magnetic fields to the junction under investigation.] It was found (6,25,32) that the absolute intensity of critical current fluctuations is nearly independent of external magnetic field (Fig. 6.9), an observation which has important consequences for the noise behavior of junctions and SQUIDs in practical situations.

6.3.5 Noise Temperature of HTS Junctions

If the normal tunneling resistance of the junction is R_n, the noise temperature T_n is defined as that temperature at which an ideal resistor $R = R_n$ would produce the same noise spectral density as the junction. The noise temperature of HTS junctions has been studied at high-frequency by examination of the Josephson radiation linewidth (34–38), by direct measurement of the power spectral density (39), and by simulation of the rounding of the Shapiro steps under microwave irradiation (10,40–42). In some cases, it was concluded that excess noise (i.e., noise in excess of the expected Johnson–Nyquist level) was present. In other cases, the noise temperature of the device was found to be close to its physical temperature; that is, there was no significant level of excess noise. The former result was obtained for step-edge grain-boundary junctions and bicrystal junctions, whereas the latter situation was found for proximity-effect SNS junctions and bicrystal junctions. It has been shown (38,42) that the high-frequency properties of a bicrystal junction are likely to be affected by the presence of low-frequency ($1/f$) fluctuations of the junction critical current and/or resistance. In an experimental demonstration of Josephson heterodyne oscillation in a two-junction resistive HTS SQUID (38), linewidth broadening was observed, which would be consistent with noise temperatures approximately two to three times higher than the physical temperature, as suggested by Koelle et al. (43) to explain excess noise in HTS dc SQUIDs (see also in Sec 6.4).

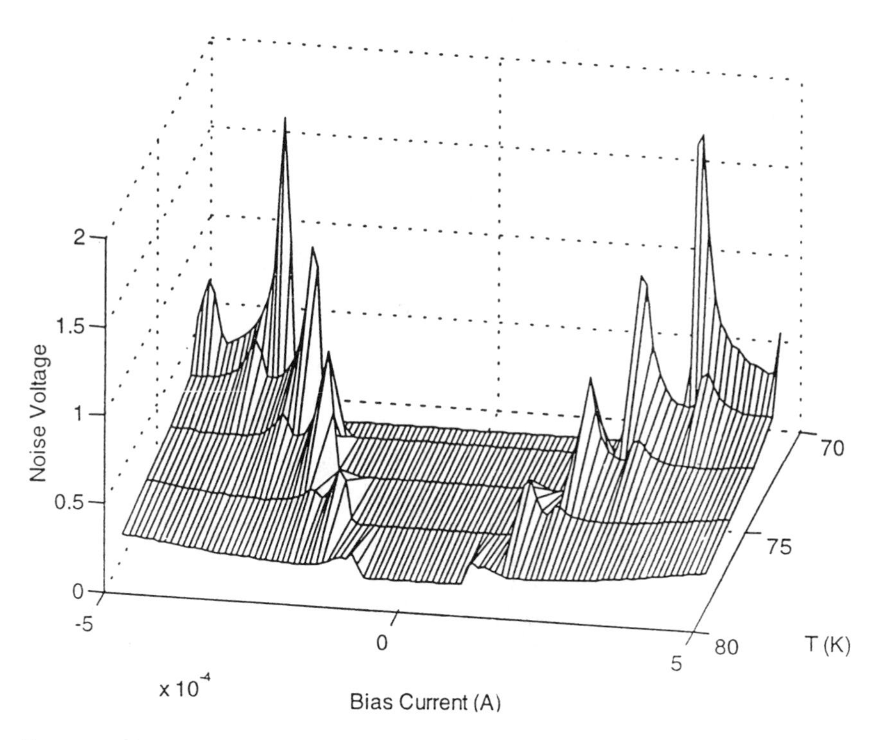

FIGURE 6.7 Dependence of voltage noise on bias current and temperature for a step-edge junction.

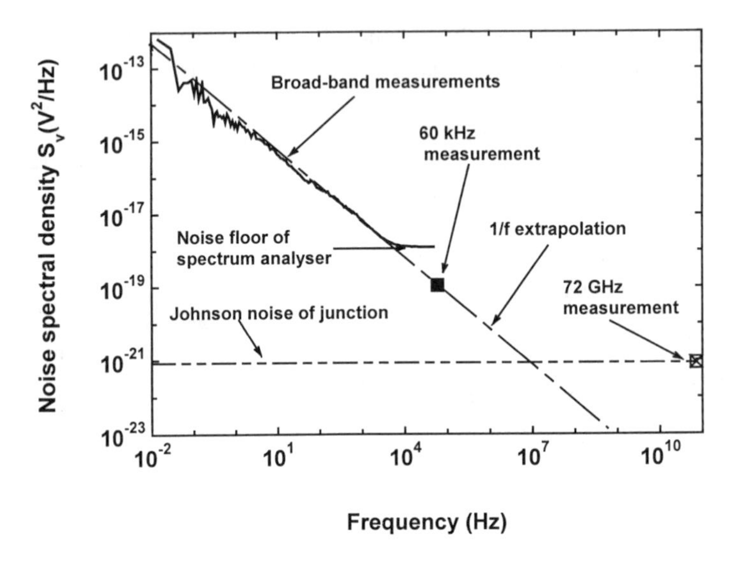

FIGURE 6.8 Combined noise data for a step-edge junction over 12 decades of frequency. The broadband low-frequency measurements extrapolate consistently to the 60-kHz high-resolution point, and the 72-GHz result (29,30) lies on the Johnson noise floor of the junction.

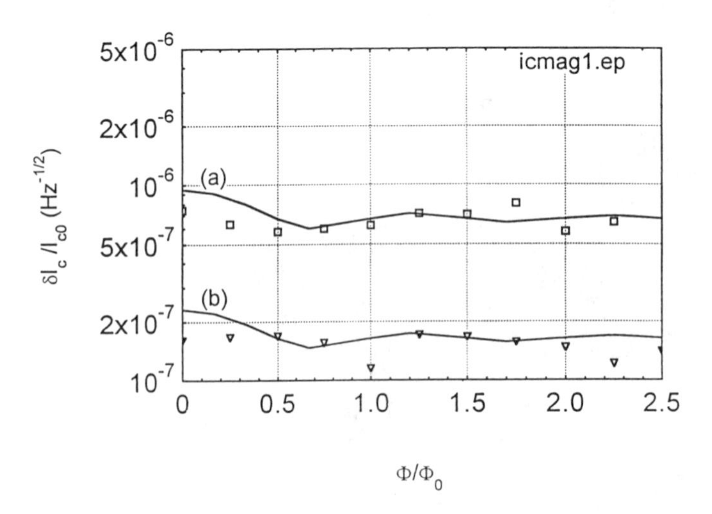

FIGURE 6.9 Magnetic field dependence of normalised critical current fluctuations: (a) MgO bicrystal junction; (b) MgO step-edge junction. Curves are calculated values; points are experimental data.

6.3.6 Comparative Data for Various Junction Types

Because the measurement technique described in Section 6.2 allows repeatable measurements to be rapidly carried out, it is possible to compile a set of representative data for a range of junction types (25). These results are presented in Table 6.1.

6.4 NOISE IN dc SQUIDs

In this final section, a concise overview of noise in HTS dc SQUIDs will be given. An excellent and recent review article by Koelle et al. (43) provides a comprehensive state-of-the-art treatment of this topic.

6.4.1 Broadband "White" Noise

The dc SQUID consists essentially of two Josephson junctions connected in parallel in a superconducting loop of inductance L. The junctions are assumed to be identical, having critical currents I_c, shunt capacitance C, and shunt resistance R. As a first step toward estimating the theoretical noise performance of an "ideal" dc SQUID, it was shown by Tesche and Clarke (44) that for $\beta_L = 1$ (where $\beta_L = 2LI_c/\Phi_0$) the (white) flux noise spectral density S_Φ is given by

$$S\phi = \frac{16k_BTL^2}{I_cR} \tag{25}$$

For a typical LTS SQUID, putting $T = 4$ K, $L = 0.01$ nH, $I_c = 10$ μA, and $R = 1$ Ω, this corresponds to

$$S_\Phi \sim 0.2\mu\Phi_0 \text{ Hz}^{-1/2} \tag{26}$$

TABLE 6.1 Summary of Measured Noise Data at Indicated Frequencies for All Junction Types Averaged over Temperatures in the Range 40–80 K

Junction type	1 Hz $\delta I_c/I_c$ ($\times 10^{-5}$ Hz$^{-1/2}$)	100 Hz $\delta I_c/I_c$ ($\times 10^{-6}$ Hz$^{-1/2}$)	60 kHz $\delta I_c/I_c$ ($\times 10^{-7}$ Hz$^{-1/2}$)	60 kHz $\delta R_n/R_n$ ($\times 10^{-7}$ Hz$^{-1/2}$)	60 kHz $p = \lvert\delta I_c/I_c\rvert / \lvert\delta R_n/R_n\rvert$
MgO step edge	9.3 ± 4.9	7.83 ± 2.3	2.6 ± 0.9	1.7 ± 0.1	1.65 ± 0.6
MgO bicrystal (θ = 24°)	3.05 ± 0.74	2.91 ± 0.96	2.9 ± 0.5	1.22 ± 0.07	2.43 ± 0.18
STO bicrystal (θ = 24°)	23 ± 2.5	23.3 ± 2.5	9.5 ± 1.0	3.8 ± 0.1	2.48 ± 0.93
YSZ bicrystal (θ = 32°)	28 ± 9	28 ± 9	11.6 ± 5.3	4.38 ± 0.7	2.55 ± 0.91
Biepitaxial (θ = 45°)	42 ± 5	42 ± 5	17 ± 2	N/A	N/A

On a straightforward scaling by temperature, an HTS SQUID of similar dimensions and critical current to the LTS device mentioned earlier, when operating at 77 K, would be expected to have a white-noise floor $\sim 4.2\mu\Phi_0\,\mathrm{Hz}^{-1/2}$. It is widely found in practice, however, that HTS SQUID noise levels are at least a factor of 2 higher than the theoretical arguments would predict. In other words, the "noise temperature" is twice the thermodynamic temperature (38,43). Other factors such as flux movements and temperature fluctuations (see below) can contribute to noise levels an order of magnitude greater than the ideal case.

There is a correlation between the flux-to-voltage transfer function $\delta V/\delta\Phi$ and the SQUID's flux noise power density S_Φ. A full theoretical treatment is provided, for example, in Ref. 43, but the general behavior can be represented by the graphs (Fig. 6.10). The inductance of the device, which is determined largely by the area and/or geometry of the SQUID loop, enters into the theory via the quantity β_L, as defined earlier. In practice, it is often more important to optimize the field noise S_B rather than the flux noise S_Φ. It then becomes necessary to consider the means whereby an external field B is coupled into the SQUID's inductance. Again, a full discussion of the evolving technology and theory is given in Ref. 43. As an illustration of typical SQUID performance, a 20-pH SQUID with $I_c = \sim 50$ μA and $R \sim 3\ \Omega$ coupled to a 50-mm^2 pickup loop would be expected to achieve a white-field-noise floor of the order 50 fT/Hz$^{1/2}$.

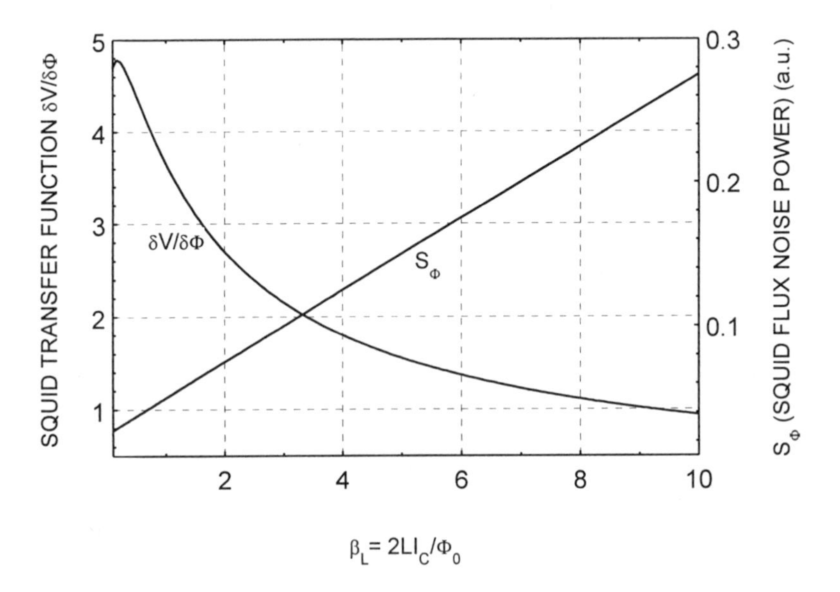

FIGURE 6.10 Dependence of the SQUID transfer function and flux noise on β_L. (Adapted from Ref. 40.)

6.4.2 Low-Frequency (1/*f*) Noise

Excess low-frequency noise in the junctions, as discussed in a previous section, will inevitably contribute to the overall noise limitations of the SQUID. Fortunately, as illustrated in Figure 6.11, techniques based on periodic switching of the bias current can minimize these effects. Nevertheless, because the two junctions are seldom identical, the bias-reversal method is unlikely to be 100% effective and junction noise remains as an undesirable feature of most HTS dc SQUIDs in use at the present time. Therefore, we make some straightforward estimates of the limits to the SQUID sensitivity based on our previous discussion of junction noise. As a basis for comparison with other work, the noise performance (without bias reversal) that would be expected of a dc SQUID incorporating bicrystal junctions on MgO is estimated. The critical current is assumed typically to be 100 μA and the low-frequency SQUID noise is assumed predominantly due to junction noise, for the purpose of this discussion. At $T = 77$ K, using the data in Table 6.1, the normalized critical current fluctuations are of the order of magnitude 3×10^{-5}, corresponding to 3 nA/Hz$^{1/2}$ at 1 Hz. With a typical SQUID loop inductance of 60 pH, this would be equivalent to a 1/*f* flux noise at 1 Hz of $9.0 \times 10^{-5} \Phi_0 \text{ Hz}^{-1/2}$, a level which is commonly reported. Similar values found in "state-of-the-art" single-layer YBCO SQUIDs have been effectively reduced by nearly two orders of

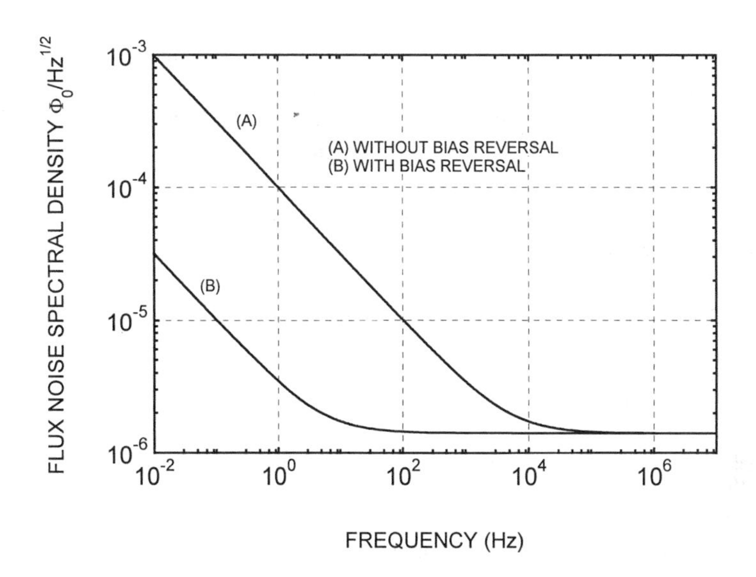

FIGURE 6.11 Example of HTS dc SQUID flux noise versus frequency (A) without and (B) with bias reversal.

magnitude with the use of bias-reversal techniques (40). (Although a treatment of rf SQUIDs is outside the scope of this section, it is nevertheless interesting to note that "intrinsic" bias reversal, inherent in the operation of the rf SQUID, has been claimed to reduce the effects of low-frequency junction noise under certain conditions.) A further observation on SQUID performance concerns the level of white noise and the "$1/f$ knee" (25). It has been shown conclusively that excess intrinsic noise for all junction types tested continues to decrease as $1/f$ all the way to 60 kHz and above (see, e.g., Fig. 6.8). By contrast, the noise in a dc SQUID under optimized bias-reversal conditions usually falls as $1/f$ from dc to around 10–20 Hz and is then "white" at higher frequencies. It is also reported that bias reversal is ineffective in removing this higher-frequency noise floor (45). Therefore, it is probable that in contrast to the low-frequency noise, the white-noise level in high-T_c SQUIDs is caused not by critical current fluctuations in the junctions but, rather, by magnetic flux noise due to movements of weakly trapped vortices in the SQUID body. This source of noise is unrelated to the junction type, because it is a property of the YBCO films rather than the junctions themselves. It is also dramatically worse after the SQUID has been exposed to background magnetic fields of a few milliTeslas. Improvements have nevertheless been achieved by incorporating various ingeniously shaped slots or "flux dams" in the SQUID washer and by reducing the width of strip-line leads. These measures are designed either to confine the trapped flux to areas remote from the sensitive area of the SQUID or to hinder the trapping of vortices in the first place. The use of very narrow linewidths in the construction of the dc SQUID has been shown to improve its noise performance, where operation in unshielded environments is required (46). In this case, residual low-frequency noise was ascribed to the effects of temperature fluctuations which caused the effective inductance of the pickup loop to fluctuate also. The best noise achieved for a SQUID which had been cooled in a static field of 64 μT at 1 Hz was 65 $fT/Hz^{1/2}$. This problem is a topic of intense research in many laboratories. It will be solved only when the mechanisms of flux penetration, trapping, and creep in thin-film structures have been more fully analyzed, and control over the HTS thin-film materials both during fabrication and operation in harsh environments has been greatly improved.

6.5 CONCLUSIONS

The universal occurrence of noise in all electronic devices, including superconducting junctions, has been reviewed. Noise measurement techniques have been illustrated with particular reference to a system that was developed in the authors' laboratory. Measurements of excess noise in HTS Josephson junctions have been described and selected results have been presented. Significant conclusions include the following:

- The theory of Likharev and Semenov (14) successfully accounts for the observed levels of thermal noise in high-T_c junctions and provides a useful reference against which excess noise can be identified (6,25).
- For noise measurements at 60 kHz, where thermal noise is not negligible, the method originated by Miklich et al. (26) for modeling the excess noise in terms of critical current and resistance fluctuations must be modified to include the Likharev–Semenov equation.
- The levels of normalized critical current and resistance fluctuations for all types of junctions tested are nearly independent of temperature in the range $T = 30$–80 K, with typical values $|\delta I_c/I_c| \sim 1 \times 10^{-5}\ \mathrm{Hz}^{-1/2}$ and $|\delta R_n/R_n| \sim 4 \times 10^{-6}\ \mathrm{Hz}^{-1/2}$ at 100 Hz.
- In most cases, the frequency dependence of the excess noise power spectral density follows the universal $1/f$ law, with some exceptions which can be understood as Lorentzian terms due to two-level or multilevel trapping processes.
- The *absolute* intensity of critical current fluctuations $|\delta I_c|$ is nearly independent of the external magnetic field (32). It follows that when the dc critical current of a junction (or a SQUID) is partly suppressed by a magnetic field in the range 0.1–1 mT, the *relative* level of current noise increases.
- The intrinsically shunted junction model (29,31) provides a clear physical picture which describes very well the observed tunneling and noise behavior of the junctions.
- The overall general conclusion about the various grain-boundary-junction types studied is that surprisingly similar levels of intrinsic fluctuations are present in all of them.
- Remarkable progress has clearly been made in recent years in the noise performance of high-T_c dc SQUIDs by technological strategies such as bias reversal, flux dams, and reduced linewidths. Effective noise temperatures of HTS Josephson junction devices nevertheless still display excess noise, such that, in many cases, their noise temperature is some two to three times higher than the physical temperature. It now appears that significant further reductions in noise levels will require a fundamental breakthrough in the understanding of the penetration and movement of magnetic flux into and through thin films of the HTS superconductors and of properties of the grain boundary on a submicrometer scale.

ACKNOWLEDGMENT

We acknowledge the expert advice and assistance of Dr. Jan Kuznik (Czech Academy of Sciences) in the design and operation of noise measurement apparatus.

REFERENCES

1. JB Johnson. Thermal agitation of electricity in conductor. Phys Rev 32:97–109, 1928.
2. H Nyquist. Thermal agitation of electric charge in conductors. Phys Rev 32:110–113, 1928.
3. W Schottky. Über Spontance Stromschwankungen in Verschiedenen elektrizitatsleitern. Ann Phys (Leipzig) 57:541–567, 1918.
4. FN Hooge. 1/*f* Noise is no surface effect. Phys Lett A 29:139–140, 1969.
5. BD Josephson. Possible new effects in superconductive tunnelling. Phys Lett 7:251–253, 1962.
6. L Hao, JC Macfarlane, CM Pegrum. Excess noise in YBCO thin film grain boundary Josephson junctions and devices. Supercond Sci Technol 9:678–687, 1996.
7. IK Harvey, JC Macfarlane, RB Frenkel. Monitoring the NSL standard of emf using the ac Josephson effect. Metrologia 8:114–124, 1972.
8. JC Gallop, BW Petley. Operational experience with a cryogenic volt-monitoring system. IEEE Trans Instrum Meas IM-23:267–270, 1974.
9. V Ambegaokar, BI Halperin. Voltage due to thermal noise in the dc Josephson effect. Phys Rev Lett 22:1364–1366, 1969.
10. RL Kautz, RH Ono, CD Reintsema. Effect of thermal noise on Shapiro steps in high-T_c Josephson weak links. Appl Phys Lett 61:342, 1992.
11. MJ Stephen. Noise in a driven Josephson oscillator. Phys Rev 186:393–397, 1969.
12. KK Likharev. Dynamics of Josephson Junctions and Circuits. New York: Gordon & Breach, 1986.
13. AJ Dahm, A Denenstein, DN Langenberg, WH Parker, D Rogovin, DJ Scalapino. Linewidth of the radiation emitted by a Josephson junction. Phys Rev Lett 22:1416–1420, 1969.
14. KK Likharev, VK Semenov. Fluctuation spectrum in superconducting point junctions. JETP Lett 15:442–445, 1972.
15. RF Voss. Noise characteristics of an ideal shunted Josephson junction. J Low Temp Phys 42:151–163, 1981.
16. RJ Soulen, RP Giffard. Josephson-effect absolute noise thermometer: resolution of unmodeled errors. Appl Phys Lett 32:531–538, 1978.
17. RH Koch, DJ Van Harlingen, J Clarke. Quantum-noise Theory for the resistively shunted Josephson junction. Phys Rev Lett 45:2132–2135, 1980.
18. RH Koch, DJ Van Harlingen, J Clarke. Measurement of quantum noise in resistively shunted Josephson junctions. Phys Rev B 26:74–87, 1982.
19. J Clarke, G Hawkins. Flicker (1/*f*) noise in Josephson tunnel junctions. Phys Rev B 14:2826–2831, 1976.
20. R Koch. 1/*f* Noise in Josephson junctions measurements and proposed model. In: M Savelli, G Lecoy, JP Nougier, eds. Noise in Physical Systems and 1/*f* Noise. Oxford: Elsevier Science, 1983, pp 377–380.
21. P Dutta, PM Horn. Low-frequency fluctuations in solids: 1/*f* noise. Rev Mod Phys 53:497–516, 1981.
22. RH Koch, J Clarke, WM Goubau, JM Martinis, CM Pegrum, DJ Van Harlingen. Flicker (1/*f*) noise in tunnel junction dc SQUIDs. J Low Temp Phys 51:207–224, 1983.

23. J Clarke. SQUIDs: Theory and practice. In: H Weinstock, RW Ralston, eds. The New Superconducting Electronics. Berlin: Kluwer Academic, 1993, pp 123–180.
24. JH Clark, GB Donaldson, RM Bowman. A fully automated pulsed laser deposition system for HTS multiplayer devices. IEEE Trans Appl Supercond AS-5:1661–1664, 1995.
25. L Hao, Excess noise in YBCO thin film grain boundary junctions and devices. PhD thesis, University of Strathclyde, Glasgow, 1995.
26. AH Miklich, J Clarke, MS Colclough, K Char. Flicker ($1/f$) noise in bi-epitaxial grain boundary junctions of YBCO. Appl Phys Lett 60:1899–1901, 1992.
27. M Kawasaki, P Chaudhari, T Newman, A Gupta. $1/f$ Noise in YBCO superconducting bicrystal grain-boundary functions. Phys Rev Lett 68:1065–1068, 1992.
28. R Gross, B Mayer. Transport processes and noise in $YBa_2Cu_3O_{7-\delta}$ grain boundary junctions. Physica C 180:235–242, 1991.
29. A Marx, U Fath, W Ludwig, R Gross, T Amrein. $1/f$ Noise in $Bi_2Sr_2CaCu_2O_x$ bicrystal grain-boundary Josephson junctions. Phys Rev B 51:6735–6738, 1995.
30. A Marx, R Gross. Scaling behavior of $1/f$ noise in high-temperature superconductor Josephson junctions. Appl Phys Lett 70:120–122, 1997.
31. R Gross. Grain boundary Josephson junctions in the high-temperature superconductors. In: SL Shinde, ed, Interfaces in High-T_c Superconducting Systems. New York: Springer-Verlag, 1994, pp 176–209.
32. L Hao, JC Macfarlane, CM Pegrum, GJ Sloggett, CP Foley. Magnetic field and microwave effects on critical current fluctuations in HTS grain-boundary Josephson junctions. IEEE Trans Appl Supercond AS-7:2840–2844, 1997.
33. R Gross, L Alff, A Beck, OM Froehlich, R Gerber, R Gerdemann, A Marx, B Mayer, D Koelle. On the nature of high-T_c Josephson junctions: Clues from noise and spatially resolved analysis. In: DHA Blank, ed. Proceedings of HTS-Workshop on Application and New Materials, Twente, The Netherlands, 1995, pp 1–8.
34. YY Divin, J Mygind, NF Pedersen. Josephson oscillations and noise temperatures in $YBa_2Cu_3O_{7-x}$ grain-boundary junctions. Appl Phys Lett 61:3053–3055, 1992.
35. YY Divin, AV Andreev, GM Fischer, J Mygind, NF Pedersen. Millimeter-wave response and linewidth of Josephson oscillations in $YBa_2Cu_3O_7$ step-edge junctions. Appl Phys Lett 62:1295–1297, 1993.
36. M Tarasov, A Shul'man, O Polyansky, E Kosarev, Z Ivanov, G Fischer, E Kollberg, T Claeson: Applied Superconductivity Conference, Pittsburg, Abstracts. New York: IEEE, 1996.
37. GM Fischer, J Mygind, NF Pedersen. IEEE Trans Appl Supercond AS-7:3654–3657, 1997.
38. JC Macfarlane, L Hao, DA Peden, JC Gallop. Linewidth of a resistively shunted high-temperature-superconductor Josephson heterodyne oscillator. Appl Phys Lett 76:1752–1755, 2000.
39. EN Grossman, LR Vale, DA Rudman. Microwave noise in high-T_c Josephson junctions. Appl Phys Lett 66:1680–1682, 1995.
40. G Kunkel, M Bode, F Wang, M Siegel, W Zander, J Schubert, AI Braginski. Analysis of Josephson radiation from $YBa_2Cu_3O_7$ thin-film Josephson junctions. Supercond Sci Technol 7:313–316, 1994.

41. R Gupta, Qing Hu, D Terpstra, GJ Gerritsma, RH Rogalla. Noise study of a high-T_c Josephson junction under near-millimeter-wave irradiation. Appl Phys Lett 64:927–929, 1994.
42. L Hao, JC Macfarlane. Estimation of the noise temperature of Y–Ba–Cu–O grain boundary Josephson junctions. Physica C 292:315–321, 1997.
43. D Koelle, R Kleiner, F Ludwig, E Dantsker, J Clarke. High transition temperature superconducting quantum interference devices. Rev Mod Phys 71:631–685, 1999.
44. C Tesche, J Clarke. Optimisation of DC SQUID voltmeter and magnetometer circuits J Low Temp Phys 37:405–420, 1979.
45. D Grundler, R Eckart, B David, O Dossel. Origin of $1/f$ noise in $Y_1Ba_2Cu_3O_{7-x}$ step-edge dc SQUIDs. Appl Phys Lett 62:2134–2136, 1993.
46. F Ludwig, D Drung. Low-frequency noise of improved direct-coupled high-T_c superconducting quantum interference device magnetometers in ac and dc magnetic fields. Appl Phys Lett 75:2821–2823, 1999.

7

High-Temperature RF SQUIDS

V. I. Shnyrkov
Institute for Low Temperature Physics and Engineering, Academy of Sciences, Kharkov, Ukraine

7.1 INTRODUCTION

Superconducting quantum interference devices (SQUIDs) are extraordinarily sensitive detectors of magnetic flux variations. These devices have numerous applications as sensors in a wide range of experiments in the fields of physics, geology, medicine, biology, and industry. The progress in technology and in understanding the origin of the noise in low-transition-temperature (T_c) SQUIDs brought a dramatic improvement in the resolution of electrical and magnetic measurements. Many rather new and nonstandard applications of SQUIDs have been reviewed in detail; see reviews and contributions in Refs. 1 and 2.

The discovery of high-T_c oxide superconductors by Bednorz and Muller (3) quickly made apparent that macroscopic quantum phenomena may be very useful for a number of electronic applications. High-T_c materials have opened new possibilities by increasing the operating temperature of superconducting instruments and sensors.

The High-T_c SQUID was the first superconducting electronic circuit employing Josephson junction cooled by liquid nitrogen. The development of superconducting magnetometers operating at 77 K holds promise of expanding the useful range of application of these devices to include remote operation in the field and space, where the availability of liquid helium as a cryogen may be limited by some factors (4). In addition, the wider range of operation temperature may permit measurement on room-temperature samples such as living tissue (5), nondestructive material evaluation with Joule–Thomson cryocooler in industry (6), and routine checks of moving samples with greater sensitivity (7), high-T_c SQUID microscopy (8), and convenience due to nitrogen cooling.

However, oxide superconductors have introduced some completely new problems. Because of the extremely short coherence length, a conventional Josephson junction structure is not possible, and various inhomogeneities and structural defects in these materials lead to the formation of parasitic weak links between regions with well-developed superconductivity, lead to increased flux creep, reduce the critical current of the junctions, and creates an excess high noise level.

In practice, twin boundaries are essentially nonsuperconducting regions in high-T_c materials and SQUIDs due to these can be observed. The values of zero-temperature coherence length $\xi(0)$ and lattice parameter c are very much different from the conventional low-T_c superconductors. Such however, is not the case for a usual superconductors:

$$\frac{(\xi_{(0)}/c)_{\text{high}} - T_c}{(\xi_{(0)}/c)_{\text{usual}}} \approx 10^2\text{–}10^3 \qquad (1)$$

This difference between a usual and high-T_c SQUIDs eliminate the main physical and technology problems toward practical application of the new superconductors: anomalously large critical current anisotropy in single cyrstals and epitaxy sensors at moderate magnetic fields, the existence of intragrain Josephson junctions and randomness, frustration effects in presence of a magnetic field, $1/f$ noise, and so forth.

At 77 K, the Josephson coupling energy in high-T_c junctions may be of the order of the thermal energy, and under this condition, thermally activated phase-slippage processes result in an observable reduction of the high-T_c SQUIDs' dynamic range and sensitivity. Except that there are specific for oxide superconductors phenomena, the ultimate sensitivity of both single- and double-junction SQUIDs is limited by the characteristic frequency R/L of the interferometer. The sensitivity of an optimized dc SQUID is limited primarily by two parameters determined by the high-T_c technology process: loop inductance L and junction characteristics. In the case of radio-frequency (RF) SQUIDs, the sensitivity is limited by pump frequency and preamplifier noise. The requirements for fabrication technology are not as strict as for DC SQUIDs. However, in external magnetic fields both dc and RF SQUIDs sensitivities are limited by the specific for oxide superconductors noise sources and make these SQUIDs competitive to each other.

Some difficulties in the technology of high-T_c SQUIDs have been overcome and excellent sensitivity is achieved in practical devices with rf sensors (9) with probably, exact condition on k^2Q in quasi-nonhysteretic mode (10). For real-instrument applications, one can make a comparison of different SQUIDs parameters: energy sensitivity, magnetic field sensitivity, bandwidth, slew rate, main SQUID electronics structure, and so on. Deviation in these parameters of high-T_c RF SQUIDs from the low-T_c RF SQUIDs are attributed to specific properties of oxide superconductors and thermal fluctuations.

In this chapter, two different types of high-T_c RF SQUID fabricated from both bulk ceramic and thin film are briefly reviewed. We focus on the most important results and on general problems in the design and fabrication of low-noise high-T_c magnetometers based on RF SQUIDs. The single-junction interferometer in the presence of large thermal fluctuations and the RF SQUIDs are discussed. Flux-creep noise in high-T_c magnetometers and "Josephson fluctuators" are analyzed. Finally, the design and pilot applications of a high-T_c SQUID are discussed.

7.2 HIGH-*Tc* SINGLE-JUNCTION INTERFEROMETER IN THE RSJ MODEL

When connecting two superconducting electrodes by a weak electric contact, the macroscopic coherence of supercoducting state results in the following fundamental expression:

$$\frac{d\varphi}{dt} = \frac{2\pi}{\Phi_0} V \tag{2}$$

which relates a rate of the phase-difference change φ of wave functions in two electrodes to the voltage across them. The quantity $\Phi_0 = h/2e \cong 2.07 \times 10^{-15}$ Wb is the flux quantum, e is the electron charge, and h is Planck's constant. The small superconducting current I through a weak link depends only on the phase difference, and for the resistively shunted junction (RSJ) model, this relation is of the form

$$T = T_c \sin \varphi \tag{3}$$

where I_c is the critical current of a high-T_c Josephson junction.

Three types of weak-link step-edge junctions (11,12), bicrystal junctions (13,14), and grain-boundary junctions in bulk materials (15) are commonly used to fabricate high-T_c RF SQUIDs. A considerable number of articles have been published dealing with these types of weak link (16) and sometimes junctions looked at as a complicated connection between chaotically located weak links with random parameters. Perhaps a separate article is needed to review all of these effects caused by the magnetic field and current flowing through complicated junction. Regular high-T_c weak-link SQUIDs seem to be more promising both for RF and dc SQUID applications in low and moderate magnetic fields. At present, the bicrystal and step-edge structures seem to be the more promising, which can be made with low intrinsic capacitance ($C \sim 10$ fF) and a high characterizing parameter $V_c = I_c R \sim 0.1$–0.5 mV, with the current–phase relation close to RSJ model (17,18), $I = I_c \sin \varphi$.

1. *Step-edge junctions:* Step-edge junctions are formed by depositing a thin high-T_c epitaxial film on substrate that has a step etched into the surface. The weak link is then formed as the single-layer film bridges the two levels. The steps usually are formed by patterning and ion milling the $SrTiO_3$ substrate. There are significant advantages to step-edge technology. A variety of large-area substrates can be employed in the step-edge process (bicrystal technology is limited to a 10-mm $SrTiO_3$ substrate). The step-edge height is highly reproducible. However, the junction parameters appear to depend greatly on film thickness, and variations observed for step-edge technology are probably a result of the film thickness and superconducting parameters variations, on the "bottom" and on the "top" of the step.
2. *Bicrystal junctions:* Bicrystal junctions are fabricated on the substrate that has a twin boundary formed by two single-crystal domains with different crystal orientations. On $SrTiO_3$ bicrystal substrates, a misorientation angle usually is an order 25°–37°. They are made by fusing two separate single crystals and then dicing substrates from the single piece. This results in a twin boundary down the center of the substrate. When the superconductor film is epitaxially grown, a twin boundary forms in the superconductor film at this interface.
   More details of the fabrication and characterization of ramp edge and step edge junction are given in Chapters 3 and 4.
3. Josephson's junctions for bulk HTS SQUIDs are conventionally manufactured by means of local impact of a pulse laser. A sample specimen is positioned into an optical cryostat, whereas, by nonstop monitoring of volt–ampere characteristics, the amplitude of pulse irradiation has been increased, up to an emerged required nonlinearity. Such sensors are very inexpensive in manufacture and are relatively low cost (about $30 each). However, said products are typical for having excessive $1/f$ noise in low-frequency area. A response of said SQUIDs can be improved only due to quality of bulk materials and to the creation of superconductive input coils.

The investigated step-edge and bicrystal junctions routinely made by many groups had a width of weak link ranging from 1 to 50 μm [$W \gg \xi_{ab}(0)$] and with a thickness of 0.1–0.3 μm. The scaling relation between $I_c R$ and J_c for both step-edge and bicrystal junctions fabricated from YBaCuO and GdBaCuO and tested at 77K is shown in Figure 1 (16). However, there is no technological process for fabricating high-T_c Josephson junctions with reproducible characteristics. How-

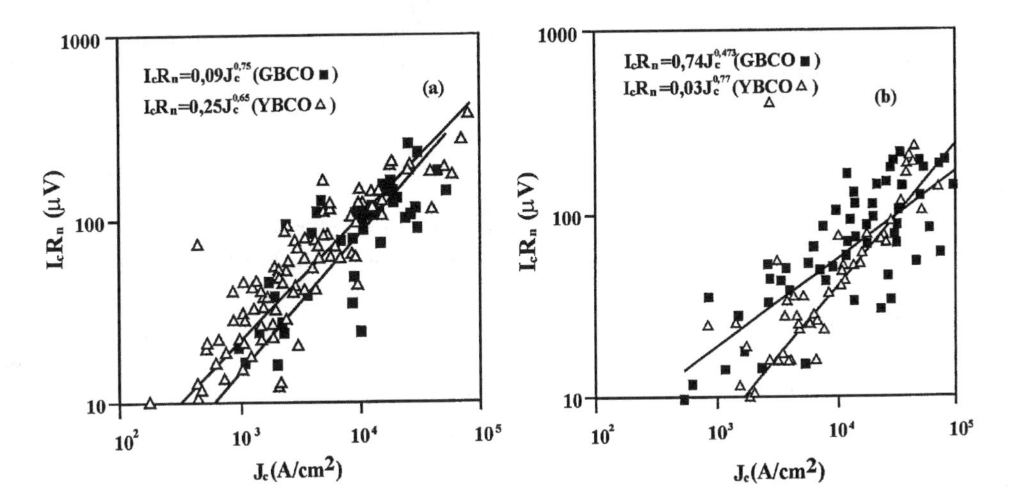

FIGURE 7.1 The scaling relation between I_cR_n and J_c for both (a) step-edge and (b) bicrystal junctions fabricated from YBaCuO and GdBaCuO and tested at 77K. (From Ref. 16.)

ever, high-quality tunnel junctions can be routinely fabricated for low-temperature superconductors.

In a RSJ model, the total current through the high-T_c Josephson junction can be considered as a sum of the superconducting current, normal current, and bias current (19,20):

$$I = I_c \sin\varphi + \frac{V}{R} + C\frac{dV}{dt} \quad (4)$$

Here R and C are the normal resistance and junction capacitance, respectively. When the junction is incorporated in the superconducting ring (Fig. 7.2), the constant voltage should occur across it, with only a time variation in the magnetic flux Φ through the ring:

$$V = \frac{d\Phi}{dt} \quad (5)$$

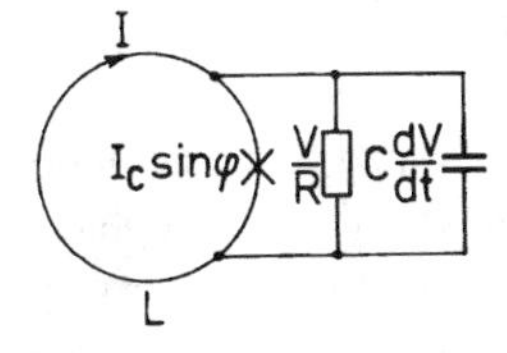

FIGURE 7.2 Superconducting quantum interferometer for rf SQUID with Josephson junction in terms of the RSJ model.

Equating Eqs. (1) and (5) and integrating with respect to the time results in an unambiguous relation between the contact-phase difference and total magnetic flux Φ through the RF SQUID loop:

$$\varphi = \frac{2\pi\Phi}{\Phi_0} \tag{6}$$

The total magnetic flux through the loop is equal to a difference between the external magnetic flux Φ_e and self-inductance flux LI due to the SQUID circulating current:

$$\Phi = \Phi_e - LI \tag{7}$$

where L is the SQUID loop inductance. Using Eqs. (4)–(6) yields for the current.

$$C\frac{d^2\Phi}{dt^2} + \frac{1}{R}\frac{d\Phi}{dt} + I_c \sin\left(2\pi\frac{\Phi}{\Phi_0}\right) = \frac{\Phi_e - \Phi}{L} \tag{8}$$

Equation (8) is equivalent to a classical equation for the motion of the particle with the mass $M = C(\Phi_0/2\pi)^2$ in the one-dimensional potential field:

$$U(\Phi, \Phi_e) = \frac{(\Phi - \Phi_e)^2}{2L} - \frac{\Phi_0 I_c}{2\pi}\cos\left(\frac{2\pi\Phi}{\Phi_0}\right) \tag{9}$$

The whole analysis of a single-junction high-T_c interferometer within the framework of the RSJ model reduces essentially to a study of the particle motion in a potential field, Eq. (9), depending on the mass of the particle, viscosity, rate of external force variation, and so forth. As mentioned earlier, the mass of the particle (the high-T_c junction's capacitance $\sim$ 10 fF) is very low, and if an external flux varies slowly in time $[(1/\Phi_0)(d\Phi/dt) \ll R/L]$, then in the low-fluctuation limit, it follows an equation describing the stationary high-T_c RF SQUID's state from Eq. (8):

$$\varphi + \ell \cdot \sin\varphi = \varphi_e \tag{10a}$$

or at high-excitation frequency $\omega \sim L/R$:

$$q\dot{\varphi} + \ell \sin\varphi + \varphi = \varphi_e \tag{10b}$$

Here, dimensionless variables have been used:

$$\varphi = \frac{2\pi\Phi}{\Phi_0}, \qquad \varphi_e = \frac{2\pi\Phi_e}{\Phi_0}, \qquad \ell = \frac{2\pi L I_c}{\Phi_0}, \qquad q = \frac{\omega L}{R} \tag{11}$$

The quantity ℓ is a fundamental RF SQUID parameter equal to the geometrical loop inductance normalized by characteristic inductance of the Josephson junction $L_J = \Phi_0/2\pi I_c$. The values ℓ and q determine the shape of the curves for the stationary SQUID characteristic and potential energy and agree upon the classification for the modes of one-contact SQUID operation in a small fluctuation limit (21).

7.3 SMALL FLUCTUATION LIMIT FOR HIGH-T_c RF SQUIDs

Superconducting rings, coils, and transformers are essential elements of all high-T_c superconducting magnetometer sensors. At 77 K, in a nonshielding environment there are some main sources of noise: Johnson noise generated by thermal energy in the normal resistance and low-frequency noise generated by magnetic flux instability (flux creep noise) and by bistable or multistable Josephson fluctuators in SQUID body. The fact is that the excess-noise amplitude of high-T_c SQUIDs decreases in a low external magnetic field and development of high-quality epitaxial film system brought a dramatic improvement in resolution. The important point about these SQUIDs is that they operate at 77 K in a small thermal fluctuation limit.

When the Josephson junction with normal resistance R is incorporated in a high-T_c superconducting ring, Johnson noise generated in R by thermal energy $k_B T$ produces a flux fluctuation spectral density in the ring inductance L:

$$\langle \Phi_n^2(\omega) \rangle = L^2 \langle I_n^2(\omega) \rangle = \frac{2 k_B T R L^2 \, d\omega}{\pi (R^2 + \omega^2 L^2)} \tag{12}$$

and total flux noise in the classical limit is

$$\langle \Phi_n^2 \rangle = \frac{2}{\pi} \int_0^{R/L} \frac{k_B T R L^2 \, d\omega}{R^2 + \omega^2 L^2} \cong k_B T L \quad at \, \frac{R}{L} \sim \infty \tag{13}$$

where k_B is the Boltzman constant 1.38×10^{-23} J/K. The integration gives a result which also follows from the equipartition theorem for a system with one degree of freedom. The fluctuation spectrum of magnetic flux noise in any closed conducting (superconducting) loop is calculated in the same way. An important practical consideration for high-T_c SQUIDs applications is to estimate the magnitude of fluctuation that will be introduced by a normal-metal enclosure surrounding the sensor, either outside or inside the dewar.

From a practical point of view, it is important to choose different parameters of a high-T_c Josephson junction, such as the critical current I_c, normal resistance R, capacitance C, and the geometrical inductance L of a SQUID loop. The capacitance is negligible for the high-T_c Josephson junction and the McCumber parameter

$$\beta_c = \frac{2\pi C R^2 I_c}{\Phi_0} < 1 \tag{14}$$

for these SQUIDs. In order to observe the magnetic flux quantization in a SQUID loop with inductance L, one needs that the uncertainity of the magnetic flux (13) must be lower than fundamental quantity of the magnetic flux quantum defined by

$$\langle \Phi_n^2 \rangle = k_B T L < \left(\frac{\Phi_0}{2} \right)^2 \tag{15}$$

The theoretical investigation of the RF SQUID interferometer in the presence of thermal fluctuations on the basis of a Fokker–Plank equation (22) has shown that a low thermal fluctuation limit is required at 77 K:

$$L \ll L_F = \left(\frac{\Phi_0}{2\pi}\right)^2 \frac{1}{k_B T} \approx 10^{-10}\, H \tag{16a}$$

In practice, SQUID's inductance in this limit is defined by

$$L < \frac{L_F}{\pi} \tag{16b}$$

where L is the temperature-dependent fluctuation inductance. In order to determine the real size of the SQUID quantization loop for a thin-film high-T_c interferometer, one can use expressions for a circular-shaped form (Fig. 7.3a),

$$L = 2\mu_0 r \qquad r \ll w \tag{17}$$

or rectangular-shaped form (Fig. 7.3b),

$$L = 1.25\mu_0 d \qquad d \ll w \tag{18}$$

where $\mu_0 = 4\pi \times 10^{-7}$ H/m, and the optimizing size of the loop for a low fluctuation limit from Eqs. (16) and (18) is about $20 \times 20\ \mu m^2$! It is very difficult for real technology because there are some "parasitic" inductances in SQUID topology, and to provide for a good coupling with the signal source, complex multiloop input circuits are required.

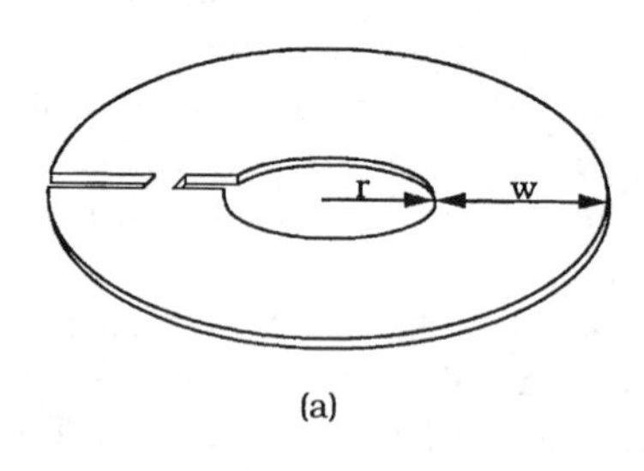

(a)

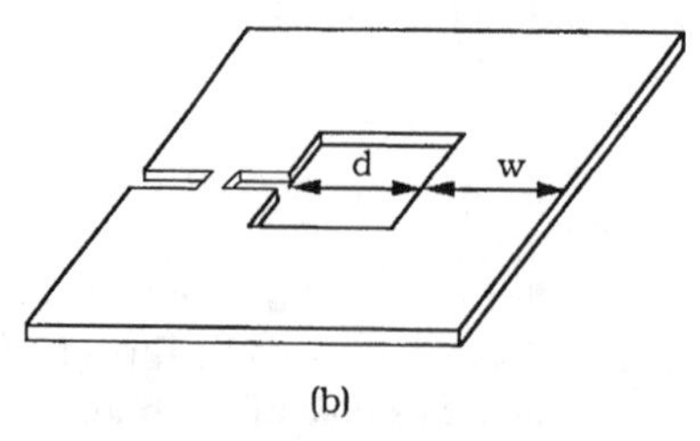

(b)

FIGURE 7.3 Schematic view of the simple structures of high-T_c thin-film RF SQUIDs, (a) circular, (b) rectangular.

The most widely used thin-film high-T_c RF SQUID is the washer type (23), where the SQUID inductance is a square washer containing a slit. This configuration allows one to simply design a good coupling; however, a relatively large inductance (about 100–300 pH) is usually used for high-T_c RF SQUIDs.

In multiloop SQUIDs topology (21,24,25) offered some 20 years ago by Zimmerman, the inductance is obtained from the parallel of many loops, can be coupled to an external (input) coil, and can be reduced to a very low value (about 20–30 pH). This type high-T_c RF SQUIDs can be a good candidate on small fluctuation limit.

For a bulk ceramic SQUID with simple cylindrical topology, one can use

$$L = \frac{\mu_0}{4\pi} df\left(\frac{h}{d}\right) \tag{19}$$

where f is a function of the height/diameter ratio for a bulk SQUID with a cylinderlike interferometer:

d/h	1.0	0.8	0.6	0.4	0.2	0.1	0.01
f	6.79	5.8	4.67	3.35	1.81	0.946	0.098

Parasitic inductance for bulk SQUIDs in practice is about 20–30 pH and more. That is why it is really not possible to have bulk ceramic RF SQUIDs in a low fluctuation limit.

The second effect in the high temperature is the rounding of the voltage–current, and the signal characteristics can be characterized in the high-T_c SQUIDs by a parameter

$$\Gamma = \frac{2\pi k_B T}{\Phi_0 I_c} \tag{20}$$

An inspection of the RF SQUID characteristics (25,26) indicates that the effect of noise is small for $\Gamma < 0.05$. Therefore, one can use for noise-free dynamic critical current of the high-T_c Josephson junction at 77K from

$$I_c \geq \frac{2\pi k_B T}{\Gamma \Phi_0} \cong 54\ \mu A \tag{21}$$

In the low fluctuation limit, the fundamental SQUID parameter in multiloop topology (with low inductance, 20 pH $< L <$ 30 pH), the dimensionless inductance ℓ can be varied from 1 to 7. The dynamics and noise characteristics of the single-junction SQUIDs depend fundamentally on whether $\ell < 1$ (nonhysteretic SQUID) or $\ell > 1$ (hysteretic SQUID). In the small fluctuation limit, RF SQUIDs have been studied intensively both theoretically and experimentally (19–21,25–27) for low-T_c SQUIDs. For the best high-T_c SQUID in the small fluctuation limit, all of this noise analysis predicted an intrinsic RF SQUID sensitivity limit with the same dependence of this noise upon SQUID ring parameters, tank circuit, temperature, and preamplifier characteristics.

7.4 HIGH-T_c RF SQUIDs IN NONHYSTERETIC MODE

High-T_c RF SQUIDs (in the small fluctuation limit $L < L_F/\pi$, $\Gamma < 0.05$) analysis is based on the results reported for low-T_c SQUIDs by many groups (19–21,26–28). The subsequent analysis has shown that because of the low transfer coefficients $dV/d\Phi_e$ and $d\vartheta/d\Phi_e$, the intrinsic energy sensitivity of the nonhysteretic high-T_c RF SQUIDs should be defined by an amplifier noise and is always worse than that of hysteretic one. For $\ell \ll 1$, the transfer function "conversion efficiency" is approximately $(\omega/k)(L_T/L)^{1/2}\ell/2$, which is $\ell/2$ times lower than that of the hysteretic high-T_c RF SQUID with $q = \omega L/R \ll 1$. It is a good approximation for the step-edge and bicrystal junctions parameters ($R \sim 2$–$3\ \Omega$ and $L \sim 30$ pH) up to $\omega/2\pi = 1$ GHz. Please note that L_T is resonant contour inductance.

However, if the condition $k^2Q\ell > 1$, $\ell < 1$ (here k is the coefficient of coupling between the interferometer and resonant circuit and $Q = \omega L_T/R_T$ the circuit quality), is satisfied, then at some points of signal characteristics, the conversion efficiency can be made extremely high, up to $\sim 10^{12}$ V/Wb. From this a possibility follows that there is a strong increase in the transfer coefficient at $k^2Q\ell > 1$, assuming that the value's sensitivity can be achieved, which are defined by resonant circuit noise and pumping frequency ω.

The oscillation equation for a resonant circuit (Fig. 7.4) is of the form

$$\ddot{\varphi}_T + Q^{-1}\dot{\varphi}_T + (1 - 2\xi_0)\varphi_T = e \cos \tau + k^2\ell\ddot{\imath} \quad (22)$$

where $\varphi_T(\tau) = 2\pi M I_T/\Phi_0$ is the normalized interferometer flux induced by the resonant circuit, $e = 2\pi M I_T/\Phi_0$ is the normalized pumping amplitude, $\xi_0 = (\omega - \omega_0)/\omega_0$ is the detuning of the generator frequency ω, and ω_0 is the resonant circuit frequency. The induced interferometer current can be found from Eqs. (7) and

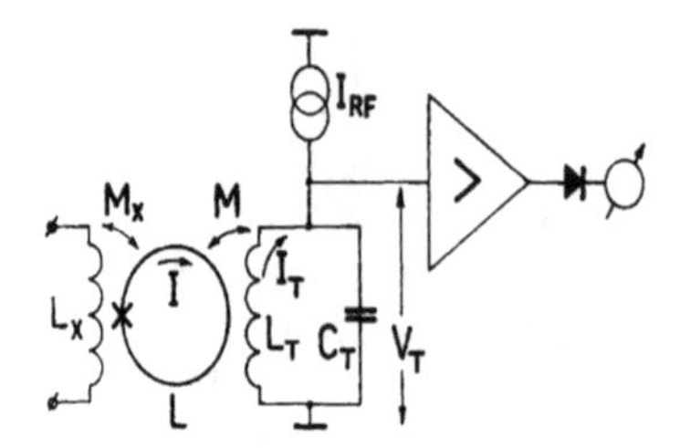

FIGURE 7.4 A basic RF SQUID circuit including the input coil and signal detection circuits.

(10b):

$$i(\tau) = \frac{I(\tau)}{I_c} = \frac{\varphi_T(\tau) + \varphi_e - \varphi}{l} = \frac{q}{\ell}\dot{\varphi} + \sin\varphi \tag{23}$$

where differentiation with respect to dimensionless time $\tau = \omega t$ is marked by a dot.

In practice, both damping and detuning are low: $Q >> 1$, $\xi_0 << 1$, and $k^2 << 1$; hence, the term $k^2\ell\ddot{i}$ should be negligible, the resonant circuit oscillations being of a quasiharmonic nature; that is,

$$\varphi_T(\tau) = a(\tau)\cos[\tau + \theta(\tau)] \tag{24}$$

where $a(t)$ and $\vartheta(t)$ are the slowly varying amplitude and phase of oscillations, respectively. In this case, Eq. (22) reduces to two differential equations:

$$\dot{a} + \frac{a}{2Q} + \frac{e}{2}\sin\theta = k^2\ell\langle i \sin\psi\rangle \tag{25a}$$

$$a\dot{\theta} + \xi_0 a + \frac{e}{2}\cos\theta = k^2\ell\langle i \cos\psi\rangle \tag{25b}$$

where $\psi = \tau + \vartheta$ and the brackets indicate the averaging over the RF oscillation period. A system of Eq. (23) along with Eqs. (25a) and (25b) permits the analysis of RF SQUID voltage–current, amplitude–frequency, and signal characteristics.

A system of reduced equations (25a), (25b) describing the stationary processes occurring in the resonant circuit which is inductively coupled with an interferometer at $\ell < 1$ is assumed to be of the form

$$2\delta(a, \varphi_e)\, a = -e\sin\theta \tag{26a}$$

$$2\xi(a, \varphi_e)\, a = -e\cos\theta \tag{26b}$$

Here, the $2\delta(a, \varphi_e)$ and $2\xi(a, \varphi_e)$ functions are effective damping and detuning, respectively, of the resonant-parametric circuit taking account of the interferometer contribution. With a due account of the terms proportional to ℓ and $1 + q^2 \sim 1$, these functions can be written as follows:

$$2\delta(a, \varphi_e) = Q^{-1} + k^2 q\left(1 - 2\ell\frac{2J_1(a)}{a}\cos\varphi_e + \ell^2\cdots\right) \tag{27a}$$

$$2\xi(a, \varphi_e) = 2\xi_0 - k^2\ell\left(\frac{2J_1(a)}{a}\cos\varphi_e + \ell\cdots\right) \tag{27b}$$

where $J_1(a)$ is the Bessel function of the first kind.

Eliminating the oscillation phase ϑ from Eqs. (26a) and (26b) yields

$$e^2 = a^2\left[(2\xi)^2 + (2\delta)^2\right] \tag{28}$$

Substitution of effective values for damping and detuning into this equation makes it possible to analyze SQUIDs characteristics numerically depending on different system parameters; see Ref. 21. Because the increase in the transfer coefficient for the nonhysteretic mode with $k^2Q\ell > 1$ results from a variation in the signal characteristics behavior over almost the same amplitude range, the $\Delta\varphi_e$ spacing markedly decreases with a great slope of $da/d\varphi_e$. This decreasing of $\Delta\varphi_e$ is proportional to an increase in $da/d\varphi_e$ and, therefore, the increase in the slope from 10^{10} to 10^{12} V/Wb changes the $\Delta\varphi_e$ range from $\Phi_0/2$ to $10^{-2}\Phi_0$. The same increase in the transfer coefficient can be achieved in dc SQUIDs with a positive feedback.

At $k^2Q\ell > 1$, the main feature of the nonhysteretic RF SQUIDs is the divergence of the transfer coefficients by amplitude and phase:

$$\eta_a = -\eta_0 \frac{(2k^2Q\ell)\, J_1(a) \sin \varphi_e}{(1 + k^2Q\ell\omega/\omega_{0J})^2 + (2Q)^2\xi\tilde{\xi}}\, 2\xi Q \tag{29}$$

$$\eta_a = -\eta_0 \frac{(2k^2Q\ell)\, J_1(a) \sin \varphi_e}{(1 + k^2Q\ell\omega/\omega_{0J})^2 + (2Q)^2\xi\tilde{\xi}} \left(1 + k^2Q\ell \frac{\omega}{\omega_{0J}}\right) \tag{30}$$

where $\tilde{\xi}(a, \varphi_e) = \xi_0 - k^2\ell J_1(a) \cos \varphi_e$ is the effective detuning at small variations of amplitude approaching those points at the voltage–current characteristics for which the values in the denominators of Eqs. (29) and (30) go to zero. In this case, the output fluctuations rise proportionally to $\eta_{a,\vartheta}$ and that is why the noise reduced to the SQUID input remains finite and defines the intrinsic nonhysteretic SQUID energy by a high-T_c RF loop and tank circuit noise. Detailed researches of the nonhysteretic regime with $k^2Q\ell > 1$ (10), have shown that by greater values, up to $k^2Q\ell > 2$, a portion of pumping energy, the RF power, has been transformed into stochastic oscillations. Therefore, an increase of the $k^2Q\ell$ value is limited in the nonhysteresis regime by the following condition:

$$1 < k^2Q\ell < 2 \tag{31}$$

7.5 HIGH-T_c RF SQUIDs IN HYSTERETIC MODE

Let us return to a basic circuit of the RF SQUID (Fig. 7.4). If the magnetic flux φ_T generated in the interferometer by a resonant circuit is less than the critical value $\varphi_{ec} \cong \ell + \pi/2$, then according to Ohm's law, the voltage V_T should be related with the generator current by the following expression:

$$V_T = \omega L_T Q I_{RF} = \omega L_T I_T \tag{32}$$

With the appearance of hysteresis losses, a well-known system of steps is formed at the I–V curves. The first step can be observed at $\varphi_x < \ell + \pi/2$ in the low fluctuation limit of the high-T_c RF SQUID. If an extra external field φ_x is applied to the SQUID, the energy absorption can be initiated under a lower voltage across the tank circuit; that is,

$$\varphi_T = \varphi_{ec} - \varphi_x = \frac{2\pi}{\Phi_0} M I_T = \frac{2\pi}{\Phi_0} M \frac{V_T}{\omega L_T} \tag{33}$$

Also, assuming linear coupling between the interferometer and resonant circuit, the magnetic flux-to-voltage transfer coefficient can be deduced from Eq. (33):

$$\eta_0 = \eta_a = \frac{dV}{DI_{RF}} = \frac{\omega L_T}{M} = \frac{\omega}{k}\sqrt{\frac{L_T}{L}} \tag{34}$$

It should be emphasized that η_0 in the hysteretic mode of RF SQUIDs is independent on the superconducting materials, junction parameters, current–phase relation, small amplitude and frequency variations (due to the form of the step), and so on. The high-T_c RF SQUID sensitivity is defined by the transfer coefficient when the main source of noise is that of the transistor. This case is typical of the most practical low-T_c SQUIDs, including commercial ones, because the transfer coefficient ~ (3–5) × 10^{10} V/Wb for a 20–30-MHz excitation frequency:

$$\varepsilon = \varepsilon_a = \frac{\delta\phi_x^2}{2L\Delta f} = \frac{1}{2L}\frac{S_A}{\eta^2\Delta f} \approx \left(\frac{k}{\omega}\right)^2 \tag{35}$$

A possibility of increasing η_0 and, respectively, ε (sensitivity) for high-T_c RF SQUIDs in a low fluctuation limit by a reduction in L to $L_F/\pi = 30$ pH, is restricted by the intrinsic resonant circuit and interferometer noise, which are dependent on the position in the I–V characteristic (step) [i.e., the probability of the "jumps" (19,20,29,30). In this mode, the φ (φ_e) dependence is multivalid, which leads to "jumps" of the phase φ (and, consequently, of the loop current) when the externally applied flux approaches one of the threshold values $\varphi_{ec} = l + \pi/2$. In the presence of the external RF pumping flux applied by the tank circuit oscillations, the repeated phase jumps lead to the appearance of the almost horizontal "plateau" regions at the RF SQUID I–V curve. This is a plateau region, where the bias point should be located for a proper SQUID operation. At the plateau, the maximum value of the external flux is slightly less than the threshold values φ_{ec}. In the presence of the junction intrinsic noise, the finite probability p exists for the phase φ to make a "jump" during the $\varphi_e(t)$ passage through the vicinity of its maximum value. In this model, the consequent phase jumps are independent of each other. This is true only if the change in oscillation due to a single jump ($\sim 4\pi k^2$) is much less than the width of the probability distribution, $\sim \ell(\Gamma/2)^{2/3}$, and the jumps can lead to the neighboring stable state [$2\pi >> l(\Gamma/2)^{2/3}$]. Hence, all of the above results are valid at

$$k^2 \ll \ell(\Gamma/2)^{2/3} \ll 2\pi \tag{36}$$

The plateau slope cannot be assumed to be constant, as it has been in the early publications. In fact, the slope strongly increases at the plateau edges and p

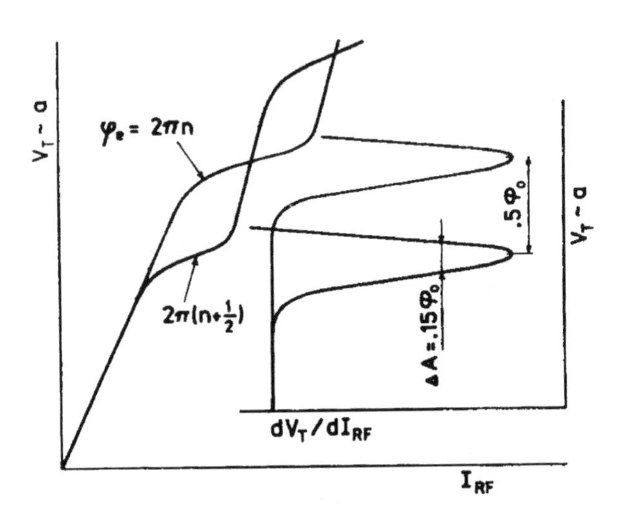

FIGURE 7.5 Current–voltage characteristics of high-T_c RF SQUID in the small fluctuation limit and its first derivatives, $dV_T/dI_{RF}(V_T)$, for two values of external flux: Φ_0 and $\Phi_0/2$.

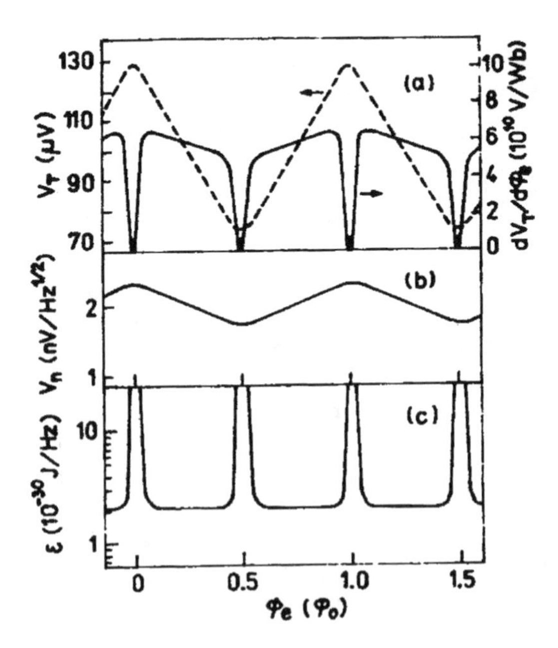

FIGURE 7.6 Signal characteristic $V_T(\Phi_e)$, transfer coefficient $dV_T/d\Phi_e$, spectral noise density V_n, and sensitivity ε as a function of magnetic flux for the high-T_c RF SQUIDs in the hysteretic mode for a small fluctuations limit.

varies linearly over the step length from 0 to 1. The density of the jumps probability $\sigma = -dp/du$ is proportional to the first derivative, $dV_T/dI_{RF}(V_T)$, with respect to I–V characteristic, which can be measured directly. $\sigma = -dp/du$ has a minimum at the optimum working point. Figure 7.5 shows I–V and $dV_T/dI_{RF}(V_T)$ characteristics of RF SQUID for two values of the magnetic flux, in the small fluctuation limit. Let us remember that a minimal value of interferometer's own noise, within these limits, is observed only in a single optimal state point and depends on the C_p parameter that follows from theory (19,29) and experiments (1). The signal characteristics, transfer coefficients, spectral noise density, and sensitivity are shown in Figure 7.6 as a function of external field for hysteretic mode operation of RF SQUIDs. In other words, within small thermal fluctuations limit, an optimization of RF high-T_c SQUIDs response and sensitivity does not differ from that used in low-T_c SQUIDs. It is a worth mentioning that due to the $L < L_F(T)$ condition, the transfer coefficient [Eq. (34)] becomes significantly greater.

7.6 SENSITIVITY OF HIGH-T_c RF SQUIDs

It would be of significant importance to estimate ultimate noise properties of two types of high-T_c RF SQUID operating in nonhysteretic and hysteretic regime in the small-signal mode operation, with nitrogen cooling. This will indicate limit of the possible sensitivity of the high-T_c RF SQUIDs for the wide-band measurements.

At first, one should avoid features in the ring and Josephson junction design which increase excess noise. This noise contribution can be made small for multiloop epitaxial thin-film interferometers with a high-quality Josephson junction. This is true; however, now it is possible only in very low external magnetic fields.

In the low fluctuation limit, sensitivity for the nonhysteretic high-T_c RF SQUIDs can be estimated as follows (19):

$$\varepsilon = \frac{1}{j^2(a)\sin^2\varphi_e}\,\frac{k_B}{\ell\omega_{0J}}\left[T + T_T\left(\frac{\omega_{0J}}{k^2Q\ell\omega}\right)\right]\frac{(1 + k^2Q\ell\omega/\omega_{0J})^2 + (2\xi Q)^2}{(2\xi Q)^2} \tag{37}$$

This formula does not take into account the preamplifier noise, whose contribution becomes negligible for a large transfer coefficient. The signal characteristics, transfer coefficients, spectral noise density, and sensitivity for RF SQUID in the nonhysteretic mode are shown in Figure 7.7. The noise of the transistor currents can be easily allowed in expression (37) by the increased noise temperature of the resonant circuit T_T. If all of the parameters a, ξ, φ_e and Q are chosen to be near an optimum operating value, Eq. (37) can be greatly simplified as follows:

$$\varepsilon_V \cong \frac{4k_B}{\omega_{0J}\ell}\left(T + T_T\frac{\omega_{0J}}{k^2Q\ell\omega}\right) \tag{38}$$

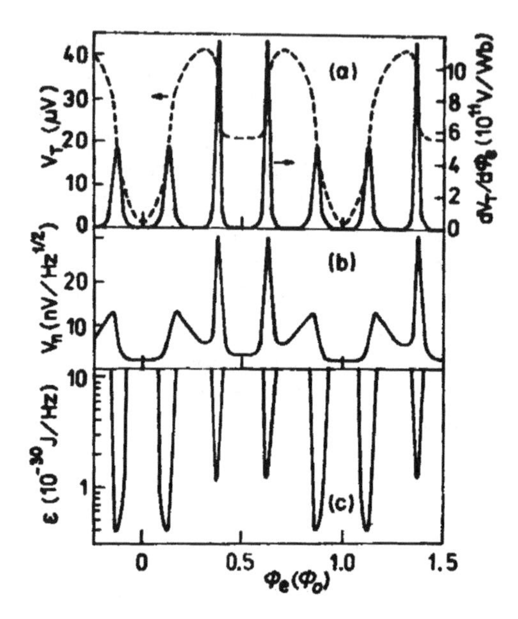

FIGURE 7.7 Signal characteristic $V_T(\Phi_e)$, transfer coefficient $dV_T/d\Phi_e$, spectral noise density V_n, and sensitivity ε as function of magnetic flux for nonhysteretic regime high-T_c RF SQUIDs in the small fluctuations limit with $k^2Ql > 1$.

and for the best interferometer parameters $\ell \cong 1$, $k^2Q\ell \cong 2$,

$$\varepsilon_V \cong 4k_B\left(\frac{T}{\omega_{0J}} + \frac{T_T}{2\omega}\right) \tag{39}$$

It is evident from this equation that the ultimate sensitivity for nonhysteretic high-T_c RF SQUID is limited by the resonant circuit noise for excitation frequency $\omega \ll \omega_{0\epsilon J}$:

$$\varepsilon_V \cong \frac{2k_BT_T}{\omega} \tag{40}$$

At liquid-nitrogen temperature $T \sim 80$ K and multiloop design, $\ell \sim 1$, for the realistic value $V_c = I_c R = 90$ μV ($I_c = 30$ μA, $R = 3\ \Omega$, $\omega_{0J} \sim 3 \times 10^{11}\ \mathrm{s}^{-1}$), the energy sensitivity [Eq. (39)] is as low as $\varepsilon_V = 10^{-30}$ J/Hz at $\omega = 2\pi$, 300 MHz, and $\varepsilon_V \sim 10^{-32}$ J/Hz at $\omega \sim \omega_{0J}$.

For hysteretic high-T_c RF SQUIDs in the small fluctuation limit, the magnetic flux-to-voltage transfer coefficient can be estimated for $20(L_0)_{\mathrm{high}\ Tc} \sim (L_0)_{\mathrm{low}\ Tc}$ as:

$$(\eta_0)_{\mathrm{high}-Tc} = \frac{\omega}{k}\left(\frac{L_T}{L_0}\right)^{1/2} \approx 4(\eta_0)_{\mathrm{low}\ Tc} \tag{41}$$

It should be emphasized that the transfer coefficient for RF SQUIDs with $\ell > 1$ is independent of the superconducting interferometer feature and is the main factor in the limitation of energy sensitivity in low-T_c RF SQUID due to the noise temperature of preamplifier. For high-T_c RF SQUIDs in the small fluctuation limit, η_0 is four times larger and the resonant circuit and interferometer are main sources of the noise.

The intrinsic interferometer and resonant circuit noise depends on the position in the "step" of the RF voltage–current SQUID characteristic (19,27) (i.e., on the probability p of the "jumps") and is considered by the expression

$$\varepsilon_V = \frac{\pi\ell}{\sigma^2}\left(\frac{\Gamma}{2}\right)^{2/3}\frac{k_B}{\omega}[p(1-p)\left(\frac{\Gamma}{2}\right)^{-1/3}T + \frac{\ell(\Gamma/2)^{2/3}}{2\pi k^2 Q}T_T] \tag{42}$$

where T and T_T are the contact and tank circuit temperatures, respectively, p varies linearly over the step length from 0 to 1 for an operating point that corresponds to $p(1-p) = 0.16$. Q is the quality of the tank circuit, σ is the density of the jump probability that is proportional to the first derivative $dV_T/dI_{RF}(V_T)$ with respect to the step of the RF voltage–current characteristic (27). An uncertainty of the magnetic flux values corresponding to the instant of decay (jump) seems to be a reason for uncertainty of flux variations which gives to the intrinsic high-T_c SQUID loop noise.

In the small fluctuation limit, for the high-T_c RF SQUID sensitivity, the tank circuit and preamplifier noise contribution is supposed to be independent of the external flux. Its value is equal to $\varepsilon \sim$ 10–30 J/Hz for a broad band of frequencies up to 100 kHz, with parameters $L = 30$ pH, $dV/d\Phi_e = 5 \times 10^{11}$ V/Wb, $I_c = 50$ μA, $T_T = T = 77$ K, noise temperature amplifier $T_a = 120$ K, and pumping frequency $\omega/2\pi = 30$ MHz. As the transfer coefficient of single-junction SQUID tends to increase linearly (and intrinsic SQUID loop noise decrease) with a frequency ω, the sensitivity can be further improved by using the ultra high-frequency (UHF) pumping, and a recent experiment (9) demonstrated achievement of very high sensitivity of high-T_c UHF SQUIDs.

In practice, quite often, this limit (small fluctuations) is not attained for high-T_c UHF and RF SQUIDs, because the intrinsic noise considerably exceeds the classical prediction due to additional phenomena such as flux creep, inhomogeneities, and structural defects of oxide superconductor material. SQUID and Josephson junction research has recently made a comeback, because of the improvement of its characteristics at nitrogen temperature. In Section 7.8, the noise in high-T_c materials is discussed. In the following section, the performance of a high-T_c RF SQUID in the presence of thermal fluctuations is discussed.

7.7 RF SQUIDS IN THE PRESENCE OF HIGH THERMAL FLUCTUATIONS

Since the discovery of macroscopic quantum interference in high-T_c materials, almost all RF SQUIDs are operating in the presence of high thermal fluctuations (22,31,32):

$$\Gamma \sim 0.5\text{–}2 \quad \text{and} \quad L \sim L_F \tag{43}$$

Generally speaking, at such values of parameters, the notion of "critical current" of the Josephson junction becomes conditional both for dc and RF SQUIDs. In stricter terms, at ultimate Γ values, a junction critical current equals zero. However, nonlinear properties of the interferometer are still present. In the subsequent discussion, we will use the term "critical current" to denote characteristics of nonlinearity.

It is very interesting, from a practical point of view, that suppression of the transfer function generated by the thermal fluctuations is smaller for a high-T_c RF SQUID than for the dc one. In addition, it is also found that the energy sensitivity of dc SQUIDs degrades more rapidly with the increase of the noise parameter Γ. On the other hand, for the RF SQUID, better $1/f$ noise properties are observed for the higher-noise parameter. Therefore, in the thermal limit, high-T_c RF SQUIDs should have a better intrinsic $1/f$ noise spectrum. These aspects are supported by experiments and represent important advantage of high-T_c RF SQUIDs (31).

Probably, the first theoretical investigation of RF SQUIDs interferometer in the presence of high thermal fluctuations traces back to the work of Khlus and Kulik (22). In this publication, it has been shown, on the basis of a Fokker–Plank equation, that thermal fluctuations result in the reduction of the critical currents and, in practice, of the nonlinearity of the system:

$$I = I_c \exp\left(-\frac{L}{2L_F}\right) \tag{44}$$

in particular, for the screening current in the high-T_c SQUIDs and dimensionless inductance ℓ will be rendered as $\ell_F = \ell \exp\left(-\frac{L}{L_F}\right)$. With a due account of this result, one can

write Eqs. (27a) and (27b) for the effective damping and detuning of the resonant circuit with high-T_c SQUID in the presence of high thermal fluctuations as

$$2\delta(a, \varphi_e) = Q^{-1} + k^2 q \left[1 - 2\ell \exp\left(-\frac{L}{2L_F}\right)\frac{2J_1(a)}{a}\cos\varphi_e + \ell_F^2 \cdots\right] \tag{45}$$

$$2\xi(a, \varphi_e) = 2\xi_0 - k^2 \ell \exp\left(-\frac{L}{2L_F}\right)\left(\frac{2J_1(a)}{a}\cos\varphi_e + \ell_F \cdots\right) \tag{46}$$

These simple formulas can be used for the first-hand determination of the high-T_c SQUID parameters in the regime $\ell \ll 1$ and $L \sim L_F$. Similar to the case of the small fluctuation limit, here also only the positions of the resonances and their widths are necessary. However, expressions (45) and (46) are quite different; see Eqs. (27a) and (27b). Recently (32), this approach was applied to the detailed study of signal characteristics of high-T_c RF SQUIDs with $\ell < 1$ and $\ell > 1$ and to the determination of SQUID parameters from voltage–frequency characteristics. For $L >> L_F$ and $\Gamma >> 1$, the averaged supercurrent of a high-T_c SQUID is $\langle I_c \rangle \sim I_c \exp(-L/L_F) \sin \varphi$ in accordance with the old theory (22). Moreover, in this work and in some experiments (33) and the theory (29), it has been shown that in the presence of high thermal fluctuations, $L >> L_F$ and $\Gamma >> 1$ can be considered from a unique standpoint, independently of the value of parameter ℓ. Roughly speaking, it is due to the above-mentioned reduction of the critical current, which makes the characteristics of all RF SQUIDs "quasi-nonhysteretic."

Some comments should be made here. Compared with the case of a small fluctuation limit, the maximum of flux modulation (i.e., the difference in the RF SQUID output voltage or phase) as a signal flux is changed from $(2n + 1)\Phi_0/2$ to $n\Phi_0$, the transfer function decreases very quickly ($\sim \exp(-L/2L_F)$. In this mode ($\ell < 1, L \sim L_F, \Gamma \sim 1$), high-$T_c$ RF SQUID sensitivity is determined by the noise temperature of the preamplifier (T_{amp}) and the pump frequency ω:

$$\varepsilon \sim \frac{k_B T_{amp}}{\tau_{opt}\{\ell^2[\exp(-L/L_F)](\omega/k)^2/(L_T/L)\}} \tag{47}$$

where $\tau_{opt} = L/R_{opt}$, with $R_{opt,}$ the optimal input impedance of the preamplifier. In contrast with small fluctuations, in this mode one has a higher intrinsic noise of high-T_c RF SQUIDs; if the pump frequency is 30 MHz, $\ell \sim 0.5$, $k \sim 0.1$, and $L/L_F \sim 1$, then $T_{amp} \sim 120$ K, corresponding to the energy sensitivity $\varepsilon_{min} \sim 2 \times 10^{-27}$ J/Hz or $\delta\Phi \sim 5 \times 10^{-3}\Phi_0/\text{Hz}^{1/2}$. These drawbacks explain why inductive ($\ell < 1$). High-T_c SQUIDs in a high fluctuation limit are rarely used in practice and we need high critical current in SQUIDs.

The screening current in the RF high-T_c SQUID inductance have been calculated recently (31,32) for higher values of $l \sim 3$ and $\Gamma > 0.3$:

$$\langle i \cos(\omega t + \vartheta) \rangle = \exp\left(-\frac{L}{2L_F}\right) g\left\{J_1(\text{a}) \cos \varphi_e - \left(\frac{f}{2}\right) J_1(2\text{a}) \cos 2\varphi_e\right\} \tag{48}$$

where g and f are coefficients that depend on both ℓ and Γ:

$$g = \exp\left(\frac{\ell 1/2}{\exp[3.5(\Gamma - 0.5)] + \ell}\right) \quad \text{and} \quad 0.1 < f < 1 \tag{49}$$

at $\Gamma = 0.5$ and $\ell = 2$, the coefficient g is equal to 0.72. Then the transfer function has the same Eq. (48) terms:

$$\eta_T \sim \eta_0 k^2 \ell g \exp\left(-\frac{L}{2L_F}\right)[J_1(a) \sin \varphi_e - fJ_1(2a) \sin 2\varphi_e] \tag{50}$$

Using a computer simulation method for dc SQUIDs, it has been found (34,35) that the transfer function is proportional to

$$\eta_{dc} \sim \exp\left(-\frac{1.75\, L}{2L_F}\right) \tag{51}$$

and suppression of the transfer function of a dc SQUID with $L > L_F$ is stronger than in the case of a high-T_c RF SQUID.

However, taking into account $\eta_{dc} \sim R/L$ and $\eta_{RF} \sim (\omega/k)(L_T/L)^{1/2}$, the total value of the transfer function and the optimal energy sensitivity can be higher for dc SQUIDs. This formula does not take into account the $1/f$ noise region. It should be pointed out that for RF SQUIDs, $1/f$ noise performance can be better due to the noise parameter Γ; see, for example, Refs. 31, 34, and 35.

However, there is no difference between the two types of macroscopic quantum device with dc or RF operation modes because, in practice, the main reasons for the sensitivity for both systems are excess noises in high-T_c materials in external magnetic fields.

7.8 EXCESS NOISE IN HIGH-T_c MATERIALS AND SQUIDs

A number of materials-related problems have hindered the development of a technology process for fabrication of high-T_c SQUID magnetometers (36). The first 1-2-3 structure discovered, YBaCuO, is one of the most commonly used materials for high-T_c electronics device fabrication. Depositing high-quality YBaCuO and GdBaCuO thin films (16) forms the development of a reproducible process for step-edge and bicrystal junctions. The critical currents displayed by the bicrystal and step-edge junctions are typically ~30–100 μA with the I_cR product of 100–200 μV at 77K for both materials with a current density 5×10^5–10^6 A/cm^2 (Fig.7. 1). The noise level of the devices (about 100 fT/Hz$^{1/2}$) does not appear to differ significantly between magnetometers fabricated from YBaCuO and GdBaCuO with step-edge and bicrystal Josephson junctions. Also, it has been noted that significant parts of these devices are generally noisier. The reason of the higher noise level of the all types (dc and RF) of high-T_c SQUIDs may be discrete or a large ensemble of Josephson fluctuators (JF) (37,38) and thermally activated flux motion in oxide superconductors (39,40).

7.8.1 Flux Noise Generated by Josephson Fluctuators

The fluctuation phenomena in high-T_c materials result in much more pronounced features than those observed in traditional low-T_c superconductors. The order-parameter depression is essential at temperatures close to T_c and it arises due to the short coherence length in the oxide superconductors, which may be of the order of interatomic distances in the direction transverse to the Cu–O layers. Various inhomogeneities and structural and orientation defects in these materials lead to the formation of weak links between regions with a high critical current density. The power spectral density of the magnetic flux noise in a simple model of a multi-contact weakly coupled system is called the "Josephson fluctuator" (JF), which reflects to some extent specific properties of real high-T_c superconductors.

Due to higher operating temperature for the SQUIDs, thermal fluctuations must be an important part of noise generation, which leads to breaking up phase coherence in the system of Josephson junctions. In macroscopic Josephson fluctuators which consist of superconducting loops closed by weak links, thermal fluctuations initiate transitions between different metastable states of the system. The existence of many metastable states in polycrystalline samples (high-T_c bulk SQUIDs, shields, flux focusers, coils) investigated in the form of ensembles of independent fluctuators, produces low-frequency magnetic noise (37,38). In the most advanced specimens of SQUIDs made from epitaxial films, the number of such Josephson fluctuators is drastically reduced, and it is possible to observe new phenomena induced by a few, or even a single, JFs (38).

In this subsection, we carry out a numerical and experimental analysis of the dependence of the noise spectral density of the magnetic flux noise $S_\Phi(\omega)$ on the temperature, magnetic field, and JF parameters. The results of experimental investigations of $S_\Phi(\omega)$ in high-T_c RF SQUIDs are discussed both in the limit of discrete fluctuators (SQUIDs manufactured from high-quality epitaxial films) and in the case of a large ensemble of statistically independent fluctuators (bulk ceramic SQUIDs or thin films with a small critical current density $J_c < 3 \times 10^5$ A/cm^2).

Under certain conditions, a system of superconducting regions connected by Josephson or Josephson-like junctions can be regarded as a quasicontinuum and can be described by certain effective parameters in analogy with type II superconductors (41,42). In the model of a quasiregular Josephson array of granules forming a cubic lattice of period a, the penetration depth of the magnetic field depends on the screening due to intergranular Josephson currents (43,44):

$$\lambda = \left(\frac{a\Phi_0}{2\pi\mu_0 S j_c} \right)^{1/2} \tag{52}$$

Here, μ_0 is magnetic permeability of the vacuum, j_c is the average transition current density, and S is the effective area of penetration of the field into intergranular space per unit cell, with allowance for London penetration depth. The effective field penetration depth for a two-dimensional system modeling a polycrystalline high-T_c film of thickness d is

$$\lambda_{\text{eff}} = \frac{\lambda^2}{\text{d}} \tag{53}$$

and from Eqs. (52) and (53), we find

$$\frac{\lambda_{\text{eff}}}{a} = \frac{\Phi_0}{2\pi L_0 I_c} \tag{54}$$

where $I_c = j_c ad$ is the transition current of a Josephson junction between adjacent granules and $L_0 = \mu_0 S/a$ is the inductance of the elementary intergranular loop of JF. This inductance, like the area S, depends on the temperature, increasing as the transition temperature T_c of the homogeneous high-T_c superconducting materials is approached. If we consider the system to be close packed and disregard the size of the nonsuperconducting regions at the grain boundaries, we find that the effective area of penetration of the magnetic field varies from $S \sim a\lambda_L(T)$ for $\lambda_L \leq d \ll a$ to $S \sim a^2$ for $\lambda_L(T) \leq (da)^{1/2}$.

The granular system can be regarded as a continuum under the condition $\lambda_{eff} \gg a$ (i.e., $\Phi_0 \ll 2\pi L_0 I_c$. This condition is always satisfied sufficiently close to the transition temperature. However, for a high intergranular transition current density and with an increase in the period a (large size of polycrystalline high-T_c materials), the depth λ_{eff} becomes of the order of the granule size and the system begins to exhibit a discrete character. In high-quality epitaxial high-T_c superconducting films, the dimensions of the grains at whose boundaries the Josephson links form, can be quite large. It is therefore necessary to go to a discrete description of the system and to consider a Hamiltonian in which the energy of the Josephson links depends on the phase difference (37,38,44):

$$\tilde{H} = \sum_{\langle i,j \rangle} E_{ij}(1 - \cos \Theta_{ij}) + E_M \tag{55}$$

where

$$\Theta_{ij} = \varphi_i - \varphi_j - \left(\frac{2\pi}{\Phi_0}\right) \int \mathbf{A}\, dr \tag{56}$$

is the gradient-invariant Josephson phase difference between adjacent crystalline granules within which the phases of the order parameter are equal to φ_i and φ_j. The integral of the vector potential of the magnetic field $\mathbf{A}$ is evaluated along a contour situated in the region of the weak link; E_M is the magnetic energy of the system. The width of the Josephson junction is assumed to be smaller than or of the same order as the penetration depth of the intergranular weak link λ_J, and the magnetic field obeys $H \ll H_{cj} \sim \Phi_0/\mu_0\lambda_J\lambda_L$. The spatial variation of the phase difference along the intergranular boundary can be disregarded under these conditions.

As mentioned earlier, the number of weak links in epitaxial high-T_c thin-film SQUID sensors is not very large, and the small number of macroscopic Josephson fluctuators determines their magnetic flux noise. The process of thermally activated transitions in such fluctuators can be described quite accurately in the approximation of two-level systems whose parameters depend on the external magnetic field. It is well known that the spectral density of the magnetic flux fluctuations in this case is characterized by Lorentzian frequency dependence:

$$S_{\Phi(\omega)} = \frac{\delta\Phi^2}{\pi}\left(\frac{\tau}{1 + (\omega\tau)^2}\right) \tag{57}$$

where

$$\delta\Phi = \frac{\Phi_0^2 \Delta\Phi}{4\cosh^2(\Delta U/2k_B T)} \tag{58}$$

is proportional to the root mean square value of the flux fluctuations, and the characteristic time τ is equal to

$$\tau = \frac{1}{\omega_0}\left(\frac{\exp(U_0/k_B T}{1 + \exp(-\Delta U/k_B T}\right) \tag{59}$$

Here, $\Delta\Phi$ is the normalized difference between the magnetic fluxes in the loop for the two switching states. U_0 is the height of the barrier that must be surmounted upon exit from a higher-energy state. ΔU is the absolute value of the difference in the energy of the states, and the frequency ω determines the relaxation rate of the system to the state of a local energy minimum. All of these quantities are calculated for a given external magnetic flux from the equations describing the metastable energy stages of the two-level system (37). The significant parameters in this case are the dimensionless quantities characterizing the elementary JF:

$$\ell(T) = \frac{2\pi L I_c}{\Phi_0} \tag{60}$$

and

$$\frac{E_L}{k_B T_c} = \frac{\Phi_0^2}{4\pi^2 k_B T_c L} \tag{61}$$

where L is the loop inductance and $I_c(T)$ is the transition current (all of the weak links are assumed to be identical in this model).

For typical intergranular weak links in epitaxial high-T_c superconducting films, one observes current–voltage characteristics corresponding to the resistive Josephson junction model in the strong damping regime (19,45). In this case, $\omega_0 \sim L/R$; that is, it is determined by the loop inductance L and the normal resistance R of the junction that is typically $\omega_0 \sim 10^{10}$–10^{11} s^{-1}.

The strong temperature dependence of the spectral density $S(\omega)$ near the transition temperature is associated with the variation of the quantity $\ell(T)$ in terms of which the JF energy characteristics in relations (5)–(7) are expressed. In numerical analysis we consider two cases corresponding to different forms of the temperature dependence of $I_c(T)$. In the first case, $I_c(T)$ is chosen in accordance with the Ambegaokar–Baratov equation, which is valid for links of the SIS type (45). For numerical calculations, a constant parameter $\ell(0) = 2\pi L I_c(0)/\Phi_0$ is specified, and the temperature dependence $\ell(T)$ is calculated on the assumption that the superconducting gap $\Delta(T)$ has a Bardeen–Cooper–Schrieffer temperature dependence. The second case is drawn on the results of measurements of the characteristics of the intergranular weak links (46) in epitaxial films of YBaCuO, which show that the transition current is proportional to $(1 - T/T_c)^2$ at temperatures close to T_c. Such dependence is inherent in weak links of the SNS type (47). In our case, $\ell(T)$ is chosen in the form

$$\ell(T) = \ell_1 \left(1 - \frac{T}{T_c}\right)^2 \quad (62)$$

where the constant ℓ_1 is used as an independent parameter in calculations and whose value is chosen from experimental data (48) on the transition current. The inductance L of the loop is estimated from its possible length range 10^{-1}–10^{-2} cm, which corresponds to inductance $L \sim 10^{-10}$–10^{-11} H. At the transition temperature $T_c = 90$ K, the ratio E_L/k_BT lies in the interval 0.1–10 for weak links in epitaxial films. We will give only the results of a numerical calculation of the temperature dependence of the spectral density of the flux fluctuations for a single JF when the external magnetic flux is close to $\Phi_0/2$. Here, the form of the potential energy of the system is such that the latter can be regarded as a two-level system (elementary fluctuator).

Figures 7.8 and 7.9 show graphs of the calculated temperature dependence of the spectral density of magnetic flux noise at a fixed observation frequency ω chosen equal to $10^{-11}\omega_0$; that is, $f = 0.1$–1 Hz, which corresponds to the frequency at which excess magnetic noise is usually observed in high-T_c films and

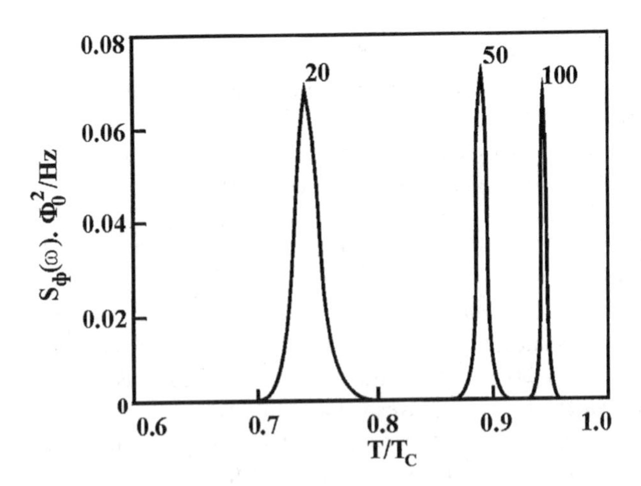

FIGURE 7.8 Spectral density of magnetic flux noise versus temperature for three Josephson fluctuators at an observation frequency $\omega = 10^{-11}\omega_0$, $E_L/k_BT_c = 1$, and $\Phi_e = 0.5\Phi_0$. The parameter of the family of curves is the normalized current $I(0) = 20, 50, 100$ with Ambegaokar–Baratov temperature dependence. (From Ref. 38.)

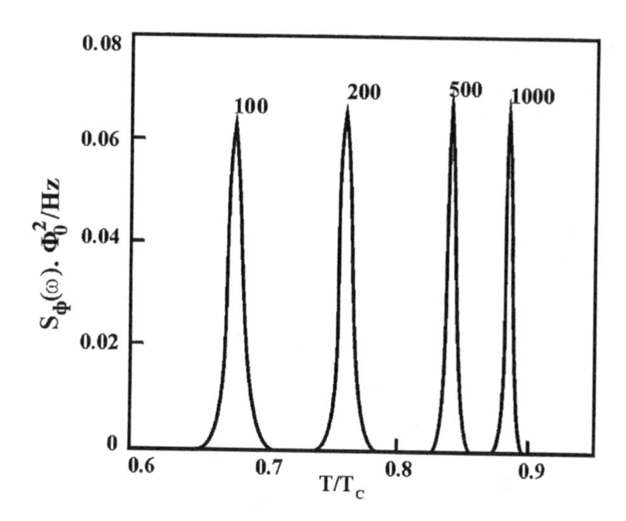

FIGURE 7.9 Temperature dependence of spectral noise density at observation frequency $\omega = 10^{-11}\omega_0$ for four fluctuators whose transition currents depend on the temperature in the form $(1\text{-}T/T_c)^2$, $E_L/k_BT_c = 1$, and $\Phi_e = 0.5\Phi_0$. The parameter of the curves is $l_1 = 100, 200, 500, 1000$. (From Ref. 38.)

SQUIDs. The external magnetic flux through the loop is fixed at the value $\Phi_0/2$. The noise intensity differs appreciably from zero in a narrow temperature interval. As the transition current of the Josephson fluctuator decreases (i.e., as the parameters $\ell(0)$ in Figure 7.8 and ℓ_1 in Figure 7.9 decrease), the $S(T)$ peak shifts toward lower temperatures and its width increases. The spectral density of the noise acquires a narrow temperature peak as a result of strong exponential temperature dependence of the quantity $\delta\Phi^2$ and the time in Eqs. (57) and (59). For a fixed inductance, the variation of the coupling weak-link energy (i.e., the transition current of Josephson fluctuators) shifts the noise peak on the temperature scale and causes its amplitude to change slightly.

The temperature dependence of spectral power of the magnetic flux fluctuations for various inductances of the Josephson fluctuator is shown in Figure 7.10, where E_L/k_BT is varied for a temperature dependence of the transition current of the form Eq. (62) and the L-independent parameter $\ell_1E_J/k_BT = \gamma_1$ is fixed. Specifically, $\gamma_1 = \Phi_0I_c/2\pi k_BT$ is a parameter characterizing the influence of thermal fluctuations on the form of I–V characteristics near the transition temperature (46,47).

If the external magnetic flux differs from $0.5\Phi_0$, the potential energy of the Josephson fluctuator is asymmetric. The total noise intensity decreases as the difference in the energy of the stationary states increases. The suppression of the tem-

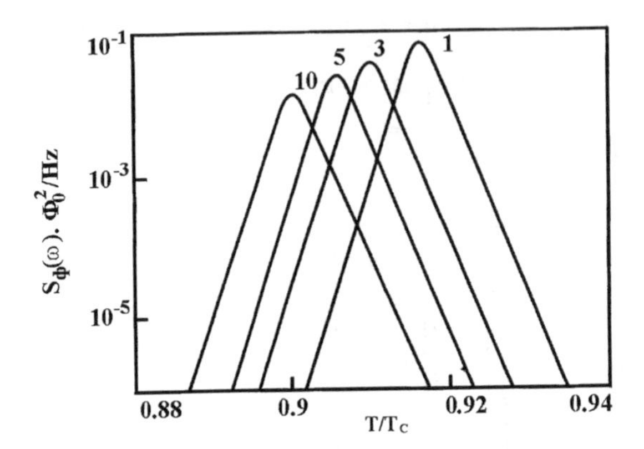

FIGURE 7.10 Temperature dependence of the spectral noise density of Josephson fluctuators at low frequencies $\omega = 10^{-11}\omega_0$. The parameter of the family is $E_L/k_BT_c = 1, 3, 5, 10$ [i.e., the inductance (or dimensions) of the JF]. Here, $\Phi_e = 0.5\Phi_0$, $I(T) = I_1(1 - T/T_c)^2$, and $\gamma = 2000$. (From Ref. 38.)

perature peak of $S(\omega)$ due to the increased asymmetry of the potential of the two-level system is far more pronounced for peaks closer to the transition temperature (i.e., for a high transition current) (Fig. 7.11).

Consequently, when several Josephson fluctuators appear in epitaxial high-T_c superconducting films, RF and dc SQUIDs, input coils, flux focusers, bolome-

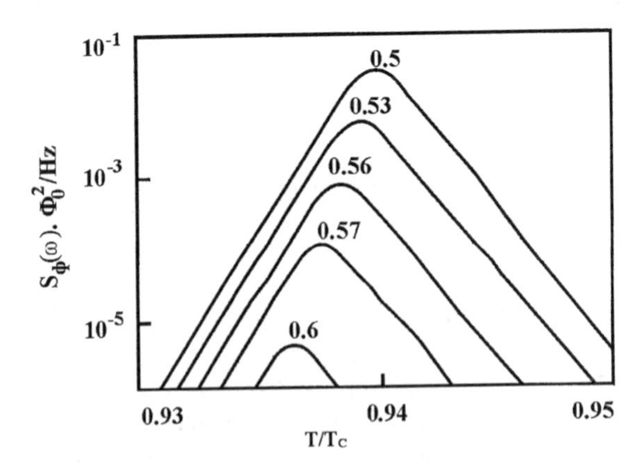

FIGURE 7.11 Temperature dependence of the spectral noise density of Josephson fluctuators at low frequencies $\omega = 10^{-11}\omega_0$. The parameter of the family of curves is the external magnetic flux $\Phi_e/\Phi_0 = 0.5, 0.53, 0.56, 0.57, 0.6$. Parameters $E_L/k_BT_c = 1$, $I_1 = 1000$, and $\gamma = 5000$ are fixed. (From Ref. 38.)

ters and SIS mixers, the spectral density of the magnetic flux noise will have a nonmonotonic temperature dependence, which can be represented by the superposition of the noise contributions from each of the Josephson fluctuators. The maximum of the noise intensity will be spaced along the temperature scale in accordance with the scatter of the critical currents, sizes of the loop, and mutual inductance between high-T_c SQUID and fluctuators. The intensity of the noise has a sharp dependence on the external magnetic field with period $\Delta B = \Phi_0/S_{eff}$, where S_{eff} is the effective area of the JF with allowance for the amplification of the magnetic flux by superconducting edges, the permeability of the medium, and so forth.

The nature of the temperature and frequency dependence of the noise spectrum changes upon transition from SQUID sensors having a small number of weak links (high-quality thin-film sensors) to systems having many junctions with statistically distributed parameters (bulk SQUIDs). Averaging over the heights of the energy barriers of the two-level systems (i.e., over the critical currents) yields a frequency dependence of the $1/f$ type. The spectral density $S_\Phi(\omega)$ as a function of the temperature has a maximum near T_c, which is caused by the existence of the above-mentioned peaks of the function $S_\Phi(T)$ for individual fluctuators with high critical current densities ($\ell >> 1$). For bulk ceramic sensors, the noise spectral density has a maximum shift toward lower temperature. As before, the main contribution to the spectral power is from those Josephson fluctuators through which the magnetic flux is close to a half-integral number of magnetic flux quantum. However, the scatter of effective areas of the current loops causes the spectral density of the magnetic noise of the disordered system to have a weak dependence on the external magnetic flux, in contrast to the strong quasiperiodic dependence associated with the individual JF.

From the experimental point of view, the magnetic fluctuations in high-T_c materials can be measured by using an apparatus in which the receiving antenna of a low-temperature SQUID is inductively coupled with the high-T_c sample (36–40). Figure 7.12 shows typical low-noise spectra for bulk polycrystalline YBaCuO sample with critical current density $J_c(T = 77\text{ K}) \sim 300\text{ A/cm}^2$. Below T_c and down to 8 K, the spectral noise density varies approximately as $1/f$ in the range 1–10 Hz. This is consistent with the predictions of the theory of the model of an ensemble of independent fluctuators. Far from T_c (see the curve for $T = 8$ K), the values of $S_\Phi(\omega)$ at frequency of 1 Hz are only an order of magnitude lower than that at $T = 88.4$ K, where this noise is a maximum. The fact that the excess noise amplitude decreases slowly with decreasing temperature in the case of bulk ceramic with $J_c = 300\text{ A/cm}^2$ is attributed to large scatter in absolute values of the transition currents of individual junctions. This noise arises from incoherent superposition of many thermally activated jumps of the supercurrent and the internal magnetic flux due to Josephson fluctuators, in low external magnetic fields up to 10^{-10} T.

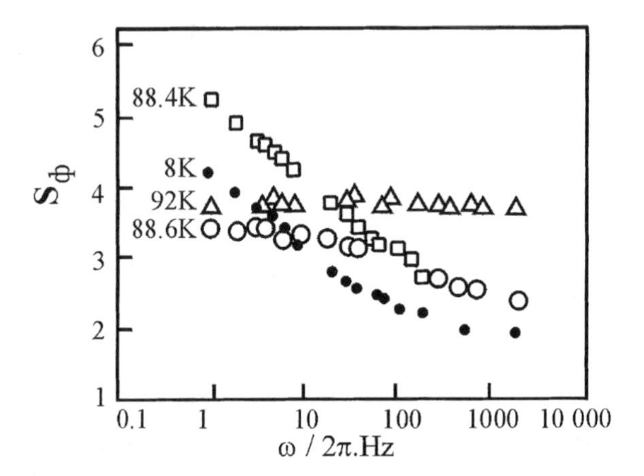

Figure 7.12 Spectral noise density of magnetic flux versus frequency for a polycrystalline sample of YBaCuO ceramic at different temperatures *T*. The external magnetic field is lower than 10^{-9} T. (From Ref. 38.)

7.8.2 Flux Creep Noise in High-T_c Superconductors

The random telegraph fluctuations of the magnetic flux are also observed in high-T_c films (39) which are being interpreted in terms of thermally activated jumps of single Abrikosov vortices between two fixed pinning sites (this is another type of elementary fluctuator). High-T_c oxide materials are type II superconductors with a relatively small lower critical field H_{c1}. A thin-film high-T_c RF SQUID, the flux focuser has a large demagnetizing factor for perpendicular fields, hence even weak fields like the Earth's cause Abrikosov flux vortices to penetrate the perimeter of the sensors. Abrikosov vortices also have other unwanted effects, including drift in SQUIDs output due to flux creep and perhaps most importantly, in some applications, an increase of flux and voltage hysteresis to field variations of the order of the Earth's magnetic field (44).

Magnetic relaxation effects arising from thermally activated flux motion were known to occur in conventional superconductors, although on a much smaller scale. Relaxation is assumed to take place via jumps of a flux line boundless from one pinning center to another. However, the assumption that the activation energy $E(T)$ for flux motion is much larger than the thermal energy k_BT is not applicable to high-T_c SQUID material. On the basis of Monte Carlo simulations (48), it is shown that, for not too long times, the relaxation of the magnetization in high-T_c materials (SQUID output signal drift) is well described by

$$V(t, T) = V_0\left[1 - \frac{k_BT}{E(T)}\ln\left(1 + \frac{t}{\tau}\right)\right] \tag{63}$$

even in the case where the activation energy $E(T)$ is comparable to k_BT. [$\sim \omega \exp[E(T)/k_BT]$. Here, ω is essentially given by the attempt frequency for a flux line to make a thermally activated jump from one pinning state to another. For a very long time [i.e., $V(t, T) < 0.1V_0$] the drift of the output voltage (due to magnetization drift) is shown to follow a simple exponential decay (48). It is evident that twin-plane pinning might be important in YBaCuO as compared to BiSrCaCuO, which does not have twin planes. Figure 7.13 shows the distribution function of activation energies for thermally activated flux motion, probably the first obtained (49) on YBaCuO superconductors used for high-T_c SQUIDs, shields, and flux focuser preparation. It is very interesting that a much narrower distribution was found in BiSrCaCuO materials.

The adverse effects of Abrikosov vortex penetration in 77 K SQUIDs, including excess low-frequency noise, hysteresis, and drift, are minimized when the critical current J_c of the superconducting material is large. Qualitatively, this is to be expected because J_c is a measure of both the material's ability to resist vortex penetration and the strength with which vortices, once present, are pinned. Because the largest current densities occur near the device edges, care must be taken during processing to ensure that minimal degradation of film-edge, bulk-edge J_c occurs during patterning of the sensors.

7.8.3 Processing of Bulk High-T_c Materials for RF SQUIDs

For most high-T_c thin-film SQUID applications, large critical current densities of the order of 10^6–10^7 A/cm^2 are required, often in the presence of magnetic fields. Critical current density is not an intrinsic property of oxide superconductors and is strongly dependent on its fabrication technology, processing conditions, mi-

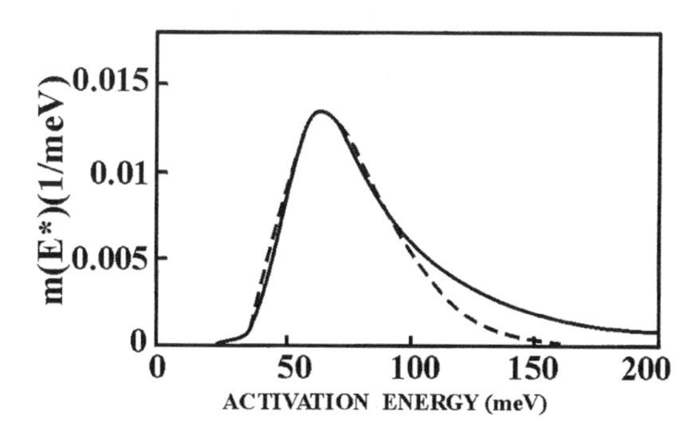

FIGURE 7.13 Distribution function of activation energy for thermally activated flux motion on a single crystal of YBaCuO. A log-normal distribution is shown as the dashed line. (From Ref. 48.)

crostructure, pinning centers, and so forth (50). The important point about bulk high-T_c RF SQUIDs is that they operate well (with good sensitivity $\sim 10^{-28}$ J/Hz) in external magnetic field with a "small" critical current density of bulk material ($\sim 10^3$ A/cm^2). In this case, penetration of the external magnetic field is described by $dx = dH/J_c$. Therefore, when J_c is high, a large magnetic field in the interior regions can be shielded by the bulk superconductor. This effect can be used for magnetic shielding by oxide superconductor high-T_c electronics circuits. Bulk high-T_c SQUIDs are low-cost, very simple to design, and can provide good sensitivity in the Earth's magnetic field. Up until now, the main differences in its characteristics are results of bulk ceramic properties (J_c = 300–1000 A/cm^2) and weak-link parameters.

Various processes have been employed for the fabrication of bulk high-T_c superconductors. Sintering is very commonly used in ceramic processing and has many advantages in preparing ceramics in precise shapes needed for practical high-T_c RF bulk SQUIDs, focusers, and shields (51). The compounds for bulk SQUIDs can be produced by solid-state reactions at relatively low temperatures. Furthermore, by controlling heat-treatment conditions, microstructure characteristics such as grain size and critical current density can be controlled. By this method, good T_c values can be easily achieved (for YBaCuO, $T_c \sim$ 89–90 K). However, J_c values at 77 K are very small in bulk sintered oxide superconductors and limited by 1000 A/cm^2 for commonly used bulk SQUID sensors. Several parameters such as density, oxygen content, cracks, homogeneity, and energy coupling at grain boundaries are considered to be very important in determining J_c in bulk sintered YBaCuO SQUIDs sensors. These parameters are strongly affected by processing conditions, powders, and sintering technology.

The starting powders are prepared by various solution techniques (52). Densification is possible without the help of liquid, if fine powders (<1 μm) with a uniform size distribution are used. After pressing under high pressure (preferably by cold isostatic pressing), bulk ceramics can be sintered below 925°C, resulting in highly dense samples without the intrusion of liquid. However, even in such samples, J_c values are low. This is partly due to the presence of a crack along the grain boundaries. Oxide superconductors have an anisotropic crystal structure, and most properties such as thermal expansion coefficients are also anisotropic. In sintered samples of a bulk RF SQUID body where grain orientations are random, the stress will be accumulated at grain boundaries during thermal cycles, leading to cracking and low values of the critical current density ($J_{c\,\max} \sim 1000$ A/cm^2 at 77 K in zero field) (see Fig. 7.14). The fact that the single crystals exhibit J_c values to two orders of magnitude larger and excess low-frequency noise two orders of magnitude lower than polycrystalline samples suggests that grain boundaries are the weak links.

As mentioned earlier, the thermally activated transitions between metastable states in macroscopic Josephson fluctuators and in a system of

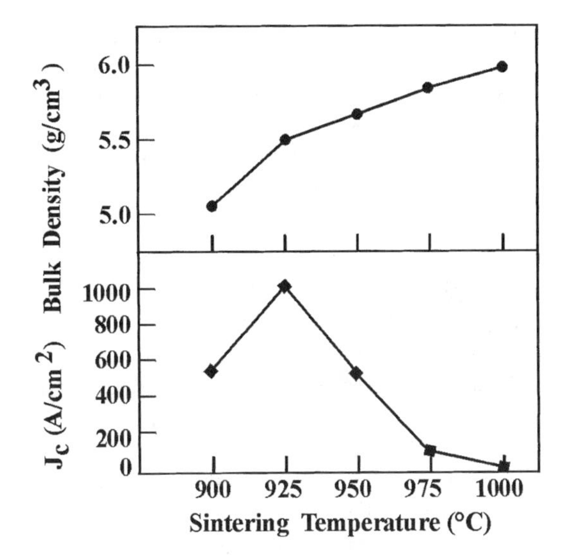

FIGURE 7.14 Effect of sintering temperature on the bulk density and J_c of bulk YBaCuO. Note that although the bulk density can be increased by increasing the sintering temperature, J_c decreases on sintering above 950°C. This is partly due to the presence of a crack along the grain boundaries. (From Ref. 51.)

Abrikosov vortexes cause low-frequency excess noise in RF SQUIDs. Both phenomena can induce the Lorentzian flux noise spectrum [see Eq. (57) for "telegraph" noise] in high-T_c materials, Josephson junctions, and RF SQUIDs. We find that for both high-quality thin-film and bulk RF SQUIDs ($\omega/2\pi = 24$ MHz), sensitivity at low frequencies is $\Delta B \sim 1.0 \times 10^{-12}$ T/Hz$^{1/2}$ at 1 Hz in the Earth's magnetic field without additional noise-compensation electronics. For best bulk SQUIDs, there are simple relations $\Delta B \sim 1/J_c$ in high-T_c material. It is an order of magnitude larger than usually obtained for the best of these high-T_c sensors in zero magnetic fields. In summary, for the reduction of low-frequency noise in high-T_c RF SQUIDs, further improvements in the processing technology are required, especially with regard to the quality of the edge (thin-film sensors), Josephson junctions, and quality (critical current density $\sim 10^4$ A/cm^2) ceramic bulk materials based on melt-textured technology or the technology UPtYBaO, which can be formed within melt-textured YBaCuO by the addition of U (53).

7.9 HIGH-T_c SQUID APPLICATIONS

Currently, the sensitivity of high-T_c RF SQUIDs reaches 40 fT/Hz$^{1/2}$ for several samples (54–56). The parameter of noise ($1/f$) begins for better samples with units of several hertz. It is obvious that such a sensitivity range is not sufficient for the application in magneto-encephalographs. On the other hand, this value is quite enough for measurements of magneto-cardiograms, myograms, iron overload in human liver, and magnetized lung contaminants (57). It is also clear that employment of nitrogen-cooled SQUIDs for solving these tasks of biomagnetism would greatly improve the cryogenic facility service at large hospitals and would give an impetus to use such instruments at medical facilities in a wide area of medical application (55,58–61).

Currently, a required sensitivity of 30–40 $fT/Hz^{1/2}$ field strength is possessed only by occasional RF SQUIDs samples, whereas the overwhelming majority of researchers' groups claim of attaining a range of 100 $fT/Hz^{1/2}$ field strength. There are no objective reasons to doubt that in several decades to come, the SQUIDs with a sensitivity of 30 $fT/Hz^{1/2}$ field strength will be manufactured for wide commercial availability. This advantage would enable one to use the high-T_c SQUIDs to win the medical-facilities market-segment worldwide (55). In the opinion of experts, neither data analysis nor modeling or signal processing are serious obstacles to the wide application of SQUID technologies at hospitals. The point is that at a regular hospital, everyday problems arise when it is necessary to employ liquid helium as a cryocoolant. For this reason, replacement of liquid helium by liquid nitrogen (or a cryocooler) would make handling of these instruments a trivial pursuit.

Another decent area for high-T_c SQUIDs application is presumably geophysics (62). As a matter of fact, the above-mentioned sensitivity, as well as dynamic range and frequency band of the RF SQUIDs (being realized today), are quite sufficient for implementing a great number of geophysical investigations. In the RF SQUID, there still exist some underdeveloped $1/f$-noise-field problems for frequencies lower than 1 Hz.

It is clear that substituting the liquid helium by liquid nitrogen would ensure a dramatic economic gain from field practice in the realm of remote geology. Recently, similar pioneering studies with high-T_c SQUID magnetometers have been done by a researchers team from Ulich (63). All of these factors contribute to the idea of high-T_c SQUID employment in the problems of paleo-magnetism measurements in remote-territory geologic expeditions.

Prospects of high-T_c RF SQUID usage in nondestructive evaluation and testing (NDE and NDT) methodologies for industrial objects (64) are very optimistic. As publications show (65–68), most of the NDE and NDT technologies (such as the search for metal cracks, chemical corrosion, magnetic impurities, or occasional body cavities) are well applicable at field-strength sensitivity values in the range 30–100 $fT/Hz^{1/2}$. For such a field sensitivity range, the high-T_c RF SQUIDs are quite competitive with rather expensive low-T_c SQUIDs.

It should be noted that there exists a specific field, such as magnetic microscopy imaging techniques, where high-T_c SQUIDs are, by far, more advantageous, contrary to their low-T_c opponents, due to the fact that the former can be physically positioned in closest vicinity to a sample surface. The matter is that a source field decreases as $1/R^3$. Therefore, reducing a detector-to-object clearance by three to four times would result in a signal-intensity gain by 1.5–2 orders of magnitude. Technically speaking, the conventionally adopted 8–10-mm "gold size" for helium dewar may be reduced to 2.5–3 mm for nitrogen cryostats.

A special focus should be concentrated on the creation of ultrasensitive electromagnetic metal detectors (69) based on high-T_c SQUIDs. These detector systems include active excitation coils of low-frequency magnetic field, whereby a signal being induced in a sample object is measured by SQUID detector. It is noteworthy that such radarlike systems (of 10^{-13} T/m /Hz$^{1/2}$ sensitivity) ensure a detection of metal objects even in conducting media, such as under sea level at depths up to a dozen meters, whereby seawater conductivity is rated as much as 4 S/m. Undoubtedly, such a radar–SQUID trend will evolve using a low-T_c SQUID also; however, high-T_c SQUIDs, in their turn, possess a definite relevant niche.

As a matter of fact, if a SQUID's input circuit (i.e., antenna) is construed to a phenomenon of stochastic resonance (70), the signal-to-noise ratio would be improved by 20 to 30 times (71). Generally speaking, stochastic resonance effect-based antenna inclusion in the above-mentioned input circuits can facilitate a fast coming-to-market for active SQUID systems for biomagnetic application, such as lung pollution tests, iron overload in liver, NDE and NDT, scanning microscopy, electromagnetic metal detectors, and so forth.

7.10 CONCLUSIONS

In the recent decade, significant progress has been achieved in the field of response and energy-sensitivity enhancement for high-T_c thin-film RF SQUIDs (6,9,23,30,56,58,59,63,72–75). Figure 7.15 shows the spectral density characteristic of the noise for high-T_c RF SQUIDs operating at two RF frequencies (76). By increasing the pumping frequency to 360 MHz from 20 MHz, a considerable improvement has been found. For 360 MHz operating frequency, flux noise is $\sim 5 \times 10^{-5}\ \Phi_0/\text{Hz}^{1/2}$ which corresponded to a field sensitivity of 6×10^{-14} T/Hz$^{1/2}$. Figure 7.16 shows the frequency dependence of flux noise of RF and dc SQUIDs for the $L > L_F$ limit (31). The parameters R and L are the same for both the SQUIDs. It shows that there are no $1/f$ noise at 1 Hz as predicted by theory (31) for RF SQUID.

It has proven to be possible to obtain in this range of inductance (of large fluctuations), practically the same energy sensitivity as that of best dc high-T_c SQUIDs that operate within the $L < L_F$ regime. This has been possible to increase in the excitation frequency (UHF high-T_c SQUIDs) and improvement of $k^2Q\ell$ (9,77).

The good results for dc and RF high-T_c SQUIDs have been obtained in the application of μ-metal shields in exterior fields which are by far lower compared to magnetic field of the Earth. With an increased exterior magnetic field, a spec-

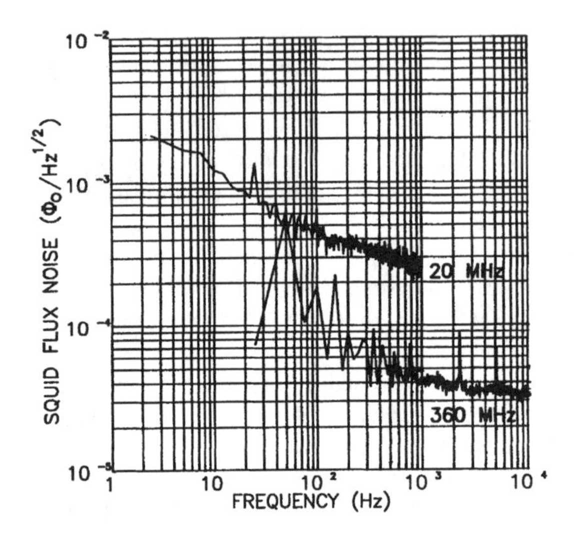

FIGURE 7.15 Flux noise versus frequency for thin-film high-T_c RF SQUIDs with two RF frequencies in a low magnetic field. (From Ref. 76.)

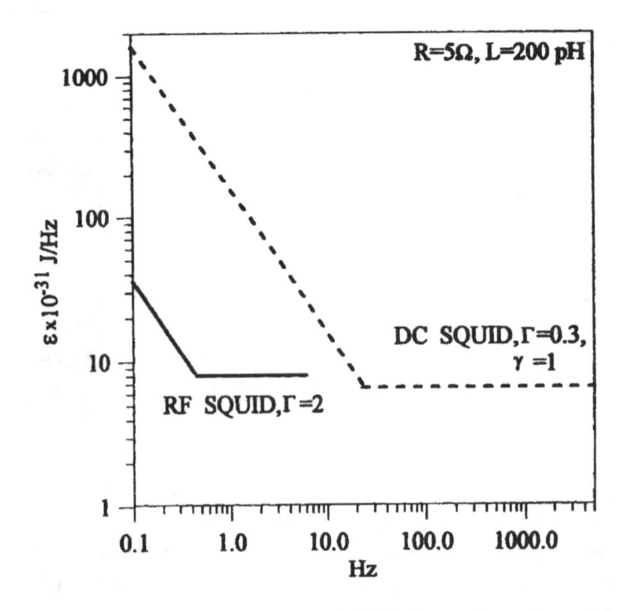

FIGURE 7.16 The energy sensitivity as a function of frequency. The parameters R and L are identical for both RF and dc SQUIDs; $L > L_F$. (From Ref. 31.)

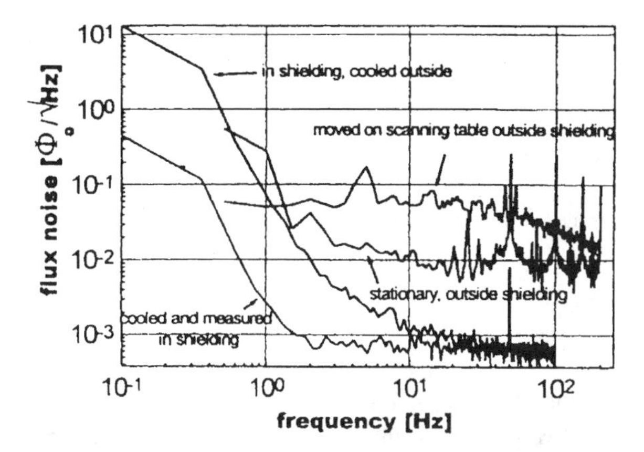

FIGURE 7.17 $1/f$ flux noise and field resolution of a high-T_c RF washer-type SQUID in external magnetic field. (From Ref. 78.)

tral density of noise at low frequencies is abruptly deteriorating (78) (see Fig. 7.17), which is also typical for high-T_c dc SQUIDs.

Progress in the bulk RF SQUIDs is considerably slow. Practically up to date, there are no energy sensitivities better that those described in Refs. 74 and 75, as yet. However, the importance of sensitivity enhancement for bulk high-T_c RF SQUIDs, for example, due to improved materials properties or an increased merit factor or a resonance circuit, input coils, and so forth is emphasized by the fact that the RF SQUIDs do not, practically, change the noise spectral density up to ~ 1 Oe within exterior magnetic fields.

As a conclusion, we list several tasks, a resolution hereof is stimulated by further progress of single-contact high-T_c SQUIDs, or, more accurately, due to SQUID applications in NDE, biomagnetism, geomagnetism and some other tasks:

- The creation of Zimmerman multiloop RF interferometer with $L < L_F$ inductance based on oxide materials of high (up to 10^4–10^5 A/cm^2) critical current density
- The creation of bulk-wire superconducting input (and output with improvement Q) circuits for magnetometers and gradiometers
- The creation of reproducible submicrometer high-T_c Josephson junctions for thin-film Zimmerman multiloop SQUIDs as well as of input and output thin-film-struture circuits (i.e., of completely integrated chips)
- The creation of input amplifiers for high-T_c RF SQUIDs based on stochastic resonance phenomena

For the best operation of the high-T_c RF SQUIDs at nitrogen temperatures, one can use interferometers in the small fluctuation limit, which were described in detail by Kurkijarvi–Webb and Danilov–Likharev (79,80) theory.

ACKNOWLEDGMENT

The author thanks A. S. Garbuz for stimulating discussions.

REFERENCES

1. B Cabrera. Fundamental physics experiments using SQUIDs. In: A Barone, ed. Principles and Applications of SQUIDs. Singapore: World Scientific, 1992, pp 345–416.
2. GB Donaldson. SQUIDs for Everything Else. In: H. Weinstock, M Nisenoff, eds. Superconducting Electronics. NATO ASI Series, Vol. F39. Berlin: Springer-Verlag, 1989, pp 175–207.
3. JG Bednorz, KA Muller. Possible high-T_c superconductivity in the BaLaCuO system. Z Phys B 64:189–193, 1986.
4. H Weinstock. A review of SQUID magnetometry applied to nondestructive evaluation IEEE Trans Magnetics M-27:3231–3236, 1991.
5. A Moya, F Baudenbacher, JP Wilkswo, FC Wellstood. Design of high resolution high-T_c SQUID magnetometers for biomagnetic imaging. IEEE Trans Appl Supercond AS-9:3511–3514, 1997.
6. R Hohmann, H-J Krause, H Soltner, Y Zhang, CA Copetti, H Bousack, AI Braginski, MI Faley. HTS SQUID system with Joule–Tomson cryocooler for eddy current nondestructive evaluation of aircraft structures. IEEE Trans Appl Supercond AS-7:2860–2865, 1997.
7. T Nagaishi, H Kugai, H Toyoda, H Itozaki. NDE of high speed fine particles by high-T_c SQUID. IEEE Trans Appl Supercond AS-:2886–2889, 1997.
8. TS Lee, YR Chemla, E Dantsker, J Clarke. High-T_c SQUID microscope for room temperature samples. IEEE Trans Appl Supercond AS-7:3147–3150, 1997.
9. Y Zhang, M Muck, K Herrmann, J Shubert, F Ruders, W Zader A.I. Braginskij, C Heiden. Low-noise YBaCuO rf SQUID magnetometer. Appl Phys Lett 60:645–647, 1992.
10. SA Bulgakov, VB Rybov, VI Shnyrkov, DM Vavriv. Effect of the magnetic-flux variations on SQUID stability. J Low Temp Phys 83:241–255, 1991.
11. AA Golubov, VM Krasnov, MY Kupriyanov. Properties of HTS step-edge SNS Junctions. IEEE Trans Appl Supercond AS-7:3204–3207, 1997.
12. JB Bulman, OO Salazar, JM Murduk. AFM analysis of step-edge Josephson junctions. IEEE Trans Appl Supercond AS-7:3650–3653, 1997.
13. OM Froehlich, P Richter, A Beck, D Koelle, R Gross. Barrier properties of grain boundary junctions in high-T_c superconductors. IEEE Trans Appl Supercond AS-7:3189–3192, 1997.
14. LR Vale, RH Ono, DA Rudman. YBaCuO Josephson junctions on bicrystal Al_2O_3 and $SrTiO_3$. IEEE Trans Appl Supercond AS-7:3193–3197, 1997.

15. Y Soutome, Y Okabe. Magnetic field modulation of critical currents in YbaCuO Coplanar. IEEE Trans Appl Supercond AS-7:2311–2314, 1997.
16. SG Haupt, DK Lathrop, R Matthews, SL Brown, R Altman, WJ Gallagher, F Milliken, JZ Sun, RH Koch. Materials issues related to the fabrication of HTS SQUIDs. IEEE Trans Appl Supercond AS-7:2319–2326, 1997.
17. E Il'ichev, V Zakosarenko, V Schultze, H-G Meyer, HE Hoenig, VN Glyntsev, A Golubov. Temperature dependence of the current–phase relation for $YBa_2Cu_3O_{7-x}$ step-edge Josephson junctions. Appl Phys Lett 72:731–733, 1998.
18. E Il'chev, V Zakosarenko, RPJ Isselsteijn, V Schultze, H-G Meyer, HE Hoenig, H Hilgenkamp, J Mannhart. Nonsinusoidal current–phase relationsip of grain boundary Josephson junctions in high-T_c superconductors. Phys Rev Lett 81:894–897, 1998.
19. VV Danilov, KK Likharev, OV Snigirev. Signal and noise parameters of SQUIDs. Proceedings of International Conference on SQUIDs, SQUID'80. 1980, pp 445–472.
20. T Ryhanen, H Seppa, R Ilomoniemi, J Knuutila. SQUID magnetometers for low-frequency applications. J Low Temp Phys 76:287–386, 1989.
21. VI Shnyrkov, VA Khlus, GM Tsoi. On quantum interference in a superconducting ring closed by weak link. J Low Temp Phys 39:477–496, 1979.
22. VA Khlus, IO Kulik. Fluctuations and quantum interference in system with weak links. Z Tehn Fiz 45:449–452, 1975.
23. M Muck, D Diehl, C Heiden. A planar gradiometer based on a microwave RF-SQUID IEEE Trans Appl Supercond AS-27:2465–2468, 1993.
24. RA Buhrman. Noise limitations of rf SQUIDs. In: HD Hahlbohm, H Lubbig eds. Superconducting Quantum Interference Devices and Their Applications. Berlin: Walter de Gruyter, 1977, pp 395–431.
25. LD Jackel RA Buhrman. Noise in the rf SQUID. J Low Temp Phys 19:201–246, 1975.
26. IM Dmitrenko, VI Shnyrkov, GM Tsoi, VV Kartsovnik. RF SQUID in the nonhysteretic regime with $k^2Q\ell > 1$. Physica B 107:739–740, 1981.
27. VI Shnyrkov, GM Tsoi. Signal and noise characteristics of RF SQUIDs. In: A. Barone, ed. Superconducting Quantum Interference Device, Singapore: World Scientific Publishing, 1992, pp. 77–149.
28. PK Hansma. Superconducting single-junction interferometers with small critical currents. J Appl Phys 44:4191–4194, 1973.
29. N Khare, AK Gupta, S Chaudhry, SK Arora, V.S. Tomar, V.N. Ojha. Fabrication and performance of high-T_c two hole bulk and single hole thick film RF SQUIDs at 77 K. IEEE Trans Magnetics M-27:3029–3031, 1991.
30. JR Buckley, N Khare, GB Donaldson, A Cochran, Z Hui. Use of bulk high-T_c magnetometer for non-destructive evaluation. IEEE Trans Magnetics M-27:3051–3054, 1991.
31. B Chesca. Theory of RF SQUIDs operating in the presence of large thermal fluctuations. J Low Temp Phys 110:963–1001, 1998.
32. YS Greenberg, Self-consistent theory of *V–I* characteristic and of intrinsic noise of a hysteretic RF SQUID. J Low Temp Phys 92:367–413, 1993.
33. SS Tinchev. Current–phase relation in high-T_c weak links made by oxygen-ion irradiation. Physica C 222:173–176, 1994.
34. K Enpuku, Y Shinomura, T Kisu. Effect of thermal noise on the characteristics of a high-T_c superconducting quantum interference device. J Appl Phys 73:7929–7934, 1993.

35. K Enpuku, G Tokita, T Maruo, T Minotani. Parameter dependencies of characteristics of a high-T_c dc superconducting quantum interference device. J Appl Phys 78:3498–3503, 1995.
36. BH Moeckly, K Char. Properties of interface-engineered high-T_c Josephson junctions. Appl Phys Lett 71:2526–2528, 1997.
37. VA Khlus, AV Dyomin. AC magnetic susceptibility and flux noise in simple Josephson-coupled superconducting clusters. Physica C 212:352–364, 1993.
38. VI Shnyrkov, VP Timofeev, GM Tsoi, VA Khlus, AV Demin. Low-frequency magnetic flux noise in high-T_c superconducting SQUIDs. J Low Temp Phys 21:470–476, 1995.
39. M Johnson, MJ Ferrari, FC Wellstood, J Clarke, MR Beasly, A Inam, XD Wu, L Nazar, T Venkatesan. Random telegraph signals in high-temperature superconductors. Phys Rev B 42:10,792–10,795, 1990.
40. P Selders A M Castellanos, M Vaupel, R Wordenweber. Reduction of $1/f$-noise in HTS SQUIDs by artificial defects. IEEE Trans Appl Supercond AS-9:2967–2970, 1999.
41. M Tinkham. Flux motion and dissipation in high temperature superconductors. IEEE Trans Magnetics M-27:828–832, 1991.
42. R Gross, L Alff, A Beck, OM Froehlich, D Koelle A Max. Physics and technology of high temperature superconducting Josephson junctions. IEEE Trans Appl Supercond AS-9:2929–2935, 1997.
43. K Nakajima, Y Savada. Numerical analysis of vortex motion in two-dimensional array of Josephson junctions. J Appl Phys 52:5732–5743, 1981.
44. G Deutscher, KA Muller. Origin of superconductive glassy state and extrinsic critical currents in high-T_c oxides. Phys Rev Lett 59:1745–1747, 1987.
45. KK Likharev. Dynamic of Josephson Junction and Circuits. New York: Gordon & Breach, 1991.
46. P Gross, P Chaudhari, D Dimos, A Gupta, G Koren. Thermally activated phase slippage in high-T_c grain boundary Josephson junctions. Phys Rev Lett 64:228–231, 1990.
47. JP Sydow, M Berninger RA Buhrman, BH Moecky. Effect of oxygen on YBCO Josephson junction structures. IEEE Trans Appl Superconduct AS-9:2993–2996, 1999.
48. CW Hagen, R Griessen. Thermally activated magnetic relaxation in high-T_c superconductors. In: AV Narlikar ed. New York Nova Science, 1989, pp 1–36.
49. RJ Wijngaarden, R Griessen, J. Fendrich, WK Kwok. Influence of twin planes in YbaCuO on magnetic flux movement and current flow. Phys Rev B 55:3268–3275, 1997.
50. AA Golubov, VM Krasnov, MY Kupriynov. Properties of HTS step-edge SNS junctions. IEEE Trans Appl Supercond AS-7:3204–3207, 1997.
51. M. Murakami. Processing of bulk YBaCuO. Supercond Sci Technol 5:185–203, 1992.
52. S Jin, JE Graebner. Processing and fabrication techniques for bulk high-T_c superconductors: A critical review. Mater Sci Eng B7:243–260, 1991.
53. A Marino, F Ichikawa, H Rodriguez, L Rinderer. Superconducting properties as function of substrate type in BSCCO thin films. Physica C 282:2277–2278, 1997.

54. AI Braginski, K Barthel, B Chesca, YS Greenberg D Koelle, Y Zhang, X Zeng. Progress in undestanding of high-transition temperature SQUIDs. Physica C 341–348:2555–2559, 2000.
55. AI Braginski. Superconducting electronics coming to market. IEEE Trans Appl Supercond AS-9:2825–2836, 1999.
56. DF He, HJ Krause, Y Zhang, M Bick, N Wolters, W Wolf, H. Bousack. HTS SQUID magnetometer with SQUID vector reference for operation in unshielded environment. IEEE Trans Appl Supercond AS-9:3684–3687, 1999.
57. JW Wikswo Jr. SQUID magnetometers for biomagnetism and nondestructive testing: Important questions and initial answers. IEEE Trans Appl Supercond AS-5:74–120, 1995.
58. DA Konotop, SS Khvostov, VP Timofeev, VI Shnyrkov, GM Tsoi. Development of high-T_c superconductor SQUID-based magnetometer. Cryogenics 33:632–635, 1993.
59. Y Tavrin, Y Zang, M Muck, AI Braginski, C Heiden. YBCO thin film SQUID gradiometer for biomagnetic measurements. Appl Phys Lett 62:1824–1826, 1993.
60. J Borgmann, AP Rijpma, HJM Brake, H Rogalla. Highly balanced gradiometer systems based on HTS-SQUIDs for the use in magnetically unshielded environment. IEEE Trans Appl Supercond AS-9:3680–3683, 1999.
61. P Seidel, F. Schmidl, S Wunderlich, T Vogt, H Schneidewind, R Weidl, S Losche, U Leder, O Solbig, H Nowak. High-T_c SQUID system for practical use. IEEE Trans Appl Supercond AS-9:4077–4080, 1999.
62. J Clarke. Geophysical applications for SQUIDs. IEEE Trans Magnetics M-19: 288–294, 1983.
63. M Bick, G Panaitov, N Wolters, Y Zhang, H Bousack, AI Braginski, U Kalberkamp, H Burkhardt, U Matzander. A HTS rf SQUID vector magnetometer for geophysical exploration. IEEE Trans Appl Supercond AS-9:3780–3785, 1999.
64. GB Donaldson. SQUIDs for everything else. In: H Weinstock, M Nisenoff eds. Superconducting Electronics. NATO ASI Series Vol. 59, Berlin: Springer-Verlag, 1989, pp 175–207.
65. V Schultze, R Stolz, R Ijssenlsteijn, V Zakosarenko, L Fritzsch, F. Thrum E Il'ichev, HG Meyer. Integrated SQUID gradiometers for measurement in disturbed environments. IEEE Trans Appl Supercond AS-7:3473–3476, 1997.
66. A Haller, Y Tavrin, H-J Krause. Eddy-current nondestructive material evaluation by high-temperature SQUID gradiometer using rotating magnetic fields. IEEE Trans Appl Supercond AS-7:2874–2877, 1997.
67. C Carr, DMcA McKirdy, EJ Romans, GB Donaldson, A Cochran. Electromagnetic nondestructive evaluation: Moving HTS SQUIDs, inducing field nulling and dual frequency measurements. IEEE Trans Appl Supercond AS-7:3275–3278, 1997.
68. F Schmidl, S Wunderlich, L Dorrer, H Specht, S Linzen, H Schneidewind, P Seidel. High-T_c SQUID system for nondestructive evaluation. IEEE Trans Appl Supercond AS-7:2756–2759, 1997.
69. PV Czipott, WN Podney, Use of superconductive gradiometer in an ultrasensitive electromagnetic metal detector. IEEE Trans Magnetics M-25:1204–1207, 1989.
70. AD Hibbs, AL Singsaas, EW Jacobs, AR Bulsara, JJ Bekkedahl, F Moss. Stochastic resonance in a superconducting loop with a Josephson junction. J Appl Phys 77:2582–2589, 1995.

71. OG Turutanov, VI Shnyrkov, YP Bliokh. Stochastic resonance-based input circuit for SQUIDs. 5th European Conference on Appl. Superconductivity (EUCAS 2001), to be published.
72. VI Shnyrkov, VP Timofeev, SS Khvostov, GM Tsoi. Mod Phys Lett B 5:1281–1286, 1991.
73. N Khare. YBCO thin film rf-SQUID using bicrystal junction. Physica C 313: 281–284, 1999.
74. VI Shnyrkov, VP Timofeev, SS Khvostov, GM Tsoi, VY Kosiev, MD Strikovsky. UHF high-T_c thin film SQUID with sensitivity 3×10^{-30} JHz. Preprint, Institute for Low Temperature Physics, Kharkov, Ukraine, 1993.
75. VI Shnyrkov, GM Tsoi, GA Kozyr, and VN Glyntsev. Nitrogen-cooled RF SQUID sensitive up to 10^{-28} J/Hz. Fiz Niz Temp Sov J Low Temp Phys 14:770–773, 1988.
76. VP Timofeev, SS Khvostov, GM Tsoi, VI Shnyrkov. UHF SQUID-magnetometer at 77 K. Cryogenics 32:517–519, 1992.
77. VI Shnyrkov, VP Timofeev, AS Garbuz, CG Kim. High-T_c rf SQUIDs for operation in magnetic fields. Effect of thermal fluctuations. Low Temp Phys 25:826–828, 1999.
78. ML Lucia, R Hohmann, H Soltner H-J Krause, W Wolf, H Bousack, MI Faley, G Sporl, A Binneberg. Operation of HTS SQUIDs with a portable cryostat: A SQUID system in conjunction with eddy current technique for non-destructive evaluations. IEEE Trans Appl Supercond AS-7:2878–2882, 1997.
79. J Kurkijarvi, WW Webb. Thermal noise in the superconducting quantum flux detectors. In: Proceedings of the Applied Superconductivity Conference, Annapolis, New York: IEEE 1972, pp 581–585.
80. VV Danilov, KK Likharev, OV Snigirev. In: H Hahlbohm, H. Lubbig eds. SQUIDs and Their Applications. Berlin: Walter de Gruyter, 1980, pp 473–507. MV Kreutzbruck, J Troll, M Muck, C Heiden, Y Zhang. Experiments on eddy current NDE with HTS rf SQUIDs. IEEE Trans Appl Superconduct AS-7:3279–3282, 1997. GJ Sloggett, CP Foley, S Lam, RA Binks, DL Dart. 77 K SQUIDs operating in the Earth's magnetic field. IEEE Trans Appl Superconduct AS-7:3044–3047, 1997.

8

High-Temperature SQUID Magnetometer

Neeraj Khare
National Physical Laboratory, New Delhi, India

8.1 INTRODUCTION

The superconducting quantum interference device (SQUID) is the most sensitive detector of the magnetic flux. This feature makes the SQUID an attractive device for a range of applications. The SQUIDs based on low-T_c superconductors have shown unsurpassed sensitivity for the measurement of current, voltage, magnetic field, and magnetic field gradient (1). Potentiality of the applications of these SQUID magnetometers in measuring the biomagnetic field (2), nondestructive testing (2), and geological prospecting (3) have been demonstrated much earlier. In spite of the commercial availability of low-T_c SQUID for several years, the SQUID-based applications did not gain widespread acceptability. This has been mainly due to the inconvenience of its operation at liquid-helium temperature.

The feasibility of the fabrication of high-T_c SQUID operating at 77 K was soon demonstrated after the discovery of high-T_c superconductors (HTS). Since then, there has been continuous progress in this area (4–6). Several novel approaches have been conceived and applied for improving the performance of the high-T_c SQUIDs. HTS SQUID magnetometers exhibiting magnetic field sensitivity $\sim$ 10–50 fT/Hz$^{1/2}$ in the white-noise region at 77K have been demonstrated (5). Several companies have started commercializing high-T_c SQUIDs (7). HTS

SQUID magnetometers and gradiometers have been successfully used in the detection of biomagnetic signals from the heart and brain (8,9), nondestructive testing of deep buried flaws in metallic specimens (10), geophysical applications (11), and several other novel applications such as in biological immunoassays (12), sentinels-lymph node biopsy (13), and so forth.

This chapter presents a review of the developments of high-T_c SQUID magnetometers and discusses their applications in different areas.

8.2 SUPERCONDUCTING QUANTUM INTERFERENCE DEVICE

The SQUID is an ultrasensitive magnetic flux sensor that converts magnetic flux into voltage. There are two types of SQUIDs: dc-SQUID and rf-SQUID (1). A dc-SQUID consists of a superconducting loop interrupted by two Josephson junc-

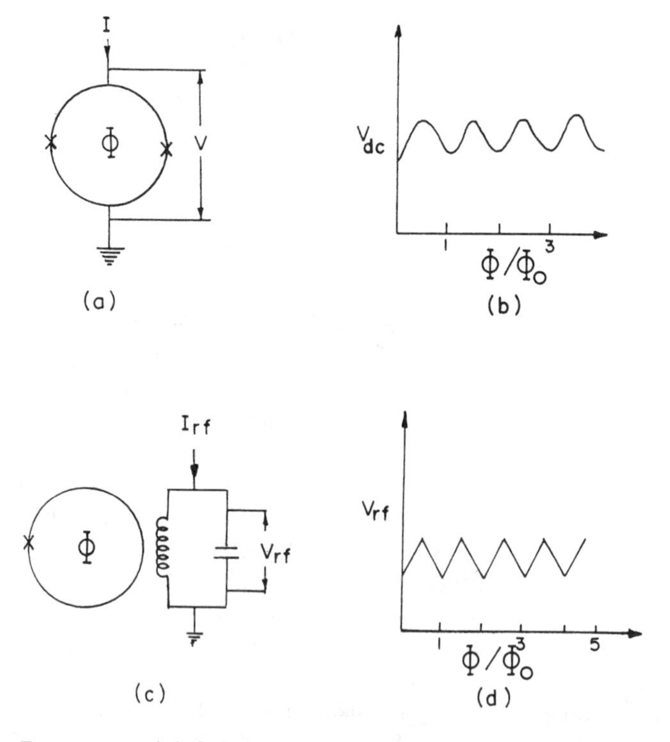

FIGURE 8.1 (a) Schematic of a dc-SQUID. Two Josephson junctions are shown by × in the superconducting ring, (b) Variation of voltage of the dc-SQUID with the applied flux (Φ) for a constant bias current, (c) Schematic of an rf-SQUID and (d) variation of peak amplitude of V_{rf} with the applied flux (Φ) for a fixed rf bias current.

tions, as shown in Figure 8.1a. The prefix dc implies that it is biased with a direct current. Both Josephson junctions in the dc-SQUID have identical characteristics. Critical current of the dc-SQUID is an oscillatory function of the applied flux with a period of one flux quantum, Φ_0. The value of one flux quantum, Φ_0 ($=h/2e$), is 2×10^{-7} G/cm^2. When the dc-SQUID is biased with a dc current $I_B \approx I_c$, where I_c is the critical current of the Josephson junction, the voltage across the SQUID shows an oscillatory function of applied magnetic flux with the periodicity of Φ_0 (Fig. 8.1b).

The rf-SQUID consists of one Josephson junction in the superconducting ring as shown in Figure 8.1c. The rf-SQUID is biased with an rf current applied to the SQUID through a inductively coupled tank circuit. Here also for an appropriate biasing, the rf voltage across the SQUID oscillates as a function of magnetic flux having a periodicity of Φ_0 (Fig 8.1d).

8.2.1 Designing of SQUID

The performance of a SQUID depends on the characteristics of the Josephson junction and inductance of the SQUID loop, L_{SQ}. For the Josephson junction, important parameters are the critical current (I_c), the capacitance (C), and the shunt resistance (R). Designing of a highly sensitive SQUID requires careful selection of the appropriate values of L_{SQ}, I_c, R, and C (14,15).

Noise due to thermal fluctuation puts an upper limit on the selection of the value of L_{SQ}. The thermal noise power in the SQUID is $\frac{1}{2}k_BT$/Hz and the mean energy per Hertz in an inductor is $\frac{1}{2}L_{SQ}I_n^2$; thus $I_n^2 = k_BT/L_{SQ}$. The corresponding equivalent flux noise is

$$\Phi_n^2 = L_{SQ}^2\, I_n^2 = L_{SQ}\, k_BT \tag{1}$$

where k_B is the Boltzman constant and T is the operating temperature of the SQUID. Due to this constraint of the thermal fluctuation noise, the SQUID effect can be observed only if Φ_n is less than $\Phi_0/2$, where Φ_0 is the flux quantum. Thus, the condition for observing the SQUID response is

$$L_{SQ} < \frac{\Phi_0^2}{4k_BT} \tag{2}$$

which imposes a condition that $L_{SQ} < 1$ nH for operating the SQUID at 77 K.

Similar to L_{SQ}, the choice of the critical current (I_c) of the junction is also very crucial. In order to observe the quantum interference effect, the junction coupling strength $I_c\Phi_0/2\pi$ must be significantly greater than the thermal energy, the (k_BT).

For taking into account Josephson coupling strength and thermal energy, the Γ parameter for the SQUID is defined as

$$\Gamma = \frac{2\pi k_BT}{I_c\Phi_0} \tag{3}$$

For observing the SQUID effect, the value of the Γ parameter should be <1 (preferably ≈ 0.1). This puts a condition that $I_c \approx 10$–$50\ \mu$A for operating the SQUID at 77 K.

Other important parameters for the SQUID design are

$$\beta_c = \frac{2\pi I_c R^2 C}{\Phi_0} \tag{4}$$

and

$$\beta_L = \frac{2 L_{SQ} I_c}{\Phi_0} \tag{5}$$

To ensure nonhysteretic characteristics of the junction, the McCumber parameter $\beta_c < 1$ and for the optimum SQUID performance $\beta_L \approx 1$. For a optimum bias current, the SQUID transfer function $V_\Phi \approx \delta V/\delta\Phi$ is given as

$$V_\Phi \approx \frac{R}{L_{SQ}} \tag{6}$$

The value of spectral density of thermal noise (S_v) is

$$S_v(f) \approx 16 k_B T R \tag{7}$$

and the thermal flux noise density is given by

$$S_\Phi = \frac{S_v}{(V_\Phi)^2} \tag{8}$$

$$S_\Phi \approx \frac{16 k_B T L_{SQ}^2}{R} \tag{9}$$

Thus, for reducing the thermal flux noise, the value of R should be large and C should be kept small to satisfy the condition for β_c. Typically, $\sqrt{S_\Phi(f)}$ is frequency independent down to a frequency below which it scales approximately as $1/f$.

In order to compare different SQUIDs, a figure of merit (energy resolution of the SQUID) is defined as

$$\varepsilon(f) = \frac{S_\Phi(f)}{2 L_{SQ}} \tag{10}$$

Substitution of S_Φ from Eq. (9) gives

$$\varepsilon(f) \approx \frac{8 k_B T L_{SQ}}{R} \tag{11}$$

thus, one should reduce T and L_{SQ} and increase R in order to improve the energy resolution of the SQUID.

8.2.2 SQUID Magnetometer

The dependence of SQUID voltage on the applied flux suggests that it can be used as a flux meter. Figure 8.1b shows that the SQUID voltage varies periodically with the applied flux. However, for several applications, one requires that the output of the SQUID should vary linearly with the applied flux even when the flux is much greater than Φ_0. This linear response is obtained by means of a feedback circuit, as shown in Figure 8.2. The dc-SQUID is biased with a constant current $I_B \approx I_c$, and an ac magnetic field ($f_{ac} \sim$ 10–100 kHz) is applied to the SQUID with a peak-to-peak amplitude $\leq \Phi_0/2$. The ac signal developed across the SQUID is detected with a lock-in at the fundamental frequency. When the external field is equal to exactly $n\Phi_0/2$, the signal from the SQUID is at twice the fundamental frequency and the lock-in output is zero. For an external flux corresponding to $(n + \frac{1}{4})\Phi_0$ and $(n + \frac{3}{4})\Phi_0$, the output from the lock-in is maximum and minimum, respectively. Therefore, the lock-in output is also oscillatory and the period of the oscillation is one flux quantum. To obtain a linear response to the external flux, the lock-in output is amplified and is fed back to the SQUID through a feedback resistor and a coil coupled to the SQUID. A flux change ($\delta\Phi$) produces a voltage across the SQUID which is fed back through the coil coupled to the SQUID, thereby generating a flux ($-\delta\Phi$) exactly equal but opposite to the applied flux. Thus, the SQUID will experience no flux and it will always be "locked" at the minima of

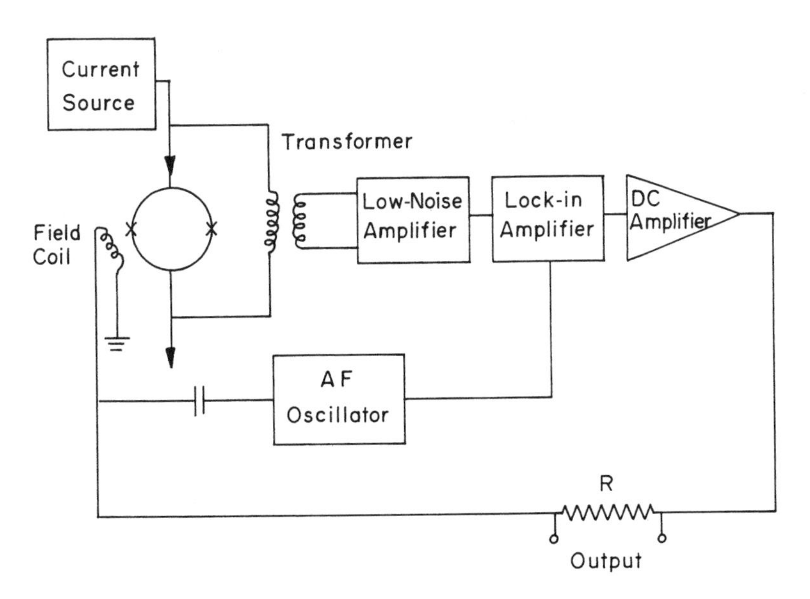

FIGURE 8.2 Arrangement for a feedback circuit to operate the dc-SQUID in flux locked loop mode.

the V–Φ characteristics. Any increase in the applied flux increases the feedback current linearly and the voltage across the feedback resistance is proportional to the applied flux. In the flux-locked loop mode, one can detect flux much less than Φ_0 as well as flux corresponding to several Φ_0's with the same sensitivity.

The overall performance of the SQUID magnetometer is also determined by the readout electronics. In recent years, there has been considerable advancement in SQUID electronics (16–18). Direct-coupled SQUID electronics with large bandwidth and high slew rate at a very low noise level have been developed. These electronics have features like automatic bias voltage tuning, background field cancellation, and bias reversal scheme.

The SQUID can also be used to measure the magnetic field in the digital mode by counting the number of oscillations that occur when the field is applied. The digital SQUID electronics uses the SQUID itself as an integral part of the analog-to-digital conversion. Digital SQUID electronics have been developed to perform software gradiometry with SQUID magnetometers (19) for measurements in an unshielded environment or in a moving platform.

8.2.2.1 Flux Transformer and Gradiometer

The magnetic field sensitivity of a SQUID depends on its effective area. The intrinsic magnetic field sensitivity of the SQUID is less due to its small geometric area. Due to the design constraints, the area of the SQUID loop cannot be made very large. For an applied field B_a, the flux coupled to the SQUID is $\Phi_{SQ} = B_a A_S$, where A_S is the area. Using a superconducting flux transformer, one can increase the magnetic field sensitivity of the SQUID. The configuration of such a superconducting transformer is shown in Figure 8.3. It is a closed superconducting loop consisting of a large pickup coil connected in series with a small area input coil that is inductively coupled to the SQUID. When a magnetic field is applied to the pickup coil, fluxoid quantization requires that the total flux in the superconducting loop remain fixed in the multiples of flux quantum. As a result, a supercurrent is gener-

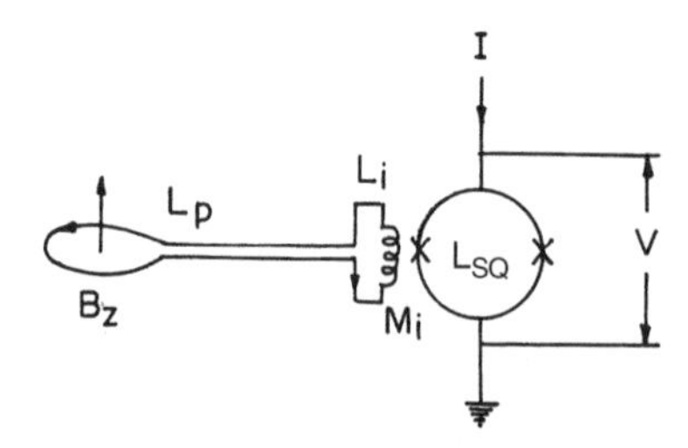

FIGURE 8.3 Superconducting flux transformer inductively coupled to a SQUID. L_p, L_i, and L_{SQ} are inductance of pickup loop, the input coil, and the SQUID loop respectively, and M_i is the mutual inductance between the input coil and the SQUID loop.

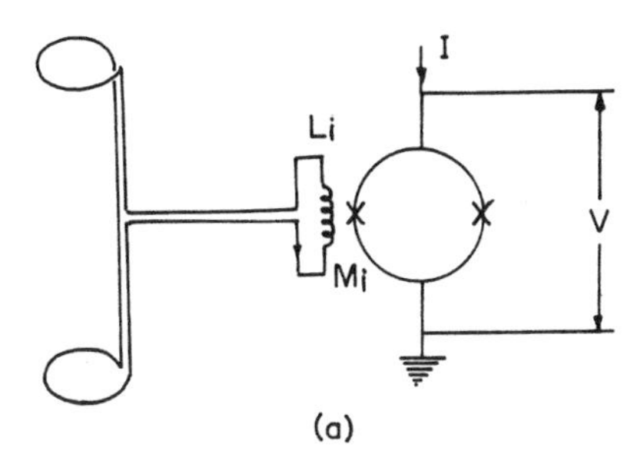

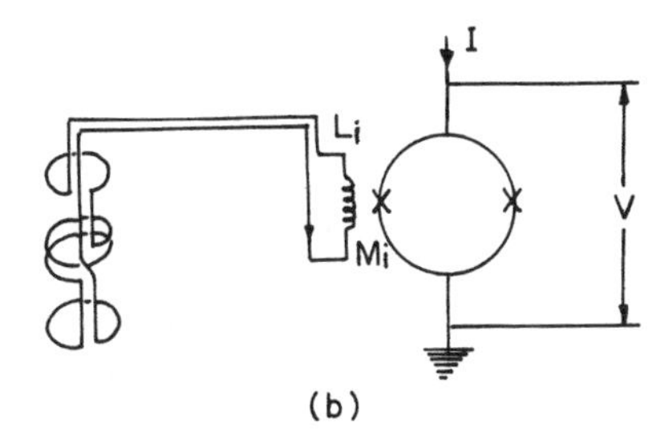

FIGURE 8.4 (a) Superconducting first-order gradiometer and (b) second-order gradiometer.

ated in the transformer with the appropriate direction and magnitude so as to counter any change in the magnetic flux within the loop. As this supercurrent flows through the input coil, it couples a magnetic field on the SQUID. The small loop that couples flux to the SQUID is magnetically shielded from the external field.

For a field B_a applied to the pickup loop, flux quantization in the flux transformer requires that (15)

$$B_a A_p - (L_p + L_i)I = 0 \tag{12}$$

where A_p is the area of the pickup loop, L_p and L_i are the inductance of pickup loop and input coil of the transformer, respectively, and I is the supercurrent in the loop. The flux coupled to the SQUID is

$$\Phi = \frac{M_i A_p B_a}{L_p + L_i} \tag{13}$$

where M_i is the mutual inductance between input coil and the SQUID inductance. It is evident that flux coupled to the SQUID through the flux transformer is much larger than the flux coupled to the SQUID without a transformer. For an efficient flux transformer, inductance of the pickup loop, L_p, should be nearly equal to the inductance of the input coil, L_i.

An important extension of flux transformer is a gradiometer that is used to measure the magnetic field gradient. Figure 8.4 shows schematic configurations of first–order and second–order gradiometers that enables one to measure minute localized magnetic signals even in the presence of large uniform background magnetic field. In the first-order gradiometer, the two superconducting pickup loops are wound in the opposite sense so that an uniform magnetic field produces no supercurrent in the flux transformer, whereas a gradient $\partial H_z/\partial z$ generates a net supercurrent that is proportional to the difference in the flux threading the two loops.

In the second-order gradiometer, two first-derivative gradiometers are wound end to end (Fig. 8.4b). This configuration measures the double derivative $\partial^2 Hz/\partial z^2$. In the design of the gradiometer, considerable care is taken to ensure that the loops are of the same size and exactly parallel to each other. In the low-T_c SQUID, a wire-wound gradiometer (20) or thin-film gradiometer (21) is used. For the high-T_c superconductor, wires of adequate quality are still not available and, thus, planar thin-film HTS flux transformers and gradiometers are used.

8.3 HIGH-T_C JOSEPHSON JUNCTIONS AND SQUIDs

In the low-T_c SQUID, S-I-S (superconductor–insulator–superconductor) tunnel junctions are usually used for the fabrication of the SQUID. In high-T_c superconductors, it is extremely difficult to prepare S-I-S-type tunnel junctions because of the short coherence length along the *c* axis and various other novel approaches have been followed. Figure 8.5 shows the schematics of some of the high-T_c junctions that have been used in the fabrication of HTS SQUIDs.

Soon after the discovery of high-T_c superconductors, it was established that natural grain boundaries (NGBs) in polycrystalline HTS samples behave as Josephson junctions. Several groups have reported the fabrication of rf- and dc-high T_c SQUIDs using NGB junctions (references cited in Ref. 6). The characteristics of NGB junctions varies from one grain boundary to the other. It has been possible to fabricate a NGB junction rf-SQUID with reasonable characteristics because only one junction is required for the rf-SQUID (22–25). However, fabrication of a NGB dc-SQUID has not been very reproducible (26). Moreover, it has also been found that the trapping of flux in the body of the SQUID causes large $1/f$ noise resulting into a poor sensitivity of the SQUID.

Soon it was realized that the flux noise in an *in situ* grown epitaxial HTS film is much smaller as compared to the polycrystalline films (27) and many efforts have been directed toward the fabrication of HTS Josephson junctions using epitaxial films. Figure 8.5 shows the step-edge junction, the bicrystal junction, the ramp-edge junction, and the biepitaxial junction that have been fabricated using epitaxial films. The fabrication of the step-edge junction and the bicrystal junction involves the preparation of single-layer HTS film, whereas for the biepitaxial junction and the ramp-edge junction, deposition of multilayers of HTS films are required. For the HTS bicrystal grain-boundary junction, the HTS film is epitaxially grown on a bicrystal substrate (28). The critical current density across the

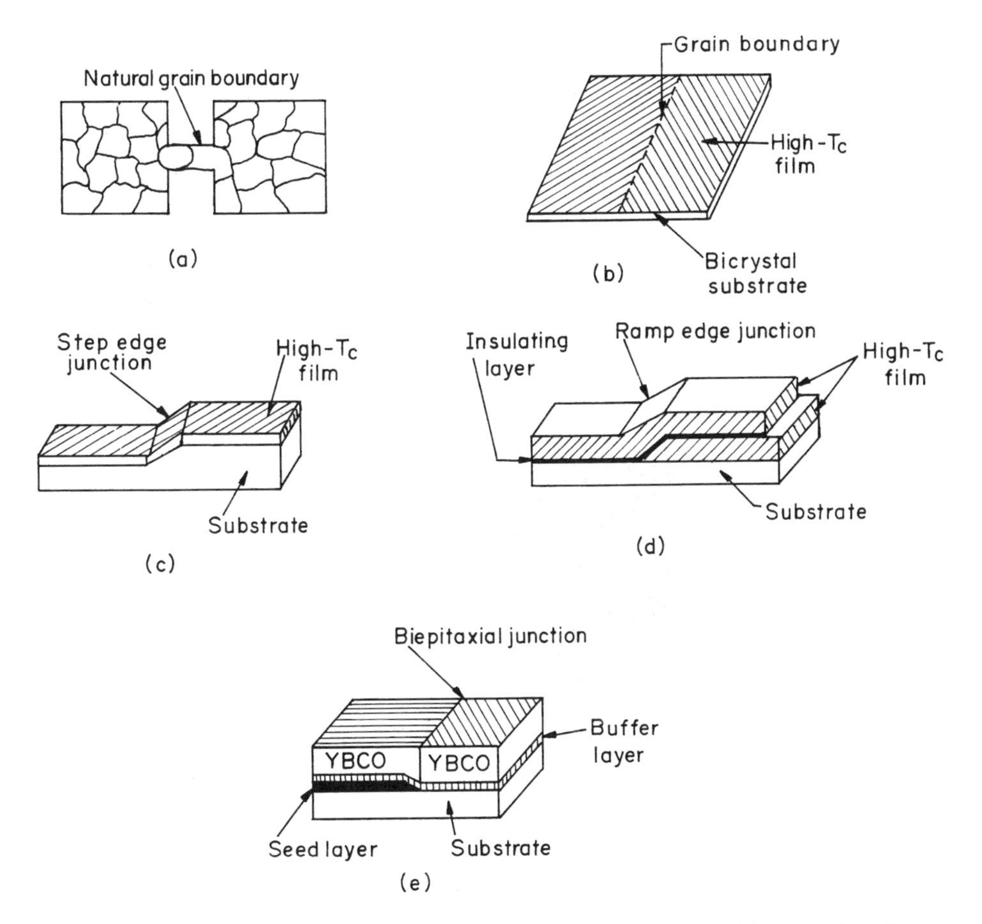

FIGURE 8.5 Various types of the high-T_c Josephson junction: (a) natural grain boundary, (b) bicrystal junction, (c) step-edge junction, (d) multilayer ramp-edge junction, and (e) biepitaxial junction.

grain boundary depends on the misorientation angle of the grain boundary (29). For the HTS step-edge junction, a sharp step is created on a single-crystal substrate using lithography and the ion-beam-milling technique (30–32). The thickness of the film is kept smaller than the step height. In the ramp-edge junction, first a layer of HTS film is deposited on a substrate and the ramp edge is created using lithography and an etching process. In the second step, a thin insulating layer and HTS film are deposited on the ramp edge (33). For the biepitaxial junction, a seed epitaxial layer of MgO is deposited on an *r*-plane sapphire and then an epitaxial buffer layer of $SrTiO_3$ is deposited. The $SrTiO_3$ film grows on MgO and sapphire

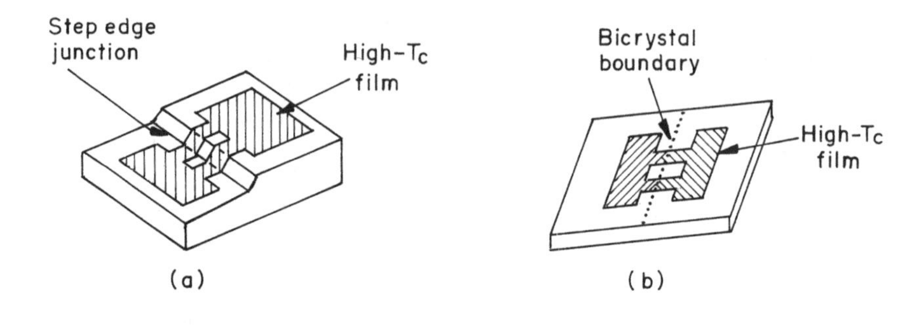

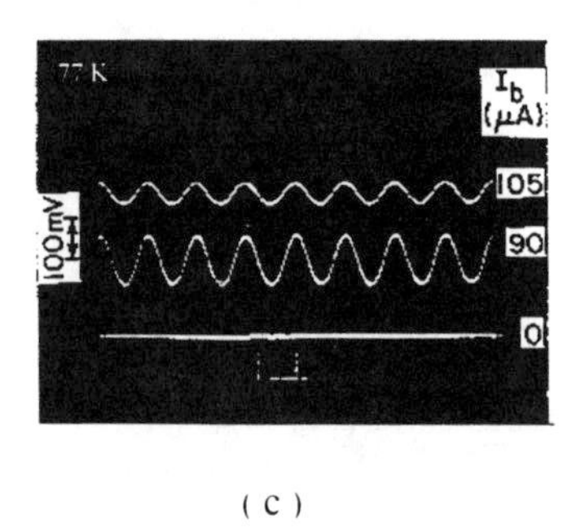

(c)

FIGURE 8.6 (a) Step-edge junction high-T_c SQUID, (b) bicrystal junction high-T_c SQUID, and (c) voltage-flux characteristics of YBCO bicrystal junction dc-SQUID at different biasing current. (Adapted from Ref. 36).

in two different orientations separated by a 45° grain boundary. The epitaxial YBCO film deposited on the $SrTiO_3$ film also has a 45° grain boundary (34,35).

In all of the above cases, a HTS junction exhibiting Josephson junction characteristics can be fabricated by patterning a microbridge across the bicrystal grain boundary, the biepitaxial boundary, a step edge, or a ramp edge. The characteristics of these artificial HTS Josephson junctions fabricated using HTS epitaxial films can be easily controlled and the reproducibility of the junction is very high. Several groups have reported on the fabrication of high-T_c SQUIDs using these artificial junctions (references cited in Refs 5 and 6).

Figures 8.6a and 8.6b show the geometry of step-edge junction and bicrystal junction dc-SQUIDs, respectively. Figure 8.6c shows typical voltage-flux characteristics for the YBCO thin-film bicrystal junction dc-SQUID (36). The SQUID characteristics depends on the geometry of the SQUID and characteristics of the junction. For designing the HTS SQUID, similar criteria as discussed in Section 8.2.1 are used. For a HTS SQUID magnetometer, the SQUID needs to be coupled with a flux transformer. Different types of configuration for HTS SQUID magnetometers have been conceived and demonstrated in recent years. The following section presents detail accounts of these developments.

8.4 HIGH T_C SQUID MAGNETOMETER

Several different approaches have been followed for improving the magnetic field sensitivity of HTS SQUIDs. These are use of a large area of the SQUID washer, a flux focuser, a single-layer flux transformer, and a flux transformer with multiturn input coil and by using a large-area shunt inductance parallel to the SQUID inductance. These approaches can be broadly classified into three categories;

1. Single-layer SQUID magnetometer
2. Flip-chip SQUID magnetometer
3. Multilayer thin-film device on a single substrate

8.4.1 Single-layer SQUID Magnetometer

Single-layer HTS SQUID magnetometers have been fabricated using large washer geometry or direct-coupled SQUID-type geometry. Figure 8.7 show schematic of a large-area washer dc-SQUID and a direct-coupled dc-SQUID. It is relatively simpler to fabricate these single-layer devices. In the large washer design, due to flux focusing, the effective area of the SQUID increases. Zhang et al (37) used a rf-SQUID with a large square washer to achieve a magnetic field sensitivity of $\sim$170 fT/Hz$^{1/2}$ at 1 Hz and 77 K.

In the direct-coupled high-T_c SQUID magnetometer, a large pickup loop is connected in parallel with a small inductance of the bicrystal junction high-T_c SQUID (Fig. 8.7b). In this geometry, the dynamics of the SQUID are controlled

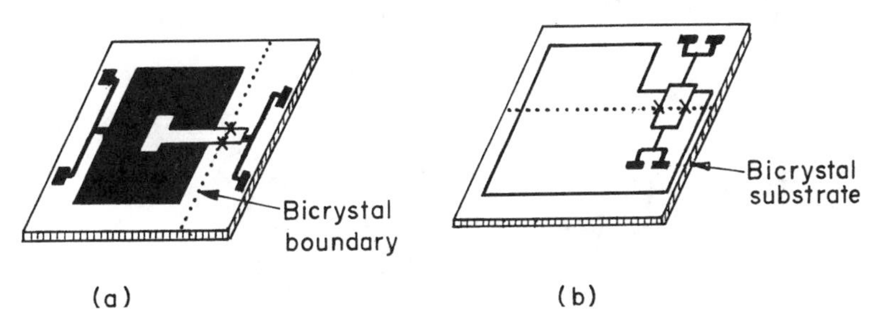

FIGURE 8.7 (a) Large washer type of geometry and (b) direct-coupled HTS SQUID magnetometer.

by the SQUID inductance, whereas flux linking to the pickup loop induces a current that couples flux to the SQUID. The effective area (A_{eff}) of a direct-coupled SQUID magnetometer is given as (38).

$$A_{eff} = A_s + \alpha A_p \left(\frac{L_{SQ}}{L_p} \right) \tag{14}$$

$$\approx \alpha A_p \left(\frac{L_{SQ}}{L_p} \right) \tag{15}$$

where A_s and A_p are the effective area of the SQUID loop and pickup loop respectively, L_p and L_{SQ} are the inductance of the pickup loop and the SQUID loop, respectively. α is the coupling constant, which is unity when the current in L_p produces a maximum possible flux in L_{SQ}.

The first HTS direct-coupled SQUID magnetometer was reported in 1991 (39). The magnetometer was fabricated using Bi-Sr-Ca-Cu-O oxide film and showed a magnetic field sensitivity of 1.5 $pT/Hz^{1/2}$ above 20 Hz and 4.2 K. A considerably improved direct-coupled HTS SQUID magnetometer have been reported by several groups in later years (38,40–46). A magnetic field sensitivity of 32 $fT/Hz^{1/2}$ at 2 Hz and 77 K was achieved for a device of 1×1 cm^2 (46). For a dc-SQUID device fabricated using YBCO film on a 2×2 cm^2 $SrTiO_3$ bicrystal substrate, a field sensitivity of 10 $fT/Hz^{1/2}$ above 10 kHz and to 26 $fT/Hz^{1/2}$ at 1 Hz (38) was demonstrated.

8.4.2 Flip-Chip Magnetometer

In the flip-chip configuration, the SQUID is fabricated on one substrate and the flux focuser or flux transformer is fabricated on another substrate. The flux focuser or flux transformer couples the flux to the SQUID through inductive coupling, which is achieved by placing the SQUID on the top of the flux transformer, as shown in Figure 8.8. The input coil of the transformer or the center of flux focuser is mechanically aligned to the SQUID loop. In the design of the thin-film flux focuser shown in Figure 8.9a, the flux over a large area gets concentrated on

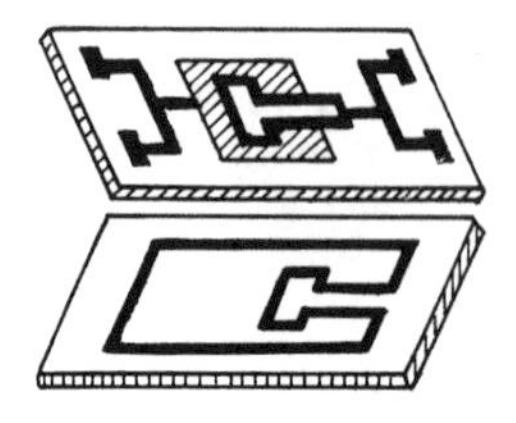

FIGURE 8.8 Schematic of flip-chip high-T_c SQUID magnetometer.

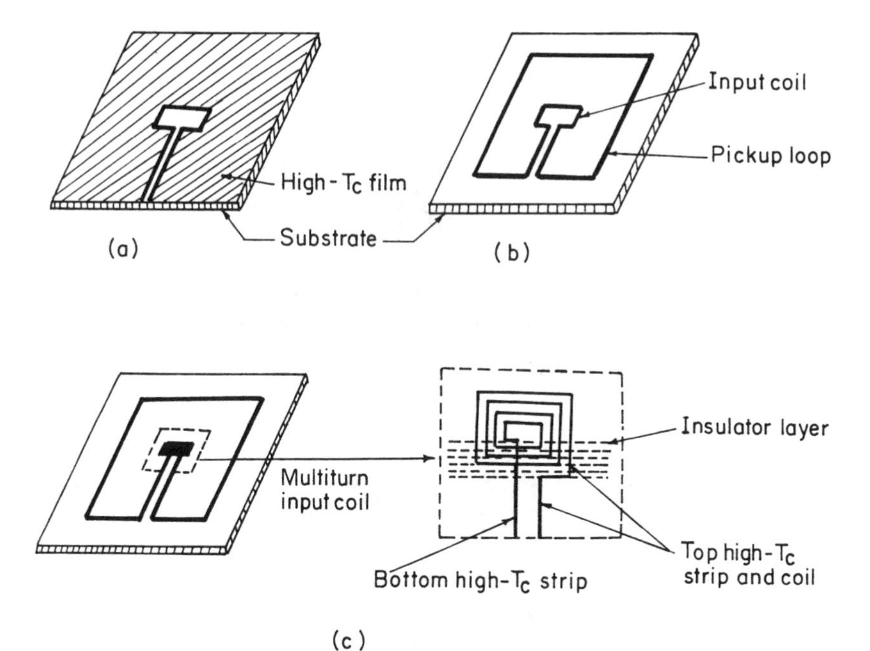

FIGURE 8.9 High-T_c thin film (a) flux focuser, (b) single-layer flux transformer, and (c) flux transformer with multiturn input coil.

the small hole at the center. Thus, a much larger flux is coupled to the SQUID, leading to an enhancement in the magnetic field sensitivity of the SQUID. Several groups have reported the enhancement of the magnetic field sensitivity of the SQUID using the flux focuser (37,47–51). A magnetic field sensitivity of 30 $fT/Hz^{1/2}$ in the white-noise region at 77 K has been achieved for a rf-SQUID with a flux focuser of diameter 13.4 mm (47).

Figures 8.9b and 8.9c show schematics of a single-layer flux transformer and multiturn input coil transformer. The first one is fabricated using a single layer of high-T_c thin film, whereas fabrication of flux transformer with multiturn input coil requires deposition of three or more layers.

When the planar flux transformer is coupled to a SQUID, the total magnetic flux induced to the SQUID is given by (52)

$$\Phi_i = B_a A_s f + \frac{B_a M(A_p - A_i)}{L_p + L_i} \tag{16}$$

where B_a is the applied field, f is the flux focusing factor of the SQUID, A_p and L_p are the area and inductance of the pickup loop, respectively, A_i and L_i are the area and the inductance of the input loop, and M is the mutual inductance between the input loop and the SQUID.

There are a few considerations for the design of a planar flux transformer, such as (1) the pickup coil should have smallest possible inductance L_p for the given area, (2) the input coil inductance, L_i should be equal to the inductance of the pickup coil, L_p, and (3) the input coil should have approximately the same size and shape as that of the SQUID in order to efficiently couple the flux in the SQUID.

Koelle et al. (41) fabricated a large single-layer flux transformer on a 50-mm-diameter substrate, and by inductively coupling of this flux transformer with a direct-coupled dc-SQUID, a magnetic field sensitivity of 39 $fT/Hz^{1/2}$ at 1 Hz and 77 K was achieved. Although it is simpler to fabricate single-layer flux transformer, but in this design, it is difficult to set $L_p \approx L_i$. Thus, the flux transformer is not very efficient.

In the multiturn flux transformer, it is possible to set $L_p \approx L_I$ by using a multiturn input coil. Thus, the flux transfer is more efficient as compared to a single-layer flux transformer. The construction of this transformer requires multilayer deposition and patterning of three or more epitaxial layers (53–58). In this flux transformer, a large-area pickup loop is in series with a small-area multiturn input coil that is inductively coupled to the SQUID hole.

Several groups have reported fabrication of the HTS thin-film multilayer flux transformer using $SrTiO_3$ or $PrBa_2Cu_3O_x$ as an insulator (53,55–58). Earlier developed high-T_c multilayer flux transformers have been found to introduce excess low-frequency noise into the SQUID. The higher $1/f$ noise in the multilayer flux transformer is attributed to the motion of the flux vortices in the portion of high-Tc film whose quality deteriorates in the fabrication of the patterned multilayer structure. Subsequent years of research and development in this area by following a more careful control of multilayer deposition conditions and patterning, it has been possible to improve the characteristics of a multilayer flux transformer (55,59–62). Dantsker et al (61) have shown that by inductively coupling a multilayer flux transformer (9×9-mm^2 pickup loop, 16-turn input coil) to a bicrystal junction dc-SQUID magnetometer in a flip-chip arrangement, a sensitivity of 27 $fT/Hz^{1/2}$ at 1 Hz and 8.5 $fT/Hz^{1/2}$ at 1 kHz for a 1×1-cm^2 device has been achieved at 77 K. Using a 30-mm-diameter $LaAlO_3$ substrate for the flux transformer, Faley et al. (62) achieved a sensitivity of 5 $fT/Hz^{1/2}$ at 1 kHz for the flip-chip magnetometer.

The HTS single-layer direct-coupled and flip-chip SQUID magnetometer show sufficiently high sensitivity, although at the expense of relatively large substrate. Such a large-area magnetometer is acceptable for the measurement of a uniform magnetic field with a relatively small number of channels. However, for the development of a multichannel HTS SQUID system, sensors of sufficiently smaller area with the same high sensitivity are required.

8.4.3 Monolithic Magnetometer

In order to achieve tighter coupling between the input coil of the flux transformer and the SQUID washer, the transformer should be integrated monolithically with the high-T_c SQUID on a single substrate. A monolithic integrated high-T_c dc-SQUID magnetometer operating at 77 K was first reported by Lee et al. (56) using template biepitaxial grain-boundary junction. The device consisted of eight epitaxial layers, including three superconducting layers and exhibited an enhancement of magnetic field response of the SQUID by a factor of 127 due to the integrated flux transformer. Several groups have reported fabrication of integrated SQUID magnetometers (8,63–67). HilgenKamp et al (63) reported a dc-SQUID magnetometer consisting of a dc-SQUID integrated with a flux transformer on a single bicrystal substrate. The magnetometer consisted of four layers including two superconducting layers and was operated up to 73 K. A magnetic field sensitivity of 490 $fT/Hz^{1/2}$ was obtained at 1 Hz and 65 K. In another report, Dilorio et al. (8) fabricated an integrated magnetometer using step-edge junction that showed a magnetic field sensitivity of 280 $fT/Hz^{1/2}$ at 1Hz and 77 K. Lee et al (66) reported a magnetometer with a slotted washer for operating it in an unshielded environment.

In another approach for increasing the sensitivity of a HTS SQUID magnetometer, a multiloop magnetometer or fractional-turn SQUID has been fabricated in which the effective area of the SQUID is increased while keeping its inductance low (65). In the multiloop SQUID magnetometer, N loops of the SQUID are connected in parallel so that the total inductance of the SQUID reduces while keeping the effective area constant (68). An integrated YBCO thin-film multiloop magnetometer has been fabricated using a YBCO/$SrTiO_3$/YBCO multilayer deposited on a $SrTiO_3$ bicrystal substrate (65). The magnetometer has 16 parallel loops with an outer diameter of 7 mm and showed a magnetic field noise of 37 $fT/Hz^{1/2}$ at 1 Hz and 18 $fT/Hz^{1/2}$ at 1 kHz when it was operated with a bias reversal scheme at 77 K. Scharnweber and Shilling (67) reported an integrated SQUID magnetometer in which multiloop pickup coil is used for matching the inductance of the input and the pickup loop of the flux transformer. A magnetic filed sensitivity of 100 $fT/Hz^{1/2}$ at 1 Hz and 77 K was demonstrated for this configuration.

8.5 OPERATION OF HTS SQUID IN UNSHIELDED ENVIRONMENT

For several applications, the SQUID needs to be operated in an unshielded or in moderately shielded environment. In general, HTS SQUIDs are found to exhibit a much enhanced value of low-frequency flux noise when operated in the presence of magnetic field of few microtesla. This increase in low-frequency flux noise is due to the thermally activated motion of flux vortices trapped in the superconducting film forming the SQUID body (69). The presence of an ac magnetic field in the unshielded environment also affects the performance of a SQUID (70). The large current which is induced inductively in the device by the external ac field can affect the critical current of the device and, thus, can cause severe deviation in the electronic properties. Use of a μ-metal shield or HTS cylinder covering the HTS SQUID sensor has been found to reduce the unwanted electromagnetic interference present in the unshielded environment.

Recently, there have been considerable efforts in improving the design of the HTS SQUID for suppressing the increase of low-frequency noise when the SQUID is cooled in a static magnetic field (71–85). The design of the HTS SQUID is modified such that either the penetration of the vortex in the superconductor is avoided or the flux motion in the body of SQUID is hindered. In order to prevent the vortex from entering the SQUID body, structures with a narrow-linewidth superconducting film have been proposed (71). Estimation for the threshold field of flux entry is given by

$$B_T = \frac{\pi\Phi_0}{4w^2} \tag{17}$$

where w is the linewidth of the film and Φ_0 is the flux quantum.

For a cooled HTS magnetometer, a change in the magnetic field induces a shielding current in the pickup loop. Movement of the magnetometer in a field gradient or even tilting the magnetometer in a static magnetic field can induce shielding currents. It has been proposed that the creation of flux dams in the pickup coil can limit the shielding current (72,73). Several groups have fabricated a HTS SQUID that has a structure with narrow-linewidth superconducting films or flux dams and showed the improvement in the low-frequency noise characteristics when the magnetometer is cooled in the presence of field of several microteslas (71–82).

Dantsker et al. (71) have shown that for a HTS SQUID in which arrays of slots and holes are patterned in the washer of the HTS dc-SQUID, leaving superconducting films of a fine linewidth of 4 μm, there was virtually no increase of $1/f$ noise on cooling the SQUID up to a field of $\sim$ 100 μT. The configurations such as narrowing of the pickup loop (74), fabrication of several narrow parallel loops in place of a wide solid pickup loop (75,76,79), creation of slits with a separation of 5 μm in the pickup loop (78), or fabrication of slotted flux dams in a bicrystal junction dc-SQUID (79,80), have been found to suppress the increase of $1/f$ noise when the SQUID was cooled in a field of several microteslas.

Selders and Wordenweber (83,84) observed that simple arrangement of a few antidots in a HTS rf-SQUID strongly reduced the flux noise arising due to vortex motion. In another report, suppression of the critical current of a YBCO step junction in a rf-SQUID due to the presence of an earth magnetic field is observed and fabrication of junctions of width in the submicrometer range is suggested for the operation of the SQUID in the earth magnetic field (85).

8.6 HIGH-T_C SQUID GRADIOMETER

In the case of a low-T_c SQUID, a wire-wound gradiometer (20) or thin-film planar gradiometer (21) is used to measure the weak magnetic field from a localized source even in the presence of large uniform background field in an unshielded environment. In high-T_c superconductors, wire is still not a viable option and only the planar thin-film structure is used to fabricate the gradiometer. Three approaches have been used for fabricating the HTS SQUID gradiometer:

1. Single-layer direct-coupled SQUID gradiometer
2. Flip-chip gradiometer
3. Electronic gradiometer

Figures 8.10a and 8.10b show direct-coupled first-order and second-order gradiometers, respectively, whereas Figure 8.10c shows a planar second-order gradiometer inductively coupled to the HTS SQUID in the flip-chip mode.

8.6.1 Single-Layer Direct-Coupled SQUID Gradiometer

In the first-order single-layer direct-coupled gradiometer (Fig. 8.10a), two symmetric pickup loops are directly connected to the SQUID. This type of planar gradiometer has been prepared using a bicrystal junction or step-edge junction SQUID (86–95). The gradient field resolution of such a device depends on the size of the substrate on which HTS film and SQUID have been fabricated. For a $10 \times 10\text{-mm}^2$ substrate, typically a baseline of 4.5 mm is achieved. Using a 30×10 mm^2 substrate, a baseline could be increased to 13 mm (89,90) and a gradient field resolution of 50 fT/cm $\text{Hz}^{1/2}$ at 1 kHz was demonstrated.

In the single-layer gradiometer device, there is a considerable response from the SQUID in the center of the structure, which can degrade the performance of the gradiometer in an unshielded environment. In order to reduce the magnetometric response of the SQUID in the single-layer gradiometer, two approaches have been proposed (93–95). In the first one, a gradiometric SQUID is fabricated which has two identical loops connected in parallel across the HTS step junction (93–94). In another approach, a two SQUID coupling scheme is used. The two SQUIDs are operated simultaneously and their voltage outputs are linearly combined, thus compensating the parasitic effective area (95).

The second-order direct-coupled SQUID gradiometer (Fig. 8.10b) has been fabricated from a single layer of a high-T_c thin film and consists of three pickup loops (96–97). The balancing condition for such a second-order gradiometer is given as (96)

$$\frac{A_c L}{A L_c} = \frac{\alpha_L + \alpha_R}{\alpha_c} = \frac{2\alpha}{\alpha_c} \tag{18}$$

where α_L and α_R are coupling coefficient of the left and the right loops, respectively. Both the left and right loops have the same value of inductance L and ef-

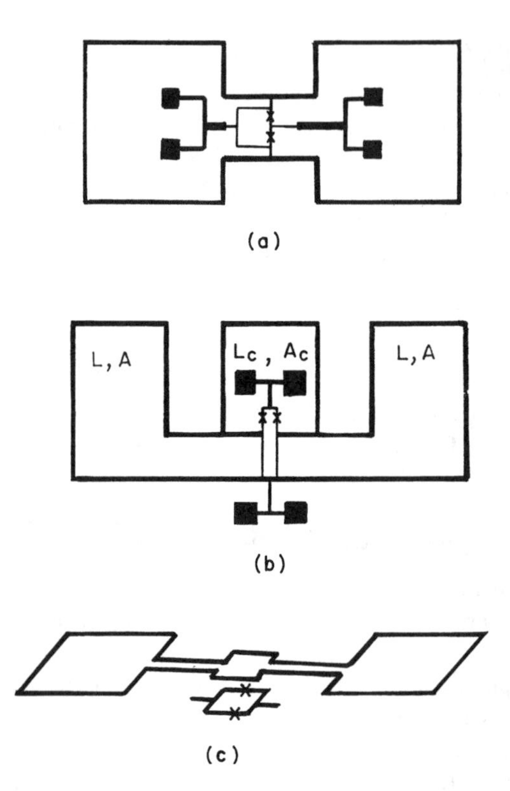

FIGURE 8.10 Schematic of (a) a single-layer HTS thin-film first-order dc-SQUID gradiometer, (b) a single-layer HTS thin-film second-order dc-SQUID gradiometer (left and right loops have the same inductance L and area A; the inductance and area of the center loop are L_c and A_c, respectively), and (c) a single-layer HTS thin-film second-order gradiometer inductively coupled to the HTS dc-SQUID.

fective area A. The effective area and inductance of the center loop are A_c and L_c, respectively. The symmetric design of the gradiometer ensures the balancing between the left and right loops, so $\alpha_L = \alpha_R = \alpha$. Because $\alpha_c = 2\alpha$, the balancing condition is

$$\frac{A_c}{L_c} = \frac{A}{L} \tag{19}$$

In this symmetric design, the flux coupled to the SQUID for a uniform field and for the first-order gradient of the field is zero. However, in practice, because of the coupling imbalance due to intrinsic inductance, the gradiometer is not fully balanced. By optimizing the inductance of the pickup loop at the center, the off-balancing factor in a uniform field of less than 7×10^{-4} has been achieved (97). This type of planar second-order gradiometer is useful for nondestructive evaluation in an unshielded environment.

The fabrication of single-layer direct-coupled gradiometer is relatively simple; however, this design has the disadvantage that the maximum baseline is limited by the size of the substrate on which SQUID gradiometer has been fabricated. The design also has extra problem if bicrystal junction SQUID is used because the pickup coil leads cross the bicrystal grain boundary.

8.6.2 Flip-Chip Gradiometer

In the flip-chip arrangement (Fig. 8.10c), the gradiometer is fabricated on one substrate and it is inductively coupled to the HTS SQUID on another substrate (62,98–101). The patterned side of the two chips are pressed together to couple the SQUID loop and the flux transformer inductively. This design enables a larger gradiometer baseline and a much higher gradient field resolution than the single-layer chip direct-coupled gradiometers. Both first- and second-order gradiometers with very high balance levels have been demonstrated using the flip-chip design. An asymmetric planar gradiometer with a base line of 48 mm has been fabricated that showed a balance with respect to a planar magnetic field of about 1 part in 3000 (99).

A planar second-order single-layer YBCO film gradiometer was fabricated within an overall length of 80 mm and inductively coupled to a direct-coupled SQUID magnetometer (100). The mutual inductance between the flux transformer and the magnetometer is adjusted mechanically to reduce the response to a uniform magnetic field applied perpendicular to the plane of the gradiometer of about 50 ppm. Tien et al. (101) reported that in a flip-chip arrangement, coupling of a direct-coupled SQUID gradiometer to a single-layer gradiometric antenna on a 50-mm Si wafer, a gradient field sensitivity of 73 fT/cm $Hz^{1/2}$ in the white-noise region at 77 K and 596 fT/cm $Hz^{1/2}$ at 1 Hz can be achieved.

By controlling the design and preparation conditions, a HTS flip-chip gradiometer having a multilayer gradiometric flux antenna on a 40-mm-diameter substrate, a gradient field sensitivity of 30 fT/cm $Hz^{1/2}$ at 1 kHz was demonstrated (62). Although a good gradient sensitivity has been achieved in the flip-chip gradiometer, the technique requires precise mechanical adjustment of the relative position of the two chips to attain a high balance level.

8.6.3 Electronic Gradiometer

The HTS electronic gradiometers have been fabricated using two or more HTS magnetometers (49–51, 102–105). Figure 8.11 is the schematic of electronic gradiometer using two HTS rf-SQUIDs. The output of the two SQUIDs is electronically subtracted to get the $\partial B_z/\partial z$ component. The high-T_c superconducting plates

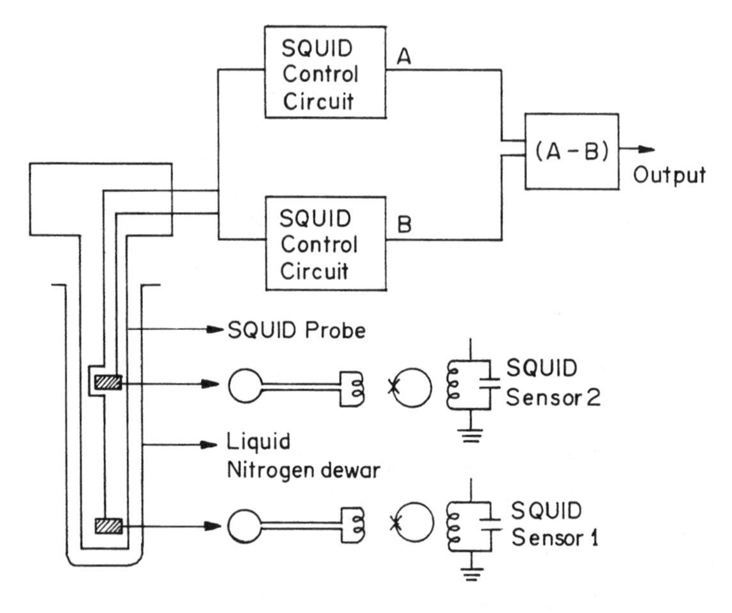

FIGURE 8.11 A HTS electronic gradiometer (two HTS rf-SQUIDs are used to construct a first-order gradiometer).

are used for tuning the common mode rejection. A field gradient sensitivity of 13 fT/cm $Hz^{1/2}$ for a baseline of 7 cm has been achieved using the two HTS rf SQUIDs (50). A first-order gradiometer using three HTS SQUIDs has also been prepared in which the third magnetometer is used to cancel ambient background noise for the other two magnetometers (51). Matlashov et al (105) reported fabrication of an electronic gradiometer using a HTS SQUID magnetometer array in which the baseline can be varied from 0.75 to 7.5 mm. One significant advantage of the electronic gradiometer is the ability to build it with a large baseline and capable of measuring any of the gradient–tensor components, providing greater flexibility than the direct coupled or flip-chip HTS gradiometer.

8.7 APPLICATIONS OF HTS SQUID MAGNETOMETER

8.7.1 Biomagnetic Measurements

Biomagnetism refers to the magnetism associated with biological processes arising from the flow of currents in neurons and muscle fibers. These biomagnetic fields are similar in many respects to the magnetic field set up by a current in a conducting wire, and the same principles of electromagnetism that hold for currents in conducting wires also apply on a much smaller scale to the currents generated by active neurons and muscle fibers. Figure 8.12 shows magnitudes of biomagnetic fields originating from electrically active tissues of different parts of the body and magnetic fields from different sources in the environment such as the Earth's magnetic field, urban noise, and so forth. It is immediately obvious that the environmental noise is many orders of magnitude higher than the biomagnetic field. The challenge of the biomagnetic measurements is in achieving the instrumentation sensitivity enough to detect weak signals such as those generated by the electrical activity of the heart or by neurons of the brain and simultaneously screening out the interference from background magnetic noise.

The potentiality of SQUID in measuring these biomagnetic field has been demonstrated earlier (2,106). Low-T_c SQUID-based multichannel SQUID systems are now commercially available for mapping the magnetic field of brain (MEG). The noise levels of the SQUID sensor in these MEG systems are typically < 10 fT (107). Magnetically shielded rooms are usually used for biomagnetic measurements using SQUIDs. In these magnetically shielded rooms, several layers of high-permeability materials are combined to reach a shielding factor of up to 10,000. The MEG field is in the range of 10 to 1000 fT. The magnetic field generated from the human heart is much larger than the magnetic field from the brain.

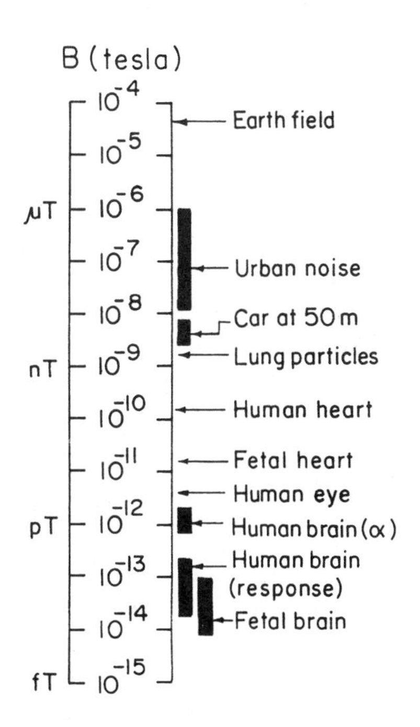

FIGURE 8.12 Comparison of biomagnetic field and background magnetic fields from an unshielded environment.

Typical signal amplitude in magnetocardiography (MCG) is ~ 10 pT. However, the physicians are more interested in the fine structure of the QRS-sequence where the mean signal amplitude reduces to roughly 2 pT. For adequate current source localization, the measurement bandwidth needs to be at least 1–100 Hz at a signal-to-noise ratio of 5. Recent clinical MCG investigations reveal several areas of cardiology where MCG using a SQUID can play a significant role (108). One undoubted advantage of MCG over electrocardiography (ECG) is the application in fetal cardiology. During the advanced stage of pregnancy, the strong attenuation and distortion of an electrical signal within various tissue layers between the fetus' heart and the mother's skin surface create problems in the use of the conventional ECG method.

Several groups have reported the measurements of magnetic signals from the human heart using a single SQUID sensor (8,37,54,64,109,110,114) and a multichannel HTS SQUID probe (7,111–113) in a magnetically shielded environment. Schilling et al. (109) reported the development of a HTS magnetometer with a hand-held cryostat for MCG measurement. However, the measurements have been carried out inside the shielding room. The magnetically shielded rooms are very expensive, and in order to make the SQUID system cost-effective for biomagnetic measurement, the HTS SQUID system should be portable and capable of operation in an unshielded environment.

Measurements of MCG in an unshielded environment using electronic first-order and second-order HTS rf-SQUID gradiometer are reported (49,50). In these measurements, the MCG signal was recorded in a bandwidth of 30 Hz using a 50-Hz digital notch filter. Yokosawa et al. (104) measured the MCG signal using an axial high-T_c first-order gradiometer in a moderately shielded room with 50-, 100-, and 150-Hz notch filters and a 0.1–80-Hz bandpass filter. Zhang et al. (115) described a second-order electronic gradiometer for recording of a MCG signal in the bandwidth of 100Hz. Ter Brake et al. (116) fabricated a electronic second-order gradiometer using three HTS SQUIDs for fetal magnetocardiography in the bandwidth of 100 Hz. Bick et al. (117) have shown that by using noise-canceling software techniques, the measurement bandwidth of electronic gradiometer can be increased from 130 to 250 Hz without using any notch filters.

The HTS thin-film gradiometers have been fabricated for monitoring the heart signal in an unshielded environment (91,118–120). Weidl et al. (118) have used a direct-coupled dc-SQUID first-order gradiometer, whereas Kouznetsov et al. (120) reported a thin-film second-order gradiometer for MCG measurement. Figure 8.13a shows a magnetocardiogram as recorded in a magnetically shielded room (8) and Figure 8.13b shows a magnetocardiogram recorded using a thin-film second-order gradiometer in an unshielded environment (120). It is evident that the HTS gradiometer eliminates background noise and the MCG data recorded in an unshielded environment is comparable to the MCG data recorded using a SQUID magnetometer in a magnetically shielded room.

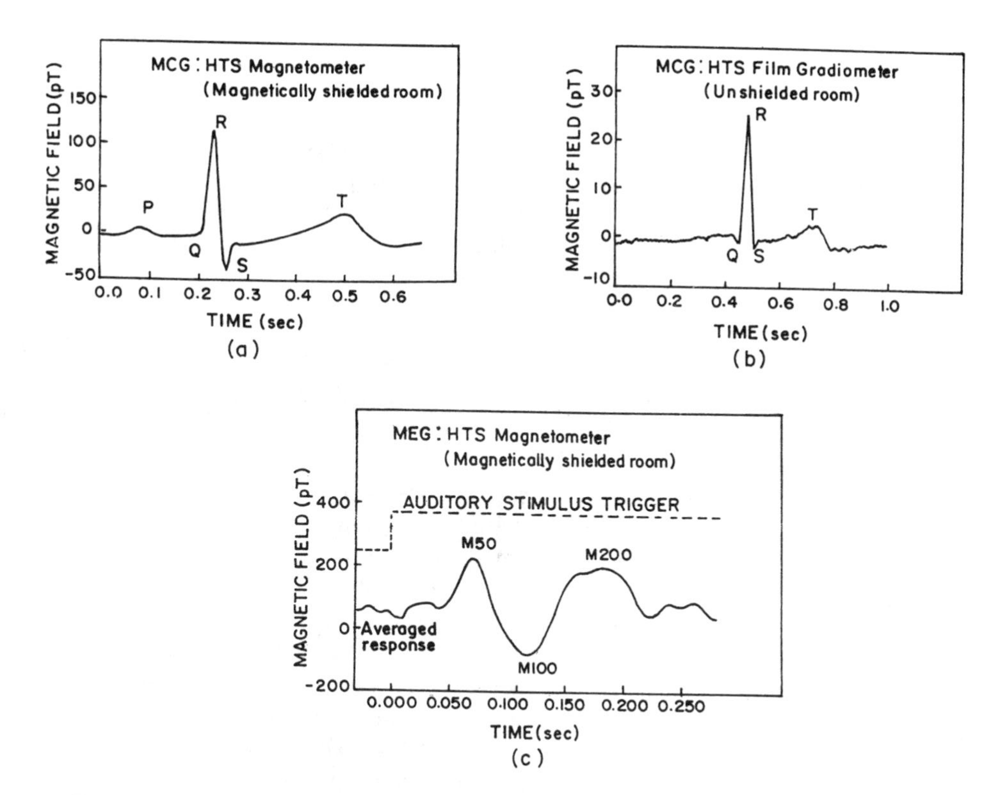

FIGURE 8.13 (a) Magnetocardiogram using a high-T_c SQUID magnetometer in a magnetically shielded room. The MCG data are averaged for 326 cardiac beats. (Adapted from Ref. 8.) (b) Magnetocardiogram using a high-T_c thin film second-order gradiometer in an unshielded environment. The MCG data are averaged for 119 beats. (Adapted from Ref. 120.) (c) Evoked brain response to auditory stimulus obtaining using a HTS magnetometer in a magnetically shielded room. (Adapted from Ref. 8.)

A handy eight-channel HTS system with encapsulated sensors have been developed for applications where a small mobile cryostat in a moderately shielded room is desired (9). The HTS SQUID sensors in this heart scanner has been designed for operation in a static magnetic field upto 50 μT without any increase in the low-frequency noise. Development of a HTS SQUID heart scanner cooled by a small stirling cryocooler is reported for operation in standard clinical environment without magnetically shielded room (121). David et al. (122) reported fabrication of a nine-channel high-T_c SQUID system for MCG measurement in an unshielded environment. In this multichannel SQUID magnetometer, a noise level of about 1 pT/Hz$^{1/2}$ for each channel was achieved using an active noise-compensation system.

Figure 8.13c shows the evoked brain response as recorded by a HTS magnetometer in a magnetically shielded room (8). There are some other reports also on the measurements of evoked magnetic signals from the brain using high-T_c SQUIDs (9,37,64). However, the sensitivity of the HTS SQUID still needs more improvement for developing MEG systems to compete with the commercially available low-T_c SQUID MEG systems.

8.7.2 Nondestructive Evaluation

Nondestructive evaluation (NDE) is the determination of defects in the specimen without damaging it. In fact, the complete absence of contact is a desirable feature of NDE. The eddy current technique is a simple and popular method for detecting hidden defects in the metallic specimens. In this technique, the eddy current generated by the field of an ac excited coil is deviated by the presence of flaws in the material. Conventionally, this distorted field is detected by a secondary coil. The sensitivity of this technique reduces considerably at the low frequency. Thus, the detection of deep buried flaws in the thick plate is not possible by this technique because the skin depth requires use of a low-frequency excitation field to penetrate deep into the material. On the other hand, the SQUID is a very suitable sensor for the detection of deep buried flaws in metallic plates due to its high sensitivity even in the low-frequency range. The SQUID is also used for NDE studies of ferromagnetic specimens. Low-T_c SQUIDs operating at 4.2 K have been used for several NDE studies (2,123,124) and the results have been very promising. The requirements for cooling and system development in the case of the low-T_c SQUID is more stringent compared to the high-T_c SQUID; thus, there seems to be great scope for high-T_c SQUID-based NDE applications.

In recent years, there have been considerable developments in using HTS SQUIDs for NDE studies (125–152). SQUID-based NDE measurements can be broadly classified into two categories: (1) observation under a direct-current (dc) field and (2) observation under an alternating-current (ac) field. In the direct-current technique, the specimen is exposed to a dc field and the SQUID is used to detect magnetic flux leakage or a stray field due to remanant magnetization. In the ac field technique, the ac field is applied to the specimen and the distribution of the magnetic field due to the eddy current is measured by SQUID. Figure 8.14 shows the high-T_c SQUID setup (developed at National Physical Laboratory, New Delhi, India) for the detection of flaws in metallic plates using an ac current technique. In this setup, the SQUID is inside a tail-type dewar at 77 K and remains stationary. The specimen is moved with the help of a computer controlled *X–Y* table. An ac field (20–200 Hz) is applied through a coil mounted at the bottom of the tail of the dewar. Normally, a circular coil or double-D-shaped coil is used for the ac

FIGURE 8.14 High-T_c SQUID setup for NDE studies of defects in metallic plates.

field excitation. This field is applied to the specimen, which induces eddy current. The lock-in technique is used to detect the signal from the sample at the eddy current frequency. The presence of any defects in the plate distorts the eddy current distribution, which is detected by the SQUID.

There are reports of the investigation of surface and subsurface defects in aluminum plates (125–133), aircraft parts (134–138), the quality of copper, aluminum, and NbTi/Cu composite wire (139), stressed steel cylinders (140), and mapping of an embedded iron rod in a concrete block (131). In other reports, the multichannel system such as an array of eleven HTS SQUIDs has been employed for investigating subsurface flaws in an aluminum plate (141) and a four-channel ramp-edge junction dc-SQUID magnetometer for magnetic inspection of prestressed concrete bridges (142).

Several of these measurements have been performed in an unshielded environment. In these studies, either a SQUID magnetometer or a gradiometer is used. The SQUID magnetometer is locally shielded with μ-metal or a HTS cylinder (130,143,144,147) to prevent the unwanted external magnetic interference.

In several practical applications, the specimen cannot be moved and the detection system has to move for scanning the defect location. There are some reports in which NDE measurement with a moving SQUID have been performed (127,128,134,145). A portable cryostat has also been developed for the investigation of the fatigue crack detection on stationary samples with a moving SQUID (145).

Considerable interests have been shown for the NDE of aircraft parts using a HTS SQUID. In the body of the aircraft, flaws present in the layers of aluminum around rivets need to be detected without the removal of the paint or disassembly of the parts. Aircraft wheels are subjected to enormous stress and breaking-generated heat during takeoff and landing. This may develop cracks that are not visible from the outside. Eddy current techniques are normally used for detecting these cracks, but due to poor sensitivity of the technique in the low-frequency region, only larger cracks can be detected. An ultrasonic technique can be employed to detect smaller flaws. However, this technique is very tedious because gel or water has to be used as a coupling medium for the ultrasonic wave into the wheel. A HTS SQUID system with a Joule–Thomson cooler has been developed to detect flaws in the thick aluminum aircraft structure and aircraft wheels (134–138). A three-layer aluminum sample (62 mm thick) from a EADS airbus was measured using a HTS rf-SQUID magnetometer (135). A defect lying at a depth to 40 mm was detected successfully by employing low-frequency (10–40 Hz) ac excitation. For aircraft wheel testing, the whole SQUID system was mounted on a robot and integrated in an automated wheel testing system (Fig. 8.15). In this setup, while the aircraft wheel is rotating, the SQUID mounted on the tip of Joule–Thomson cooler is brought close to the wheel's contour by the robot. Figure 8.16 shows the result of the SQUID measurement for the testing of a Boeing 737 wheel. An inner flaw penetrating even 10% of the wall thickness has also been detected successfully by scanning the outside surface of the rim (138).

A demonstration of capability of high-T_c SQUID NDE in the detection of deep buried defects gives a very optimistic picture for the potentiality of this tech-

FIGURE 8.15 A HTS SQUID system with a Joule–Thompson cooler mounted on a robot for aircraft wheel testing. While the aircraft wheel is rotated, the SQUID is brought closer to the wheel's contour by the robot. (From Ref. 138, © 2001 IEEE.)

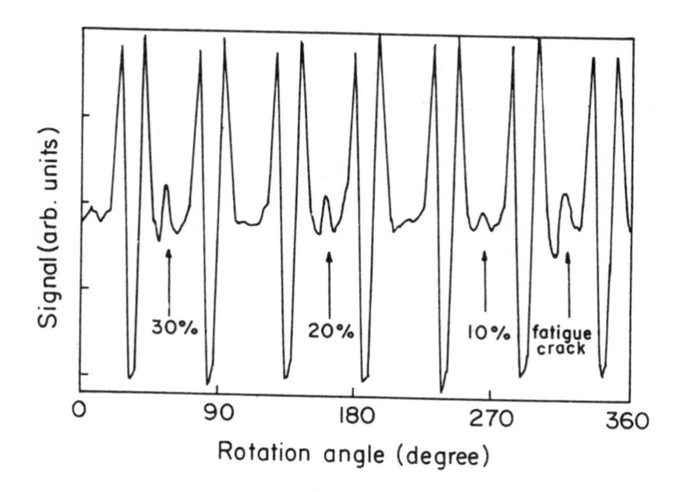

FIGURE 8.16 The NDE of Boeing 737 wheel using a HTS SQUID. The three sawcuts with 30%, 20%, and 10% wall penetration are clearly visible along with the signals from the seven keys. (From Ref. 138, © 2001 IEEE.)

nique. Several methods have been proposed for obtaining the exact depth information about the subsurface flaws. These are measurements by changing the frequency of the flowing current in the metal (146) or use of the dual-frequency eddy current technique (128). Another method is the analysis of the amplitude and phase of the signals (143,147–150) in which the amplitude of the signal is found to correspond to the size of the flaw, whereas the phase is correlated to the depth of the flaw. However, an accurate determination of depth and location of deep buried flaw is still a challenging task for NDE studies.

8.7.3 HTS SQUID Microscope

A magnetic microscope is an instrument for imaging the spatial distribution of magnetic flux in a sample on a microscopic scale. The SQUID is extremely sensitive magnetic field sensor and can be of very small dimension. Thus, the SQUID-based magnetic microscope can offer an unprecedented combination of magnetic field sensitivity and the spatial resolution (153). There is another advantage of the SQUID microscope in that the data from a scanning SQUID are highly quantitative and can be used for comparison with theoretical models.

A scanning SQUID microscope (SSM) consists of a SQUID and its electronics, *X–Y* positioning assembly, a cryogenic dewar, stepper motors for controlling the *X–Y* positioning stage, and a computer to control stepper motors and record SQUID output. The sample positioning stage is made of nonmagnetic and nonconducting material so that the movement of the position does not perturb the magnetic field of the sample. The SQUID is operated in the flux locked loop mode and its output is recorded in the computer as a function of sample position. Spatial resolution of the SQUID microscope depends on the size of a SQUID ring and the distance between the SQUID and the sample. If the SQUID is operated very close to the sample, then the spatial resolution is limited by the size of the SQUID's hole. If the distance between the SQUID and the sample is more than the hole of the SQUID, then the spatial resolution is not better than the distance between the SQUID and the sample. Thus, for increasing the spatial resolution, the SQUID should be designed to have a small hole and it should be very close to the sample. For a SQUID microscope, the spatial resolution and sensitivity are optimized when the SQUID is separated from the sample by not exceeding its own hole diameter.

Low-T_c SQUID-based magnetic microscopes have been demonstrated to have a spatial resolution of ~ 4–10 μm (154–156). These microscopes have been used in several basic and applied studies (155) such as in imaging of half-flux quantum in a tricrystal superconducting ring of YBCO for determining the pairing symmetry (157), magnetic vortices in superconducting networks and clusters (156), moat-guarded superconducting electronic circuits (158), in the detection of trapped flux quanta in superconducting film (159), and in the detection of vortex motion in a high-T_c grain boundary (160).

Wellstood and co-workers (161–163) have pioneered the development of a high-T_c scanning SQUID microscope. Later, several other groups have also reported the fabrication of a high-T_c SSM (164–171). These microscopes can be broadly classified into two categories. In the first one, the SQUID and the specimen both remain at 77 K (161,162,164), whereas in the second type of microscope, the HTS SQUID remains at 77 K and the sample is kept at room temperature (165,167,171–177). For the SQUID and specimen at 77 K, a resolution of 15 μm has been demonstrated using a bicrystal junction dc-SQUID (178). Tzalenchuk et al. (168) have developed a variable sample temperature microscope with a high-T_c step junction dc-SQUID in which both the sample and SQUID are thermally linked to the cold bath and the sample temperature can be varied from 5 to 100 K. For the investigation of a room-temperature sample, HTS SQUID microscopes have been developed using thin sapphire and Si_xN_y windows between the SQUID at low temperature and the sample at room temperature (165,171). The SQUID is mounted at the end of a sapphire point that is plugged into the end of a cold finger. The sapphire point is maintained at 77 K by a direct contact with liquid nitrogen. The distance between the sample and the SQUID can be as low as 15 μm in this arrangement (171). Neocera Inc. has developed a commercial version (MAGNA-C1) of the HTS microscope for investigating a room-temperature sample which can image current flow and make failure analysis in an integrated circuit. Tanaka et al. (169) constructed a high-T_c SSM with a sample chamber isolated by a shutter in which the sample can be investigated at room temperature and also at 77 K. Daibo et al. (170) reported the fabrication of a laser HTS SQUID microscope for photoinduced magnetic field imaging of a *p-n* junction at room temperature. A spatial resolution of 20 μm has been achieved that mainly depends on the spot size of the laser beam. In another novel design, a sharp soft magnetic needle is attached to a HTS SQUID for investigating a room-temperature sample (173–175). The needle serves as a magnetic flux guide (MFG) through which a magnetic field is transferred for a point at the sample to the SQUID. For such a combined sensor design, the spatial resolution of the SSM can be greatly improved because it is determined by the diameter of the needle tip and the distance between the tip and the sample surface.

The high-T_c SQUID microscope has been used to image a static magnetic field from domains in ferromagnetic samples, small currents in fine wires, and trapped flux and diamagnetic susceptibility in a superconducting thin film (161,166,167,169). There are other reports of imaging of the distribution of eddy current at a frequency below about 1 MHz (162), a microwave source of frequency up to 12.5 GHz at 77 K (178), and a microwave source at room temperature (176). For an imaging microwave field at 77 K, a spatial resolution of 15 μm has been achieved (178), whereas for the microwave source at room temperature, the SQUID-to-sample separation was 80 μm (176). The high-T_c SQUID microscope has also been used for different investigations of samples at room temperature, such as to image defect present in Cu-clad Nb–Ti wire (177), magnetized ion particles (167), and in the detection of motion of magnetotactic bacteria (171).

8.7.4 Geophysical Applications

Sensitive magnetometers are required for several geophysical explorations. Conventionally, a large induction coil, a flux gate magnetometer, proton magnetometers, and optically pumped magnetometers are used for this purpose. During the early 1980s, there was considerable interest in using a low-temperature SQUID magnetometer for geophysical applications (3). The potentiality of the SQUID for the studies related to electromagnetic survey, an airborne magnetic survey, rock magnetism, paleomagnetism, tectonomagnetism, and mapping hydro fracture have been demonstrated. However, these applications could not be adopted on a widespread level due to the complexity of the system and the requirement of liquid helium. With the development of a high-T_c SQUID with sufficiently high sensitivity, the interest of using SQUIDs for the geophysical applications have been rejuvenated. Several groups have used the HTS SQUID for ground-based and airborne geophysical transient electromagnetic (TEM) exploration (11,179–182). Initially, a single HTS SQUID was used for this measurement. However, more recently, three axes HTS magnetometer have been used (11,179).

The TEM exploration method is based on the measurement of the weak secondary magnetic field induced by eddy currents flowing in the ground after switching off the primary magnetic field of a large transmitter coil. The primary magnetic field is generated by passing a square-wave current pulse through a large transmitter coil that is situated on or above the surface of the Earth. The secondary field is measured through a magnetometer during the time ranging from a few tens of microseconds to a few seconds after the primary field pulse is switched off. The magnetometer in such a system is exposed to a rather large primary field ($\approx$100 nT) and should be able to measure a small secondary field. For the adequate measurement of a secondary field, the receiver magnetometer should have a slew rate $>$ 1 ms, a dynamic range $>$ 120 dB, and a bandwidth $\approx$1 Hz–20 kHz. Conventionally, an induction coil is used for measuring the secondary field. High-T_c SQUIDs having a high sensitivity, and a wide dynamic range and bandwidth are very suitable for measuring the weak secondary magnetic fields in TEM explorations.

The advantage of the HTS SQUID over the induction coil has been demonstrated specially for the late-time TEM measurements in terms of improved signal-to-noise ratio. A 64-times averaged SQUID signal yields a comparable result to a 256-times averaged coil signal (179). Another advantage of the HTS SQUID magnetometer over the induction coil method is the overall smaller size and weight of the system. The SQUID output signal is directly proportional to the magnetic field rather than its time derivative, as in the case of induction coil. Thus, the depth investigation using the SQUID is independent of the overlying ground conducting layer. The airborne TEM exploration studies using the HTS SQUID is found to give better results compared to the induction coil system, but there is still need for improving the system by increasing Q and reducing the noise due to the motion of the system (11).

The HTS SQUID electronic gradiometers have been constructed for use in mapping of the gradient of the Earth's magnetic field in surface exploration and archaeometry (183,184). A comparison of the results of a SQUID gradiometer and a cesium vapor magnetometer showed that the SQUID system has superior noise characteristics. It can measure the vector component directly and has lower power consumption (184). The performance of the SQUID gradiometer system can be further improved by reducing the enhanced noise of the system due to motion.

A prototype for three SQUID sensor gradiometers has been developed (185) and used for magnetic anomaly detection of an underwater target in a mobile survey. The system noise was 0.8 pT/m-Hz$^{1/2}$ at 1 Hz on stationary position, and when the system is in motion, the sensitivity was 5 pT/m Hz$^{1/2}$ at 1 Hz.

A HTS rf-SQUID-based spinning rock magnetometer has been developed for measuring the intensity and direction of the remanant magnetization of rock specimens collected in paleomagnetic surveys (186). Samples are rotated around two orthogonal axes to facilitate the calculation of the three-component remanance vectors with a minimum of operator intervention. Measurement of remanance values ranging from 10 to 10^{-5} A/m with errors of 1×10^{-5} A/m has been achieved in a measurement span of 20 s. A portable high-T_c SQUID for the measurement of remanant field of rock that can be operated for more than 17 hours for one filling of liquid nitrogen is also reported (187).

A portable high-T_c SQUID system has been developed for measuring magnetic field changes in the ultralow-frequency band related with earthquakes (188). The measurements were carried out at Mount Bandai, an active volcano and some anomalous magnetic field changes were recorded before and after the volcanic earthquake.

8.7.5 Other Applications

Recently high-T_c SQUIDs have been used in several novel applications such as in biological immunoassays (189), the detection method for DNA molecules (190), for sentinel–lymph node biopsy (13,191), and for the detection of a chemomagnetic field (192). Apart from directly measuring the magnetic field and field gradient, HTS SQUIDs have also been used in several other applications such as a picovoltmeter (193–195), an amplifier (196–198) and an oscillator (199).

In the biological immunoassays (12,189), an antibody that is labeled with a magnetic marker made of a γ-Fe_2O_3 nanoparticle is used to detect an antigen such as pathogenic bacteria, cancer cell, or environmental injuries. The binding reaction between the antigens and its antibody can be measured with the magnetic field from the marker. It has been demonstrated that the HTS SQUID system can detect the antigen–antibody reaction 10 times more sensitive than the conventional optical marker (189). By using a SQUID gradiometer and increasing the applied field or using the remanance of the marker, the SQUID system for immunoassays can be developed that is 100 times more sensitive compared to the conventional optical system.

In solid-state combustion and the liquid–solid reaction, the moving reaction zone, ion convection, and ion diffusion causes a macroscopic ionic current which generates electric and magnetic fields. The HTS SQUID has been used for the detection of these camomagnetic fields (192). Studies of the time series of these chemomagnetic fields is expected to throw more light on the dynamics of the chemical reaction.

Figure 8.17 is a schematic of a SQUID-based picovoltmeter. The basic components of a picovoltmeter are the SQUID, the preamplifier, and current feedback. The voltage from a source with resistance R_s is connected in series with a inductance L and a known resistance R_0. The SQUID is operated in flux locked loop mode. The current generated by the voltage source through the input circuitry is cancelled by the feedback current through R_0. In this circuit, the SQUID output voltage is proportional to the input voltage. A sensitivity of 2.3 $pV/Hz^{1/2}$ at 1 Hz is demonstrated for the HTS SQUID-based picovoltmeter (193). The HTS SQUID picovoltmeter has been used as a preamplifier for a Rogowski-type current sensor (194).

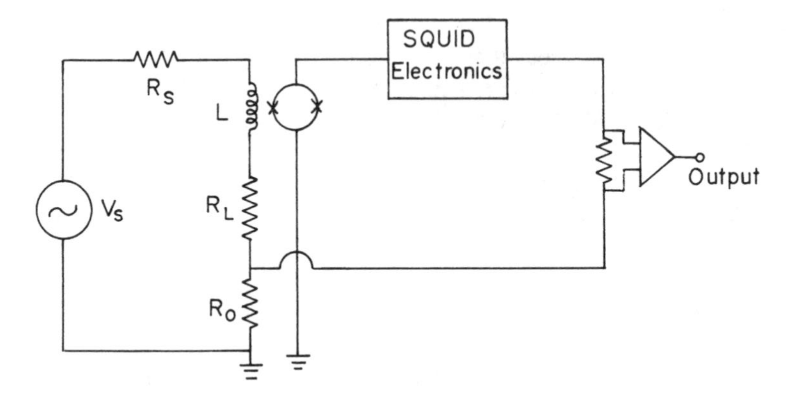

FIGURE 8.17 Block diagram of a high-T_c SQUID-based picovoltmeter.

A HTS SQUID-based amplifier combined with a HTS bandpass filer is developed for use in a front end receiver subsystem for a mobile base station (196). The combination of a HTS SQUID amplifier and a HTS bandpass filter reduces the band noise and optimizes the amplifier performances. Discharge measurement using a HTS SQUID-based amplifier is also demonstrated (197). A concept of a rf amplifier based on a direct-coupled bicrystal junction dc-SQUID is proposed (198) which can be a very promising device for a number of practical low-noise applications where small size and extremely low-power dissipation are important factors.

Hao et al. (199) reported fabrication of a tunable HTS rf Josephson effect oscillator using a YBCO–Au–YBCO-resistive SQUID in which the frequency in the range of 1–50 MHz can be very accurately controlled. In another report, application of a HTS SQUID in ion beam measurement is demonstrated (200).

8.8 CONCLUSION

There has been considerable progress in the developments of high-T_c SQUID magnetometers and gradiometers. As a result, SQUID magnetometers with a magnetic field sensitivity of 10 fT/Hz$^{1/2}$ or better have been achieved. The feasibility of the applications of high-T_c SQUIDs in MCG measurement, NDE studies, and geophysical measurements have been demonstrated. Multichannel HTS SQUIDs systems have been developed for MCG measurement and NDE studies. Recent use of HTS SQUIDs in biological immunoassay, sentinels–lymph node biopsy, detection of DNA, and so forth shows its potentiality for providing unique results in several other areas.

Using high-T_c SQUIDs in field applications poses many challenging problems. A practical SQUID-based system should be portable and requires its operation in an unshielded environment. In several studies, the SQUID system needs to be moved while recording the data. Efforts are in progress to develop a SQUID system for such measurements. There have been some progress in the development of a HTS SQUID system with a portable liquid-nitrogen dewar or HTS SQUID system with cryocooler assembly and battery-powered electronics.

The design of a high-T_c SQUID sensor has been improved to prevent the enhancement of excess low-frequency noise when the SQUID is cooled in the presence of a static magnetic field. HTS gradiometers with different configurations such as single-layer direct-coupled, flip-chip, and electronic gradiometers have been developed for measurement in an unshielded environment.

The development of high-T_c SQUID microscope for the study of a room-temperature samples with high spatial resolution has been a remarkable advancement. Further development in the increase of spatial resolution and sensitivity of a SQUID and cryocooler-based HTS SQUID microscope will open new areas of application.

REFERENCES

1. J Clarke. Low-frequency applications of superconducting quantum interference devices. IEEE Trans Magnetics 61:8–19, 1973.
2. JP Wikswo Jr. SQUID magnetometers for biomagnetism and nondestructive testing: important questions and initial answers. IEEE Trans Appl Supercond 5:74–120, 1995.
3. J Clarke. Geophysical applications of SQUIDS. IEEE Trans Magnetics 19:288–294, 1983.
4. K Enpuku, T Minotani. Progress high T_c superconducting quantum interference device (SQUID) magnetometer. IEICE Trans Electron E83-C:34–43, 2000.
5. D Koelle, R Kleiner, F Ludwig, E Dantsker, J Clarke. High-transition-temperature superconducting quantum interference devices. Rev Mod Phys 71:631–686, 1999.
6. AK Gupta, N Khare. High T_c SQUIDs : a review. In: AV Narlikar, ed. Studies of High Temperature Superconductors, Vol. 12. New York: Nova Science Publishers, 1994, pp. 43–94.
7. H Itozaki. Development and commercialization of high T_c SQUID. Physica C 357–360:7–10, 2001.
8. MS Dilorio, KY Yang, S Yoshizumi. BioMagnetic measurements using low-noise integrated SQUID magnetometers operating in liquid nitorgen. Appl Phys Lett 67:1926–1928, 1995.
9. HJ Barthelmess, M Halverscheid, B Schienfenhovel, E Heim, M Schilling, R Zimmermann. Low-noise biomagnetic measurements with a multichannel dc-SQUID system at 77K. IEEE Trans Appl Supercond 1:657–660, 2001.
10. HJ Krause, MV Kreutzbruck. Recent developments in SQUID NDE. Physica C 368:70–79, 2002.
11. CP Foley, KE Leslie, R Binks, C Lewis, W Murray, GJ Sioggett, S Lam, B Sankrithyan, N Savvides, A Katzaros, KH Muller, EE Mitchell, J Pollock, J Lee, DL

Dart, RR Barrow, M Asten, A Maddever, G Panjkovic, M Downey, C Hoffman, RR Turner. Field trials using HTS SQUID magnetometers for ground-based and airborne geophysical applications. IEEE Trans Appl Supercond 9:3786–3792, 1999.

12. K Enpuku, M Hotta, A Nakahodo. High-T_c SQUID system for biological immunoassays. Physica C 357–360:1462–1465, 2001.
13. S Tanaka, H Ota, Y Kondo, Y Tamaki, S Noguchi, M Hasegawa. Position determine system for lymph node relating breast cancer using a high-T_c SQUID. Physica C 368:32–36, 2002.
14. CD Tesche, J Clarke. dc SQUID: Noise and optimization. J Low Temp Phys 29:301–331, 1977.
15. TV Duzer, CW Turner. Principles of Superconductive Devices and Circuits. New York: Elsevier–North Holland, 1981.
16. T Hirano, T Nagaishi, H Itozaki. Direct readout flux locked loop circuit with automatic tuning of bias current and bias flux for high-T_c SQUID. Supercond Sci Technol 12:759–761, 1999.
17. D Drung, S Bechstein, KP Franke, M Scheiner, T Schurig. Improved direct-coupled dc SQUID read-out electronics with automatic bias voltage tunning. IEEE Trans Appl Supercond 11:880–883, 2001.
18. D Drung. High-performance DC SQUID read-out electronics. Physica C 368:134–140, 2002.
19. C Ludwig, C Kessler, AJ Steinfort, W Ludwig. Versatile high performance digital SQUID electronics. IEEE Trans Appl Supercond 11:1122–1125, 2001.
20. UA Overweg, MJ Walter-Peters. The design of a system of adjustable superconducting plates for balancing a gradiometer. Cryogenics 18:529–534, 1978.
21. MB Ketchen, WM Goubau, J Clarke, GB Donaldson. Superconducting thin-film gradiometer. J Appl Phys 49:4111–4116, 1978.
22. N Khare, AK Gupta, S Chaudhry, SK Arora, VS Tomar, VN Ojha. Fabrication and performance of high-T_c 2-hole bulk and single-hole thick film SQUIDs at 77 K. IEEE Trans Magnetics MAG-27:3029–3031, 1991.
23. S Chaudhry, N Khare, AK Gupta, VS Tomar. Performance of Bi-Sr-Ca-Cu-O thick film rf SQUIDs at liquid nitrogen temperatures. J Appl Phys 72:1172–1174, 1992.
24. VN Polushkin, BV Vasiliev. The investigation of rf-SQUIDs at liquid nitrogen temperature. Physica C 162–164:397–398, 1989.
25. N Khare, AK Gupta, HK Singh, ON Srivastava. RF-SQUID effect in Hg(Tl)-Ba-Ca-Cu-O high-T_c thin film up to 121 K. Supercond Sci Technol 11:517–519, 1998.
26. RH Koch, CP Umbach, GJ Clarke, P Chaudhari, RB Laibowitz. Quantum interference devices made from superconducting oxide thin films. Appl Phys Lett 51:200–202, 1987.
27. MJ Ferrari, M Johnson, FC Wellstood, J Clarke, A Inam, XD Wu, L Nazar, T Venkatesan. Low magnetic flux noise observed in laser-deposited in situ films of $YBa_2Cu_3O_y$ and implications for high-T_c SQUIDs. Nature 341: 723–725, 1989.
28. P Chaudhari, RH Koch, RB Laibowitz, TR McGuire, RJ Gambino. Critical-current measurements in epitaxial films of $YBa_2Cu_3O_{7-x}$ compound. Phys Rev Lett 58:2684–2686, 1987.
29. D Dimos, P Chaudhari, J Mannhart. Superconducting transport properties of grain boundaries in $YBa_2Cu_3O_7$ bicrystals. Phys Rev B 41:4038–4049, 1990.

30. J Luine, J Bulman, J Burch, K Daly, A Lee, C Pettiette-Hall, S Schwarzbek. Characteristics of high performance $YBa_2Cu_3O_7$ step-edge junctions. Appl Phys Lett 61:1128–1130, 1992.
31. JZ Sun, WJ Gallagher, AC Callegari, V Foglietti, RH Koch. Improved process for high-T_c superconducting step-edge junctions. Appl Phys Lett 63:1561–1563, 1993.
32. F Lombardi, ZG Ivanov, GM Fischer, E Olsson, T Claeson. Transport and structural properties of the top and bottom grain boundaries in $YBa_2Cu_3O_{7-\delta}$ step-edge Josephson junctions. Appl Phys Lett 72:249–251, 1998.
33. CL Jia, MI Faley, U Poppe, K Urban. The effect of chemical and ion-beam etching on the atomic structure of interfaces in $YBa_2Cu_3O_7/PrBa_2Cu_3O_7$ Josephson junctions. Appl Phys Lett 67:3635–3637, 1995.
34. K Char, MS Colclough, SM Garrison, N Newman, G Zaharchuk. Bi-epitaxial grain boundary junctions in $YBa_2Cu_3O_7$. Appl Phys Lett 59:733–735, 1991.
35. K Char, MS Colclough, LP Lee, G Zaharchuk. Extension of the bi-epitaxial Josephson junction process to various substrates. Appl Phys Lett 59:2177–2179, 1991.
36. N Khare, P. Chaudhari. Operation of bicrystal junction high-T_c direct current–SQUID in a portable microcooler. Appl Phys Lett 65:2353–2355, 1994.
37. Y Zhang, M Much, K Herrmann, J Schubert, W Zander, AI Braginski, C Heiden. Sensitive rf-SQUIDs and magnetometers operating at 77. IEEE Trans Appl Supercond 3:2465–2468, 1993.
38. LP Lee, M Teepe, V Vinetskiy, R Cantor, MS Colclough. Key elements for a sensitive 77 K direct current superconducting quantum interference device magnetometer. Appl Phys Lett 66:3059–3061, 1995.
39. M Matsuda, Y Murayama, S Kiryu, N Kasai, S Kashiwaya, M Koyanagi, T Endo, S Kuriki. IEEE Trans Magnetics MAG-27:3043–3046, 1991.
40. D Koelle, AH Miklich, F Ludwig, E Dantsker, DT Nemeth, J Clarke. dc SQUID magnetometers from single layers of $YBa_2Cu_3O_{7-x}$. Appl Phys Lett 63:2271–2273, 1993.
41. D Koelle, AH Miklich, E Dantsker, F Ludwig, DT Nemeth, J Clarke, W Ruby, K Char. High performance dc SQUID magnetometers with single layer $YBa_2Cu_3O_{7-x}$ flux transformers. Appl Phys Lett 63:3630–3632, 1993.
42. R Cantor, LP Lee, M Teepe, V Vinetskiy, J Longo. Low noise single-layer $YBa_2Cu_3O_7$ dc-SQUID magnetometers at 77 K. IEEE Trans Appl Supercond 5:2927–2930, 1995.
43. PRE Petersen, YQ Shen, MP Sager, T Holst, BH Larsen, JB Hansen. Directly coupled YBCO dc SQUID magnetometers. Supecond Sci Technol 12:802–805, 1999.
44. A Tsukamoto, Y Soutome, T Fukazawa, K Takagi. Dual bias current operation of 2-SQUID directly coupled magnetometers. IEEE Trans Appl Supercond 11:1106–1109, 2001.
45. JH Chen, KL Chen, HW Yu, MJ Chen, CH Wu, JT Jeng, HE Horng, HC Yang. Effects of modulation schemes on the performance of directly coupled high-T_c dc SQUID magnetometers. IEEE Trans Appl Supercond 11:1110–1113, 2001.
46. J Beyer, D Drung, F Ludwig, T Minotani, K Enpuku. Low-noise $YBa_2Cu_3O_{7-x}$ single layer dc superconducting quantum interference device (SQUID) magnetometer based on bicrystal junctions with 30° misorientation angle. Appl Phys Lett 72:203–205, 1998.

47. Y Zhang, W Zander, J Schubert, F Ruders, H Soltner, M Banzet, N Wolters, XH Zeng, AI Braginski. Operation of high-sensitivity radio frequency superconducting quantum interference device magnetometers with superconducting coplanar resonators at 77 K. Appl Phys Lett 71:704–706, 1997.
48. Y Zhang. Evolution of HTS rf SQUIDs. IEEE Trans Appl Supercond 11:1038–1041, 2001.
49. Y Tavrin, Y Zhang, M Muck, AI Braginski, C Heiden. $YBa_2Cu_3O_7$ thin film SQUID gradiometer for biomagnetic measurements. Appl Phys Lett 62:1824–1826, 1993.
50. J Borgmann, P David, G Ockenfuss, R Otto, J Schubert, W Zander, AI Braginski. Electronic high-temperature radio frequency superconducting quantum interference device gradiometers for unshielded environment. Rev Sci Instrum 68:2730–2734, 1997.
51. RH Koch, JR Rozen, JZ Sun, WJ Gallagher. Three SQUID gradiometer. Appl Phys Lett 63:403–405, 1993.
52. B Oh, RH Koch, WJ Gallagher, RP Robertazzi, W Eidelloth. Multilevel YBaCuO flux transformers with high T_c SQUIDs: A ptototype high T_c SQUID magnetometer working at 77 K. Appl Phys Lett 59:123–125, 1991.
53. FC Wellstood, JJ Kingston, MJ Ferrari, J Clarke. Superconducting thin-film flux transformers of $YBa_2Cu_3O_{7-x}$. Appl Phys Lett 57:1930–1932, 1990.
54. D Grundler, B David, R Eckart, O Dossel. Highly sensitive $YBa_2Cu_3O_7$ dc SQUID magnetometer with thin-film flux transformer. Appl Phys Lett 63:2700–2702, 1993.
55. MN Keene, SW Goodyear, NG Chew, RG Humphreys, JS Satchell, JA Edwards, K Lander. Low-noise $YBa_2Cu_3O_7$–$PrBa_2Cu_3O_7$ multiturn flux transformers. Appl Phys Lett 64:366–368, 1994.
56. LP Lee, K Char, MS Colclough, G Zaharchuk. Monolithic 77 K dc SQUID magnetometer. Appl Phys Lett 59:3051–3053, 1991.
57. AH Miklich, FC Wellstood, JJ Kingston, J Clarke, MS Colclough, AH Cardona, LC Bourne, WL Olson, MM Eddy. High T_c-thin-film magnetometer. IEEE Trans Magnetics MAG-27:3219–3222, 1991.
58. MD Strikovski, F Kahlmann, J Schubert, W Zander, V Glyantsev, G Ockenfuss, GL Jia. Fabrication of $YBa_2Cu_3O_7$ thin-film flux transformers using a novel microshadow mask technique for in situ patterning. Appl Phys Lett 66:3521–3523, 1995.
59. F Ludwig, D Koelle, E Dantsker, DT Nemeth, AH Miklich, J Clarke, RE Thomson. Low noise $YBa_2Cu_3O_{7-x}$–$SrTiO_3$ –$YBa_2Cu_3O_{7-x}$ multilayers for improved superconducting magnetometers. Appl Phys Lett 66:373–375, 1995.
60. J Ramos, V Zakosarenko, R IJsselsteijn, V Schultze, HG Meyer. Mutual inductance and noise of high-T_c SQUIDs with flip-chip and integrated input coils. IEEE Trans Appl Supercond 11:1118–1121, 2001.
61. E Dantsker, F Ludwig, R Kleiner, J Clarke, M Teepe, LP Lee, NM Alford, T Button. Addendum: Low noise $YBa_2Cu_3O_{7-x}$–$SrTiO_3$ –$YBa_2Cu_3O_x$ multilayers for improved superconducting magnetometers. Appl Phys Lett 67:725–726, 1995.
62. MI Faley, U Poppe, K Urban, DN Paulson, TN Starr, RL Fagaly. Low noise HTS dc-SQUID flip-chip magnetometers and gradiometers. IEEE Trans Appl Supercond 11:1383–1386, 2001.

63. JWM Hilgenkamp, GCS Brons. JG Soldevilla, RPJ IJsselsteijn, J Flokstra. Four layer monolithic integrated high T_c dc SQUID magnetometer. Appl Phys Lett 64:3497–3499, 1994.
64. D Drung, F Ludwig, W Muller, U Steinhoff, L Trahms, H Koch, YQ Shen, MB Jensen, P Vase, T Holst, T Freltoft, G Curio. Integrated $YBa_2Cu_3O_{7-x}$ magnetometer for biomagnetic measurements. Appl Phys Lett 68:1421–1423, 1996.
65. F Ludwig, E Dantsker, R Kleiner, D Koelle, J Clarke, S Knappe, D Drung, H Koch, NM Alford, TW Button. Integrated high-T_c multiloop magnetometer. Appl Phys Lett 66:1418–1420, 1995.
66. HJ Lee, WK Park, SM Lee, JD Park, SH Moon, O Oh. Integrated multilayer high-T_c SQUID magnetometers with slotted washer. IEEE Trans Appl Supercond 11:1327–1330, 2001.
67. R Scharnweber, M Schilling. Integrated $YBa_2Cu_3O_7$ magnetometer with flux transformer and multiloop pick-up coil. Appl Phys Lett 69:1303–1306, 1996.
68. JE Zimmerman. Sensitivity enhancement of superconducting quantum interference devices through the use of fractional-turn loops. J Appl Phys 42:4483–4487, 1971.
69. MJ Ferrari, M Johnson, FC Wellstood, JJ Kingston, TJ Shaw, J Clarke. Magnetic noise in copper oxide superconductors. J Low Temp Phys 94:15–61, 1994.
70. HJ Barthelmess, B Schiefenhovel, M Schilling. DC-SQUID magnetometers from $YBa_2Cu_3O_{7-\delta}$ in AC fields for application in multichannel systems. Physica C 368:37–40, 2002.
71. E Dantsker, S Tanaka, J Clarke. High-T_c super conducting quantum interference devices with slots or holes: low $1/f$ noise in ambient magnetic fields. Appl Phys Lett 70:2037–2039, 1997.
72. RH Koch, JZ Sun, V Foglietta, WJ Gallagher. Flux dam, a method to reduce extra low frequency noise when a superconducting magnetometer is exposed to a magnetic field. Appl Phys Lett 67:709–711, 1995.
73. FP Milken, SL Brown, RH Koch. Magnetic field-induced noise in directly coupled high T_c superconducting quantum interference device magnetometers. Appl Phys Lett 71:1857–1859, 1997.
74. MP Sager, PRE Petersen, T Holst, YQ Shen, JB Hansen. Low-frequency flux noise in YBCO dc SQUIDs cooled in static magnetic fields. Supercond Sci Technol 12:798–801, 1999.
75. V Schultze, N Oukhanski, V Zakosarenko, R IJsselsteijn, J Ramos, A Chwala, HG Meyer. HTS dc SQUID behavior in external magnetic fields. IEEE Trans Appl Supercond 11:1319–1322, 2001.
76. F Ludwig, J Beyer, D Drung, S Bechstein, T Schurig. High-performance high-T_c SQUID sensors for multichannel systems in magnetically disturbed environment. IEEE Trans Appl Supercond 9:3793–3796, 1999.
77. F Ludwig, ABM Jansman, D Drung, MO Lindstrom, S Bechstein, J Beyer, J Flokstra, T Schurig. Optimization of direct-coupled high-T_c SQUID magnetometers for operation in magnetically unshielded environment. IEEE Trans Appl Supercond 11:1315–1318, 2001.
78. K Yokosawa, H Oyama, S Kuriki, D Suzuki, K Tsukada, M Matsuda. Effect of a static magnetic field on a slotted high-T_c SQUID magnetometer without a flux dam. IEEE Trans Appl Supercond 11:1335–1338, 2001.

79. H Oyama, S Kuriki, M Matsuda. Effects of flux dam on low-frequency noise in high-T_c SQUID magnetometers. IEEE Trans Appl Supercond 11:1331–1334, 2001.
80. H Oyama, S Hirano, M Matsuda, S Kuriki. Suppression of field-induced low-frequency noise in high-T_c SQUID magnetometers using slotted flux dams. Physica C 357–360:1451–1454, 2001.
81. M Matsuda, S Ono, K Kato, T Matsuura, H Oyama, A Hayashi, S Hirano, S Kuriki, K Yokosawa. High-T_c SQUID magnetometers for use in moderate magnetically-shielded room. IEEE Trans Appl Supercond 11:1323–1326, 2001.
82. HM Cho, YT Andresen, J Clarke, MS DiIorio, KY Yang, S Yoshizumi. Low-frequency noise in high-transition-temperature superconducting multilayer magnetometers in ambient magnetic fields. Appl Phys Lett 79:2438–2440, 2001.
83. P Selders, R Wordenweber. Low-frequency noise reduction in $YBa_2Cu_3O_{7-\delta}$ superconducting quantum intereference devices by antidots. Appl Phys Lett 76:3277–3279, 2000.
84. P Selders, R Wordenweber. Low-frequency noise reduction in Y-Ba-Cu-O SQUIDs by artificial defects. IEEE Trans Appl Supercond 11:928–931, 2001.
85. M Bick, J Schubert, M Fardmanesh, G Panaitov, M Banzet, W Zander, Y Zhang, HJ Krause. Magnetic field behavior of YBCO step-edge Jospehson junctions in rf-washer SQUIDs. IEEE Trans Appl Supercond 11:1339–1342, 2001.
86. S Knappe, D Drung, T Schurig, H Koch, M Klinger, J Hinken. A planar $YBa_2Cu_3O_7$ gradiometer at 77 K. Cryogenics 32:881–884, 1992.
87. V Zakosarenko, F Schmidl, H Schneidewind, L Dorrer, P Seidel. Thin-film dc SQUID gradiometer using a single $YBa_2Cu_3O_{7-x}$ layer. Appl Phys Lett 65:779–780, 1994.
88. LR Bar, GM Daalmans, KH Barthel, L Ferchland, M Selent, M Kuhnl, D Uhl. Single-layer and integrated YBCO gradiometer coupled SQUIDs. Supercond Sci Technol 9:A87–A91, 1996.
89. CM Pegrum, A Eulenburg, EJ Romans, C Carr, AJ Millar, GB Donaldson. High-temperature single-layer SQUID gradiometers with long baseline and parasitic effective area compensation. Supercond Sci Technol 12:766–768, 1999.
90. A Eulenburg, EJ Romans, C Carr, AJ Millar, GB Donaldson, CM Pegrum. Highly balanced long-baseline single-layer high-T_c superconducting quantum interference device gradiometer. Appl Phys Lett 75:2301–2303, 1999.
91. A Tsukamoto, K Yokosawa, T Fukazawa, D Suzuki, K Tsukada, T Miyashita, A Kandori, K Takagi. Noise properties of highly balanced $YBa_2Cu_3O_y$ directly coupled gradiometers and MCG measurement in magnetically unshielded environment. Physica C 368:41–44, 2002.
92. IS Kim, JM Kim, HR Lim, YK Park. Development of a high-T_c first-order gradiometer system. IEEE Trans Appl Supercond 11:1359–1362, 2001.
93. AJ Millar, EJ Romans, C Carr, E Eulenburg, GB Donaldson, P Maas, CM Pegrum. High-T_c gradiometric superconducting quantum interference device and its incorporation into a single-layer gradiometer. Appl Phys Lett 76:2445–2447, 2000.
94. AJ Millar, EJ Romans, C Carr, A Eulenburg, GB Donaldson, CM Pegrum. Step-edge Josephson junctions and their use in HTS single-layer gradiometers. IEEE Trans Appl Supercond 11:1351–1354, 2001.

95. C Carr, EJ Romans, AJ Millar, A Eulenburg, GB Donaldson, CM Pegrum. First-order high-T_c single-layer gradiometers: Parasitic effective area compensation and system balance. IEEE Trans Appl Supercond 11:1367–1370, 2001.
96. SG Lee, Y Hwang, BC Nam, JT Kim, IS Kim. Direct-coupled second-order superconducting quantum interference device gradiometer from single layer of high temperature superconductor. Appl Phys Lett 73:2345–2347, 1998.
97. Y Hwang, JR Ahn, SG Lee, JT Kim, IS Kim, YK Park. Balancing of the single-layer second-order high-T_c SQUID gradiometer. IEEE Trans Appl Supercond 11: 1343–1346, 2001.
98. W Eidelloth, B Oh, RP Robertazzi, WJ Gallagher, RH Koch. $YBa_2Cu_3O_{7-\delta}$ thin-film gradiometers: Fabrication and performance. Appl Phys Lett 59:3473–3475, 1991.
99. E Dantsker, OM Froehlich, S Tanaka, K Kouznetsov, J Clarke, Z Lu, V Matijasevic, K Char. High-T_c superconducting gradiometer with a long baseline asymmetric flux transformer. Appl Phys Lett 71:1712–1714, 1997.
100. A Kittel, KA Kouznetsov, R McDermott, B Oh, J Clarke. High T_c superconducting second-order gradiometer. Appl Phys Lett 73:2197–2199, 1998.
101. YJ Tian, S Linzer, F Schmidl, L Dorrer, R Weidl, P Seidel. High-T_c directly coupled direct current SQUID gradiometer with flip-chip flux transformer. Appl Phys Lett 74:1302–1304, 1999.
102. Y Tavrin, Y Zhang, W Wolf, AI Braginski. A second-order SQUID gradiometer operating at 77 K. Supercond Sci Technol 7:265–268, 1994.
103. DF He, HJ Krause, Y Zhang, M Bick, H Soltner, N Wolters, W Wolf, H Bousack. HTS SQUID magnetometer with SQUID vector reference for operation in unshielded environment. IEEE Trans Appl Supercond. 9:3684–3687, 1999.
104. K Yokosawa, H Oyama, S Kuriki, D Suzuki, K Tsukada. Axial high-temperature superconducting-quantum-interference-device gradiometer composed of magnetometers with a monolithic feedback and compensation coil. Appl Phys Lett 78:2745–2747, 2001.
105. A Matlashov, M Espy, RH Kraus Jr., KR Ganther, LD Snapp. Electronic gradiometer using HTc SQUIDs with fast feedback electronics. IEEE Trans Appl Supercond 11:876–879, 2001.
106. J Clarke. SQUIDs, brains and gravity waves. Phys Today 39:36–44, 1986.
107. J Vrba. Magnetoencephalography: the art of finding a needle in a haystack. Physica C 368:1–9, 2002.
108. H Koch. SQUID Magnetocardiography: Status and Perspectives. IEEE Trans Appl Supercond 11:49–59, 2001.
109. M Schilling, S Krey, R Scharnweber. Biomagnetic measurements with an integrated $YBa_2Cu_3O_7$ magnetometer in a hand-held cryostat. Appl Phys Lett 69:2749–2751, 1996.
110. HR Lim, IS Kim, YK Park, DH Kim. Noise properties of $YBa_2Cu_3O_{7-x}$ step-edge junction dc SQUID magnetometers prepared on sapphire substrates. IEEE Trans Appl Supercond 11:1355–1358, 2001.
111. S Tanaka, H Itozaki, H Toyoda, N Harada, A Adachi, K Okajima, H Kado, T Nagaishi. Four-channel $YBa_2Cu_3O_{7-x}$ dc SQUID magnetometer for biomagnetic measurements. Appl Phys Lett 64:514–516, 1994.

112. HJM ter Brake, N Janseen, J Flokstra, D Veldhuis, H Rogalla. Multichannel heart scanner based on high-T_c SQUIDs. IEEE Trans Appl Supercond 7:2545–2548, 1997.

113. KY Sai, YQ Zhen, Y Uchikawa, M Kotani. Analysis of MCG measured by high-T_c SQUID with spatial filter method. IEEE Trans Magnetics MAG-36:3718–3720, 2000.

114. S Kuriki, A Hayashi, T Washio, M Fujita. Active compensation in combination with weak passive shielding for magnetocardiographic measurements. Rev Sci Instrum 73:440–445, 2002.

115. Y Zhang, G Panaitov, SG Wang, N Wolters, R Otto, J Schubert, W Zander, HJ Krause, H Soltner, H Bousack, AI Braginski. Second-order, high-temperature superconducting gradiometer for magnetocardiography in unshielded environment. Appl Phys Lett. 76:906–908, 2000.

116. HJM ter Brake, AP Rijpma, JG Stinstra, J Borgmann, HJ Holland, HJG Krooshoop, MJ Peters, J Flokstra, HWP Quartero, H Rogalla. Fetal magnetocardiography: Clinical relevance and feasibility. Physica C 368:10–17, 2002.

117. M Bick, K Sternickel, G Panaitov, A Effern, Y Zhang, HJ Krause. SQUID gradiometry for magnetocardiography using different noise cancellation techniques. IEEE Trans Appl Supercond 11:673–680, 2001.

118. R Weidl, S Brabetz, F Schmidl, F Klemm, S Wunderlich, P Seidel. Heart monitoring with high-T_c d.c. SQUID gradiometers in an unshielded environment. Supercond Sci Technol 10:95–99, 1997.

119. P Seidel, F Schmidl, R Weidl, S Brabetz, F Klemm, S Wunderlich, L Dorrer, H Nowak. Development of a heart monitoring system based on thin film high-T_c DC-SQUIDs. IEEE Trans Appl Supercond 7:3040–3043, 1997.

120. KA Kouznetsov, J Borgmann, J Clarke. High T_c second-order gradiometer for magnetocardiography in an unshielded environment. Appl Phys Lett. 75:1979–1981, 1999.

121. PJ van den Bosch, HJM ter Brake, JJ Holland, MA de Boer, JFC Verberne, H Rogalla. Cryogenic design of a high-T_c SQUID-based heart scanner cooled by small stirling cryocoolers. Cryogenics 37:139–151, 1997.

122. B David, O Dossel, V Doormann, R Eckart, W Hoppe, J Kruger, H Laudan, G Rabe. The development of a high-T_c magnetocardiography system for unshielded environment. IEEE Trans Appl Supercond 7:3267–3270, 1997.

123. H Weinstock. Prospects on the application of HTS SQUID magnetometery to nondestructive evaluation (NDE). Physica C 209:269–272, 1993.

124. RJP Bain, GB Donaldson, S Evanson, G Hayward, in HD Hahlbolhm, H Lubbig, eds. SQUID'85 Proc. 3rd International Conference on Superconducting Quantum Devices. Berlin: deGruyter, 1985, pp 841–846.

125. A Cochran, JC Macfarlane, LNC Morgan, J Kuznik, R Westson, L Hao, RM Bowman, GB Donaldson. Using a 77 K SQUID to measure magnetic fields for NDE. IEEE Trans Appl Supercond. 4:128–135, 1994.

126. CX Fan, DF Lu, KW Wong, Y Xin, B Xu, NS Alzayed, M Chester, DE Knapp. High temperature RF SQUID gradiometer applied to non-destructive testing. Cryogenics 34:667–670, 1994.

127. C Carr, A Cochran, J Kuznikt, DM McKirdly, GB Donaldson. Electronic gradiometry for NDE in an unshielded environment with stationary and moving HTS SQUIDs. Cryogenics 36:691–695, 1996.

128. C Carr, DM McKirdy, EJ Romans, GB Donaldson, A Cochran. Electromagnetic nondestructive evaluation: Moving HTS SQUIDs, inducing field nulling and dual frequency measurements. IEEE Trans Appl Supercond 7:3275–3278, 1997.
129. Y Tavrin, HJ Krause, W Wolf, V Glyantsev, J Schubert, W Zander, H Bousack. Eddy current technique with high temperature SQUID for non-destructive evaluation of non-magnetic metallic structures. Cryogenics 36:83–86, 1996.
130. N Tralshawala, JR Claycomb, JH Miller Jr. Practical SQUID instrument for nondestructive testing. Appl Phys Lett 71:1573–1575, 1997.
131. SG Han, JH Kang, KW Wong. Application of a two-hole HTS RF-SQUID gradiometer to non-destructive evaluation. Supercond Sci Technol 10:516–520, 1997.
132. H Nakane, R Kabasawa, H Adachi. Non-destructive testing using a HTS SQUID. IEEE Trans Appl Supercond 11:1291–1294, 2001.
133. DF He, M Yoshizawa. Detection of metal deformation using a high-T_c rf SQUID planar gradiometer. Physica C 357–360:1466–1468, 2001.
134. MV Kreutzbruck, J Troll, M Muck, C Heiden, Y Zhang. Experiments on eddy current NDE with HTS rf SQUIDs. IEEE Trans Appl Supercond 7:3272–3282, 1997.
135. MV Kreutzbruck, K Allweins, G Gierelt, HJ Krause, S Gartner, W Wolf. Defect detection in thick aircraft samples using HTS SQUID magnetometers. Physica C 368:85–90, 2002.
136. R Hohmann, HJ Krause, H Soltner, Y Zhang, CA Copetti, H Bousack, AI Braginski, MI Faley. HTS SQUID system with Joule–Thomson cryocooler for eddy current nondestructive evaluation of aircraft structures. IEEE Trans Appl Supercond 7:2860–2865, 1997.
137. R Hohmann, M Maus, D Lomparski, M Gruneklee, Y Zhang, HJ Krause, H Bousack, AI Braginski. Aircraft wheel testing with machine-cooled HTS SQUID gradiometer system. IEEE Trans Appl Supercond 9:3801–3804, 1999.
138. R Hohmann, D Lomparski, HJ Krause, MV Kreutzbruck, W Becker. Aircraft wheel testing with remote eddy current technique using a HTS SQUID magnetometer. IEEE Trans Appl Supercond 11:1279–1282, 2001.
139. H Weinstock, N Tralshawala, JR Claycomb. Nondestructive evaluation of wires using high-temperatures SQUIDs. IEEE Appl Supercond 9:3797–3800, 1999.
140. F Schmidl, S Wunderlich, L Dorrer, H Specht, S Linzen, H Schneidewind, P Seidel. High-T_c DC-SQUID system for nondestructive evaluation. IEEE Trans Appl Supercond 7:2756–2759, 1997.
141. MA Espy, AN Matlashov, JC Mosher, RH Kraus. Non-destructive evaluation with a linear array of 11 HTS SQUIDs. IEEE Trans Appl Supercond 11:1303–1306, 2001.
142. HJ Krause, W Wolf, W Glaas, E Zimmermann, MI Faley, G Sawade, R Mattheus, G Neudert, U Gampe, J Krieger. SQUID array for magnetic inspection of prestressed concrete bridges. Physica C 368:91–95, 2002.
143. JT Jeng, SY Yang, HE Horng, HC Yang. Detection of deep flaws by using a HTS-SQUID in unshielded environment. IEEE Trans Appl Supercond 11:1295– 1298, 2001.
144. JA Lobera-Serrano, JR Claycomb, JH Miller Jr, K Salama. Hybrid double-D sheet-inducer for SQUID-based NDT. IEEE Trans Appl Supercond 11: 1283–1286, 2001.

145. ML Lucia, R Hohmann, H Soltner, HJ Krause, W Wolf, H Bousack, MI Faley, G Sporl, A Binneberg. Operation of HTS SQUIDs with portable cryostat: a SQUID system in conjunction with eddy current technique for non-destructive evaluation. IEEE Trans Appl Supercond 7:2878–2881, 1997.
146. Y Hatsukade, N Kasai, H Takashima, R Kawai, F Kojima, A Ishiyama. Development of an NDE method using SQUIDs for the reconstruction of defect shapes. IEEE Trans Appl Supercond 11:1311–1314, 2001.
147. JT Jeng, HE Horng, HC Yang. High-T_c SQUID magnetometers and gradiometers for NDE applications. Physica C 368:105–108, 2002.
148. JT Jeng, HE Horng, HC Yang, JC Chen, JH Chen. Simulation of the magnetic field due to defects and verification using high-T_c SQUID. Physica C 367:298–302, 2002.
149. HE Horng, JT Jeng, HC Yang, JC Chen. Evaluation of the flaw depth using high-T_c SQUID. Physica C 367:303–307, 2002.
150. HC Yang, JT Jeng, HE Horng, SY Wang, SY Hung, JC Chen, JH Chen. Noise characteristics of high-T_c $YBa_2Cu_3O_y$ Physica C 367:290–294, 2002.
151. AK Gupta, N Khare, S Khare. Development of high-T_c superconducting thick film rf-SQUID sensors and their application in non-destructive testing. Measure Sci Technol 8:111–114, 1997.
152. P Seidel, S Wunderlich, F Schmidl, L Dorrer, S Linzen, F Schmidt, F Schrey, C Steigmeier, K Peiselt, S Muller, A Forster, S Losche, S Gudochnikov. Improvement of spatial and field resolution in NDE systems using superconducting sensors. IEEE Trans Appl Supercond 11:1176–1179, 2001.
153. J Kirtley. Imaging magnetic fields. IEEE Spectrum Dec: 41–48, 1996.
154. JR Kirtley, MB Ketchen, KG Stawiasz, JZ Sun, WJ Gallagher, SH Blanton, SJ Wind. High-resolution scanning SQUID microscope. Appl Phys Lett 66:1138– 1140, 1995.
155. JR Kirtley. SQUID microscopy for fundamental studies. Physica C 368:55–65, 2002.
156. LN Vu, MS Wistrom, DJ Van Harlingen. Imaging of magnetic vortices in superconducting networks and clusters by scanning SQUID microscopy. Appl Phys Lett 63:1693–1695, 1993.
157. CC Tsuei, JR Kirtley, CC Chi, LS Yu-Jahnes, A Gupta, T Shaw, JZ Sun, MB Ketchen. Pairing symmetry and flux quantization in a tricrystal superconducting ring of $YBa_2Cu_3O_{7-\delta}$. Phys Rev Lett 73:593–596, 1994.
158. M Jeffery, TV Duzer, JR Kirtley, MB Ketchen. Magnetic imaging of moat-guarded superconducting electronic circuits. Appl Phy Lett 67:1769–1771, 1995.
159. K Tanaka, T Morooka, A Odawara, Y Mawatari, S Nakayama, A Nagata, M Ikeda, K Chinone, M Koyanagi. Study of trapped flux in a superconducting thin film—Observation by scanning SQUID microscope and simulation. IEEE Trans Appl Supercond 11:230–233, 2001.
160. S Hirano, H Oyama, M Matsuda, T Morooka, S Nakayama, S Kuriki. Direct detection of vortex motion in high-T_c grain boundary junctions. IEEE Trans Appl Supercond 11:924–927, 2001.
161. RC Black, A Mathai, FC Wellstood, E Dantsker, AH Miklich, DT Nemeth, JJ Kingston, J Clarke. Magnetic microscopy using a liquid nitrogen cooled $YBa_2Cu_3O_7$ superconducting quantum interference device. Appl Phys Lett 62:2128–2130, 1993.

162. RC Black, FC Wellstood, E Dantsker, AH Miklich, JJ Kingston, DT Nemeth, J Clarke. Eddy current microscopy using a 77-K superconducting sensor. Appl Phys Lett 64:100–102, 1994.
163. FC Wellstood, Y Gim, A Amar, RC Black, A Mathai. Magnetic microscopy using SQUIDs. IEEE Trans Appl Supercond 7:3134–3138, 1997.
164. SA Gudoschnikov, II Vengrus, KE Andreev, OV Snigirev. Magnetic microscope based on YBCO bicrystal thin film dc SQUID operating at 77 K. Cryogenics 34:883–886, 1994.
165. TS Lee, E Dantsker, J Clarke. High transition temperature superconducting quantum interference device microscope. Rev Sci Instrum 67:4208–4215, 1996.
166. KE Andreev, AV Bobyl, SA Gudoshnikov, SF Karmanenko, SL Krasnosvobodtsev, LV Matveets, OV Snigirev, RA Suris, II Vengrus. Magnetic field maps of YBCO thin films obtained by scanning SQUID microscopy for HTSC microelectronics. Supercond Sci Technol 10:366–370, 1997.
167. T Nagaishi, H Itozaki. High T_c SQUID microscope head for room temperature sample. Supercond Sci Technol 12:1039–1041, 1999.
168. AY Tzalenchuk, ZG Ivanov, S Pehrson, T Claeson, A Lohmus. A variable temperature scanning SQUID microscope. IEEE Trans Appl Supercond 9:4115–4118, 1999.
169. S Tanaka, O Yamazaki, R Shimizu, Y Saito. High-T_c SQUID microscope with sample chamber. Supercond Sci Technol 12:809–812, 1999.
170. M Daibo, T Kotaka, A Shikoda. Photo-induced magnetic field imaging of p-n junction using a laser SQUID microscope. Physica C 357–360:1483–1487, 2001.
171. TS Lee, YR Chemla, E Dantsker, J Clarke. High-T_c SQUID microscope for room temperature. IEEE Trans Appl Supercond 7:3147–3150, 1997.
172. EF Fleet, S Chatraphorn, FC Wellstood, C Eylem. Determination of magnetic properties using a room-temperatue scanning SQUID microscope. IEEE Trans Appl Supercond 11:1180–1183, 2001.
173. SA Gudoshnikov, YV Deryuzhkina, PE Rudenchik, YS Sitnov, SI Bondarenko, AA Shablo, PP Pavlov, AS Kalabukhov, OV Snigirev, P Seidel. Magnetic flux guide for high-resolution SQUID microscope. IEEE Trans Appl Supercond 11:219–221, 2001.
174. SA Gudoshnikov, BY Liubimov, LV Matveets, AP Mikhailenko, YV Deryuzhkina, YS Sitnov, OV Snigirev. Flux guide for high-T_c SQUID microscope with high spatial resolution. Physica C 368:66–69, 2002.
175. T Nagaishi, K Minamimura, M Itozaki. HTS SQUID microscope head with sharp permalloy rod for high spatial resolution. IEEE Trans Appl Supercond 11:226–229, 2001.
176. S Chatraphorn, FE Fleet, RC Black, FC Wellstood. Microwave electric-field imaging using a high-T_c scanning superconducting quantum interference device. Appl Phys Lett 73:984–986, 1998.
177. E Fleet, A Gilberston, S Chatraphorn, N Tralshwala, H Weinstock, FC Wellstood. Imaging defects in Cu-clad NbTi wire using a high-T_c scanning SQUID microscope. IEEE Trans Appl Supercond 11:215–218, 2001.
178. RC Black, FC Wellstood, E Dantsker, AH Miklich, D Koelle, F Ludwig, J Clarke. Imaging radio-frequency fields using a scanning SQUID microscope. Appl Phys Lett 66:1267–1269, 1995.

179. M Bick, G Panaitov, N Wolters, Y Zhang, H Bousack, AI Braginski, U Kalberkamp, H Burkhardt, U Matzander. A HTS rf SQUID vector magnetometer for geophysical exploration. IEEE Trans Appl Supercond 9:3780–3785, 1999.
180. A Chwala, R Stolz, J Ramos, V Schultze, HG Meyer, D Kretzschmar. An HTS dc SQUID system for geomagnetic prospection. Supercond Sci Technol 12: 1036–1038, 1999.
181. G Panaitov, M Bick, Y Zhang, HJ Krause. Effect of repetitive transmitter signals on SQUID response in geophysical TEM. IEEE Trans Appl Supercond 11:888–891, 2001.
182. V Zakosarenko, A Chwala, J Ramos, R Stolz, V Schultze, H Lutjen, J Blume, T Schuler, HG Meyer. HTS dc SQUID systems for geophysical prospection. IEEE Trans Appl Supercond 11:896–899, 2001.
183. A Chwala, R Stolz, R Ijsselsteijn, V Schultze, N Ukhansky, HG Meyer, T Schuler. SQUID gradiometers for archaeometry. Supercond Sci Technol 14:1111–1114, 2001.
184. CP Foley, DL Tilbrook, KE Leslie, RA Binks, GB Donaldson, J Du, SK Lam, PW Schmidt, DA Clark. Geophysical exploration using magnetic gradiometry based on HTS SQUIDs. IEEE Trans Appl Suprcond 11:1375–1378, 2001.
185. TR Clem, DJ Overway, JW Purpura, JT Bono, RH Koch, JR Rozen, GA Keefe, S Willen, RA Mohling. High-T_c SQUID gradiometer for mobile magnetic anomaly detection. IEEE Trans Appl Supercond 11:871–875, 2001.
186. KE Leslie, RA Binks, CJ Lewis, MD Scott, DL Tilbrook, J Du. Three component spinner magnetometer featuring rapid measurement times. IEEE Trans Appl Supercond 11:252–255, 2001.
187. S Tanaka, R Shimizu, Y Saito, K Shin. Development of a high-T_c SQUID cryosystem for the measurement of a remanent magnetic field of rock. IEICE Trans Electron E83-C:44–48, 2000.
188. N Kasai, Y Fujinawa, H Iitaka, K Nomura, Y Hatsukade, S Sato, H Nakano, T Doi, T Nemoto, A Ishiyama. Development of the HTS-SQUID system for measuring ULF band magnetic field changes related with earthquakes. Supercond Sci Technol 14:1135–1139, 2001.
189. K Enpuku, T Minotani, M Hotta, A Nakahodo. Application of High T_c SQUID magnetometer to biological immunoassays. IEEE Trans Appl Supercond 11:661–664, 2001.
190. S Katsura, T Yasuda, K Hirano, A Mizuno, S Tanaka. Development of a new detection method for DNA molecules. Supercond Sci Technol 14:1131–1134, 2001.
191. S Tanaka, A Hirata, Y Saito, T Mizoguchi, Y Tamaki, I Sakita, M Monden. Application of high T_c SQUID magnetometer for sentinel-lymph node biopsy. IEEE Trans Appl Supercond 11:665–668, 2001.
192. J Claycomb, M Nersesyan, D Luss, JH Miller Jr. SQUID detection of magnetic fields produced by chemical reactions. IEEE Trans Appl Supercond 11:863–866, 2001.
193. AH Miklich, D Koelle, F Ludwig, DT Nemeth, E Dantsker, J Clarke. Picovoltmeter based on a high transition temperature SQUID. Appl Phys Lett 66:230–232, 1995.
194. T Eriksson, J Blomgren, D Winkler. An HTS SQUID picovoltmeter used as preamplifier for Rogowski coil sensors. Physica C 368:130–133, 2002.

195. J Blomgren, T Eriksson, D Winkler. An HTS SQUID picovoltmeter with a flip-chip flux transformer. IEEE Trans Appl Supercond 11:892–895, 2001.
196. AS Kalabukhov, MA Tarasov, EA Stepantsov, S Gevorgian, A Deleniv, ZG Ivanov, OV Snigirev, OG Vendik, OA Mukhanov. A high-T_c L-band SQUID amplifier combined with superconductive thin-film filters. Physica C 368:171–175, 2002.
197. T Eriksson, J Blomgren, D Winkler. Discharge measurements using a HTS-SQUID based amplifier system. IEEE Trans Appl Supercond 11:256–259, 2001.
198. GV Prokopenko, SV Shitov, IV Borisenko, J Mygind. HTS dc SQUID based rf amplifier: Development concept. Physica C 368:153–156, 2002.
199. L Hao, DA Peden, JC Gallop, JC Macfarlane. Tunable HTS rf Josephson-effect oscillator based on YBCO-Au-YBCO resistive SQUID. Supercond Sci Technol 14:1119–1123, 2001.
200. L Hao, JC Macfarlane, JC Gallop. Ion beam measurement with a high-temperature superconductor SQUID and current comparator. IEEE Trans Instrument Measure. 50:302–305, 2001.

9

High-Temperature Superconducting Digital Circuits

Mutsuo Hidaka
NEC Corporation, Ibaraki, Japan

9.1 INTRODUCTION

Superconducting digital circuits have two advantages compared with their competitive semiconductor circuits, such as Josephson junctions and superconducting microstrip transmission lines. The Josephson junction can switch its zero-voltage state to a finite-voltage one within a few picoseconds and power dissipation of the switching is extremely low because the voltage state is less than a few millivolts. The superconducting microstrip transmission line is able to transfer picosecond waveforms over virtually any interchip distance with a speed approaching that of light and low attenuation and dispersion. Superconducting microstrip lines can be laid out densely because there is little cross-talk between them, and the junctions can be impedance matched with the strip lines to ensure the ballistic transfer of the generated waveforms along the lines.

There have been many efforts to develop circuits for exploring the advantages of ultrahigh-speed processing system by superconducting digital circuits using metallic superconductor materials, such as Pb and Nb. Two examples of these efforts are the IBM project (1969–1983) (1) and the Japanese MITI project (1981–1991) (2). Successful demonstrations for the low-T_c superconductor (LTS) circuits have been made, such as a 4-kbit RAM that has 42,000 junctions and operates at 620 MHz (3) and a computer-communication-network logic circuit that has 4300 junctions and operates at 2 GHz (4). It has, nevertheless, become clear that the first-generation superconducting digital circuits, so-called "latching logic" circuits using the zero- and a finite-voltage states for logical "0" and "1" states, cannot compete with high-speed semiconductor circuits after paying their cooling penalty. The main drawback of the "latching logic" is that it is clocked by large radio-frequency (RF) current from outside of the chip. The operation frequency is restricted to a few gigahertz, because a large amount of current (e.g., several amperes) cannot be supplied at a higher-frequency.

Much attention has thus been directed to the single-flux-quantum (SFQ) logic, which codes the binary information not by using the dc voltage, but by using single quanta of magnetic flux ($\Phi_0 = h/2e = 2.07 \times 10^{-15}$ Wb). Superconducting digital circuits using the SFQ logic were originally proposed by Nakajima and Onodera in 1976 (5), and since 1985 have been dramatically improved by the Moscow State University group, represented by Likharev and Semenov (6). Their SFQ circuits, called rapid single-flux-quantum (RSFQ) circuits, have become the most popular SFQ circuits and are expected to operate at a frequency greater than 100 GHz. Several high-speed RSFQ circuits based on tunnel-type LTS Josephson junctions have been reported, and the highly important of these is an analog-to-digital converter circuit, which was made by Semenov et al. and has thousands of junctions and operates at frequency up to 11 GHz (7).

High-T_c superconducting (HTS) digital circuits are more suitable for use in SFQ circuits than LTS ones, because HTS Josephson junctions are naturally overdamped, which means that their I–V curves do not show hysteresis, and the junctions in SFQ circuits must be overdamped junctions. The tunnel-type LTS Josephson junctions, on the other hand, are underdamped ones and require some shunt resistance between the two electrodes of each junction. This makes the characteristic voltage (I_cR_n product) values lower, which results in lower operating speeds, and also complicates the layout and the fabrication process. The I_cR_n product of HTS junctions can also be expected to be larger than that of LTS junctions because it intrinsically depends on the gap voltage of the superconductor. A number of tests of the RSFQ circuits using HTS Josephson junctions have been reported, but most of the circuits that have been reported are small-scale circuits because the fabrication technology for HTS junctions is still in a primitive stage.

9.2 OPERATING PRINCIPLE OF SFQ DIGITAL CIRCUITS

Magnetic flux is quantized in a superconducting closed loop and the minimum unit is a SFQ. Figure 9.1 shows the simplest loop for the SFQ circuit, which is a superconducting closed loop including a Josephson junction. As magnetic-flux crossing of superconducting lead is forbidden by the Meissner effect, the Josephson junction plays the role of a "gate" for going in and out of the loop. When the Josephson junction switches to a voltage state, magnetic flux goes in and out

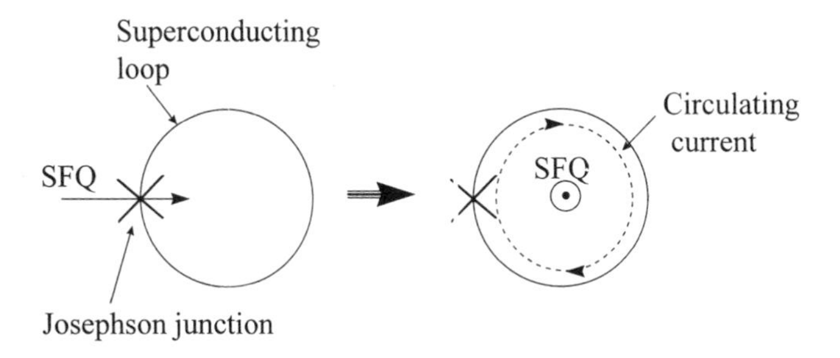

FIGURE 9.1 An explanation of a SFQ storage in a superconducting loop including a Josephson junction.

through the junction. If the product of the junction critical current I_c and loop inductance L is $\Phi_0 < LI_c < 2\Phi_0$, only a SFQ can exist in this loop after resetting the junction to superconducting state and the SFQ makes a circulating current I_{cir} in the loop. Figure 9.2 shows another explanation of SFQ storage and release in the superconducting loop. When a dc bias current I_b is supplied to a superconducting loop including a Josephson junction, almost all of the current goes through the junction because of inductance L in another branch (Fig. 9.2a). Here, I_b is smaller than the critical current I_c of the junction. If a signal current I_s is then supplied to the junction and the sum of I_b and I_s is larger than I_c, the junction switches to a voltage state and I_b and I_s flow through the inductance branch (Fig. 9.2b). After resetting the junction to the superconducting state and I_b and I_s turning off, the current flowing through the inductance branch is preserved in the loop (Fig. 9.2c). The preserved current I_{cir} is Φ_0/L when the L and I_c values satisfy $\Phi_0 < LI_c < 2\Phi_0$. The I_{cir} preservation in the loop corresponds to SFQ going in the loop. The I_{cir} is released by supplying I_s in the opposite direction. The currents I_s and I_{cir} are

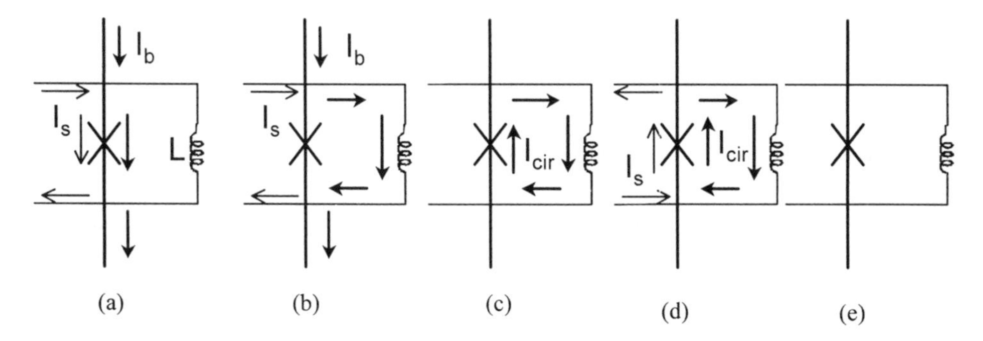

FIGURE 9.2 The basic operations of a SFQ gate.

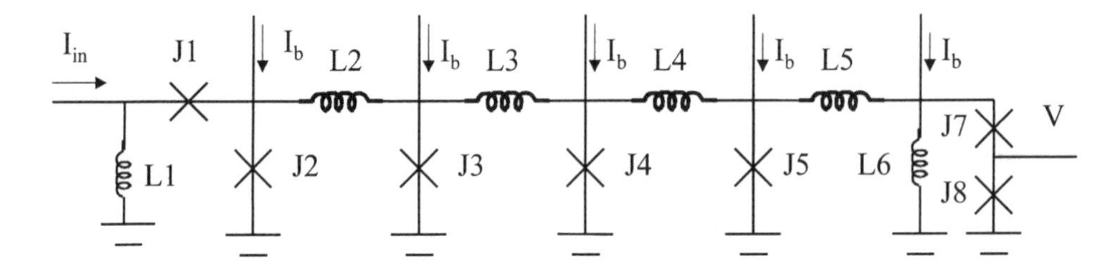

FIGURE 9.3 A series of SFQ circuits: DC/SFQ, JTL, and SFQ/DC.

added because they flow in the same direction at the junction and their sum exceeds I_c (Fig. 9.2d). Then, the junction turns on and I_{cir} dissipates at the junction (Fig. 9.2e). This corresponds to a SFQ going out of the loop.

Figure 9.3 shows a series of SFQ circuits: a DC/SFQ converter, a Josephson transmission line (JTL) and a SFQ/DC converter. The DC/SFQ converter, which consists of junctions J1 and J2 and inductance L1, makes a SFQ pulse by a dc current I_{in} input. If I_{in} increases beyond a threshold value, a SFQ pulse is generated by J2 turning on and is transferred to the right direction in Figure 9.3. The DC/SFQ converter resets to its initial state when I_{in} falls below a certain value. The reset of the circuit is accompanied by the generation of a SFQ pulse across J2, which does not propagate to the right. The JTL consists of three superconducting loops including junctions J3–J5 and inductances L2–L4. Because the product of I_c and L for each superconducting loop is less than Φ_0 in the JTL, the SFQ pulse propagates through the JTL without being stored in these loops. The SFQ/DC converter contains a SFQ storage loop with J5, L5, and L6, in which the LI_c product is larger than Φ_0, and a readout SQUID consisting of junctions J7 and J8 and a voltage output terminal between them. The SFQ from the JTL is stored in this loop, and the stored SFQ is converted to dc voltage by the readout SQUID.

Because this circuit is biased by dc current I_b shown in Figure 9.3, the rf bias current indispensable to latching circuits for their reset operations is unnecessary in SFQ circuits. This is the main reason that SFQ circuits are so much faster than latching circuits. Any logic functions and memory operations can be implemented using SFQ circuits by combining $LI_c > \Phi_0$ loops and $LI_c < \Phi_0$ loops. A detailed explanation of the RSFQ circuits can be found in Ref. 6.

In the SFQ circuits, binary information is propagated as very short voltage pulses instead of dc voltage in the superconducting latching circuits as well as in all semiconductor circuits. The voltage pulse $V(t)$ has a quantized area given by

$$\int V(t)\,dt = \Phi_0 \equiv \frac{h}{2e} \approx 2.07 \text{ mV} \times \text{ps} \tag{1}$$

The switching speed τ of the simple SFQ loops like those in Figures 9.1–9.3 is restricted by the characteristic frequency of the ac Josephson effect. Using a critical current of Josephson junction I_c and its normal resistance R_n, we can represent τ is as follows:

$$\tau = \frac{\Phi_0}{I_c R_n} \qquad (2)$$

The $I_c R_n$ product is one of the most important parameters for evaluating the Josephson junctions used in SFQ circuits. If the $I_c R_n$ product is 1 mV, which is a reasonable value for HTS Josephson junctions, τ can be as little as 2 ps.

The power consumption for one switching of a Josephson junction in a SFQ gate is about $I_c^2 R_n \tau = I_c \Phi_0$ and that required in switching at frequency f is $n I_c \Phi_0 f$ when the number of Josephson junctions in the gate is n. Using such typical values as $I_c = 0.4$ mA and $n = 4$, we can estimate the power consumption of an HTS SFQ circuit operating at 100 GHz to be 0.33 μW. On the other hand, the power consumption of a complimentary-metal oxide semiconductor (C-MOS) gate with a 3-V signal level and a 7-fF capacitance is 62 μW, even when its operation frequency is only 1 GHz. Thus, the power consumption of the HTS SFQ gate is two orders of magnitude smaller than that of the two orders of magnitude slower C-MOS gate.

9.3 BASIC ISSUES IN HTS SFQ CIRCUITS

9.3.1 Circuit Parameters

9.3.1.1 Josephson Junctions

Overdamped Josephson junctions, which have dc I–V curves with no hysteresis, are used in SFQ circuits. Junction damping is represented by a McCumber–Stewart factor $\beta_c = (2\pi/\Phi_0) I_c R_n^2 C$, where C is a junction capacitance (8,9). A McCumber–Stewart factor for a junction whose dc I–V curve shows no hysteresis is $\beta_c < 1$. The Nb/AlO_x/Nb Josephson junction (10) used in LTS SFQ circuits is a tunnel junction and its β_c is much larger than 1. Therefore, the β_c of the junction has to be reduced by adding a shunt resistance across its tunnel barrier (6). The shunt resistance used in so-called "NEC standard process" is 3–5 Ω (11). The adding of the shunt resistance reduces the $I_c R_n$ product at 4.2 K from 1.7 mV to 0.3 mV and lengthens the switching time from 1.2 ps to 6.7 ps. The β_c of HTS Josephson junctions, on the other hand, is less than 1 without additional shunt resistance. This is because the HTS junctions are not tunnel junctions but are weak links and, therefore, are characterized by smaller R_n values. The intrinsic $I_c R_n$ products of HTS Josephson junctions can be expected to be larger than that of LTS junctions because of the larger energy gaps of HTS materials. The development of superior-quality Josephson junctions with high $I_c R_n$ products is one of the most important research issues related to HTS SFQ circuit applications. Ramp-edge HTS junctions with using a Ga-doped $PrBa_2Cu_3O_x$ barrier and having an I_cR_n product of 8 mV at 4.2 K were reported by the Twente University group (12) and ramp-edge HTS junctions using a Co-doped $YBa_2Cu_3O_x$, barrier and having a I_cR_n product of 0.8 mV at 65 K were reported by the Northrop Grumman group (13).

The relations between I_c spread in a chip and circuit integration level are discussed (14,15). To illustrate the effect of critical current spread on circuit yield, we consider a one-junction gate with a designed critical current of I_c. We further assume that the fabrication process will yield junctions whose distribution of critical currents is a Gaussian distribution with a standard deviation of σ.

The probability of a given junction falling within the circuit margin Δ is given by

$$P = 2\int_{\mu}^{\mu+\Delta} \frac{1}{\sqrt{2\pi\sigma}} \exp\left[-\frac{1}{2}\left(\frac{x-\mu}{\sigma}\right)^2\right] dx \tag{3}$$

The critical current of each junction in a circuit consisting of N such junctions must be between $I_c - \Delta$ and $I_c + \Delta$. The total circuit yield will be P^N. Figure 9.4 shows the σ values required in the production of circuits with a given junction count (15). The smallest critical current spread achieved to date is $1\sigma = 8\%$ for 100 ramp-edge junctions by a modified interface barrier (16). As shown as Figure 9.4, this critical current spread corresponds to a yield of 50% for circuits with a few hundred junctions.

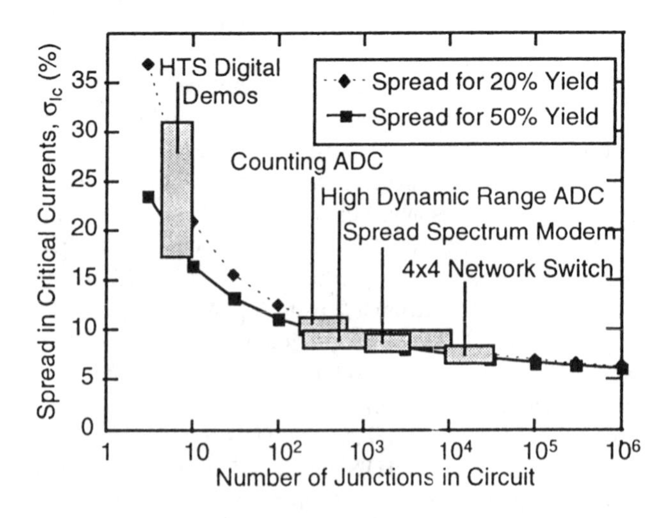

FIGURE 9.4 The spread σ in junction critical currents required for the production of circuits with a given junction count. (From Ref. 15.)

Spreads of R_n and I_cR_n product are less critical for circuit yields than the spread of I_c. Moreover, the spreads of R_n and I_cR_n product are usually smaller than the spread of I_c (17).

9.3.1.2 SFQ Loops

SFQ circuits contain two kinds of SFQ loops. One is a storage loop for which $\Phi_0 < \beta_L = LI_c < 2\Phi_0$. The other is a transfer loop for the JTL for which $\beta_L < \Phi_0$. These typical values are 1.5 Φ_0 and 0.5 Φ_0, respectively. Here, L is the inductance of the SFQ loop and I_c is the critical current of a junction including the loop.

The inductance L of a superconducting microstrip line like that whose cross section is shown in Figure 9.5 is given by

$$L = \frac{\mu_0}{Kw}\left[d + \lambda_L \coth\left(\frac{t_G}{\lambda_L}\right) + \lambda_L \coth\left(\frac{t_s}{\lambda_L}\right)\right] l \tag{4}$$

where w is the width of the line, l is the length of the line, λ_L is a superconducting penetration depth of the ground plane and the line, t_G and t_s are the thicknesses of the ground plane and the line, respectively, d is the thickness of the insulation layer, μ_0 is the permeability of free space, and K is a fringing factor (1). Because the λ_L of HTS materials is larger than that of LTS materials, the L per square ($L_\square$) value for a HTS microstrip line is larger than that for a Nb microstrip line with the same insulator thickness.

It is difficult to lay out a small L loop because of the large $L_\square$ value. Moreover, as we will explain in detail in Section 9.3.2.1, I_c cannot be decreased too much because of thermal noise. The difficulty of making smaller β_L loops is one of the most serious problems in HTS SFQ circuits.

9.3.1.3 Resistance

Three kinds of resistance are required in a HTS SFQ chip. Resistances of less than a few ohms are placed in some SFQ gates. Some SFQ gates do not require these

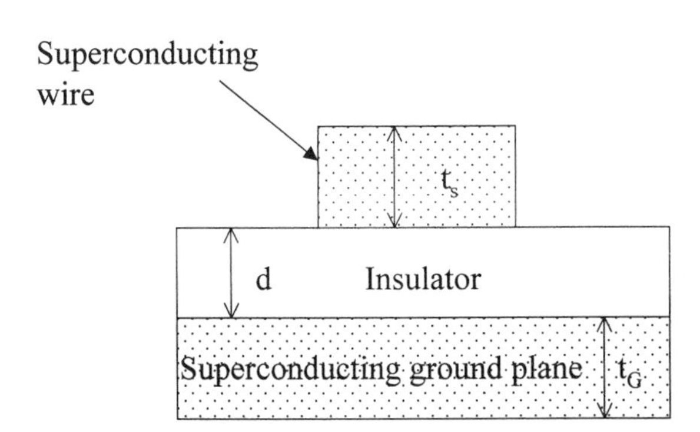

FIGURE 9.5 Cross section of a superconducting microstrip line over a superconducting ground plane.

small resistance, but some gates utilize the resistance for damping (6) and Sigma–Delta modulators, which are used for main parts of a kind of analog-to-digital converter, are indispensable to the resistance. Resistance used for dividing bias current to each SFQ loop in parallel has a value of a few tens of ohms. This resistance is required for preventing faulty operations caused by current reflected from other switched junctions. The third kind of resistance is the matching resistance in a high-speed I/O line. It is needed for impedance matching to a 50-Ω external signal line.

9.3.2 Factors Limiting HTS SFQ Circuit Operations

9.3.2.1 Thermal Noise

Unfortunately, digital circuits based on Nb need to be cooled to the temperature of liquid helium. The use of HTS materials will reduce cooling costs as well as increase the operating frequencies because of the higher $I_c R_n$ products of HTS Josephson junctions, but a higher operating temperature results in more thermal noise. The energy barrier between two flux states in a SFQ gate is very low. A rough estimation (18) shows that for typical critical currents of the order of 10^{-4} A, this energy barrier is the order of 10^{-19} J. Thus, some fluctuations, not accounted for in the yield estimation described in the previous section, may increase the spontaneous switching between the flux states. The probabilities of SFQ gates errors caused by thermal noise have been investigated theoretically and experimentally.

A balanced comparator using two junctions (Fig. 9.6) is the basic component of RSFQ logic gates and the SFQ-counting analog-to-digital converter (6). The State University of New York group (18–21) investigated the effects of thermal noise on SFQ gates theoretically by analyzing the operations of the balanced comparature. When an external driver gate sends a SFQ pulse to the balanced

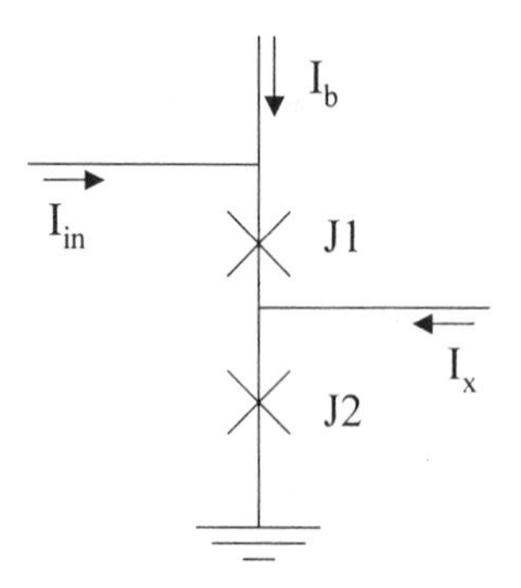

FIGURE 9.6 Equivalent circuit of a balanced comparator.

comparator, one of the junctions switches. Which junction switches is determined by the additional current I_x fed into the central node of the device. J2 switches when $I_x > 0$, and J1 switches when $I_x < 0$. However, the unavoidable fluctuations create, a gray zone around $I_x = 0$, where each of the junctions has a probability $0 < P(I_x) < 1$ of being switched. The effective width ΔI_x of this gray zone, which is defined as $\Delta I_x = (dp/dI_x)^{-1} \mid_{p=1/2}$, reduces the parameter margins of RSFQ logic gates. Results of this theory are approximately as follows:

$$P(I_x) = 0.5\left[1 + \text{erf}\left(\pi^{1/2}\frac{I_x - I_t}{\Delta I_x}\right)\right] \tag{5}$$

where $I_t = (2e/\hbar)k_BT \approx (0.042\ \mu\text{A}) \times T\,(\text{K})$ and T is a temperature. In the thermal fluctuation limit, $\hbar(\kappa\omega_c)^{1/2} << k_BT$.

$$\Delta I_x = (I_cI_t)^{1/2}\left(\frac{32\pi\kappa}{\omega_c}\right)^{1/4} \tag{6}$$

In the opposite, quantum limit, the rate dependence is different,

$$\Delta I_x = \left(\frac{8I_cI_Q\kappa}{\omega_c}\right)^{1/2} \tag{7}$$

and the thermal current unit I_t is replaced by the quantum current unit $I_Q = (2e/\hbar)eI_cR_n$.

Without considering the thermal and quantum fluctuations, the balanced comparator operates normally with $0 < I_x/I_c < 1$. The operation margin becomes narrow as a result of the fluctuations. The deterministic operation margins shown in Figure 9.4 have to be revised by taking into account these noises.

Satchell (22) and Jeffery et al. (23) simulated the bit error rate (BER) of various SFQ gates, and their results were in good agreement with the theoretical predictions. Satchell concluded that operation at temperatures above 40 K is possible only for those circuits which have good noise tolerance, and Jeffrey concluded that the toggle–flip-flop (T-FF) operating temperature should be below 40 K in order to obtain bit error rates less than 10^{-6} at gigahertz speeds.

Oelze et al. studied the effect of thermal noise on a balanced comparator made of bicrystal Josephson junctions experimentally (24), and the relation between bias current and the ΔI_x measured at 40 K is shown in Figure 9.7. The $\Delta I_x/I_c$ ratios with various bias conditions were estimated from 6% to 17%. A static error occurs when a SFQ loop loses a stored flux quantum because of thermal noise, and the static error rate of a SFQ storage loop fabricated from HTS multiplayer bicrystal Josephson junctions was measured by Chong et al. (25). A stack configuration of two HTS dc SQUIDs was used in this experiment; one serving as a storage loop for flux quanta and the other serving as a readout to detect the flux state of the storage SQUID. The stable times measured for both the "+I" and "−I" SFQ in the storage loop near the threshold bias current are shown in Figure 9.8. A decrease

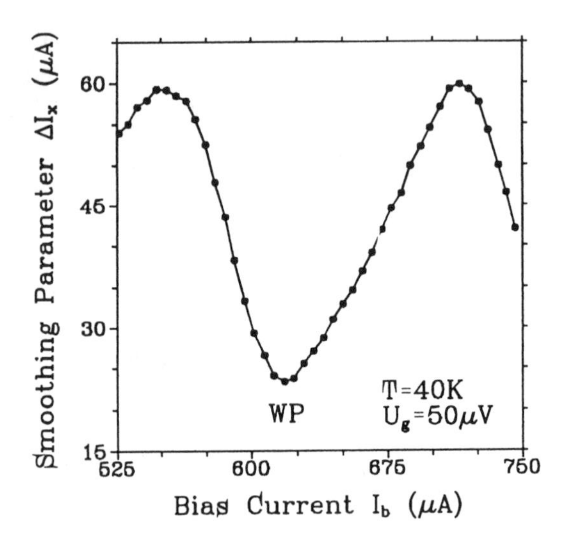

FIGURE 9.7 Dependence of ΔI_x on bias current I_b. (From Ref. 24.)

of about 6–7 μA of the bias current increased the stable time by one order of magnitude.

Ruck et al. measured the dynamic error in a SFQ loop. This is a switching failure of the comparator configuration: The wrong junction switches its phase in response to an incoming SFQ pulse (26). These dynamic errors would be domi-

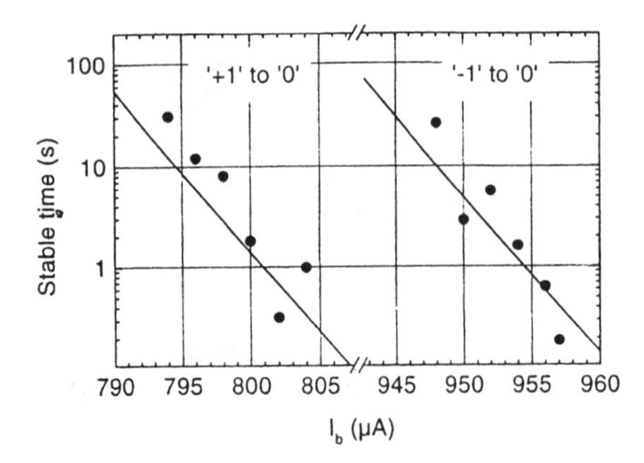

FIGURE 9.8 Measured stable time for both the "+1" and "−1" states near the threshold bias current. The solid lines show results of the model calculation. (From Ref. 25.)

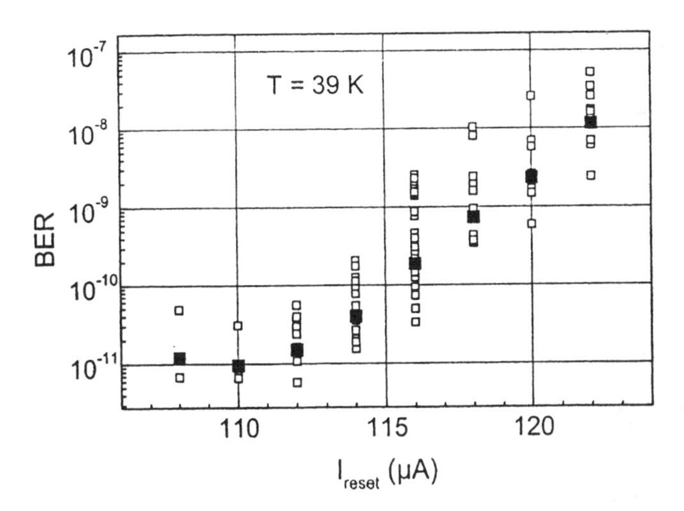

FIGURE 9.9 Bit-error-rate dependence on the applied reset current. Open squares are the measured points; solid squares the corresponding averages. (From Ref. 26.)

nant in practical RSFQ circuits. A balanced comparator consisting of focused-electron-beam-irradiated junctions was used in this experiment. Figure 9.9 shows the BER dependence on the applied reset current, which is the same as I_x. The temperature in this experiment was 39 K and the incoming SFQ pulse frequency was approximately 1 GHz. A BER of less than 10^{-11} was obtained, showing that SFQ circuits can operate at 39 K. Their measurements also indicated, however, that a significant reduction of the circuit parameter margins must be taken into account when the operation temperature is above 4.2 K. They suggested that the operating temperature of HTS SFQ circuits cannot be raised much above 40 K.

Maintaining constant noise margins at elevated temperatures would require junction I_c values to increase in proportion to operating temperature and the circuit inductance values to decrease, keeping the β_L constant. The long penetration depth of HTS materials, however, makes such low-inductance values impractical. Also, values of I_c are likely to be restricted to ~0.4 mA. The circuits must either be restricted to relatively low operating temperatures or operate with reduced noise margins.

The maximum divided voltage V_d of a RSFQ T-FF determines the maximum operation frequency f_{max} of the TFF: $f_{max} = V_d/\Phi_0$. The I_cR_n product of Josephson junctions in a T-FF and the V_d of that T-FF were compared and their temperature dependence was examined by Saito et al. (27). The temperature dependences of the I_cR_n product and V_d are shown in Figure 9.10. The V_d is clearly smaller than the I_cR_n product, although their temperature dependences are quite

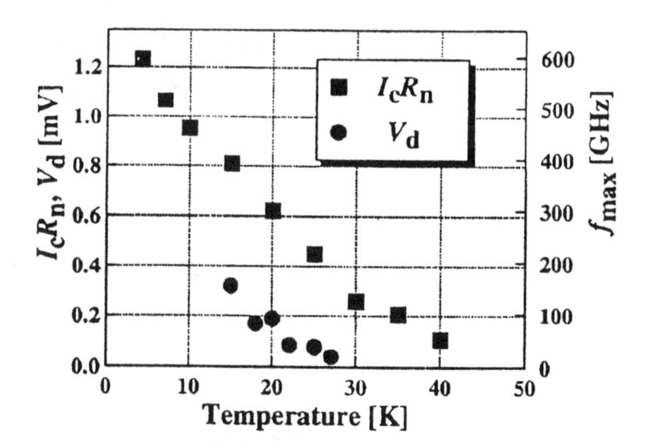

FIGURE 9.10 Temperature dependence of the I_cR_n product and maximum divided voltage V_d. (From Ref. 27.)

similar. The maximum V_d at 15 K corresponds to a f_{max} of 155 GHz. An evaluated limit factor γ, defined as $f_{max} = \gamma I_cR_n/\Phi_0$, was $0.4 > \gamma > 0.1$ for 15 K $< T <$27 K. Saito et al. assumed that thermal noise affects T-FF operation, and so they included thermal noise in their circuit simulations. Circuit simulation results and experimental results agreed quite well. These results indicate that the thermal noise affects T-FF logic operation and suppresses the maximum frequency. They speculated that Josephson junctions for which the I_cR_n product is greater than 1 mV are necessary if operation speeds over 100 GHz are to be obtained at 30 K.

9.3.2.2 Parasitic Inductance

Parasitic inductance is unavoidable in practical layouts arranging Josephson junctions and contacts in SFQ digital circuits. The line inductance in HTS circuits is twice as large as that in LTS circuits. Moreover, smaller inductance elements are used in HTS circuits for keeping the β_L against larger I_c. The parasitic inductance is, therefore, a more serious problem for HTS SFQ circuits than for LTS circuits.

Satchell (22) and Jeferry et al. (23) simulated the parasitic inductance influence to circuit yield as well as a thermal noise one. In Figure 9.11, the simulation results by Jeferry in which yield results for an SFQ T-FF with various process conditions and operation frequencies are described with (Fig. 9.11a) and without (Fig. 9.11b) parasitic inductances. These results show quantitatively that parasitic inductance can have a significant effect on the probability of obtaining HTS SFQ circuit working at ultrahigh-speeds. Hidaka et al. estimated the parasitic inductance of their sampler circuit experimentally (28). The parasitic inductance in a SFQ loop was 2.4 pH, whereas the necessary loop inductance was 5.0 pH. This

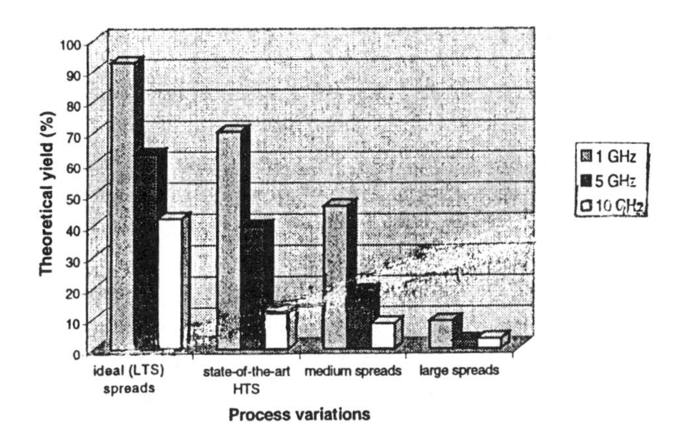

(a)

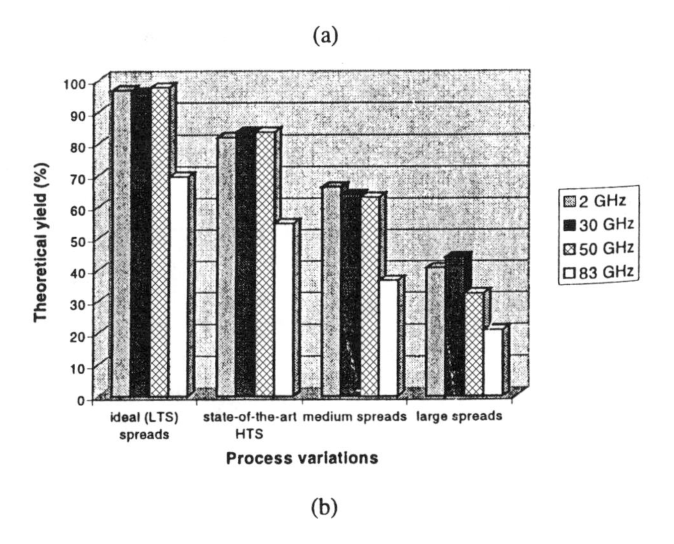

(b)

FIGURE 9.11 Monte Carlo yield results for a T-FF (a) with and (b) without parasitic inductances. (From Ref. 23.)

parasitic inductance value is large enough to narrow the operating margin of the sampler circuit.

9.3.3 Fabrication of SFQ Circuits

9.3.3.1 Materials

The superconductor that has been most widely used in HTS digital circuits is a $YBa_2Cu_3O_x$ (YBCO) thin film because of the reproducibility of T_c, robustness to water, and large critical current density. There are a number of suitable substrates on which YBCO films can be grown: $SrTiO_3$ (STO), MgO, $LaAlO_3$, NdGaO3, YSZ (yttria-stabilized zirconia), Sr_2AlTaO_6 (SAT), Sr_2AlNbO_6 (SAN), and $(La_{0.3}Sr_{0.7})(Al_{0.65}Ta_{0.35})O_x$ (LSAT). Among of them, STO has so far been the most popular substrate material for SFQ digital circuits because its lattice constant and thermal expansion coefficient are close to those of YBCO. The choice of the superconductor and substrate restricts the choice of insulators. It is highly desirable that the insulator can be deposited using the same technique used to deposit YBCO and at temperatures not significantly higher. The obvious choices for insulators are the substrate materials. High-quality STO films can be grown at temperatures similar to those used for depositing YBCO. Layers of MgO, $LaAlO_3$, $NdGaO_3$, SAT, SAN, and LSAT can also be deposited at temperatures similar to these used for depositing YBCO, and the resistivity of these materials is high enough for digital circuit applications. In YBCO-based multilayer circuits, it is necessary to diffuse oxygen through an epitaxial insulator film to fully reoxidized one or more buried YBCO layers. Oxygen diffuses more easily through epitaxial STO films than through films of other insulators (e.g., SAT and SAN) (29). The main drawback with STO is its large dielectric constant, approximately 10^3 in films at 77 K, which makes it unsuitable for high-frequency applications. CeO_2 is one of the candidate insulators for high-frequency HTS digital circuits because of its relatively low dielectric constant and its high-permeability to oxygen.

Ruck et al. tested various materials in order to evaluate their suitability for resistors in SFQ circuits like Ag/Au, Cr, Pd, and Pd/Au (30). Among of them, Pd/Au had a much smaller temperature coefficient and higher resistance, especially at low-temperature, than Pd. The sheet resistance of 400-nm-thick Pd/Au was about 0.6 Ω from 4.2 K to 77 K. They suggested that Pd/Au was the most suitable material for SFQ circuits. Forrester et al. used Au for resistors with Ti adhesion layer in a Sigma–Delta modulator (31) and Mo film was used for a 1-Ω-per-square resistor by Miller et al. (32). The contact resistance between these resistor layers and a YBCO layer is undesirably large compared with their sheet resistances, so it is important that ways to reduce the contact resistance be investigated.

The YBCO-based fabrication technologies available before 1993, including materials, deposition techniques, and patterning techniques, were reviewed by Wellstood et al. (33).

9.3.3.2 Josephson Junctions

Of the various types of Josephson junction developed so far, the ramp-edge-type junctions (34), shown schematically in Figure 9.12a, seem to be the most promising for digital circuit applications because of their small dimensions, the potential controllability of junction critical current and junction resistance values, and the ease of superconducting wiring. Many materials have been tried as the artificial barrier layer of the ramp-edge junctions, but only $PrBa_2Cu_3O_x$ (PBCO), doped

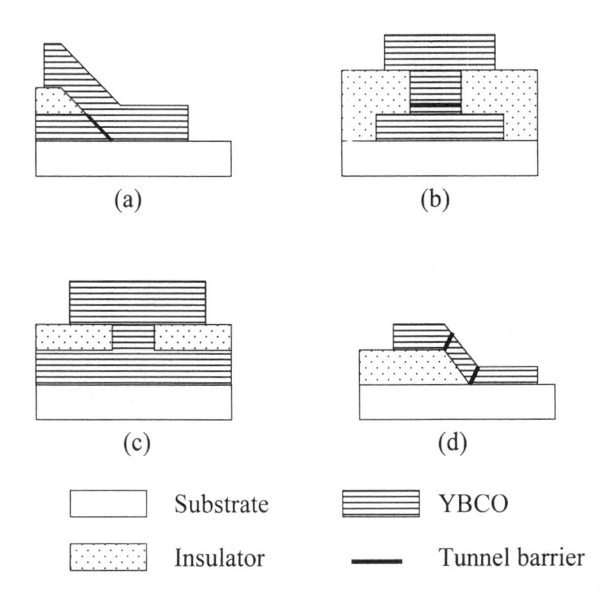

FIGURE 9.12 Schematic cross sections of HTS Josephson junctions: (a) ramp-edge junction, (b) stacked junction, (c) *c*-axis microbridge junction, and (d) step-edge grain-boundary junction.

YBCO, and doped PBCO barriers achieved a usable level in small-scale circuits including less than a few tens junctions. Satoh et al. developed an in situ edge preparation process for the PBCO ramp-edge junctions and obtained a I_c spread of $1\sigma = 10\%$ for 12 junctions with an I_cR_n product of 2 mV at 4.2 K (35). Co-doped YBCO acts as barrier at temperature above 50 K, and the 20 ramp-edge junctions with the Co-doped barrier that were made by Mallison et al. had an I_cR_n product of 0.3 mV at 50 K and showed an I_c spread of 12% (1σ) (36). Another ramp-edge junction with a Co-doped YBCO barrier, whose base electrode YBCO contains 5% La, exhibited I_cR_n products of 0.5–0.8 mV at 65 K with a 1σ I_c spreads down to 12%, was reported by Hunt et al. (13). It was found that by Ga doping, the R_n was systematically increased while I_c remained constant. Vanhoven et al. found that by Ga doping, the I_cR_n products were increased up to 8 mV at 4.2 K (12).

Interface-engineered ramp-edge junctions (IEJ) developed by Moeckly et al. (37) attracted much attention because the reproducible fabrication process is quite suitable for digital circuit applications. In this process, no barrier deposition is carried out: The barrier is formed just by structural modification using ion bombardment and vacuum annealing. Satoh et al. modified the process, in which the edge of the base YBCO film was formed by normal ion milling and then the film was heated to a deposition temperature for the counterelectrode in O_2 without being exposed to air. Their modified interfaced junctions (MIJ) also showed reproducible I_c with a 1σ spread in I_c of less than 8% for 100 junctions (16).

A drawback of using the ramp-edge junctions in SFQ circuits is that it would be difficult reduce the loop inductance. One way to reduce the inductance of the SFQ loop is by using a vertical structure. A vertical loop can be constructed by using stacked junctions (Fig. 9.12b) and *c*-axis microbridge (CAM) junctions (Fig. 9.12c). As the configuration of the stacked junction is the same as that of the Nb/AlO_x/Nb junctions used in LTS SFQ circuits, the development of stacked junctions for HTS SFQ circuits had been expected from the first stage of HTS junction development. The I_cR_n product and I_c spread of stacked junctions, however, had been too poor for these junctions to be used in SFQ circuits. Recently, an overdamped stacked junction with an I_cR_n product of 2.1 mV and a 10% 1σ spread of I_c at 4.2 K was reported by Maruyama et al. (38). These characteristics are similar to these of ramp-edge junctions and are promising with regard to SFQ circuit applications.

A CAM is simply a short, columnlike superconducting structure without an intentional barrier forming a weak-link connection between two YBCO layers. This junction has an I_cR_n product as large as 1.2 mV at 60 K. However, because the critical current of the conventional (2-μm-diameter) CAM technology are too high, a CAM diameter of 0.5 μm is required for a target critical current of 0.5 mA at 40–60 K (39). It seems that achieving good uniformity of I_c would be difficult owing to its small area.

Step-edge-grain boundary (SEGB) junctions, which are formed by discontinuities in crystal orientation where a HTS film covers a step in the substrate (Fig. 9.12d), are more easily integrated with multilayers than the ramp-edge junctions (32,40). The disadvantages of the SEGB junctions, however, are that their I_c is spread rather wide and their I_cR_n product is small. Focused-electron-beam-irradiated (FEBI) junctions on a single YBCO layer are defined by irradiating the edges with high electron doses, which renders them purely resistive. Thus, it was possible to accurately define different critical currents for the FEBI junctions (41). The FEBI junctions are not suitable for use in large-scale circuits because too much irradiation time is required for making each junction. Several HTS digital circuits have been fabricated using grain-boundary junctions, which are made by depositing an epitaxial YBCO film on a bicrystal substrate (42), because they are easy to make and have a relatively large I_cR_n product. The use of bicrystal grain-boundary junctions is limited in small-scale circuits, because flexibility of their positions is extremely restricted because they have to be arranged on a line.

9.3.3.3 Stacked Ground Plane

In order to implement high-performance HTS SFQ circuits, development a circuit process which integrates reproducible Josephson junctions into epitaxial multilayers is important. In particular, a superconducting ground plane is required to keep circuit inductance low enough that a SFQ pulse can generate enough current in a load inductor and the β_L in a SFQ loop can be kept within a designed range.

Missert et al. were the first to report the fabrication of SEGB junctions over a ground plane (43). This device operated as a SQUID only up to 20 K. The operation temperature of a SQUID that consisted of SEGB junctions with a 200-nm-thick ground plane was increased to above 77 K by Forrester et al. (44). They measured the temperature dependence of $L_\square$ and found it to be in good agreement with the theory according to which the temperature dependence of the penetration depth is represented, using the Corter–Casimir form, by $\lambda(t) = \lambda_0/[1 - (T/T_c)^4]^{1/2}$, where λ_0 is the penetration depth at $T = 0$.

Ramp-edge junctions with a 450-nm-thick stacked ground plane were first made by Miura et al. (45). Although a SQUID using the junctions operated at temperatures up to 50 K, the use of the thicker ground plane makes the surface of the base electrode YBCO rougher, causing excess current in the ramp-edge junctions. Hunt et al. (13) and Mallison et al. (36) fabricated YBCO/Co-doped YBCO/YBCO junctions over a 200-nm-thick YBCO ground plane. Both layered structures are the same except that Hunt et al. used La-doped YBCO for the base electrode and Mallison et al. used SAN for the insulator. Hunt et al. reported an $L_\square$ of 1.0 pH and large I_cR_n products of 0.5–0.8 mV at 65 K. Mallison reported an $L_\square$ of 1.2 pH at 70 K. These measured inductances are low enough to begin high-speed tests of small-scale circuits, but even lower inductances of $L_\square \approx 0.8$ pH, were obtained by Ruck et al. (30) and Henrici et al. (46). As in the structure shown in Figure 9.13, the base electrode YBCO in the structure of Ruck et al. acts as a ground plane for the counterelectrode YBCO. Therefore, this structure requires no additional ground plane. The same structure was used reported in Refs. 47 and 48. CAM technology used by Henrici naturally results in a low inductance because of the presence of a ground plane.

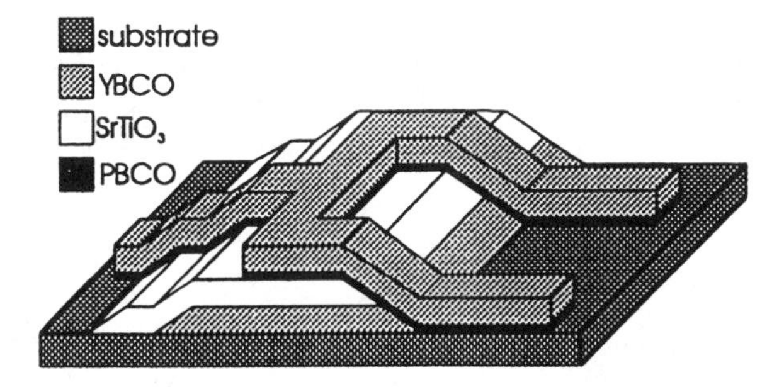

FIGURE 9.13 Drawing of a dc SQUID in which the base electrode YBCO acts as a ground plane for the counterelectrode YBCO. (From Ref. 30.)

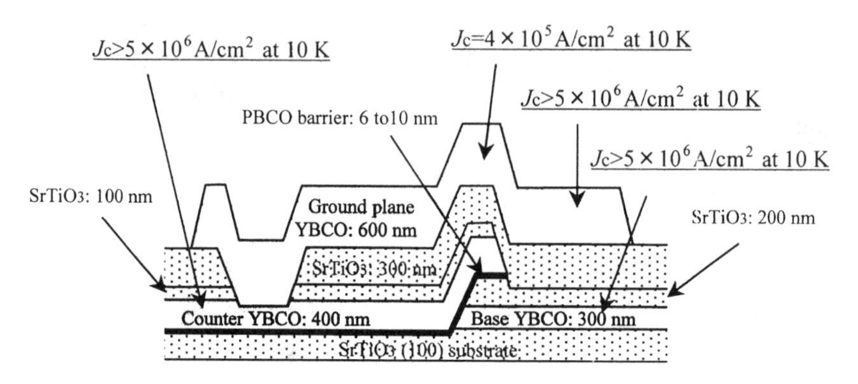

FIGURE 9.14 Schematic cross section of YBCO/PBCO/YBCO ramp-edge junction integrated with the upper ground plane "HUG structure." (From Ref. 49.)

The above-described ground planes are buried under the junctions, and the thickness of the buried ground planes had to be kept below 200 nm because thicker ground planes resulted in more surface roughness, which reduced junction quality. To overcome this drawback of the buried ground planes, Terai et al. (49) developed a multilayer structure, called a HUG (HTS circuit with an upper-layer ground plane) structure, where the YBCO ground plane is located in the top layer, as shown in Figure 9.14. The advantage of this is that higher-quality junctions can be made directly on the smooth substrate than over a thick ground plane. Moreover, the upper ground plane can be made as thick as one wants. Each YBCO layer of the HUG structure was verified to have current density close to that of an as-grown YBCO film. The resistance of the trilayer across a 400-nm-thick STO film was measured to be over 1 MΩ in a range from 4.2 K to 300 K over an area of 100 × 100 μm, which is sufficient for circuit operation. The high-temperature process used in forming the ground plane does not affect the junction quality, such as the I_cR_n product and excess current. At 30 K, experimentally estimated $L_\square$ values were around 1 pH with a 600-nm-thick ground plane, whereas the $L_\square$ without a ground plane was 2.8 pH. Figure 9.15 shows the temperature dependences of $L_\square$ in the HUG structure. The temperature dependences can be fitted by a strip-line model. This model is slightly different from Corter–Casimir form, $\lambda(t) = \lambda_0/[1 - (T/T_c)^2]^{1/2}$.

9.4 IMPLEMENTED HTS DIGITAL CIRCUITS

Although the scale of HTS digital circuits has been restricted to less than 30 junctions by the limitation of the available fabrication processes, many HTS digital circuits have been implemented. Those reported to date are listed in Table 9.1. Circuits whose operation has not been reported are not included in the table.

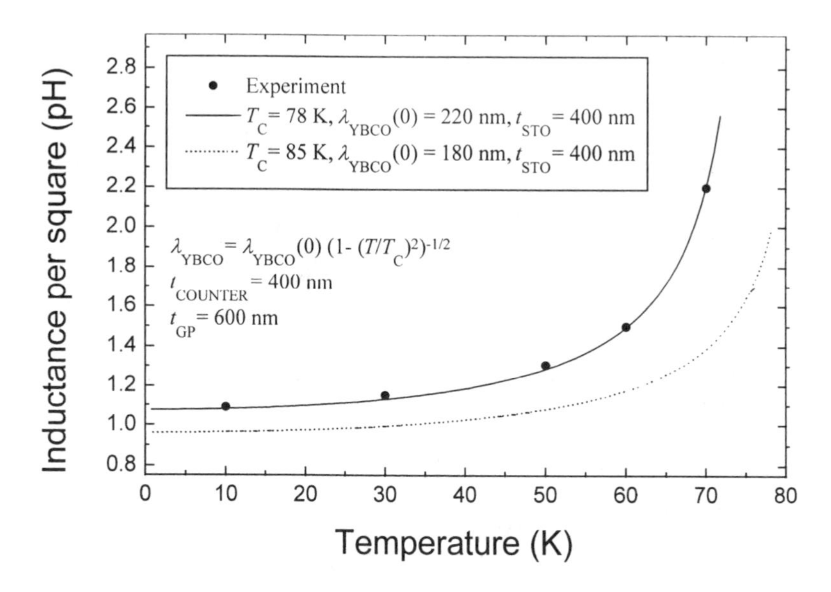

FIGURE 9.15 Temperature dependence of $L_{\square}$ observed in a SQUID integrated with an upper ground plane. (From Ref. 49.)

9.4.1 Elementary RSFQ Circuits

Several simple RSFQ circuits were fabricated and tested at low frequencies in order to test the basic SFQ operations of flux storage and controlled flux motion and so test the applicability of a specific fabrication process.

The first demonstrated operation of a HTS SFQ circuit was by Ivanov et al. (50), who demonstrated the operation of a circuit consisting of a truncated reset–set (RS) flip-flop (FF) (without the buffer junctions in the reset channel) complemented by the necessary input and output circuits, using grain-boundary junctions in a YBCO thin film. The use of an LTS (lead-alloy) ground plane, though, has limited the circuit operation to 4.2 K.

Forrester et al. reported the simple two-stage shift register with a magnetically coupled READ SQUID, as shown in Figure 9.16a (51). This circuit was fabricated by single-layer YBCO with five SEGB junctions. Figure 9.16b shows that the shift resistor loaded and shifted SFQ data on command at 65 K. Note that there was an error near 130 s, when the flux quantum shifted in the absence of a shift command. Although the efficiency of the coupling between the READ SQUID and the first DATA SQUID was only about 4%, both the storage of SFQ and their motion in response to applied signals were demonstrated in a HTS circuit.

A series RSFQ gate that includes two DC/SFQ converters, two JTLs, a complete RS-FF, and a SFQ/DC converter (readout SQUID), was implemented with

TABLE 9.1 Implemented HTS Digital Circuits

First author	Organization	Year	Kind of gate	Kind of JJ	No. of JJs	No. of HTS layers	Maximum operating temperature (K)
Ivanov	Chalmers Univ.	1994	RS-FF	Bicrystal		1	4.2
Forrester	Westinghouse	1994	2-bit shift resistor	SEGB	5	1	65
Weigerink	Twente Univ.	1995	QOS	Ramp edge	2	2	4.2 (?)
Shokhor	SUNY	1995	RS-FF	FEBI	14	1	30
Kaplunenko	Charmers Univ.	1995	Voltage divider	Bicrystal	11	1	30
Oelze	KFA	1996	Comparator	Bicrystal	8	1	50
McCambridge	Northrop Grumman	1996	1-bit ADC	SEGB	10	3	65
Gerritsma	Twente Univ.	1996	4-bit ADC	Ramp edge	8	2	25
Umezawa	TERATEC	1996	QOS	SEGB	2	1	77
Oelze	KFA	1997	3-bit shift resistor	Bicrystal	26	1	50
Hidaka	NEC	1997	Sampler	Ramp edge	5	3	50
Ruck	KFA	1998	Ring oscillator	FEBI	15	1	39
Hirst	DERA	1998	RS-FF	CAM	16	2	45
Kim	KIST	1998	RS-FF	Bicrystal	4	1	71
Hashimoto	Toshiba	1998	Voltage divider	Ramp edge	9	3	12.5
Sun	TRW	1998	T-FF	Ramp edge	14	3	65 (?)
Sonnenberg	Twente Univ.	1999	Comparator	Ramp edge	8	3	30
Forrester	Northrop Grumman	1999	Σ–Δ modulator	Ramp edge	15	3	65
Ruck	KFA	1999	Σ–Δ modulator	Bicrystal	10	2	33
Saito	Hitachi	2000	Voltage divider	Ramp edge	11	2	27

14 in-plane Josephson junctions formed by FEBI by Shokhor et al. (52). Low-frequency testing has shown that this dc-current-biased circuit operates correctly and reliably at 30 K, a few degrees below the effective critical temperature of the junctions. A three-bit SFQ shift register consisting of a shift register, two DC/SFQ converters, one readout SQUID serving as a SFQ/DC converter, and three JTLs (Fig. 9.17) was developed by Oelze et al. (53). The circuit consists of 26 bicrystal Josephson junctions, which is the largest number in any of the HTS circuits developed to date, and the correct operation of all circuit components has been confirmed by low-frequency testing at 50 K. Operating margins of the circuit were $\pm 3\%$ for a clock current and $\pm 5\%$ for a bias current of the shift register. These

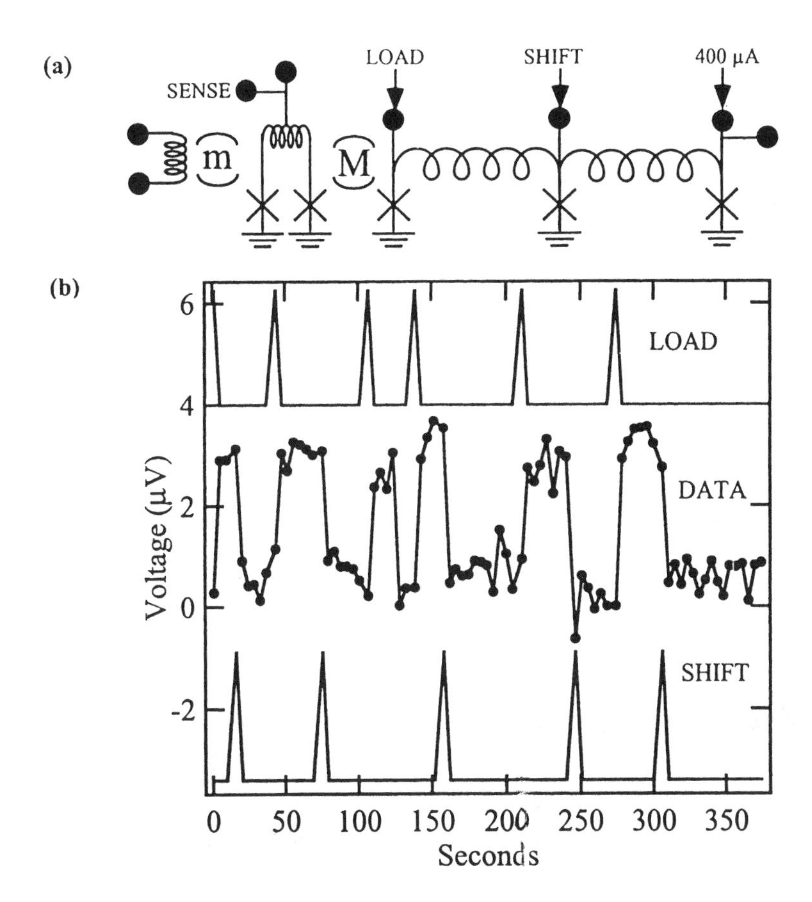

FIGURE 9.16 Schematic for (a) the two-stage shift register and (b) the shift register loaded and shifted SFQ data on command at 65 K. Note the error near 130 s. (From Ref. 51.)

narrow margins possibly resulted from a considerable spread of the critical currents of the Josephson junctions. Once the bias currents were fixed, no errors were observed during a 2-h measurement period.

An RS-FF with 16 CAM junctions was made by Hirst et al. and operated at 45 K (39). Its design was similar to that reported previously by Shokhor et al. (52). The CAM junctions have advantages that make them especially suitable for making vertical SFQ loops with low inductance and reducing the parasitic inductance. Kim et al. tested the operations of RS-FF with four bicrystal junctions at 71 K (54). SFQ stored in the RS-FF was read out by using a magnetically coupled SQUID.

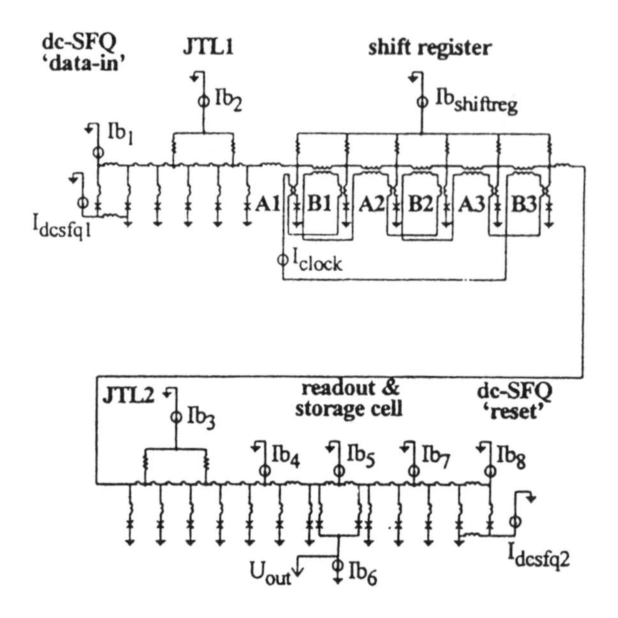

FIGURE 9.17 Equivalent circuit of a three-bit SFQ shift register. (From Ref. 53.)

9.4.2 Balanced Comparator

A balanced comparator, in which two overdamped Josephson junctions are connected in series, is not only one of the important elements of RSFQ circuits but is also useful when one is investigating the switching probability of Josephson junctions. Oelze et al. (24) and Ruck et al. (26) measured "smoothing parameters" of switching the gray zone and BER using the comparator, respectively, described in Section 9.3.2.1. Ruck's comparator was a part of a ring oscillator including 15 FEBI junctions (Fig. 9.18). A SFQ can circulate in the ring oscillator and its circulation frequency can be calculated, according to the ac Josephson relation, from the voltage across the ring oscillator. A maximum stable circulation frequency of 6 GHz was obtained. This corresponds to a delay of 17 ps per junction.

Sonnenberg et al. tested a balanced comparator in a three-HTS-layer technology. Eight ramp-edge junctions and inductances were located on a buried ground plane in order to reduce the inductance values (55). The correct operation of the balanced comparator was confirmed by dc measurement of the switching properties. The gray zone of switching was measured as a function of temperature (4.2–30 K) and operation frequency (2.5–80 GHz). For each temperature, the gray zone width has a minimum at low pulse rates of 10–15 μA, where the I_c of each junction was around 100 μA. The width of the gray zone increased with increasing pulse rate and increased more rapidly at 30 K than at 4.2 K.

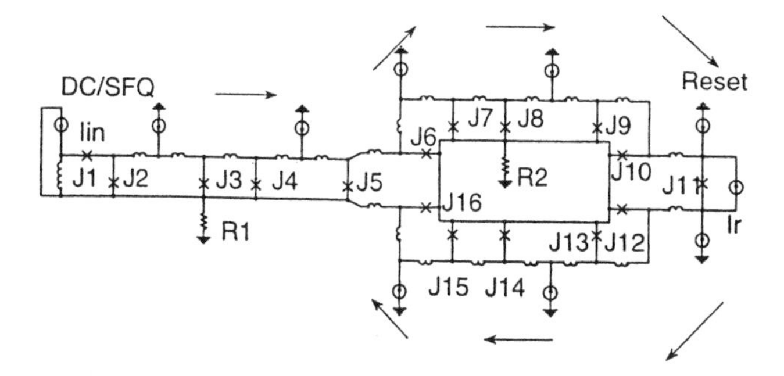

FIGURE 9.18 Equivalent scheme of a SFQ ring oscillator with a balanced comparator. (From Ref. 26.)

9.4.3 Voltage Divider

Single elements of RSFQ logic circuits can be operated at frequencies up to the characteristic Josephson frequency: $f_{\max} = I_c R_n / \Phi_0$. Kapulnenko suggested that the actual limit is $0.3 f_{\max}$ when many of these elements are connected together by means of a JTL, and this makes it necessary to use Josephson junctions with higher $I_c R_n$ products (56). The high-frequency limitations of RSFQ elements may be experimentally found by simple dc measurements using the Josephson relation between the average voltage V_{dc} across the junction and the Josephson oscillation frequency $f = \Phi_0^{-1} V_{ds}$. The T-FF passes every second fluxon from the input to its output, so the output voltage V_{out} is one-half of V_{in}. A simple dc measurement of V_{in} and V_{out} allows us to check the operation of the T-FF at high-frequency: When the frequency limit is not exceeded, V_{out} will be equal to $V_{\text{in}}/2$.

Kaplunenko et al. were the first to test a voltage divider using HTS materials (57). They used single-layer YBCO and a peculiar design. Small inductances of the SFQ loop, of about 10 pH, are formed as narrow slits of 0.4 μm width that are comparable to the London penetration depth $\lambda \sim 0.15$ μm of the superconducting YBCO film. Two slits separated by a 0.8-μm bridge provide a strong coupling between the two SFQ loops (58,59). The equivalent circuit and the layout of the T-FF circuit, which included 11 bicrystal junctions, are shown in Figures 9.19a and 9.19b. As shown in Figure 9.19c, proper operation was observed up to 0.82 mV at 4.4 K, giving $V_{\text{out}} = V_{\text{in}}/2$ within the experimental accuracy which corresponds to a Josephson frequency of about 400 GHz.

A voltage divider using nine ramp-edge junctions with a stacked ground plane was fabricated by Hashimoto et al. (60). At 12.5 K, the maximum voltage at which $V_{\text{out}} = V_{\text{in}}/2$ was 0.4 mV. This value corresponds to ~200 GHz. Saito et al. fabricated a voltage divider utilizing 11 IEJ-type ramp-edge junctions and operated it at frequencies up to 155 GHz at 15 K and 19 GHz at 27 K (27).

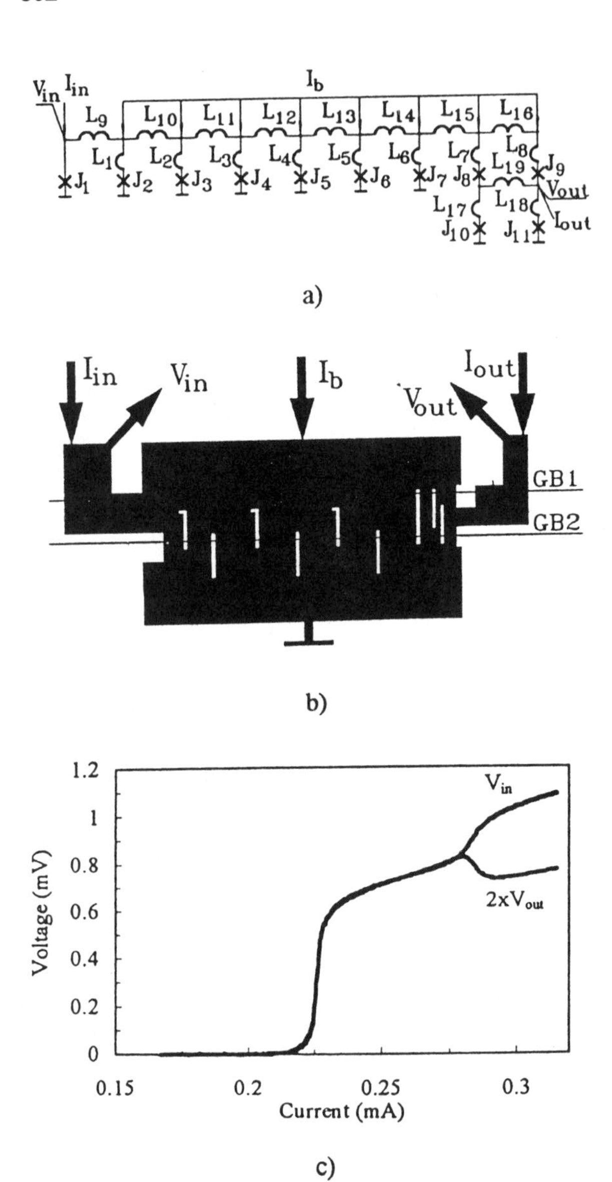

FIGURE 9.19 (a) Equivalent circuit, (b) layout, and (c) measured input, V_{in}, and output, V_{out}, voltages of a voltage divider based on submicron-slit inductances. (From Ref. 57.)

9.4.4 Flush-Type Analog-to-Digital Converters

The periodic nature of a SQUID permits the construction of a flush-type n-bit analog-to-digital (AD) converter with only n comparators, rather than the $2^n - 1$ single-threshold comparators used in the semiconductor flush-type AD converter. The circulating current in a SQUID loop is a periodic function of the flux applied with a periodicity of Φ_0. This forms the basis for 1 bit of an AD converter with extremely good differential and integral linearity. The dynamic range of such a converter is limited only by how much magnetic flux can be applied to the SQUID without suppressing the critical currents of the Josephson junctions in it. As proposed by Ko, the amount of magnetic flux can be increased dramatically by using a comparator based on a quasi-one-junction SQUID (QOS) (61).

Figure 9.20a shows the equivalent circuit of the basic QOS-based comparator. As indicated in Figure 9.20b, the current through the digitizing junction J_0 in this circuit is a periodic function of the analog input current I_a. The critical current of J_0 has to be much smaller than that of the sampling junction J_s if the influence of J_s on the behavior of the QOS is to be small. The β_L of this loop must be less than 1, because the periodic curve has to be a single value for all values of I_a. For high-frequency operation, the value of β_L should be less than 0.5. For a 4-bit AD converter, the periodic curve should contain at least four full periods. When a sampling pulse I_p is applied with a properly adjusted amplitude of I_p, the sampling junction J_s switches for 50% of the value of I_a. Each switching of the sampling junction results in a voltage at the output node of the circuit, and this state is considered to be a logical "1." When a sampling pulse I_p does not result in an output voltage, the state is considered to be a logical "0." Several LTS QOS comparators have been reported (e.g., in Ref. 62).

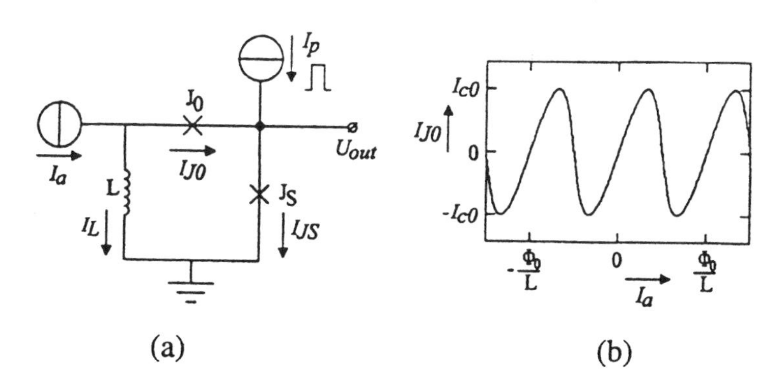

FIGURE 9.20 (a) Equivalent circuit of QOS-based periodic threshold comparator and (b) the periodic dependence of the QOS current I_{J0} on the analog input current I_a. (From Ref. 48.)

A QOS implementation using HTS materials was first reported by Wiegerink et al. (48). In their circuit, the two Josephson junctions in the QOS were constructed at two parallel ramp edges, which allows realization of a very low loop inductance. The operation of that comparator was demonstrated at slow sampling speeds, although the suppression of the junction critical currents due to the magnetic field associated with the input current I_a was clearly observed. The QOS threshold had been improved by changing its layout and up to eight full periods were observed without the suppression (63). At temperatures near 40 K, several 4-bit AD converter chips using the improved QOS were fully functional when operating at a low-frequency (1 kHz).

Umezawa et al. also fabricated a 1-bit QOS with a superconducting pickup loop using SEGB junctions and a basic AD conversion test showed that at 77 K, it operated at a frequencies from 10 kHz to 1 MHz (64).

The balanced comparator described in Section 9.4.2 can also be used as a comparator in a flush-type AD converter. Kidiyarova-Shevchenko et al. designed such a converter using three-layer HTS tricrystal junctions and simulated its operation (65).

9.4.5 Counting-Type AD Converters

Figure 9.21 shows a block diagram of a counting-type AD converter. The voltage controlled oscillator (VCO) continuously follows the analog input, generating pulses at a rate proportional to its voltage. The ac Josephson effect in a single junction affords a nearly perfect voltage-to-frequency converter, because the Josephson relation gives $f = 2eV/h$. When the "gate" is high, SFQ pulses are passed from the VCO to the counter; when the "gate" is low, SFQ pulses are prevented from passing to the counter. The total number of pulses passed in a fixed gate-open time is proportional to the time average of the analog input over the gate-open time. One critical advantage of this architecture for building an HTS AD converter is the relatively low level of circuit complexity required. A complete 12-bit NbN AD converter circuit using this architecture required only 52 Josephson junctions.

The SFQ counter shown in Figure 9.21 consists of a series of SFQ T-FFs. McCambridge et al. (66) and Sun et al. (67), in a first-step toward building a

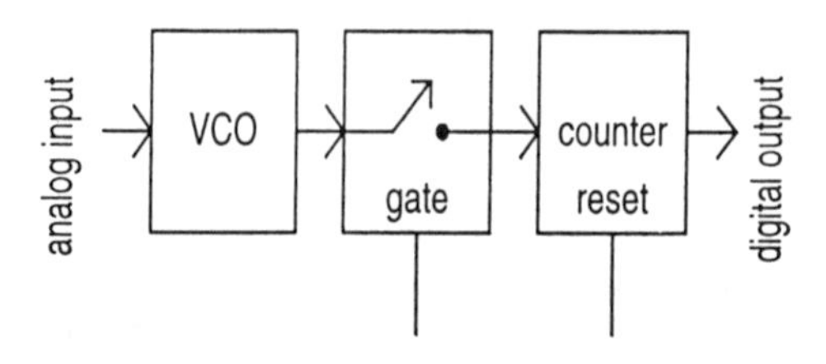

FIGURE 9.21 Block diagram of a counting-type AD converter. (From Ref. 67.)

counting AD converter, fabricated and tested single T-FF with a DC/SFQ converter, JTL, and a readout gate. McCambridge et al. used a three-layer YBCO structure, in which 10 ramp-edge junctions with a Co-doped YBCO barrier on a buried YBCO ground plane are utilized. Their circuit operated at 65 K and at low-speeds. Sun's circuit consisted of 14 ramp-edge junctions whose layered structure was same as that of the junctions in a McCambridge circuit. The readout voltage of Sun's circuit was shown to approach 40% of the junction I_cR_n product, and at 65 K, it was as high as 0.1 mV.

9.4.6 Sigma–Delta AD Converters

The Sigma–Delta (Σ–Δ) architecture is the preferred archtecture for AD converters with a high dynamic range (68). This oversampling approach implemented by semiconductor devices is used in audio applications where signals at kilohertz frequencies are sampled at megahertz frequencies by a Σ–Δ modulator and the resulting bit stream digitally filtered, to provide a resolution of 18–20 bits. The semiconductor Σ–Δ AD converter are is limited to megahertz sampling and digital filtering. A superconductor Σ–Δ AD converter, however, can perform gigahertz sampling and apply the advantages of digital filtering to megahertz-bandwidth signals. Moreover, flux quantization in a superconducting loop provides a quantum mechanically precise feedback mechanism unavailable with other technologies. Precision feedback is essential to the performance of Σ–Δ AD converters (69,70).

Figure 9.22 shows the block diagram of a first-order Σ–Δ modulator. An input signal is integrated and the result is sampled by a clocked comparator. The digitized signal is subtracted from the input signal in a feedback loop. This type of feedback causes the averaged output of the comparator to be exactly the input signal. The deviation of the quantized output of the comparator from the analog input signal can be described as noise and is called quantization noise. The quantization noise is shifted to higher frequencies. It is obvious that the signal-to-noise ratio can be improved tremendously by filtering the frequencies above the signal bandwidth.

The integrator, comparator, and feedback loop of the first-order Σ–Δ modulator is easily implemented by using Josephson junctions as shown in

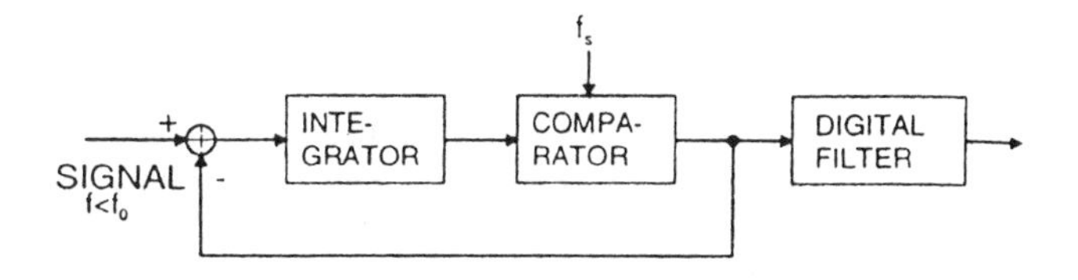

FIGURE 9.22 Block diagram of a first-order Σ–Δ modulator with a digital filter. (From Ref. 71.)

Figure 9.22. Forrester et al. fabricated a simple HTS Σ–Δ modulator, with 15 Co-doped YBCO barrier ramp-edge junctions, in an epitaxial multilayer process utilizing 3 YBCO layers, 2 epitaxial insulators, and an integrated Au resistor (31). They measured its performance at 35 K by inputting a 5.01-MHz signal and sending the output bit stream to a spectrum analyzer that measured the relative amplitude of the unwanted harmonics which determine the spur-free dynamic range (SFDR). With a 27-GHz sampling rate, a SFDP of greater than 75 dB was measured. This value is comparable to that of an LTS modulator (70).

Another HTS Σ–Δ modulator was reported by Ruck et al. (71). The circuit was fabricated on a STO bicrystal substrate. The YBCO/STO/YBCO trilayer was fabricated by laser deposition. The bottom layer served as a superconducting ground plane and the Josephson junctions were formed in the upper layer. Pd/Au thin film was used for the integrator resistance. The circuit consists of a DC/SFQ converter, a JTL, a comparator, a L/R integrator, and an output stage with 10 junctions. The correct operation of the modulator was confirmed by dc measurements at 34 K. The linearity of the modulator was studied by measuring the harmonic distortions of a 19.5-kHz sine-wave input signal, and a minimum resolution of 5 bits could be estimated from the recorded spectrum. This accuracy was limited by the noise of the preamplifier. The correct operation of the current feedback loop was demonstrated by cutting the feedback inductance.

9.4.7 Sampler

All of the above-described circuits are elements for consisting of larger RSFQ circuits or complete AD converters, except a ring oscillator and voltage dividers, which were implemented for investigating the operation frequency of SFQ circuits. These circuits themselves therefore cannot be put to practical uses. A sampler reported by Hidaka et al. (72,73), however, can measure repeated unknown electrical waveforms with high-resolutions by itself and can be used in practical measurement systems. The sampler circuit consists of five ramp-edge junctions with a stacked ground plane and is based on SFQ operations. The modified interface junctions (37) and a top-layer ground plane (HUG structure) (49) were introduced into the fabrication process.

Figure 9.23 shows the circuit design of the sampler circuit. A SFQ current pulse I_p is generated by JJ1 and JJ2 the moment the trigger current I_{tr} rises, and this pulse travels to JJ3 where it combines with a signal current $I_s(t)$ and a feedback current I_f at a given moment determined by the I_p generation. When the sum of the three currents exceeds a threshold value, the SFQ is stored in the superconducting loop that contains JJ3 and L3. Then, the stored SFQ induces an output voltage at the readout SQUID, which consists of JJ4 and JJ5. At the end of each sampling cycle, the stored SFQ in the loop is reset using negative reset currents I_r.

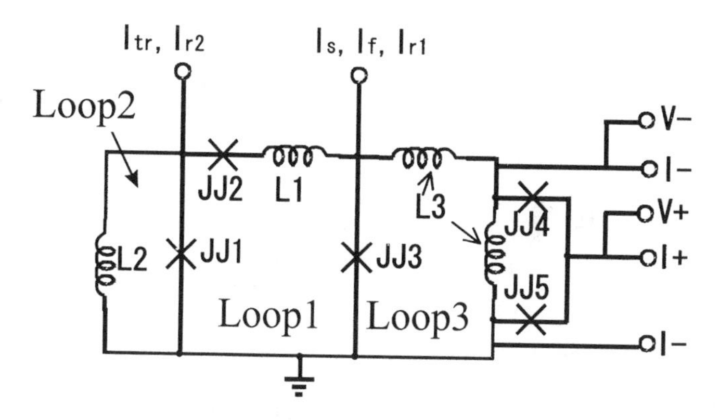

FIGURE 9.23 Circuit diagram of the HTS sampler. (From Ref. 73.)

The value of $I_{f\min}(I_s)$, which is defined as the minimum I_f required to store an SFQ in this loop for an I_s, can be determined by repeating the above operation with various I_f values. Comparing $I_{f\min}(I_s)$ with $I_{f\min}(0)$, which is the $I_{f\min}$ for $I_s = 0$, we can obtain the $I_s(t)$ value. The whole I_s waveform is measured by detecting the $I_s(t)$ values for various I_p generation times.

Figure 9.24 shows a sampler chip for high-time-resolution measurement. The chip is relatively small (2.5 mm square) so that the high-speed current lines on a STO substrate (which has a huge dielectric constant) are short. Cross-talk between current lines is kept low by the ground pads on both sides of the high-speed current pads. The signal current I_s measured by this sampler is generated by the signal trigger current I_{str} input at an on-chip Josephson signal generator (JG) and is propagated to the sampler through a 4-μm-wide, 400-μm-long YBCO line without any matching resistance.

Figure 9.25 shows one of the results with a 1-ps delay time between every sampling point. The dip structure around 160 ps in Figure 9.25a was remeasured in detail and is shown in Figure 9.25b. In Figure 9.25b, the maximum time differential of the measured waveform was 12 μA/ps, which fell 60 μA with a 5-ps time interval with a 2.5-μA current sensitivity.

The HTS sampler is able to measure current waveforms directly with picosecond and microampere resolutions. Semiconductor samplers and electro-optic (E-O) samplers are well known for characterizing the temporal shape of high-speed electrical signals. However, the semiconductor samplers measure voltage and the E-O samplers observe electrical field. In order to measure current using these samplers, the electrical impedance of the measured part has to be known. As the operation frequency of semiconductor large-scale integrated circuits (LSIs) increases toward the gigahertz frequency regime, the demand for current measure-

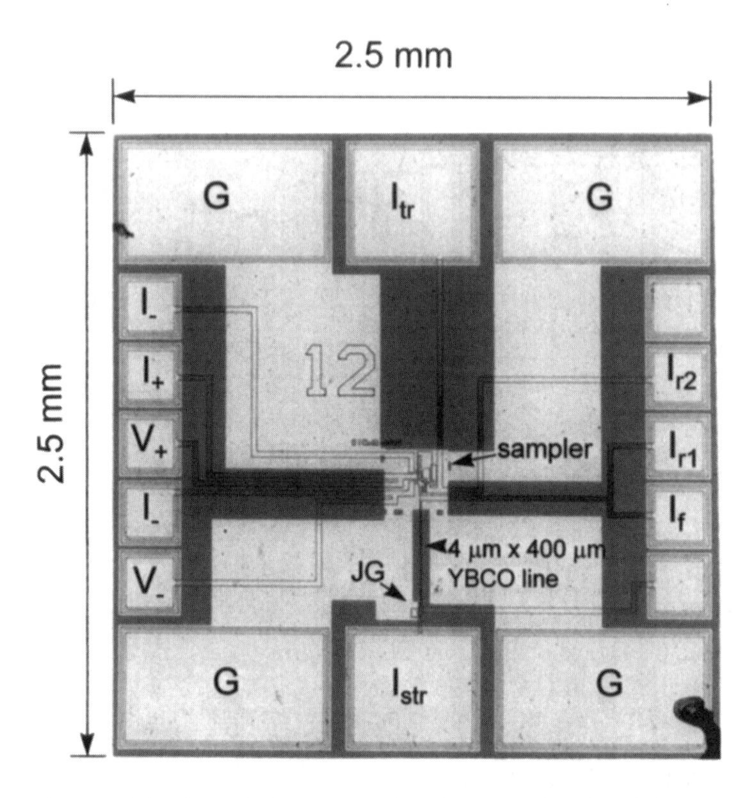

FIGURE 9.24 Optical close-up of the HTS sampler chip. The sampler circuit is in the center and the Josephson signal generator is below the sampler. (From Ref. 73.)

ments increases from the viewpoints of circuit design and electromagnetic compatibility (EMC) technology. However, because the impedance of a wiring in LSI under test is generally unknown because of its complex layered structures and via holes, current flowing through the wiring cannot be measured by using the semiconductor or E-O samplers. The HTS sampler is able to observe the current in the LSI with high-resolution. We expect the HTS sampler to be very useful for studying some transient phenomena, cross-talk, and EMC in high-speed LSI circuits.

9.4.8 Delay-Line Memory

Hattori et al. developed a HTS delay-line memory for asynchronous transfer mode (ATM) switching systems (74). This memory itself is not a digital circuit, but it is an interesting digital application of an HTS device. It is a hybrid device with high-speed semiconductor switches.

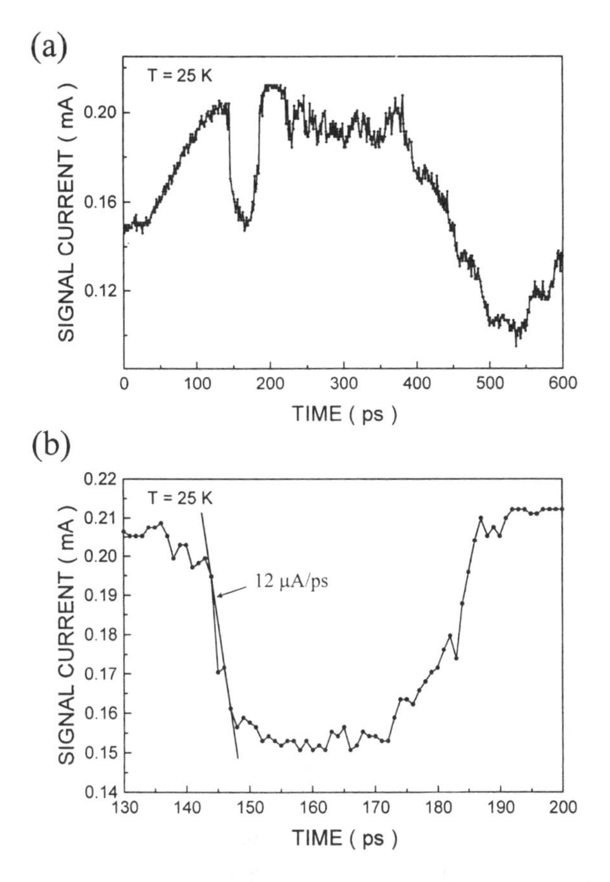

FIGURE 9.25 Signal current waveform measured by the HTS sampler: (a) waveform measured from 0 to 600 ps and (b) remeasured waveform between 130 ps and 200 ps. (From Ref. 73.)

The rapid growth in telecommunication traffic has resulted in the need for faster ATM switching systems. The upper limit of the system clock rate is currently determined by the maximum clock rate of conventional semiconductor memory devices, such as the register files used in ATM cell buffer storage. This is because the maximum clock rate of these register files is restricted by the propagation delay between each register stage. Because a re-entrant superconducting delay-line memory avoids this restriction using an analog delay given by the superconducting delay line, this memory should be used in high-speed ATM cell buffer storage.

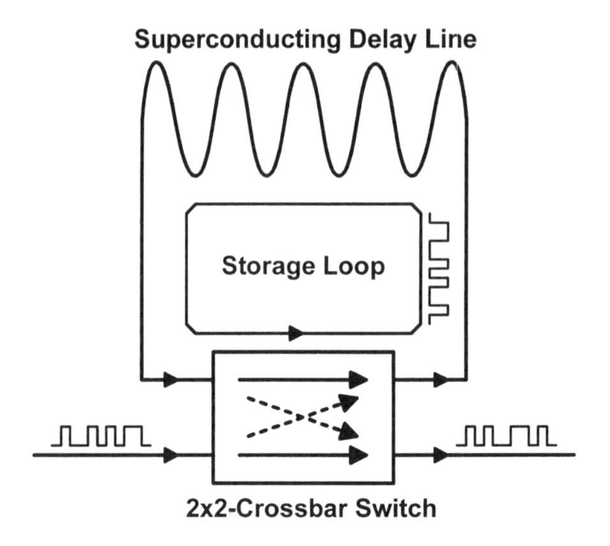

FIGURE 9.26 Configuration of the re-entrant superconducting delay-line memory. This memory consists of a superconducting delay line and a 2×2 crossbar switch. (From Ref. 74.)

The configuration of the superconducting reentrant delay-line memory is illustrated in Figure 9.26. The memory has recirculation loop storage for a fixed-length data packet and consists of a superconducting delay line and a semiconductor 2×2 crossbar switch. This delay line gives input data a fixed delay that corresponds to its line length. The crossbar switch enables cross or parallel connection between two input ports and two output ports. The delay line feeds back from an output port of the switch to its input port and forms a storage loop. This loop has a duration that corresponds to a fixed data packet length. Because signals in the delay line cannot be amplified and reshaped, the attenuation and distortion in the delay line must be extremely low despite the high clock rate and long delay. This is impossible for a planar transmission line made of normal metal, because of its surface resistance. This is why the superconducting delay line is used.

A YBCO coplanar delay line 10 μm wide and 37 cm long was fabricated. This line had a delay of approximately 2.8 ns and was used, along with commercially available semiconductor integrated circuits, to make a superconducting delay-line memory. As shown in Figure 9.27, this memory operated as a 32-bit buffer storage at a clock rate of 10 GHz at 46 K, which is several times faster than a semiconductor register file operates. This result shows that the superconducting delay-line memory is a powerful candidate for high-speed ATM cell buffer storage.

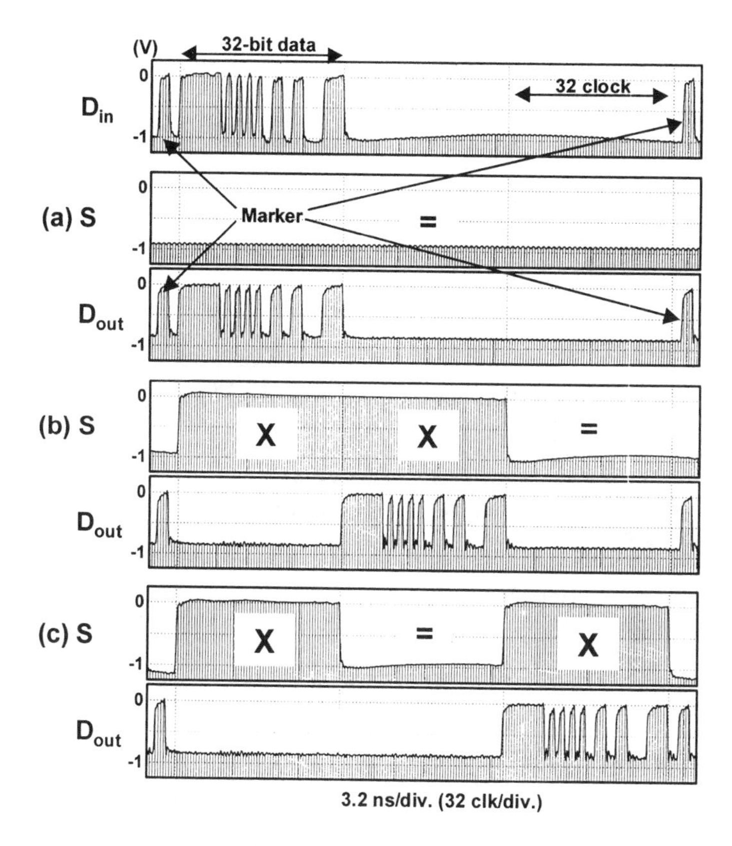

FIGURE 9.27 Thirty-two-bit 10-GHz buffering operation of the delay-line memory: (a) unbuffering operation and (b) buffering operation with a packet length (32 clock) and (c) buffering operation during double-packet length (64 clock). (From Ref. 74.)

REFERENCES

1. W Anacker. Josephson computer technology: An IBM research project. IBM J Res Dev 24:107–252, 1980.
2. H Kroger. Josephson devices and technology. In: Japanese Assessment. Park Ridge, NJ: Noyes Data Corporation, 1986, pp 250–306.
3. S Nagasawa, H Numata, Y Hashimoto, S Tahara. High-frequency clock operation of Josephson 256-word × 16-bit RAMs. IEEE Trans Appl Supercond As-9:3708–3713, 1999.
4. S Yorozu, Y Hashimoto, H Numata, M Koike, M Tanaka, S Tahara. Full operation of a three-node pipeline-ring switching chip for a superconducting network system. IEEE Trans Appl Supercond As-9:3590–3593, 1999.

5. K Nakajima, Y Onodera. Logic gate of Josephson network. J Appl Phys 47:1620–1627, 1976.
6. KK Likharev, VK Semenov. RSFQ logic/memory family: A new Josephson-junction technology for sub-terahertz-clock-frequency digital systems. IEEE Trans Appl Supercond As-1:3–28, 1991.
7. VK Semenov, YA Polyakov, D Schneider. Implementation of oversampling analog-to-digital converter based on RSFQ logic. Extended Abstracts of the 6th International Superconductive Electronics Conference, Berlin, Germany, H. Koch and S. Knappe, PTB, June 25–28, 1997, Vol. 1, pp 41–43.
8. WC Stewart. Current–voltage characteristics of superconducting tunnel junctions. Appl Phys Lett 12:277–280, 1968.
9. DE McCumber. Effects of ac impedance on dc voltage–current characteristics of superconductor weak-link junctions. J Appl Phys 39:3113–3118, 1968.
10. M Gurvitch, MA Washington, HA Huggins. High-quality refractory Josephson tunnel junctions utilizing thin aluminum layers. Appl Phys Lett 42:472–474, 1983.
11. H Numata, M Tanaka, Y Kitagawa, S Tahara. Investigation of SFQ integrated circuits using Nb fabrication process. Extended Abstracts of the 7th International Superconductive Electronics Conference, Berkeley, USA, T Van Duzer, June 21–25, 1999, pp 272–274.
12. MA Vanhoeven, GJ Gerritsma, H Rogalla, AA Golubov. Ramp-type junction parameter control by Ga doping of Pr $Ba_2Cu_3O_x$ barriers. Appl Phys Lett 69:848–850, 1996.
13. BD Hunt, MG Forrester, J Talvacchio, JD McCambridge, RM Young. High-T_c superconductor/normal-metal/superconductor edge junctions and SQUIDs with integrated ground planes. Appl Phys Lett 68:3805–3807, 1996.
14. DL Miller, JX Przybysz, JH Kang. Margins and yields of SFQ circuits in HTS materials. IEEE Trans Appl Supercond As-3, 2728–2731, 1993.
15. J Talvacchio, MG Forrester, BD Hunt, JD McCambridge, RM Young. Materials basis for a six-level epitaxial HTS digital circuit process. IEEE Trans Appl Supercond As-7:2051–2055, 1997.
16. T Satoh, M Hidaka, S Tahara. High-temperature superconducting edge-type Josephson junctions with modified interface. IEEE Trans Appl Supercond As-9:3141–3144, 1999.
17. T Satoh, M Hidaka, S Tahara. Study of *in-situ* preparation high-temperature superconducting edge-type Josephson junctions. IEEE Trans Appl Supercond As-7:3001–3004, 1997.
18. TV Filippov, YuA Polyakov, VK Semenov, KK Likharev. Signal resolution of RSFQ components. IEEE Trans Appl Supercond As-5:2240–2243, 1995.
19. TV Filippov, VK Kornev. Sensitivity of the balanced Josephson junction comparator. IEEE Trans Magnetics M-27:2452–2455, 1991.
20. TV Filippov. Current resolution of the Josephson balanced comparator at helium temperatures. Superconductivity (Moscow) 8:147–163, 1994.
21. VK Semenov, TV Filippov, YuA Polyakov, KK Likharev. SFQ balanced comparator at a finite sampling rate. IEEE Trans Appl Supercond As-7:3617–3621, 1997.
22. J Satchell. Limitations on HTS signal flux quantum logic. IEEE Trans Appl Supercond As-9:3841–3844, 1999.

23. M Jeferry, PY Xie, SR Whiteley, T Van Duzer. Monte Carlo and thermal noise analysis of ultra-high-speed high-temperature superconductor digital circuits. IEEE Trans Appl Supercond As-9:4095–4098, 1999.
24. B Oelze, B Ruck, E Sodtke, T Filippov, A Kidiyarova-Shevchenko, M Kupriyanov, W Prusseit. Investigation of the signal resolution of a high-T_c balanced comparator. IEEE Trans Appl Supercond As-7:3450–3453, 1997.
25. Y Chong, B Ruck, R Dittmann, C Horstmann, A Engelhardt, G Wahl, B Oelze, E Sodtke. Measurement of the static error rate of a storage cell for single magnetic flux quanta, fabricated from high-T_c multiplayer bicrystal Josephson junctions. Appl Phys Lett 72:1513–1515, 1998.
26. B Ruck, B Oelze, R Dittmann, A Engelhardt, E Sodtke, WE Booij, MG Blamire. Measurement of the dynamic error rate of a high-temperature superconductor rapid single flux quantum comparator. Appl Phys Lett 72:2328–2330, 1998.
27. K Saito, Y Soutome, F Tokuumi, Y Tarutani, K Takagi. Voltage divider operation using high-T_c superconducting interface-engineered Josephson junctions. Appl Phys Lett 76:2606–2608, 2000.
28. M Hidaka, H Terai, T Satoh, S Tahara. Multilayer high-T_c superconductor sampler circuit. Appl Supercond 6:615–619, 1998.
29. J Talvacchio, RM Young, MG Forrester, BD Hunt. Oxidation of multiplayer HTS digital circuits. IEEE Trans Appl Supercond As-9:1990–1993, 1999.
30. B Ruck, Y Chong, G Wahl, R Dittmann, C Horstmann, A Engelhardt, B Oelze, E Sodtke. Investigation of basic elements and devices in multiplayer technology for HTS digital RSFQ circuits. Extended Abstracts of the 6th International Superconductive Electronics Conference, Berlin, Germany H Koch and S Knappe, PTB, June 25–28, 1997, Vol 2, pp 326–328.
31. MG Forrester, BD Hunt, DL Miller, J Talvacchio, RM Young. Analog demonstration of a high-temperature superconducting sigma–delta modulator with 27 GHz sampling. Extended Abstracts of the 7th International Superconductive Electronics Conference, Berkeley, USA, T Van Duzer, June 21–25, 1999, pp. 24–31.
32. DL Miller, MG Forrester, JX Przybysz, BD Hunt, J Talvacchio. Single-flux-quantum circuits based on YBCO step-edge-grain-boundary junctions. Extended Abstracts of the 5th International Superconductive Electronics Conference, Nagoya, Japan, H. Hayakawa, September 18–21, 1995, pp 40–42.
33. FC Wellstood, JJ Kingston, J Clarke. Thin-film multiplayer interconnect technology for $YBa_2Cu_3O_{7-x}$. J Appl Phys 75:683–702, 1994.
34. J Gao, YuM Boguslavskij, BBG Klopman, D Terpstra, R Wijbrans, GJ Gerritsma, H Rogalla. $YB_2Cu_3O_x/PrBa_2Cu_3O_x/YBa_2Cu_3O_x$ Josephson ramp junctions. J Appl Phys 72:575–583, 1992.
35. T Satoh, M Hidaka, S Tahara. Edge-type Josephson junctions for HTS sampler. Extended Abstracts of the 1997 International Workshop on Superconductivity (The 3rd Joint ISTEC/MRS Workshop), Hawaii, USA, S Tanaka, ISTEC, June 15–18, 1997, pp 235–238.
36. WH Mallison, SJ Berkowitz, AS Hirahara, MJ Neal, K Char. A multilayer YBa_2 Cu_3O_x Josephson junction process for digital circuit applications. Appl Phys Lett 68:3808–3810,1996.

37. BH Moeckly K Char. Properties of interface-engineered high-T_c Josephson junctions. Appl Phys Lett 71:2526–2528, 1997.
38. M Maruyama, K Yoshida, T Furitani, Y Inagaki, M Horibe, M Inoue, A Fujimaki, H Hayakawa. Interface-treated Josephson junctions in trilayer structures. Jpn J Appl Phys 39:L205–L207, 2000.
39. PJ Hirst, TG Henrici, LL Atkin, JS Satchell, J Moxey, NJ Exon, MJ Wooliscoft, TJ Horton, RG Humphreys. *c*-Axis microbridges for rapid single flux quantum logic. IEEE Trans Appl Supercond AS-9:3833–3836, 1999.
40. KP Daly, JM James, M Murduck, CL Pettiette-Hall, M Sergant. Integration of step-edge grain boundary Josephson junctions with YBCO multilayers for electronics applications. IEEE Trans Appl Supercond AS-5:3131–3134, 1995.
41. WE Booij, AJ Pauza, EJ Tarte, DF Moore, MG Blamire. Proximity coupling in high-T_c Josephson junctions produced by focused electron beam irradiation. Phys Rev B 55:14,600–14,609, 1997.
42. D Dimos, P Chaudhari, J Manhart, FK LeGoues. Orientation dependence of grain-boundary critical current in $YBa_2Cu_3O_{7-\delta}$ bicrystals. Phys Rev Lett 61:219–222, 1988.
43. N Missert, TE Harvey, RH Ono, CD Reinstsema. High-T_c multiplayer step edge Josephson junctions and SQUIDs. Appl Phys Lett 63:1690–1692, 1993.
44. MG Forrester, A Davidson, J Talvacchio, JR Gavaler, JX Przybysz. Inductance measurements in multilevel high-T_c step edge grain boundary SQUIDs. Appl Phys Lett 65:1835–1837, 1994.
45. S Miura, W Hattori, T Satoh, M Hidaka, S Tahara, JS Tsai. Properties of a YBCO/insulator/YBCO trilayer and its application to a multiplayer Josephson junction. Supercond Sci Technol 9:A59–A61, 1996.
46. TG Henrici, SW Goodyear, JS Satchell, NG Chew, MJ Wooliscroft, K Lander, RG Humphreys. Inductance measurements and simulations for HTS SFQ logic circuits incorporating c-axis microbridges. Extended Abstracts of the 6th International Superconductive Electronics Conference, Berlin, Germany, H Koch and S Knappe, PTB, June 25–28, 1997, Vol 2, pp 260–262.
47. C Stolzel, W Wilkens, A Marx, U Fath, K Petersen, J Sollner, H Adrian. Preparation of ramp junctions and multilayers for superconducting camparators. Physica C 235–240:3349–3350, 1994.
48. RJ Weigerink, GJ Gerritsma, EMCM Reuvekamp, MAJ Verhoven, H Rogalla. An HTS quasi-one junction SQUID-based periodic threshold comparator for a 4-bit superconductive flash A/D converter. IEEE Trans Appl Supercond AS-5:3452–3458, 1995.
49. H Terai, M Hidaka, T Satoh, S Tahara. Direct-injection high-T_c dc-SQUID with an upper $YBa_2Cu_3O_{7-\delta}$ ground plane. Appl Phys Lett 70:2690–2692, 1997.
50. ZG Ivanov, VK Kaplumenko, EA Stepantsov, E Wikborg, T Claeson. An experimental implementation of high-T_c-based RSFQ set-reset trigger at 4.2 K. Supercond Sci Technol 7:239–241, 1994.
51. MG Forrester, JX Przybysz, J Talvacchio, J Kang, A Davidson, JR Gavaler. A single flux quantum shift resister operating at 65 K. IEEE Trans Appl Supercond AS-5:3401–3404, 1995.

52. S Shokhor, B Nadgomy, M Gurvitch, V Semenov, Yu Polyakov, K Likharev. All-high-T_c superconductor rapid-single flux-quantum circuit operating at ~30K. Appl Phys Lett 67:2869–2871, 1995.
53. B Oelze, B Ruck, E Sodtke, AF Kirchenko, MYu Kupriyanov, W Prusseit. A 3 bit single flux quantum shift resister based on high-T_c bicrystal Josephson junctions operating at 50 K. Appl Phys Lett 70:658–670, 1997.
54. YH Kim, JH Kang, JH Sung, JH Park, JM Lee, KR Jung, CH Kim, TS Hahn, SS Choi. Digital and analog measurements of HTS SFQ RS flip-flops and shift resister circuits. IEEE Trans Appl Supercond AS-9:3817–3820, 1999.
55. AH Sonnenberg, GJ Gerritsma, H Rogalla. Balanced comparator fabricated in ramp edge technology. Physica C 326–327:12–15, 1999.
56. VK Kaplunenko. Fluxon interaction in an overdamped Josephson transmission line. Appl Phys Lett 66:3365–3367, 1995.
57. VK Kaplunenko, ZG Ivanov, EA Stepantsov, T Claeson. Voltage divider based on submicron slits in a high-T_c superconducting film and two bicrystal grain boundaries. Appl Phys Lett 67:282–284, 1995.
58. VK Kaplunenko, ZG Ivanov, EA Stepantsov, T Claeson, T Holst, ZJ Sun, R Kromann, YQ Shen, V Vase, T Frelton, E Wikborg. Novel design of rapid single flux quantum logic based on a single layer of a high-T_c superconductor. Appl Phys Lett 67:138–140, 1995.
59. VK Kaplumenko, ZG Ivanov, A Bogdanov, EA Stepantsov, T Claeson, T Holst, ZJ Sun, R Kromann, YQ Shen, V Vase, T Frelton, E Wikborg. A new design approach for high-T_c based RSFQ logic. IEEE Trans Appl Supercond 5:2835–2838, 1995.
60. T Hashimoto S Inoue, T Nagano, J Yoshida. Design and fabrication of a voltage divider utilizing high-T_c ramp-edge Josephson junctions with a ground plane. IEEE Trans Appl Supercond 9:3821–3824, 1999.
61. H Ko, T Van Duzer. A new high-speed periodic threshold comparator for use in a Josephson AD converter. IEEE J Solid State Circuits 23:1017–1021, 1988.
62. P Bradley, H Dang. Design and testing of quasi-one junction SQUID-based comparators at low and high-speed for superconductive flush AD converters. IEEE Trans Appl Supercond AS-1:134–139, 1991.
63. GJ Gerritsma, MAJ Verhoeven, RJ Wiegerink, and H Rogalla. A high-T_c periodic threshold analog-to digital converter. IEEE Trans Appl Supercond As-7:2987–2992, 1997.
64. T Umezawa, T Fujita, Y Higashino. High-T_c superconducting flush A/D converter using planar quasi-one junction superconducting quantum interference devices. Jpn J Appl Phys 35:L981–L984, 1996.
65. AYu Kidiyarova-Shevchenko, DE Kirichenko, Z Ivanov, F Komissinsky, EA Stepancov, MM Khapaev, T Claeson. Single flux quantum comparators for HTS AD converters. Physica C 326–327:83–92, 1999.
66. JD McCambridge, MG Forrester, DL Miller, BD Hunt, J Pryzbysz, J Talvacchio, RM Young. Multilayer HTS SFQ analog-to-digital converters. IEEE Trans Appl Supercond As-7:3622–3625, 1997.
67. AG Sun, DJ Durand, JM Murduck, SV Rylov, MG Forrester, BD Hunt, J Talcvacchio. HTS SFQ T-flip flop with directly coupled readout. IEEE Trans Appl Supercond As-9:3825–3828, 1999.

68. JC Candy, GC Temes. Oversampling A/D Converters. New York: IEEE, 1992.
69. JX Przybysz, DL Miller, EH Naviasky, JH Kang. Josephson Sigma–Delta modulator for high dynamic range A/D converter. IEEE Trans Appl Supercond As-3:2732–2735, 1993.
70. DL Miller, JX Przybysz, DL Meier, JH Kang, A Hodge. Characterization of a superconductive Sigma–Delta analog to digital converter. IEEE Trans Appl Supercond As-5:2453–2456, 1995.
71. B Ruck, Y Chong, R Dittmann, M Siegel. First order sigma–delta modulator in HTS bicrystal technology. Physica C 326–327:170–176, 1999.
72. M Hidaka, JS Tsai. Circuit design for a high-T_c Josephson sampler. IEEE Trans Appl Supercond As-5:3353–3356, 1995.
73. M Hidaka, T Satoh, M Koike, S Tahara. High-resolution measurement by a high-T_c superconductor sampler. IEEE Trans Appl Supercond As-9:4081–4086, 1999.
74. W Hattori, T Yoshitake, S Tahara. A reentrant delay-line memory using a $YBa_2Cu_3O_{7-\delta}$ coplanar delay-line. IEEE Trans Appl Supercond As-9:3829–3832, 1999.

10

High-Temperature Superconductor Microwave Devices

Neeraj Khare
National Physical Laboratory, New Delhi, India

10.1 INTRODUCTION

High-temperature superconductors have been used for microwave applications since their discovery in 1986 and there has been a continuous progress in this area. Several high-T_c superconductors based microwave devices such as resonators, filters, multiplexer, receivers, delay line, antenna, phase shifter, and so forth with superior performances have been demonstrated (1–5). High-T_c microwave components and subsystems are currently being commercialized by several companies. The use of high-T_c superconductors in microwave passive devices has an advantage over normal conductors such as copper and silver in terms of low insertion loss and high gain due to their lower surface resistance. Also, because of lower losses in superconductors, a reduction in size of the devices is an added advantage in using high-T_c superconductors.

In recent years, high-T_c superconductor-based filters and subsystems have been considered for application in mobile communication as well as for satellite and some specific radio astronomy applications (3,5–12). The use of high-T_c filters in a cellular base station is being investigated for improved sensitivity. Nearly 1000 high-T_c filter subsystems have been deployed worldwide with millions of hours of cumulative operations (3). Microwave technology based on high-T_c superconductors offer the potential of considerable miniaturization of pay-load electronic equipment in space systems, leading to an overall cost reduction and accelerating the development of small satellite systems (10).

This chapter reviews the progress in the development of various high-T_c superconductor microwave passive devices such as resonators, filters, antenna, delay lines, phase shifters, tunable devices, and microwave subsystems.

10.2 HIGH-T_c SUPERCONDUCTOR MATERIALS

Several high-T_c superconductors exhibiting superconductivity above the liquid-nitrogen temperature (77 K) are listed in Chapter 1. Two high-T_c materials frequently used for microwave applications are $YBa_2Cu_3O_{7-\delta}$ (YBCO) and $Tl_2Ba_2CaCu_2O_8$ (TBCCO). The superconducting transition temperature (T_c) of YBCO is 92 K and that of TBCCO is 108 K. YBCO has been used in the form of epitaxial thin film as well as textured thick film, whereas TBCCO has been used only in the form of thin film. Earlier high-T_c microwave devices such as cylindrical cavities have been fabricated using YBCO in the form of bulk. YBCO films are deposited on both sides of a suitable substrate in situ by the laser ablation technique or magnetron sputtering, whereas for TBCCO films, first the precursor material is deposited on both sides of the substrate and then annealing is performed for obtaining a superconducting thin film. In the range of operating temperatures from 60 to 77 K that is practical for high-T_c microwave devices, high-quality YBCO and TBCCO films exhibit similar properties inspite of the higher T_c of TBCCO as compared to YBCO.

10.2.1 Substrates for High-T_c Microwave Devices

For high-T_c film microwave devices, the substrate should not only support growth of good quality epitaxial high-T_c films but also its microwave properties such as the dielectric constant (ϵ_r) and dielectric loss tangent (tan δ) should be in a desired range (13). For example, for a microwave circuit with a high-quality factor (Q), tan δ needs to be very small. The value of ϵ_r is related to the length of electromagnetic wave in the substrate material. Thus, for the operating frequency range of 1–10 GHz, the substrates with ϵ_r ~20–25 are suitable, whereas for the operating frequencies larger than 10 GHz, ϵ_r should be ~10. It is an added advantage if the dielectric constant is isotropic in the plane of the film and it has a low dispersion for wide-band devices. In addition to the above properties, the substrate should be strong and capable of being thinned to a desired thickness, as required by the application. Substrates such as $SrTiO_3$ and ZrO_2, which can support the growth of very good quality high-T_c films, are not suitable for microwave application due to their high value of loss tangent.

Table 10.1 gives a list of some of the substrates which have been used for high-T_c microwave devices. $LaAlO_3$ is found to be the first suitable substrate

TABLE 10.1 Substrates for High-T_c Superconductor Microwave Circuits

Substrate	Crystal structure	Lattice constants (Å)	Melting point (°C)	Dielectric constant (ϵ_r)	tan δ	Ref.
MgO	Cubic	$a = 4.21$	2825	9.6–10	6.2×10^{-6} [77 K, 10 GHz]	14, 15
Sapphire	Hexagonal	$a = 4.759$ $c = 12.97$	2049	9.4–11.6	1.5×10^{-8} [77 K, 9 GHz]	14–16
$LaAlO_3$	Rhombohedral	$a = 3.79$ $\alpha = 90°5'$	2100	24	7.6×10^{-6} [77 K, 10 GHz]	13,15
$YAlO_3$	Orthorhombic	$a = 3.66$ $b = 3.77$ $c = 3.69$	1875	16	1.2×10^{-5} [77 K, 10 GHz]	13,15
$GdAlO_3$	Orthorhombic	$a = 3.731$ $b = 3.724$	1940	19.5	$<10^{-4}$ [300 K, 40 GHz]	13
$NdAlO_3$	Rhombohedral	$a = 3.750$ $\alpha = 90°22'$	2070	22.5	5×10^{-5} [300 K, 40 GHz]	13
$NdGaO_3$	Orthorhombic	$a = 5.43$ $b = 5.50$ $c = 7.70$	1670	23	3.2×10^{-4} [77 K, 10 GHz]	15
$SrLaGaO_4$	Tetragonal	$a = 3.84$ $c = 12.68$	—	22	1.5×10^{-5} [77 K, 10 GHz]	15
$SrLaAlO_4$	Tetragonal	$a = 3.77$ $c = 12.5$	1650	27	1×10^{-4} [5 K, 8.5 GHz]	17
$CaNdAlO_4$	Tetragonal	$a = 3.69$ $c = 12.15$	1820	20	10^{-3} [100 K, 100 GHz]	18

which has been used for high-T_c microwave devices (19). It is easily available in a large size. Another substrate, MgO, shows isotropy of dielectric properties and low microwave losses. The drawback of MgO is the lack of mechanical strength and high hygroscopity. Apart from these, the MgO substrates are not available in a large size. Sapphire, on the other hand, exhibits high mechanical strength, high thermal conductivity, and low microwave losses and are also available in a large size. However, for the growth of good quality high-T_c films, preparation of buffer layer of MgO or CeO_2 on a sapphire substrate is required (20,21).

10.3 COMPLEX CONDUCTIVITY AND SURFACE IMPEDANCE

In a superconductor, resistance is zero for direct current and the current flows without any dissipation. However, for an alternating current, the superconductor shows a resistance, although the value of the resistance is very small. A phenomenological two-fluid model has been used to explain the general behavior of superconductors at radio frequency (RF) and microwave frequency. The two-fluid model (22) treats the current carriers in the superconductors as being of two distinct types: a supercarrier fraction with density n_s, which carries current without any dissipation, and a normal fraction n_n, which exhibit resistive scattering similar to the electrons in normal metals. Both n_s and n_n are strong functions of temperature below the transition temperature, T_c. The conductivity σ is complex and given as (22)

$$\sigma = \sigma_1 - j\sigma_2 \tag{1}$$

with

$$\sigma_1 = \frac{n_n e^2 \tau}{m} \tag{2}$$

$$\sigma_2 = \frac{1}{\omega \mu_0 \lambda_L^2} \quad j^2 = -1 \tag{3}$$

$$\lambda_L = \left(\frac{m}{\mu_0 n_s e^2} \right)^{1/2} \tag{4}$$

where m is the effective mass of the charge carriers, e is the charge of the carrier, τ is the scattering time for quasiparticle, λ_L is the London penetration depth, ω is the angular frequency, and μ_0 is the vacuum permeability. The σ_1 and σ_2 are proportional to the density of normal carriers and cooper pairs, respectively. The total carrier density, $n = n_s + n_n$, is related to the normal-state conductivity of the material by

$$\sigma_n = \frac{n e^2 \tau}{m} \tag{5}$$

The temperature dependences of σ_1 and σ_2 are expressed as

$$\sigma_1 = \sigma_n t^4 \tag{6}$$

$$\sigma_2 = \frac{1 - t^4}{\mu_0 \omega \lambda_L^2(0)} \tag{7}$$

where $t = T/T_c$ is the reduced temperature, σ_n is the normal conductivity just above the transition temperature, and $\lambda_L(0)$ is the London penetration depth at 0 K. The surface impedance, Z_s, is defined as the ratio of the electromagnetic electric field (E_y) to the magnetic field (H_x) at the surface

$$Z_s = \frac{E_y}{H_x} \tag{8}$$

where the z axis has been chosen to be normal to the superconductor surface. The formula for surface impedance of a good conductor is

$$Z_s = \left(\frac{j\omega\mu_0}{\sigma} \right)^{1/2} \tag{9}$$

Using the two-fluid relation for the complex conductivity, the expression for surface impedance of a superconductor may be written as

$$Z_s = R_s + jX_s \tag{10}$$

where R_s and X_s are real and imaginary parts of the impedance, respectively:

$$R_s = \frac{\mu_0^2 \omega^2 \lambda_L^3 n_n e^2 \tau}{2m} \tag{11}$$

$$X_s = \omega \mu_0 \lambda_L \tag{12}$$

The surface resistance R_s for a superconductor is proportional to the square of the frequency, whereas for a normal conductor, the surface resistance is proportional to the square root of the frequency.

Both the resistive and reactive components of surface impedance of a superconductor play an important role in determining the performance of filters, resonators, and other microwave devices. The surface resistance R_s determines the quality factor Q of the resonator, whereas the reactive component X_s determine the sensitivity to temperature variation of the wavelength of a transmission line and the long-term stability of the device.

10.3.1 Measurement of Surface Resistance

The measurement of surface resistance of high-T_c superconductors is important in order to determine the suitability for its application in microwave devices. Several techniques have been employed for the measurement of surface resistance; these are listed in Table 10.2. In the cavity resonating structure, the cylindrical cavity structure is very popular for the surface resistance measurement due to its high Q value and convenient shape. A cavity made totally out of a superconductor is ex-

TABLE 10.2 Techniques Used for the Measurement of Surface Resistance of High-T_c Superconductors

Cavity resonant structure (for bulk, crystals, and films)	1. Full cavity made of high-T_c superconductors
	2. Cavity end-plate replacement
	3. Cavity end-plate substitution
	4. Cavity perturbation
	5. Coaxial cavity
Dielectric resonator	1. Parallel-plate dielectric resonator
Patterned resonant structure (for films)	1. Microstrip line resonator
	2. Microstrip ring resonator
	3. Stripline resonator
	4. Coplanar line resonator

pected to provide a more precise measurement of surface resistance. Cylindrical cavities totally made of YBCO high-T_c superconductors have been fabricated for measuring surface resistance (23). Figure 10.1a is a schematic of a high-T_c cylindrical cavity. A value of surface resistance of 70 mΩ at 77 K was obtained for the YBCO bulk superconductor using this type of resonator.

It is not always possible to make the entire cavity using high-T_c superconductor materials and several other techniques are used for surface resistance measurement. In the cavity end replacement technique, cylindrical cavities made of copper or niobium is used in which one end of the cavity is replaced by a high-T_c sample (24). Figure 10.1b is a schematic of this arrangement. By measuring the Q of an all-copper (or superconductor) cavity and then by measuring the Q of the cavity with the end plate replaced by a high-T_c sample, one can estimate the value of the surface resistance. At low frequency, the size of the cylindrical cavity increases and it may not be convenient to prepare a large-size sample for the re-

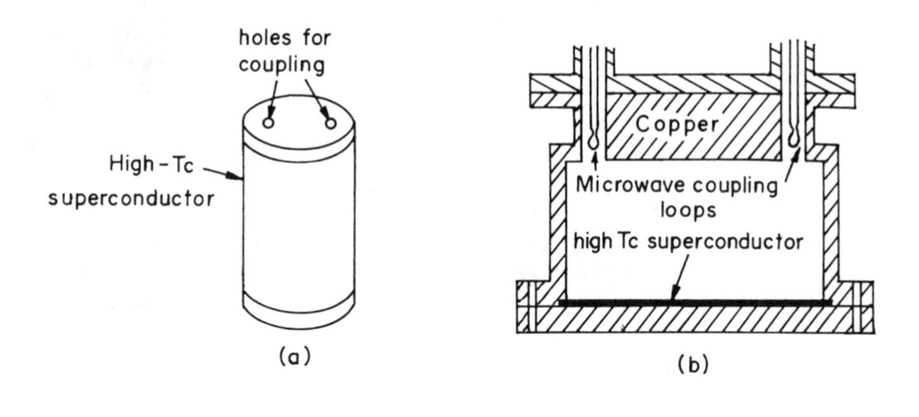

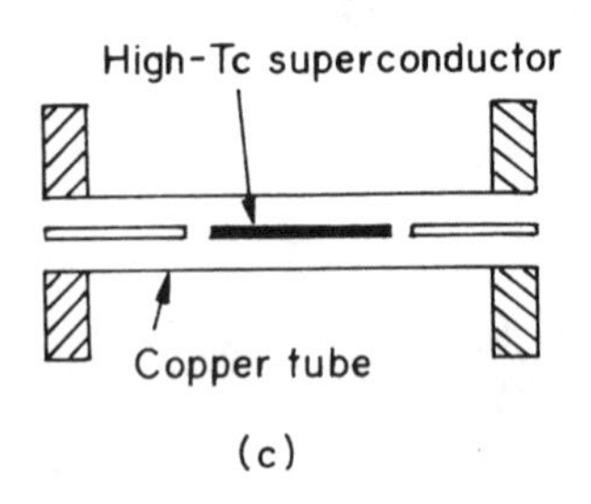

FIGURE 10.1 Schematic diagram of (a) a high-T_c cylindrical cavity resonator (adapted from Ref. 23), (b) a copper cylindrical cavity with a high-T_c end plate (adapted from Ref. 24), and (c) a high-T_c coaxial cavity resonator (adapted from Ref. 28).

placement of the full end plate. In such a case, the cavity end-plate substitution technique is preferred. This technique is also useful for a small-size sample even at high-frequency measurements. In this technique, a small high-T_c sample is placed in the center of the end plate, maintaining the circular symmetry of the cavity. The cavity perturbation technique is used for measuring the surface resistance of a smaller-size high-T_c samples, such as single-crystal or thin films (25–27). A superconductor niobium high-Q cavity is used in this measurement. The sample is mounted on a sapphire rod and placed in the cavity at a maximum magnetic field location. During the measurement, the whole cavity remains at 4.2 K and the sample temperature can be varied. The sensitivity of the measurement is high due to the high Q value of the superconductor cavity.

Figure 10.1c is a schematic of the coaxial cavity arrangement for surface resistance measurement. Coaxial cavity method has been used for measuring the surface resistance of a high-T_c wire and a rod-shaped bulk sample (28,29). Using different TM (Transverse Magnetic) and TE (Transverse Electric) modes, it was possible to estimate the directional dependence of the surface resistance (29,30). A dielectric resonator technique is very popular for measuring the surface resistance of high-T_c films (31–33). In the parallel-plate arrangement, a low-loss, high-dielectric-constant crystal is sandwiched between the two high-T_c films. The high dielectric constant of the crystal causes most of the electromagnetic energy to be confined within the crystal. This technique provides a high sensitivity for surface resistance measurement due to its very high Q value.

The surface resistance measurement of thin films by the cavity technique or dielectric resonator technique gives a value averaged over the entire surface. For the fabrication of thin-film microwave devices, different structures are patterned. The surface resistance of the patterned thin films has been measured after fabricating microstrip line, coplanar, or stripline structure (33–38).

Figure 10.2 shows variation of the surface resistance with the microwave frequency for YBCO bulk, thick, and thin films (39). The variation of surface resistance of copper is also shown in Fig. 10.2 for comparison. The surface resistance of YBCO is lower than the surface resistance of copper for the frequency $f \leq f^*$, where f^* is the crossover frequency. The value of f^* is highest for the YBCO thin film. The surface resistance of the YBCO thin film is minimum in comparison to YBCO bulk and thick films.

Microwave measurements of high-T_c superconductors have been carried out to understand the nature of the symmetry of the order parameter. Section 1.5.8 of Chapter 1 gives a detailed account of these measurements. The presence of s- or d-wave symmetry of the order parameter of a high-T_c superconductor will have affect on the microwave characteristics. For d-wave symmetry, the ultimate achievable value of R_s will be higher and this will affect the ultimate achievable Q and frequency stability of the high-T_c superconductor resonator. The design of the filter will also be more complex due to the constraint of d-wave symmetry.

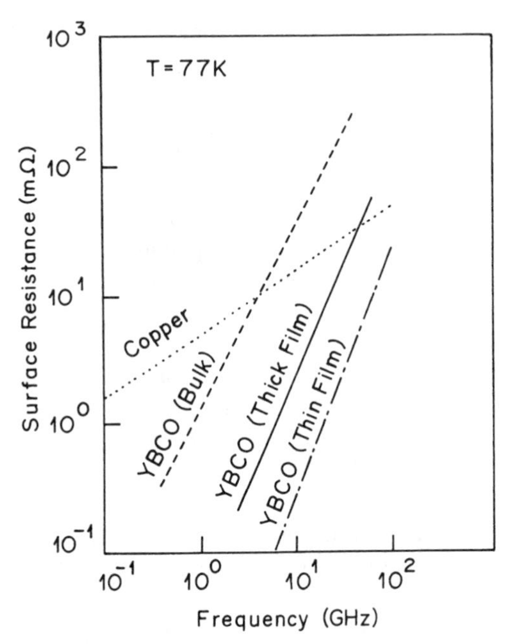

FIGURE 10.2 Frequency dependence of the surface resistance of YBCO bulk, thick film, thin film, and copper at 77 K. (Adapted from Ref. 39, © 1996 IEEE)

10.4 RESONATORS

Resonators have been fabricated in a three-dimensional (3D) structure (cavity, dielectric resonator) and in a planar structure. Resonators are specified by a fundamental characteristic, the quality factor, Q, which is defined as

$$Q = \frac{f_0}{\Delta f} \tag{13}$$

where f_0 is the resonance frequency and Δf is the 3-dB frequency bandwidth of the resonator response.

Prior to the discovery of high-T_c superconductors, planar structure resonators based on conventional metals had limited use due to its low Q value. The low value of the surface resistance of high-T_c superconductors has made it possible to realize high-Q planar resonators. Table 10.3 gives the characteristics of some of the high-T_c resonators.

10.4.1 Cavity Resonators

Cavities with high Q are required for a number of applications such as elements of filters or as frequency standards. The highest-Q resonators made with conventional

TABLE 10.3 Characteristics of High-T_c Resonators

Resonator geometry	High-T_c superconductor	Substrate	Frequency	Q	Ref.
Cylindrical cavity	YBCO bulk	—	13 Ghz	1×10^4 [77 K]	23
	YBCO thick film	YSZ (poly crystalline	5.66 GHz	7.16×10^5 [77 K]	40
Coaxial cavity	YBCO bulk	—	6–12 GHz	10^3–10^4	30
Helical resonator	YBCO bulk	—	420 MHz	9×10^3 [77 K]	41
	YBCO thick film	Zirconia	250 MHz	7×10^2 [77 K]	41
	YBCO bulk	—	355 MHz	1.6×10^4 [77 K]	42
Dielectric resonator ($LaAlO_3$)	YBCO film	$LaAlO_3$	5.6 GHz	10^5–10^6 [63 K]	43
Dielectric resonator (alumina)	YBCO film	YSZ (poly crystalline	10 GHz	1×10^5 [77 K]	44
Dielectric resonator (sapphire)	TBCCO film	$LaAlO_3$	5.55 GHz	3×10^6 [80 K]	45
Dielectric resonator ($LaAlO_3$)	YBCO film	$LaAlO_3$ (cylinder)	6.3 GHz	1.2×10^4 [77 K]	46
Microstrip resonator	YBCO film	$LaAlO_3$	5 GHz	1.37×10^4 [77 K]	47
	TBCCO film	$LaAlO_3$	33 GHz	2.74×10^3 [77 K]	48
Microstrip resonator (with Nb shield)	YBCO film	MgO	5.7 GHz	1.1×10^5 [4.2 K]	49
Microstrip resonator	YBCO film	MgO/sapphire	10.2 GHz	2×10^3 [77 K]	50
	YBCO film	CeO_2/sapphire	10.2 GHz	6×10^2 [77 K]	50
Microstrip disk resonator	TBCCO	$LaAlO_3$	4.7 GHz	3×10^4 [60 K]	51
Coplanar resonator	YBCO	MgO	2.36 GHz	4.5×10^4 [12 K]	38
Ring resonator	YBCO film	$LaAlO_3$	35 GHz	3.5×10^3 [20 K]	52
Stripline resonator	YBCO film	$LaAlO_3$	1.5 GHz	2.5×10^4 [77 K]	53
"T" resonator	YBCO film	YSZ/silicon	3.8 GHz	2×10^4 [50 K]	54
Disk resonator	TBCCO	$LaAlO_3$	1 GHz	5×10^5 [77 K]	55

metallic superconductors consist of bulk metallic cylindrical cavities. Similar types of construction have been tried using bulk and thick film of high-T_c superconductors (23,39,40). For a YBCO thick-film cavity resonator operating at 5.66 GHz in TE_{011} mode, the Q value of 715,688 at 77 K has been demonstrated (40). Measured results of cavities operating at 10 and 7.5 GHz have also been reported (39). The size of TE_{011} cavities will become too bulky if it is to be used at frequencies below 1 GHz. At lower frequencies, resonator structures such as coaxial cavity (28–30), helical cavity (41,42) and split resonators (56) have been developed.

A coaxial cavity resonator consists of an outer copper tube and a superconductor wire at the center (Fig. 10.1c). The length of the cavity usually corresponds to an integral number of half-wavelengths and this decides the frequency limit for the operation. The coaxial cavity has been operated in the frequency range 1–20 GHz (28). The unloaded Q of the operating frequency range 6–12 GHz was found to be 10^3–10^4 (30). Thick-film coaxial resonators have been found to have limited use because of the requirement of coating all the surfaces of the cavity for achieving a reasonably high Q value. Figure 10.3 shows a helical cavity resonator consisting of a helix-shaped superconducting wire placed inside a cylindrical cavity. The length of the central cable is large in this case as compared to the coaxial cavity resonator. Thus, it can be operated at a further lower frequency. A number of high-T_c helical resonators have been built and tested (41,42). These have been made from either thick-film or bulk polycrystalline material. A YBCO wire-based helical resonator has been fabricated that showed a Q of 16,000 at 77 K and an operating frequency of 355 MHz (42).

10.4.2 Dielectric–High-T_c Film Resonators

Figure 10.4a is a schematic of a parallel-plate dielectric resonator. It consists of a low-loss dielectric cylinder with superconducting plates placed on the top and bot-

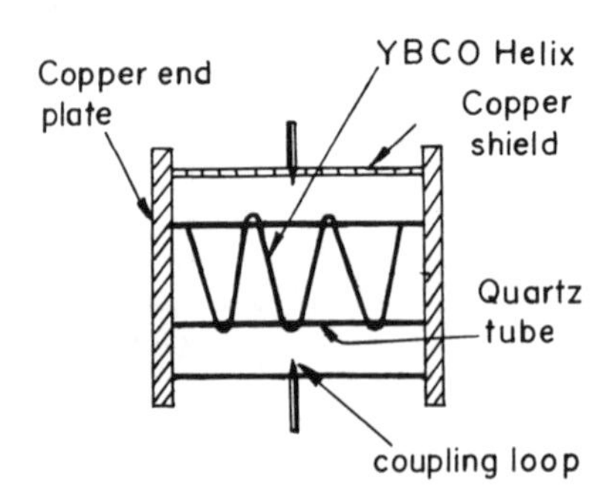

FIGURE 10.3 Schematic diagram of a high-T_c helical resonator. (Adapted from Ref. 41.)

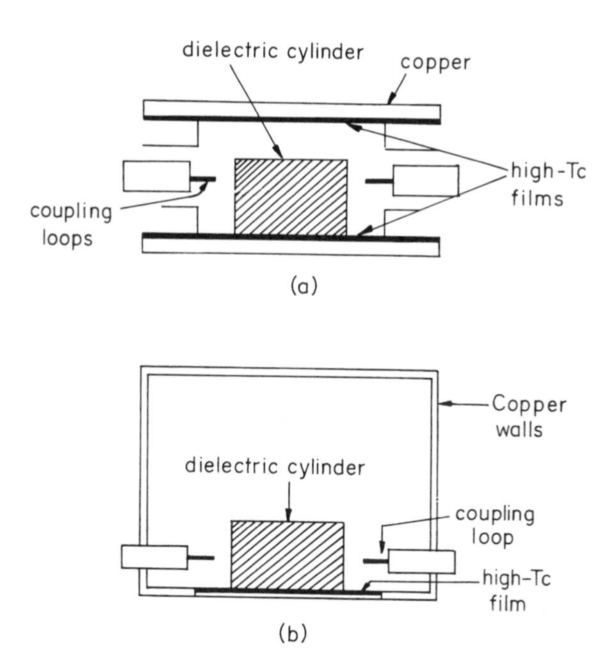

FIGURE 10.4 Schematic diagram of (a) a parallel-plate dielectric resonator and (b) a single-plate shielded dielectric resonator.

tom. Sapphire, $LaA1O_3$, and rutile exhibit sufficiently low loss and are usually used for this type of resonator. The structure is held together with copper beryllium springs and the microwave coupling is through the side walls, with a coaxial cable having loop antenna termination. Figure 10.4b is a schematic of an open-ended dielectric resonator. In this case, only one high-T_c superconducting plate is used and the dielectric cylinder is placed on the superconducting plate. The walls of the cavity is made of normal metal and are at sufficiently large distance from the dielectric so as not to affect the fields appreciably. The open-ended cavity has the advantage that only one superconducting film is required. The value of Q can also be higher because there is no loss from the second film. However, there can be problem of mechanical stability, and due to its large size, it would be inconvenient to use it in all of the applications. On the other hand, the parallel-plate resonator has a smaller size and it is much less susceptible to vibration noise because of its more rigid construction.

High-T_c film dielectric resonators have been demonstrated to have very high Q value at 77K (45,57,58). The values for unloaded Q of 3×10^6 at 80 K and 1.4×10^7 at 4.2 K and 5.55 GHz have been demonstrated using TBCCO films and a sapphire resonator (45). For a YBCO film–$LaA1O_3$ resonator, the values of Q

have been obtained as 4.5×10^5 and 1.3×10^5 at 10 K and 77 K, respectively, for the operating frequency of 11.6 GHz (59). Changing the distance of the upper superconducting plate from the dielectric cylinder can modify the resonance frequency of the parallel-plate resonator (59). Characteristic of some other high-T_c film, dielectric resonators (43–46) are described in Table 10.3. High-T_c film dielectric resonators have been demonstrated to have potentiality for applications in low-phase noise oscillators, in high-power filters, and for the frequency standard (43,58,60).

10.4.3 Planar Resonators

Figure 10.5 is a schematic of the main geometries of planar waveguides: stripline, microstrip, and coplanar. High-T_c superconducting planar resonators have been fabricated utilizing any of these geometries. The advantage of using these planar resonators over the three-dimensional cavity resonator is their smaller size and their ability to integrate into conventional microwave circuitry.

Figure 10.6 is a schematic of microstrip, coplanar, ring, and stripline high-T_c thin film resonators. In a microstrip resonator (Fig. 10.6a), a microstrip pattern is created on one of the high-T_c film, whereas the high-T_c film on the other substrate is used as a ground plane (47,48). A YBCO microstrip resonator operating at 5 GHz has been fabricated using TBCCO film as a ground plane which showed a Q value of 16,000 at 77 K (47). In the configuration of the microstrip resonator shown in Fig. 10.6a, use of a separate film for the ground plane leads to problems such as an air gap existing between the two substrates causing uncertainty of the resonance frequency and the complex fringing effect. By depositing high-T_c films on both sides of the substrate and then patterning a microstrip structure on the one side, a more compact microstrip resonator can be fabricated, as shown in Fig. 10.6b. A value of the Q for the double-sided YBCO film microstrip resonator is

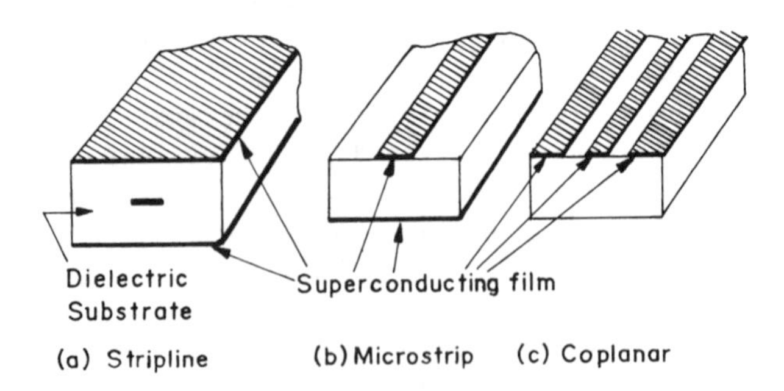

FIGURE 10.5 Schematic of planar waveguide: (a) stripline; (b) microstrip; (c) coplanar structures.

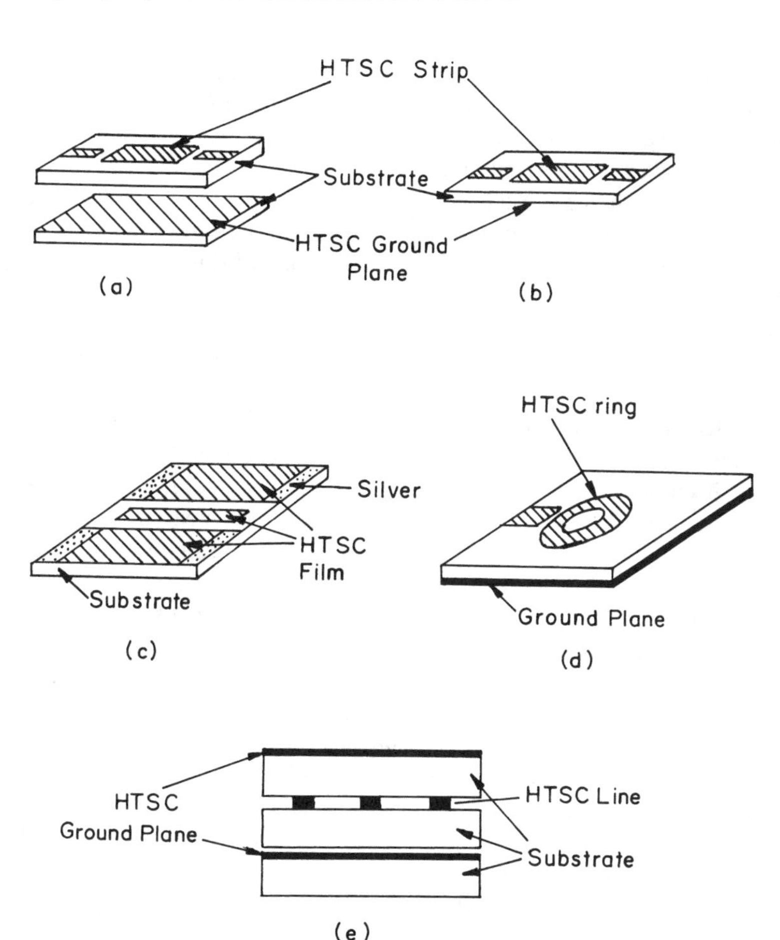

FIGURE 10.6 Schematic of (a) a planar high-T_c microstrip resonator with a separate high-T_c film for the ground plane, (b) a microstrip resonator fabricated from double-side deposited high-T_c film on the substrate, (c) a coplanar resonator (d) a ring resonator, and (e) a stripline resonator.

found to be 2.6×10^4 at 77 K and 5.7 GHz (49). It has also been found that use of a Nb superconducting shield enhanced Q to a further high value of 1.1×10^5 at 4.2 K. YBCO thin-film planar resonators on silicon and sapphire are also reported using buffer layers (50,54).

A coplanar resonator (Fig. 10.6c) consists of two ground planes and a central strip of superconductor deposited on the substrate (38). The central strip is an open circuit at each end and forms a resonant section of the coplanar transmission line. By depositing good quality YBCO film on the MgO substrate, a coplanar resonator is fabricated which shows a value of $Q = 45{,}000$ at 2.36 GHz and 12 K. A microstrip ring resonator (Fig. 10.6d) with YBCO film on a $LaAlO_3$ substrate and gold ground plane has been fabricated and tested (52). The resonator is operated at 35 GHz and showed six to seven times higher Q as compared to the identical gold resonator at 20 K. In the stripline resonator (Fig. 10.6e), high-T_c films on one substrate is patterned (central conductor) and high-T_c films on two separate substrates are used as ground plane. A value of $Q = 25{,}000$ at 77 K is reported for the YBCO film stripline resonator (53).

The microstrip resonator has a simple structure, but it requires a two-sided deposition of high-T_c thin films on the substrate or a separate film for the ground plane. On the other hand, for the fabrication of the coplanar resonator, a high-T_c film is deposited only on one side of the substrate. The stripline resonator is more complex. However, it has the advantage that the propagation mode is pure TEM (Transverse Electromagnetic) and is, therefore, dispersion free. In the stripline resonator, the field is totally enclosed within the dielectric, whereas in the case of the microstrip resonator, some of the field is external to the substrate.

The above-discussed microstrip resonator works well for low microwave input power. Due to a high Q, a large current circulates on the edges of the strips. Planar disk resonators based on high-T_c thin films that show reasonably high-power handling capability are being developed and tested (51,55). Large-area substrates coated on both sides of high-T_c films are used for fabricating this type of resonator. Different planar shapes such as octagonal, ellipse, and so forth are fabricated. Planar disk resonators operating in the TM_{010} mode show very good power handling capability. In the TM mode, only radial current flows, so that the current crowding problem is largely avoided. An extremely high-quality factor ($Q \approx 5 \times 10^5$ at 77 K) as well as excellent power handling and intermodulation characteristics are obtained for these resonators (55).

10.5 FILTERS

Filters are important components of microwave circuits. In addition to their evident application of providing frequency selectivity, they are necessary for matching networks, which form an integral part of multiplexers and amplifiers. Resonators are the most common basic elements used for making filters. The main requirements for the resonators that can be used for filters are a high quality factor as well as high power handling capability. In the design of planar filters, resonators can consist of distributed elements or the lumped elements. The circuits of distributed elements take up more space compared to the lumped element. However, the Q value of the distributed-element circuit is larger than the lumped elements.

In recent years, there has been considerable development in high-T_c superconductor-based filters. The motivation has been to develop higher-order high-T_c

filters that show very low insertion loss, high power handling capability, and miniaturization. Different types of high-T_c filter have been fabricated and tested. These include 3D structure using dielectric resonators and 2D planar structure fabricated using high-T_c thin films. The performance of some high-T_c superconductor filters is summarized in Table 10.4.

TABLE 10.4 Performance of High-T_c Superconducting Filters

Filter design	High-T_c superconductor	Substrate	Center frequency	Bandwidth (%)	Insertion loss	Ref.
Dielectric resonator based two-pole filter	YBCO	$LaAlO_3$	6.3 GHz	0.95	0.26 dB [77 K]	46
Microstrip two-pole bandpass filter	YBCO	$LaAlO_3$	11.2 GHz	5	1.5 dB	61
Microstrip three-pole bandpass filter	YBCO	$LaAlO_3$	10.5 GHz	5	0.304 dB [77 K]	62
Microstrip five-pole bandpass filter	YBCO	$LaAlO_3$	2 GHz	1.2	0.35 dB [77 K]	63
Eight-pole bandpass filter using meander-shaped open-loop resonator	YBCO	$LaAlO_3$	1.777 GHz	0.84	–	64
Cross-coupled eight-pole microstrip line bandpass filter	YBCO	$LaAlO_3$	1.949 GHz	1.04	–	65
	YBCO	$LaAlO_3$	1.950 GHz	1.02	–	65
Microstrip nine-pole bandpass filter	YBCO	$LaAlO_3$	900 MHz	2.8	0.27 dB [77 K]	63
Ten-pole bandpass filter	YBCO	CeO_2/ sapphire	1.92 GHz	0.4	–	66
Eleven-pole bandpass filter	YBCO	$LaAlO_3$	1.778 GHz	0.6	0.6 dB [65 K]	67
Cross-coupled bandpass filter	YBCO	MgO	5 GHz	4	0.3 dB [77 K]	4
Bandpass filter using "H"-type resonator	YBCO	MgO	16.44 GHz	6.86	0.4 dB [77 K]	68
Coplanar seven-pole band stop filter	YBCO	$LaAlO_3$	1.53 GHz	94.7	1.2 dB [20 K]	11
Microstrip band stop filter	YBCO	$LaAlO_3$	1.623 GHz	0.32	1.0 dB [63 K]	12
Three-pole bandpass filter with planar octagonal resonator	TBCCO	$LaAlO_3$	6.04 GHz	1.3	–	69
Lumped-element bandpass filter	YBCO	MgO	1.777 GHz	0.84	–	70
Interdigital microstrip bandpass filter	TBCCO	MgO	2.4 Ghz	4	0.4 dB [77 K]	71
Three-pole lumped-element bandpass filter	YBCO	$LaAlO_3$	2 GHz	2	0.1 dB [77 K]	72
Three-pole lumped-element bandpass filter	YBCO	$LaAlO_3$	15 MHz	0.1	1 dB [13 K]	73

10.5.1 3D Filters

A high-T_c cavity or dielectric resonator can be used as a narrow-band filter. A resonator passes a range of frequencies within a bandwidth Δf_0 around the resonance frequency f_0, where $\Delta f_0 = f_0/Q$ and Q is the loaded quality factor of the resonator. For a high-T_c dielectric resonator, the Q value is very high, thus extremely narrow-bandpass filters are possible. In order to broaden the band, a number of resonators with slightly different resonance frequencies need to be coupled. The loss tangent of the dielectric and the surface resistance of the high-T_c material mainly determine the insertion loss of these filters. The loss tangent of dielectric material improves at lower temperature, which helps to improve the insertion loss of the filter. Dielectric resonator–high-T_c film filters have been shown to handle extremely high power levels in the range 50–100 W (74). Using $LaAlO_3$ single-crystalline cylinders covered on the end faces by YBCO films, two-pole bandpass filters have been fabricated (46). At 77 K, the resonators in the filter showed a Q of 12,000 and the filter showed an insertion loss of 0.26 dB at the operating frequency of 6.3 GHz. An eight-pole dielectric resonator–high-T_c film bandpass filter has also been demonstrated (75). The central frequency of the filter was 3.8 GHz, with a bandwidth of 1%. The size of this high-T_c bandpass filter is much smaller than that of the conventional dielectric resonator filter. The filter has been incorporated successfully into a four-channel input multiplexer (74).

The main advantage in this dielectric–high-T_c film filter is that no etching of the film is required and the filter can be easily tuned using conventional tuning screws. It is also possible that by using a low-loss dielectric with a high dielectric constant, the size of the filter can be made smaller. The major drawback of this type of filter is the mechanical design complexity. The filter has to be thermally stable to ensure performance repeatability as the temperature changes from cryogenic temperature to room temperature and then back to cryogenic temperature.

10.5.2 Planar Filters

The emergence of high-T_c technology has created the opportunity for much more innovation in planar filter configurations. Several novel configurations have been proposed which allow the realization of filters with quite advanced functions. The simplest form of a planar high-T_c superconducting filter is based on a set of parallel-coupled transmission line resonators. The length of these lines is approximately half a wavelength long. In some cases, the linear resonance elements are folded back on themselves into a hairpinlike geometry for minimizing the surface area of the substrate. Figure 10.7 shows schematics of layout of a parallel coupled microstrip filter and a hairpin type of filter. The design of the hairpin type of filter is more compact. Filters based on a meander-shaped open-loop resonator (5), a "H"-type of resonator (68), and a "J"-type of resonator (76) have also been fabricated which lead to more compact high-T_c filters. High-T_c filters up to 32 poles have been

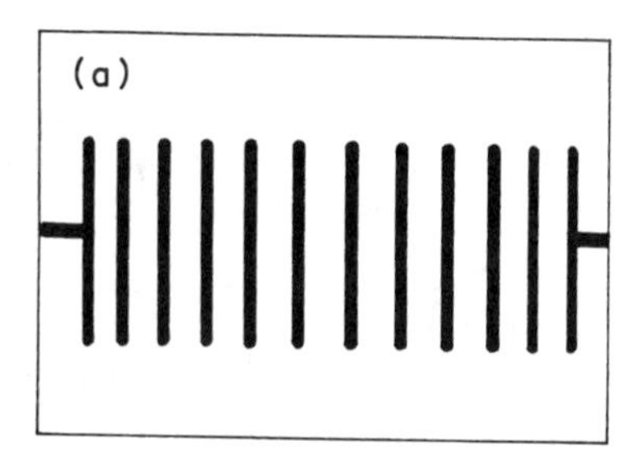

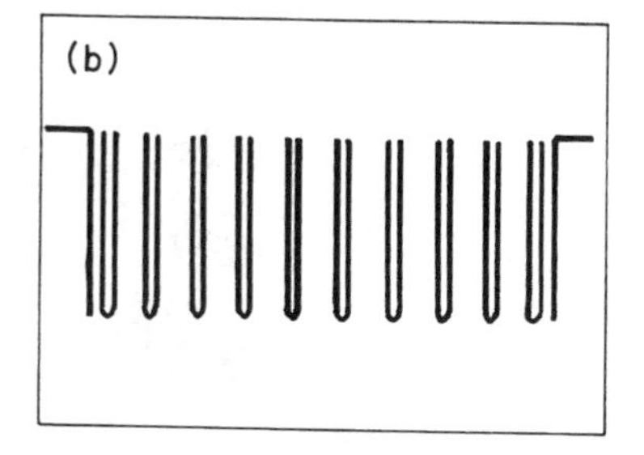

FIGURE 10.7 Layout of (a) a forward coupled microstrip filter and (b) a 10-pole hairpin filter.

demonstrated (76). The performance of a high-T_c microstrip filter with elliptical and quasielliptical functions have also been examined (4,64,77,78). It is known that elliptic filters can exhibit a sharp skirt property as compared to that of a Chebyshev filter, even with a small number of resonators. A 10-pole elliptic-function self-equalized planar filter has been developed (78). The basic building block of this filter is a dual-mode lumped-element resonator that makes it easy to create elliptic and self-equalization functions. High-T_c preselect receive filters have been designed using YBCO film to have a quasielliptic function response in order to provide low insertion loss and a very steep rolloff at the filter band edges (64).

Figure 10.8a shows layout of an eight-pole microstrip bandpass filter using a meander-shaped open-loop resonator. The filter has been fabricated on YBCO film deposited on both sides of a MgO substrate. Figure 10.8b shows the frequency response of the filter at 55 K (64). The filter shows the characteristics of the quasielliptical response with two diminishing transmission zeros near the pass band edges, which improves the selectivity of the filter. The filter has been integrated with a low-noise amplifier (LNA) into an RF module for encapsulation (64). High-band stop filters have been developed for radioastronomy applications (11,12).

The size of the filter can be considerably reduced if the filters are based on lumped elements. The very low loss of high-T_c materials enabled lumped-element

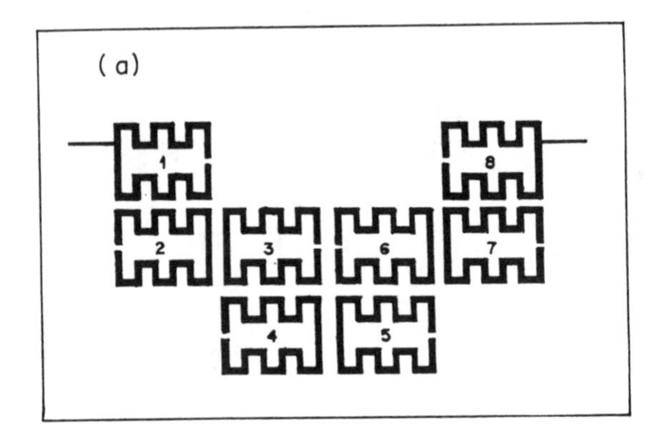

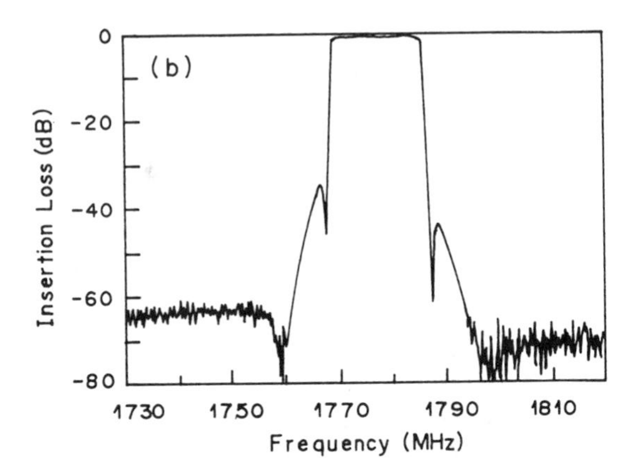

FIGURE 10.8 (a) Layout of an eight-pole bandpass filter using meander-shaped open-loop resonators; (b) frequency response of the band pass high-T_c filter shown in (a) at 55 K. (Adapted from Ref. 64, © 2000 IEEE)

filters to be designed at a microwave frequency in the form of planar structures (70–73,79,80). Highly miniaturized third-order high-T_c superconducting microwave filters have been fabricated using lumped elements (70). The filter uses very compact resonators consisting of interdigital capacitors and meander line inductors. Use of lumped-element circuits has also made it possible to design a planar high-T_c filter operating at low frequency. A three-pole bandpass filter operating at 15 MHz has been fabricated using lumped elements on YBCO films (73). This filter has a narrow bandwidth of 15 kHz with an insertion loss of less than 1 dB at 13K. A lumped-element switched band stop filter at 5.8 GHz with 36 dB isolation at 80 K for switching current of 25 mA has been demonstrated. The switching time of the device was observed to be less than 2 μs (81).

Several planar high-T_c filters show a low value of insertion loss and high out-of-band rejection only at low microwave power. In order to use high-T_c filters for telecommunication, the high-power handling capability of the device is also required apart from the low insertion loss and sharp skirts. At higher power, current crowding at the patterned film edge and associated nonlinear effect occurs. This leads to the generation of signal harmonics and intermodulation distortions. Several attempts have been made to improve the power handling capability of the microwave devices. Parallel-coupled feed lines have been used to avoid the current crowding in a narrow-band microstrip filter at 2 GHz (7). The filter showed an insertion loss of less than 0.2 dB at 45 K and showed no degradation in the characteristics up to 20 W of microwave input power. High-T_c filters using several planar shapes such as square, octagonal and elliptical resonator structures have been fabricated (69,82) for avoiding the current crowding effect. As a result, improvement in the power handling capability of the device has been noted. A TBCCO planar three-pole filter with a central frequency of 6.04 GHz and 1.3% bandwidth showed no measurable degradation for the transmitted power level up to 74 W at 77 K (69). For a two-pole high-T_c planar filter using this design, no degradation was observed even up to 115 W at 77 K. One of the problems with the high-T_c filters based on patch resonator is that the size is considerably larger as compared to the conventional high-T_c filters. In the design of split resonators the peak current density at the outer edge can be reduced also, so that the overall power handling capability of the device can be improved. A high-T_c thin-film filter employing the split resonator concept is demonstrated to show good power handling capability (74).

10.6 ANTENNA

The antenna plays an important role in transmitting and receiving signals through free space. Performance of any antenna system is governed by the antenna ohmic loss resistance R_l, the radiation resistance R_t, and the antenna reactance X_r. For a short antenna, the dipole length is much less than the radiation wavelength λ. The radiation efficiency η of a short antenna is given by

$$\eta = \frac{R_t}{R_t + R_l} \tag{14}$$

In many applications, R_t is much large than R_l and so the choice of antenna material has little effect on the overall efficiency. However, in applications such as superdirective antenna arrays or small antennas, the choice of material would affect the efficiency. For such antennas, use of high-T_c superconductor material in place of a normal conductor leads to an increase of the efficiency due to the smaller value of R_l (83). Loss in the matching network, which connects the active device to the antenna, is particularly important for the short-dipole antenna and this may be significantly reduced by the use of high-T_c superconductors with low surface resistance.

The first high-T_c superconducting antenna was demonstrated by the Birmingham Group (U.K.) in 1988 (84). This consisted of a short dipole (20 mm) made out of bulk polycrystalline YBCO material, as shown in Fig. 10.9a. The antenna was operated at 550 MHz and enhancement in the gain was 12 dB compared to the copper antenna operating at room temperature. Since then, several other groups have successfully fabricated different types of high-T_c antenna such as the loop antenna (85,88), the helical antenna (89,90), the patch or microstrip antenna (86,87,91,92), and the meander line antenna (93). Figure 10.9 is a schematic of the high-T_c superconductor loop antenna, "H"-type patch, and triangular patch antenna, along with the dipole antenna. The loop antenna is made from YBCO thick

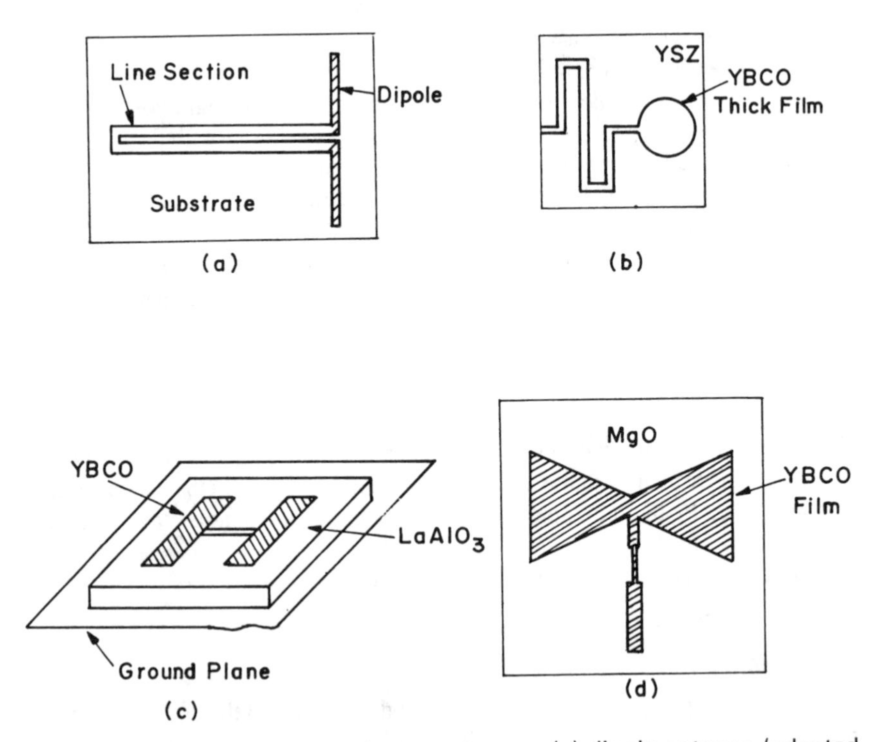

FIGURE 10.9 High-T_c superconductor antennas: (a) dipole antenna (adapted from Ref. 84); (b) loop antenna (adapted from Ref. 85); (c) "H" microstrip antenna (adapted from Ref. 86); (d) triangular patch antenna (adapted from Ref. 87).

film printed on a YSZ (yttria stabilized zirconia) substrate (85). This loop antenna operates at 410 MHz with a 1.4% bandwidth and a radiation efficiency of 80%. One of the most widely quoted superconducting patch antenna is the "H" antenna. It consists of three sections of a microstrip transmission line. The two wide sections behave as capacitors and the thin narrow section forms an inductive section of the transmission line. The "H"-type antenna has been fabricated from epitaxially grown YBCO film on a $LaAlO_3$ substrate (86). The antenna operates at 2.45 GHz and has a bandwidth of about 3 MHz. The efficiency of this YBCO antenna was ~55–65%, whereas for a copper antenna of similar type, the efficiency was only 1–6%. There is also considerable miniaturization. For a 2.4-GHz antenna, the conventional patch would have a length of 42 mm on RT-duroid ($\epsilon_r = 2.2$), whereas the YBCO "H" antenna on a $LaAlO_3$ substrate ($\epsilon_r = 24$) is 6 mm long.

The high value of Q and small substrate height in the high-T_c superconducting antenna results in a narrow bandwidth between 0.85% and 1.1%. An increase of substrate height can broaden the operation bandwidth of the antenna. However, it is not possible to increase the thickness beyond a critical value. A meander line antenna has been fabricated in order to increase the bandwidth (93). A bandwidth of 4% and radiation efficiency of 50% is achieved for a meander line high-T_c antenna. In another report, a high-T_c antenna consisting of two triangular patches has been fabricated using a YBCO film on a MgO substrate (87). The antenna operates around 20 GHz and has a 6.7% bandwidth. The high-T_c superconductor four-element patch array antenna has been developed for satellite communication that showed a 3.6-dB higher gain compared to the copper array (91). Superdirective array excitation produces a directive gain that is substantial larger than a conventional uniform or tapered excitation. Superdirective antenna arrays have been fabricated using high-T_c superconductors (89,94,95). For a 16-element YBCO high-T_c superdirective array, a supergain of 4.2 has been achieved (94). The whole array was fitted on a 100×100 mm^2 YSZ substrate, representing 0.38λ at the resonant frequency of 1.14 GHz and has a maximum gain at 45° to the axis of the antenna.

The use of high-T_c superconductors in the conventional antenna array can also have a definite advantage. As the array becomes large, its feed network becomes more complicated. This leads to a long, narrow transmission line, which becomes increasingly lossy. Due to very small value of the surface resistance of high-T_c films, the development of very large arrays is possible (96–98). Antennas of the 4-element array (96) and 16-element array (97,98) have been constructed to test the technical feasibility.

10.7 DELAY LINES

Delay lines are useful devices in electronic warfare (EW) systems, satellite communication systems, and in realizing transversal filters for analog signal processing applications. In EW systems, signals are required to be delayed for a short time while the processing of the earlier input occurs. Similarly, in satellite communication systems, signals are required to be delayed while switching takes place. Typical delays needed for these applications are in the range of 100–300 ns. Current technologies for delay lines include surface acoustic wave (SAW) devices and transmission coaxial cables in the form of a coil. Whereas the use of SAW devices are limited to low-frequency applications, the coaxial cable delay lines are bulky and have a high insertion loss. Superconducting delay lines offer the highest bandwidth with much lower insertion loss in a highly compact design. Several high-T_c planar transmission lines such as stripline, microstrip, and coplanar lines could be used to realize delay lines. Using YBCO thin film deposited on both sides of a 50-mm-diameter, 25-μm-thick $LaAlO_3$ wafer, a nondispersive delay line with 20 GHz of bandwidth has been fabricated that produces a delay of 22.5 ns with an insertion loss of 5dB (99). The high dielectric constant of the $LaAlO_3$ substrate allows a significant amount of delay to be placed on a single substrate and the low loss of high-T_c film keeps insertion loss to a minimum. High-T_c devices producing a delay of 1–100 ns have been fabricated (2).

10.8 PHASE SHIFTERS

Low-loss microwave phase shifters are important devices for phase array radar and communication systems. Conventionally, a ferrite-loaded waveguide phase shifters are employed using ferrites that are large and bulky. High-T_c superconductor-based phase shifters have been fabricated which are compact in size and have low losses (100,101). A meander line structure is patterned on high-T_c film, which is kept in close proximity to magnetized ferrite for obtaining the phase shift. In order to optimize the performance of the phase shifter, invasion of magnetic flux to the high-T_c superconductor should be minimized (100,101). Large phase shifts (over 1000°/dB) have been demonstrated in the X band using magnetized ferrite in proximity with a meander-shaped YBCO circuit (101).

10.9 TUNABLE RESONATORS AND FILTERS

There are many microwave applications in which the performance of resonators and filters needs to be tuned in situ. Tunable high-T_c resonators and filters have been fabricated by incorporating ferroelectric and ferrite-based thin films with high-T_c planar resonators and filters. A magnetically tunable YBCO resonator on a $Y_3Fe_5O_{12}$ (YIG) substrate has been fabricated that showed tuning of 500 MHz for a 500-Oe magnetic field without appreciable degradation of the Q of the resonator (102). Fabrication of a seven-pole microstrip filter with a central frequency of 14 GHz and a bandwidth of 4% is also reported whose frequency can be shifted up to 12 MHz for an applied magnetic field of 0.1 mT at 40 K (103). The frequency shift is attributed to the presence of vortices modifying the microwave penetration depth and, therefore, the kinetic inductance of the strip. A magnetically tunable YBCO three-pole bandpass filter with a 1% bandwidth and a 1-dB insertion loss at 10 GHz has been also reported in which 13% tuning has been achieved (104). Electrically tunable resonators and filters have been fabricated using a $SrTiO_3$ (STO) layer (105–107). Tunability in these devices is achieved through the nonlinear dc electrical field dependence of the relative dielectric constant of STO. A large tunability of greater than 10% is reported for the YBCO/STO/$LaAlO_3$ microstrip bandpass filter with a central frequency of 19 GHz at 77 K (107). Fabrication of a three-pole bandpass filter using YBCO/1.5-μm layer of STO with <2% bandwidth at 2.5 GHz has been reported (108) in which the pass band could be tuned by more than 15% by an external bias voltage at 77 K. Although the insertion loss in these electrically tunable filters are large due to the high loss tangent of STO, the extent and the ease of tunability is a very significant development. An electromechanical approach of active control of tuning of high-T_c resonators and filters is also reported (109).

10.10 HIGH-T_c MICROWAVE SUBSYSTEMS

High-T_c superconductor microwave subsystems are expected to offer the advantage of minimization of mass and volume by substitution of bulky and heavy waveguide components by extremely compact planar devices and minimization of losses. The major challenge in the development of high-T_c superconductor-based subsystems are not just high-T_c components alone but also the associated cryopackaging and cryocooler integration of the subsystems. In recent years, there has been much interest in developing high-T_c superconductor-based subsystems for wireless and communication satellite applications (3,6,8–10,110–112). Several research projects are in progress for the development of high-T_c microwave components and subsystems for use in a base transceiver station of digital mobile communication systems of the second generation (GSM, DAMPS/IS-136, PDC, etc.) and for the forthcoming third-generation (W-CDMA, CDMA 2000). Narrow-band and compact high-T_c receiver bandpass filters have been developed for mobile communication applications (5,63,64,110,111). The low loss and high selectivity of the planar high-T_c multipole filters can induce an enhancement of the base station sensitivity and capacity.

A high-T_c superconductor duplexer has been developed for cellular base station applications (5,112). The component of the duplexer have been fabricated using double-sided YBCO thin film on a $LaAlO_3$ substrate and this is comprised of a pair of high-T_c–3-dB tandem couplers and a pair of band stop filters. Figure 10.10 shows the configuration of the high-T_c duplexer. The measured insertion loss was less than 0.3 dB, both from the antenna to receiver ports over a receive band of 1770–1785 MHz and from the transmitter to antenna ports over a trans-

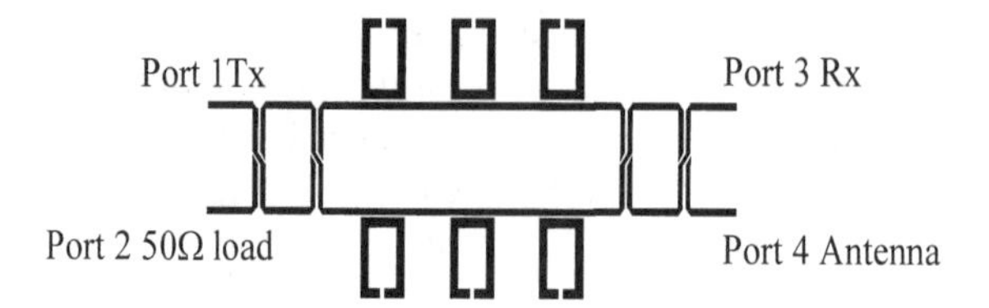

FIGURE 10.10 Schematic of the configuration of a high-T_c duplexer. (Adapted from Ref. 5, © 1999 IEEE)

mit band of 1805–1880 MHz. The isolation between the transmitter and receiver was measured to be greater than 35 dB.

An integrated receiver front-end unit with a high-T_c noise reduction filter and low-noise amplifier (LNA) has been developed by several research groups (9,64,111). It utilizes high-Q, high-T_c superconducting filters, a low-noise cryogenic amplifier, and a highly reliable cooler. Figure 10.11a shows a three-channel high-T_c bandpass filter with a low-noise amplifier in an RF connector ring for en-

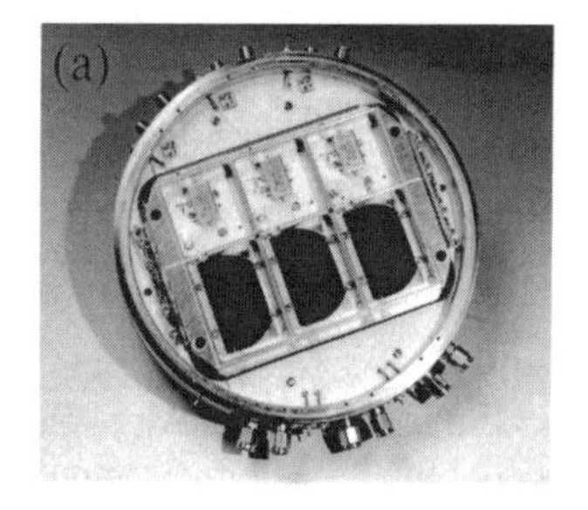

FIGURE 10.11 (a) Three-channel high-T_c filter with a low-noise amplifier assembled in an RF connector ring for encapsulation. (Adapted from Ref. 64, © 2000 IEEE) (b) An encapsulated microwave subsystem with compressor and stirling cooler (M. J. Lancaster, Birmingham University, U.K., personal communication, 2002).

capsulation and Fig. 10.11b shows the encapsulated microwave subsystems with a stirling cryocooler and compressor.

More than 800 base stations worldwide have high-T_c filter subsystems permanently installed for testing. Replacement of the cryogenic receiver front end by the conventional receiver of the base station typically results in a 2-dB improvement (9) in the receiver noise figure. The overall size of the high-T_c transceiver chains, including the associated cryocooler, is considerably less than the equivalent conventional transceiver.

A 60-channel superconducting input multiplexer integrated with a pulse tube cryocooler has been developed (78). Introduction of this high-T_c C-band multiplexer in INTELSAT-8 in place of the conventional dielectric-resonator-based input multiplexer will result in a 50% reduction in mass and a 65% reduction in size. A high-power output multiplexer has also been fabricated using hybrid dielectric resonator–high-T_c thin-film filter and planar high-T_c filter for space applications (74). The planar high-T_c filters used in this multiplexer are five-pole filters with a 1% bandwidth having a quasi elliptic function.

The high-T_c filter system has been developed for filtering of signals from satellites in a ground-based satellite communication receiver (10). This system includes two 8-MHz-wide-band high-T_c filters and LNA for the 3.7-GHz range. Use of the system allows the 2.4-m-diameter dish to be replaced with a much smaller array of two dishes having a 1-m diameter. A switch module with a switching time much less than 1 μs was also developed to allow switching between two adjacent 8-MHz bands for demonstrating the high-T_c filter's antijam capabilities (10).

For radioastronomy applications, the L-band is heavily utilized for deep-space exploration. Because the spectrum is crowded at these frequencies, suppression of spurious signals is necessary to avoid degradation of the probing frequency band of interest. Therefore, band stop filters that offer small size, low loss, and ease of integration with the receiving subsystem are highly desirable. Because the radio telescope receiver is already cooled to low temperature for reducing the noise in the semiconductor electronics, high-T_c superconductors are very attractive for this application. The development of high-T_c superconductor coplanar waveguide band stop filter (11) and microstrip band stop filter (12) has been demonstrated for this application.

10.12 CONCLUSION

There has been considerable progress in the development of high-T_c superconductor-based microwave devices. Advancement in the growth of high-T_c epitaxial films on a large-area substrate has enabled the realization of planar high-T_c microwave devices with much superior performances. It has been established that the technology based on high-T_c superconductors offers the potential of a large reduction in mass and volume of the microwave equipment. High-T_c film–dielectric resonators showed a very high quality factor and large power handling capability. A breakthrough in the high spectral purity microwave oscillators and in frequency standard may be achieved using these high-T_c resonators. Oscillators exhibiting outstanding short-term frequency stability will be utilized in radar to detect slow-moving or low observable targets. Several novel designs of filters have been proposed that led to the development of small-size, high-order, high-T_c filters with a narrow bandwidth, low insertion loss, and steep skirts. Demonstration of the feasibility of high-T_c microwave circuits using lumped elements is a very significant step toward miniaturization. Further developments in antennas, tunable microwave devices, and phase shifters are required.

The development in high-T_c microwave subsystems such as a microwave receiver front end, multiplexer, switching circuits, and so forth has been remarkable. Successful installation of a high-T_c receiver front end in several mobile base stations has demonstrated that high-T_c technology is reliable and provides a significant performance advantage for wireless telephony systems. The commercial viability of the high-T_c microwave subsystems will also depend on the reliability and cost of small cryocoolers.

REFERENCES

1. RR Mansour. Microwave superconductivity. IEEE Trans Microwave Theory Tech MTT-50:750–759, 2001.
2. MJ Lancaster. Passive Microwave Device Applications of High-Temperature Superconductors. Cambridge: Cambridge University Press, 1997.
3. BA Willemsen. HTS filter subsystems for wireless telecommunications. IEEE Trans Appl Supercond 11:60–67, 2001.
4. S Ohshima, K Ehata, T Tomiyama. High-temperature superconducting microwave passive devices, filter and antenna. IEEE Trans Electron E83-C:2–6, 2001.
5. JS Hong, MJ Lancaster, RB Greed, D Voyce, D Jedamzik, JA Holland, HJ Chaloupka, JC Mage. Thin film HTS passive microwave components for advanced communication system. IEEE Trans Appl Supercond 9:3893–3896, 1999.
6. RB Greed, DC Voyce, D Jedamzik, JS Hong, MJ Lancaster, M Reppel, HJ Chaloupka, JC Mage, B Marchilhac, R Mistry, HU Hafner, G Auger, W Rebernak. An HTS transceiver for third generation mobile communications. IEEE Trans Appl Supercond 9:4002–4005, 1999.
7. GC Liang, D Zhang, CF Shih, ME Johansson, RS Withers, AC Anderson, DE Oates. High-power high-temperature superconducting microstrip filters for cellular base-station applications. IEEE Trans Appl Supercond 5:2652–2655, 1995.
8. D Jedamzik, R Menolascino, M Pizarroso, B Salas. Evaluation of HTS sub-systems for cellular basestations. IEEE Trans Appl Supercond 9:4022–4025, 1999.
9. M Klauda, T Kasser, B Mayer, C Neumann, F Schnell, B Aminov, A Baumfalk, H Chaloupka, S Kolesov, H Piel, N Klein, S Schornstein, M Bareiss. Superconductors and cryogenics for future communication systems. IEEE Trans Microwave Theory Tech MTT-48:1227–1238, 2000.

10. ER Soares, JD Fuller, PJ Marozick, RL Alvarez. Applications of high-temperature-superconducting filters and cryo-electronics for satellite communication. IEEE Trans Microwave Theory Tech MTT-48:1190–1197, 2000.
11. S Wallage, JL Tauritz, GH Tan, P Hadley, JE Mooij. High T_c superconducting CPW bandstop filters for radio astronomy front ends. IEEE Trans Appl Supercond 7:3489–3491, 1997.
12. EM Saenz, G Subramanyam, FWV Keuls, C Chen, FA Miranda. Fixed-frequency and frequency-agile (Au, HTS) microstrip bandstop filters for L-band applications. IEEE Trans Appl Supercond 11:395–398, 2001.
13. EK Hollmann, OG Vendik, AG Zaitsev, BT Melekh. Substrates for high T_c superconductor microwave integrated circuits. Supercond Sci Technol 7:609–622, 1994.
14. J Talvacchio, RG Wagner, SH Talisa High T_c film development for electronic application, Microwave July: 105–114, 1991.
15. T Konaka, M. Sato, H Asano, S. Kubo. Relative permittivity and loss tangents of substrate materials for high T_c superconducting thin films. J Supercond 4:283–287, 1991.
16. VB Braginsky, VS Ilchenko, SK Bagdassarov. Experimental observation of fundamental microwave absorption in high quality dielectric crystals. Phys Lett 120A:300–305, 1987.
17. R Brown, V Pendrick, D Kalokitisis, BHT Chai. Low loss substrate for microwave application of high temperature superconductor film. Appl Phys Lett 57:1351–1353, 1990.
18. M Berkowski, A Pajaczkowska, P Gierlowski, SJ Lweandowski, R Sobolewski, BP Gorshunov, GV Kozlov, DB Lyudmirski, OI Sirotinsky, PA Saltykov, H Soltner, U Poppe, C Buchal, A Lubig. $CaNdAlO_4$ perovskite substrate for microwave and far-infrared applications of epitaxial high T_c superconducting thin films. Appl Phys Lett 57:632–634, 1990.
19. RW Simon, CE Platt, AE Lee, GS Lee, KP Daly, MS Wire, JA Luine, M Urbanik. Low loss substrate for epitaxial growth of high temperature superconductor. Appl Phys Lett 53:2677–2679, 1988.
20. SY Lee, JH Lee, J Lim, J Lee, JH Yun, SH Moon, B Oh. Microwave and structural properties of large $YBa_2Cu_3O_{7-\delta}$ films: a study on the homogeneity. Supercond Sci Technol 14:921–928, 2001.
21. H Schneidewind, M Manzel, G Bruchols, K Kirsch. TIBaCaCuO-(2212) thin films on lanthanum aluminate and sapphire substrates for microwave filters. Supercond Sci Technol 14:200–212, 2001.
22. TV Duzer, CW Turner. Principles of Superconductive Devices and Circuits. New York: Elsevier–North Holland, 1981.
23. MJ Lancaster, Z Wu, TSM Maclean, NM Alford. High temperature superconducting cavity for measurement of surface resistance. Cryogenics 30:1048–1050, 1990.
24. G Muller, DJ Brauer, R Eujen, M Hein, N Klein, H Piel, L Ponto, U Klein, M Peiniger. Surface impedance measurements on high Tc superconductors. IEEE Trans Magnetics MAG 25:2402–2405, 1989.
25. JC Gallop, PG Quincey, WJ Radcliffe. Surface impedance properties of in situ HTS thin films measured in a superconducting cavity. Superconductor Sci Technol 4:574–576, 1991.

26. S Sridhar, WL Kennedy. Novel technique to measure the microwave response of high T_c superconductors between 4.2 and 200 K. Rev Sci Instrum 59:531, 1988.
27. N Exon, CE Gough, A Porch, MJ Lancaster. Measurements of the surface impedance of high temperature superconducting single crystals. IEEE Trans Appl Supercond 3:1442–1445, 1993.
28. P Woodall, MJ Lancaster, TSM Maclean, CE Gough, NM Alford. Measurements of the surface resistance of $YBa_2Cu_3O_{7-\delta}$ by the use of a coaxial resonator. IEEE Trans Magnetics M-27:1264–1267, 1991.
29. JC Gallop, WJ Radcliffe, TW Button, NM Alford. Microwave surface impedance in a coaxial cavity as a material characterisation technique. IEEE Trans Magnetics MAG 27:1310–1312, 1991.
30. WJ Radcliffe, JC Gallop, CD Langham, NM Alford, TW Button. Multi-mode microwave measurements on a coaxial cavity with high-temperature superconductor centre conductor. Supercond Sci Technol 3:151–154, 1990.
31. C Wilker, ZY Shen, VX Nguyen, MS Brenner. A sapphire resonator for microwave characterisation of superconducting thin films. IEEE Trans Appl Supercond 3:1457–1460, 1993.
32. R Fletcher, J Cook. Measurement of surface impedance versus temperature using a generalised sapphire resonator technique. Rev Sci Instrum 65:2658–2666, 1994.
33. M Mishra, ND Kataria, P Pinto, M Tonouchi, GP Srivastava. Sensitivity of Rs-measurement of HTS thin films by three prime resonant technique: cavity resonator, dielectric resonator and microstrip resonator. IEEE Trans Appl Supercond 11:4128–4135, 2001.
34. R Pinto, N Goyal, SP Pai, PR Apte, LC Gupta, R Vijayaraghavan. Improved performance of Ag-doped $YBa_2Cu_3O_{7-\delta}$ thin film microstrip resonators, J Appl Phys 73:5105–5108, 1993.
35. T Yoshitake, H Tsuge, T Inui. Effects of microstructure on microwave properties in Y-Ba-Cu-O Microstrip resonators. IEEE Trans Appl Supercond 5:2571–2574, 1995.
36. HY To, GJ Valco, KB Bhasin. 10 GHz $YBa_2Cu_3O_{7-\delta}$ superconducting ring resonators on $NdGaO_3$ substrates. Supercond Sci Technol 5:421–426, 1992.
37. AN Reznik, AA Zharov, MD Chernobrovtseva. Non-linear thermal effects in the HTSC microwave stripline resonator. IEEE Trans Appl Supercond 5:2579–2582, 1995.
38. A Porch, MJ Lancaster, RG Humphreys, NG Chew. Surface impedance measurements of $YBa_2Cu_3O_7$ thin films using coplanar resonators. IEEE Trans Appl Supercond 31:1719–1722, 1993.
39. TW Button, PA Smith, G Dolman, C Megss, SK Remillard, JD Hodge, SJ Penn, NM Alford. Properties and applications of thick film high temperature superconductors. IEEE Trans Microwave Theory Tech MTT-44:1356–1360, 1996.
40. TW Button, NM Alford. High Q $YBa_2Cu_3O_x$ cavities. Appl Phys Lett 60:1378–1380, 1992.
41. A Porch, MJ Lancaster, TSM Maclean, CE Gough, NM Alford. Microwave resonators incorporating ceramic YBaCuO helices. IEEE Trans Magnetics M-27:2948–2951, 1991.
42. GE Peterson, RP Stawicki, NM Alford. Helical resonators containing high T_c ceramic superconductors. Appl Phys Lett 55:1798–1800, 1989.

43. IS Ghosh, N Tellmann, D Schemion, A Scholen, N Klein. Low phase noise microwave oscillators based on HTS shielded dielectric resonators. IEEE Trans Appl Supercond 7:3071–3074, 1997.
44. SJ Penn, NM Alford, TW Button. High Q dielectric resonators using $YBa_2Cu_3O_x$ thick films and polycrystalline dielectrics. IEEE Trans Appl Supercond 7:3500–3503, 1997.
45. ZY Shen, C Wilker, P Pang, WL Holstein, D Face, DJ Kountz. High T_c superconductor-sapphire microwave resonator with extremely high Q values up to 90 K. IEEE Trans Microwave Theory Tech MTT-40:2424–2432, 1992.
46. H Kittel, M Klauda, C Neumann, J Dutzi, YR Li, R Smithey, E Brecht, R Schneider, J Greek, J Keppler, K Klinger. Resonators for a 2 pole filter fabricated from YBCO coated $LaAlO_3$ cylinders. IEEE Trans Appl Supercond 7:2784–2787, 1997.
47. KH Young, GV Negrete, RB Hammond, A Inam, R Ramesh, DL Hart, Y Yonezawa. Clear correlations observed between $YBa_2Cu_3O_{7-\delta}$ thin-film properties and GHz microwave resonator performance. Appl Phys Lett 58:1789–1791, 1991.
48. MS Schmidt, RJ Forse, RB Hammond, MM Eddy, WL Olson. Measured performance at 77 K of superconducting microstrip resonators and filters. IEEE Trans Microwave Theory Tech MTT-39:1475–1479, 1991.
49. S Miura, T Yoshitake, H Tsuge, T Inui. Properties of shielded microstrip line resonators made from double-sided $Y_1Ba_2Cu_3O_x$ films. J Appl Phys 76:4440–4442, 1994.
50. CH Mueller, FA Miranda, S Toncich, KB Bhasin. YBCO X-band microstrip linear resonators on (1I02) and (1I00)-oriented sapphire substrates. IEEE Trans Appl Supercond 5:2559–2562, 1995.
51. AP Jenkins, KS Kale, DJ Edwards, D Dew-Huges, AP Bramley, CRM Grovenor, SV Kale. Microstrip disk resonators for filters farbicated from TBCCO thin films. IEEE Trans Appl Supercond 7:2793–2796, 1997.
52. CM Chorey, KS Kong, KB Bhasin, JD Warner, T Itoh. YBCO Superconducting ring resonators at millimeter frequencies. IEEE Trans Microwave Theory Tech MTT-39:1480–1487, 1991.
53. DE Oates, AC Anderson, DM Sheen, SM Ali. Stripline resonator measurements of Z_s versus H_{rf} in thin films. IEEE Trans Microwave Theory Tech MTT-39:1522–1529, 1991.
54. YA Vlasov, JM Vargas, P Brown, GL Larkins. $YBa_2Cu_3O_7$ on Y-stabilized ZrO_2 buffered (100) Si-"T" resonator microwave characteristics. IEEE Trans Appl Supercond 11:385–387, 2001.
55. T Dahm, D Scalapino, BA Willemsen. Microwave intermodulation of a superconducting disk resonator. J Appl Phys 86:4055–4057, 1999.
56. RD Lithgow, JM Peters. Electromagnetic resonant filter comprising cylindrically curved split ring resonators. US Patent 6616540A, 1997.
57. M Manzel, S Huber, H Bruchios, S Bornmann, P Gornert, M Klinger, M Stiller. High Q-value resonators for the SHF-Region based on TBCCO-films. IEEE Trans Microwave Theory Tech MTT-44:1382–1384, 1996.
58. N Klein, A Scholen, N Tellmann, C Zuccaro, KW Urban. Properties and applications of HTS-shielded dielectric resonators: a state-of-the-art report. IEEE Trans Microwave Theory Tech MTT-44:1369–1372, 1996.

59. N Tellmann, N Klein, U Dahne, A Scholen, H Schulz, H Chaloupka. High Q-$LaAlO_3$ dielectric resonator shielded by YBCO-films. IEEE Trans Appl Supercond 4:143–148, 1994.
60. J Gallop, L Hao, F Abbas, CD Langham. Frequency stability of dielectric loaded HTS microwave resonators. IEEE Trans Appl Supercond 7:3504–3507, 1997.
61. B Oh, HT Kim, YH Choi, SH Moon, PH Hur, M Kim, SY Lee, AG Denisov. A compact two-pole X-band high temperature superconducting microstrip filter. IEEE Trans Appl Supercond 5:2667–2670, 1995.
62. JP Hong, JS Lee. Performance of microstrip bandpass filters using high-T_c superconducting $YBa_2Cu_3O_{7-\delta}$ thin films on $LaAlO_3$. Appl Phys Lett 68:3034–3036, 1996.
63. D Zhang, GC Liang, CF Shih, RS Withers, ME Johansson, AD Cruz. Compact forward-coupled superconducting microstrip filters for cellular communication. IEEE Trans Appl Supercond 5:2656–2659, 1995.
64. JS Hong, MJ Lancaster, D Jedamzik, RB Greed, JC Mage. On the performance of HTS microstrip quasi-elliptic function filters for mobile communications applications. IEEE Trans Microwave Theory Tech MTT-48:1240–1246, 2000.
65. T Kinpara, M Kusunoki, M Mukadia, S Ohshima. Design of cross-coupled microstrip bandpass filter. Physica C 357–360:1503–1506, 2001.
66. I Vendik, A Deleniv, A Svishchev, M Goubina, A Lapshin, A Zaitsev, R Schneider, J Greek, R Aidam. Narrow band 10-pole Y-Ba-Cu-O filter on sapphire substrate. IEEE Trans Appl Supercond 11:361–364, 2001.
67. HT Kim, B Min, YH Choi, SM Moon, S Lee, B Oh, J Lee, I Park, C Shin. A compact narrowband HTS microstrip filter for PCS applications. IEEE Trans Appl Supercond 9:3909–3912, 1999.
68. DC Chung, BS Han. HTS microstrip filters using H-type resonators. IEEE Trans Appl Supercond 11:388–391, 2001.
69. ZY Shen, C Wilker, P Pang, DW Face, CF Carter, CM Harrington. Power handling capability improvement of high-temperature superconducting microwave circuits. IEEE Trans Appl Supercond 7:2446–2453, 1997.
70. HT Su, F Huang, MJ Lancaster. Highly miniature HTS microwave filters. IEEE Trans Appl Supercond 11:349–352, 2001.
71. K Huang, D Hyland, A Jenkins, D Edwards, D Dew-Hughes. A miniaturized interdigital microstrip bandpass filter. IEEE Trans Appl Supercond 9:3889–3892, 1999.
72. M Reppel, H Chaloupka, S Kolesov. Highly miniaturised superconducting lumped-element bandpass filter. Electron Lett 34:929–930, 1998.
73. H Xu, E Gao, S Sahba, JR Miller, QY Ma, JM Pond. Design and implementation of a lumped-element multipole HTS filter at 15 MHz. IEEE Trans Appl Supercond 9:3886–3888, 1999.
74. RR Mansour, S Ye, V Dokas, B Jolley, WC Tang, CM Kudsia. Feasibility and commercial viability issues for high-power output multiplexers for space applications. IEEE Trans Microwave Theory Tech 48:1199–1207, 2000.
75. RR Mansour, B Jolley, S Ye, FS Thomson, V Dokas. On the power handling capability of high temperature superconductive filters. IEEE Trans Microwave Theory Tech MTT-44:1322–1338, 1996.

76. G Tsuzuki, M Suzuki, N Sakakibara. Superconducting filter for IMT-2000 band. IEEE Trans Microwave Theory Tech MTT S Digest: 669–673, 2000.
77. KSK Yeo, MJ Lancaster, JS Hong. The design of microstrip six-pole quasi-elliptic filter with linear phase response using extracted-pole technique. IEEE Trans Microwave Theory Tech MTT-49:321–327, 2001.
78. RR Mansour, S Ye, B Jolley, G Thomson, SF Peik, T Romano, WC Tang, CM Kudsia, T Nast, B Williams, D Frank, D Enlow, G Silverman, J Soroga, C Wilker, J Warner, S Khanna, G Seguin, G Brassard. A 60-channel superconductive input multiplexer integrated with pulse-tube cryocoolers. IEEE Trans Microwave Theory Tech MTT-48:1171–1180, 2000.
79. MJ Lancaster, J Li, A Porch, NG Chew. High temperature superconductor lumped element resonator, Electron Lett 29:1728–1729, 1993.
80. MJ Lancaster, F Huang, A Porch, B Avenhaus, JS Hong, D Hung. Miniature superconducting filters. IEEE Trans Microwave Theory Tech MTT-44:1339–1346, 1996.
81. B Avenhaus, A Porch, F Huang, MJ Lancaster, P Woodall, F Wellhofer. Switched $YBa_2Cu_3O_7$ lumped element bandstop filter. Electron Lett 31:985–986, 1995.
82. K Setsune, A Enokihira. Elliptic-disc filters of high-T_c superconducting films for power-handling capability over 100 W. IEEE Trans Microwave Theory Tech MTT-48:1256–1264, 2000.
83. RJ Dinger, DR Bowling, AM Martin. A survey of possible passive antenna applications of high-temperature superconductors. IEEE Trans Microwave Theory Tech MTT-39:1498–1507, 1991.
84. SK Khamas, MJ Mehler, TSM Maclean, CE Gough. High-T_c superconducting short dipole antenna. Electron Lett 24:460–461, 1988.
85. LP Ivrissimtzis, MJ Lancaster, TSM Maclean, NM Alford. On the design and performance of electrically small printed thick film $YBa_2Cu_3O_{7-x}$ antennas. IEEE Trans Appl Supercond 4:33–40, 1994.
86. H Chaloupka, N Klein, M Peiniger, H Piel, A Pishke., G Splitt. Miniaturised high temperature superconductor microstrip patch antenna. IEEE Trans Microwave Theory Tech MTT-39:1513– 1521, 1991.
87. DC Chung. Broadband HTS microstrip antennas for satellite communication. IEEE Trans Appl Supercond 11:107–110, 2001.
88. MJ Lancaster, TSM Maclean, J Niblett, NM Alford, TW Button. YBCO thick film loop antenna and matching network. IEEE Trans Applied Supercond 3:2903–2905, 1993.
89. MJ Lancaster, Z Wu, Y Huang, TSM Maclean, X Zhou, CE Gough, NM Alford. Superconductor antennas. Supercond Sci Technol 5:277–279, 1992.
90. K Itoh, O Ishii, Y Nagai, N Suzuki, Y Kimachi, O Michikami. High T_c superconducting small antennas. IEEE Trans Appl Supercond 3:2836–2839, 1993.
91. MI Ali, K Ehata, S Ohshima. Superconducting patch array antenna on both-side YBCO thin film for satellite communication. IEEE Trans Appl Supercond 9:3077–3080, 1999.
92. K Ehata, K Sato, M Kusunoki, M Mukaida, S Ohshima, Y Suzuki, K Kanao. Miniaturized cooling systems for HTS antennas. IEEE Trans Appl Supercond 11:111–114, 2001.
93. HJ Chaloupka. High-temperature superconductor antennas: utilisation of low rf losses and non-linear effects. J Supercond 5:403–416, 1992.

94. LP Ivrissimtzis, MJ Lancaster, TSM Maclean, NM Alford. High gain printed dipole array made of thick film high-T_c superconducting material. Electron Lett 30:92–93, 1994.
95. LP Ivrissimtzis, MJ Lancaster, NM Alford. A high gain YBCO antenna array with integrated feed and balun. IEEE Trans Appl Supercond 5:3199–3202, 1995.
96. MA Richard, KB Bhasin, PC Claspy. Superconducting microstrip antennas: an experimental comparison of two feeding methods. IEEE Trans Antennas Propag AP-41:967–974, 1993.
97. JS Herd, D Hayes, JP Kenney, LD Poles, KG Herd, WG Lyons. Experimental results on a scanned beam microstrip antenna array with a proximity coupled YBCO feed network. IEEE Trans Appl Supercond 3:2840–2843, 1993.
98. JS Herd, LD Poles, JP Kenney, JS Derov, MH Champion, JH Silva, M Davidovitz, KG Herd, WJ Bocchi, SD Mittleman, DT Hayes. Twenty-GHz broadband microstrip array with electromagnetically coupled high T_c superconducting feed network. IEEE Trans Microwave Theory Tech MTT-44:1384–1389, 1996.
99. SH Talisa, MA Janocko, DL Meier, C Moskowitz, RL Grassel J. Talvacchio, P LePage, DC Buck, RS Nye, SJ Pieseski, GR Wagner. High-temperature superconducting wide band delay lines. IEEE Trans Appl Supercond 5:2291–2294, 1995.
100. KSK Yeo, MJ Lancaster. High temperature superconducting ferrite phase shifter with new latching structure. IEEE Trans Appl Supercond 11:430–432, 2001.
101. GF Dionne, DE Oates, DH Temme, JA Weiss. Ferrite-superconductor devices for advanced microwave applications. IEEE Trans Microwave Theory Tech MTT-44:1361–1368, 1996.
102. DE Oates, A Pique, KS Harshavardhan, J Moses, F Yang, GF Dionne. Tunable YBCO resonators on YIG substrates. IEEE Trans Appl Supercond 7:2338, 1997.
103. A Trotel, M Pyee, B Lavigne, D Chambonnet, P Lederer. Magnetically tunable YBaCuO microstrip resonators and bandpass filters. Appl Phys Lett 68:2559–2561, 1996.
104. DE Oates, GF Dionne. Magnetically tunable superconducting resonators and filters. IEEE Appl Supercond 9:4170, 1999.
105. H Fuke, Y Terashima, H Kayano, M Yamazaki, F Aiga, R Katoh. Tuning properties of 2 GHz superconducting microstrip-line filters. IEEE Trans Appl Supercond 11:434–437, 2001.
106. SS Gevorgian, EF Carsson, S Rudner, U Helmersson, EL Kollberg, E Wikborg, OG Vendik. HTS/ferroelectric devices for microrwave applications. IEEE Trans Appl Superconduct 7:2458–2461, 1997.
107. G. Subramanyam, WV Keuls, FA Miranda. A K-band-frequency agile microstrip bandpass filter using a thin-film HTS/ferroelectric/dielectric multilayer configuration. IEEE Trans Microwave Theory Tech MTT-48:525–530, 2000.
108. AT Findikoglu, QX Campbell, XD Wu, DW Mombourqette, D McMurry. Electrically tunnable coplanar transmission line resonators using $YBa_2Cu_3O_7/SrTiO_3$ bilayers. Appl Phys Lett 66:3674–3476, 1995.
109. H Xu, E Gao, QY Ma. Active tuning of high frequency resonators and filters. IEEE Trans Appl Superconduct 11:353–356, 2001.
110. HT Kim, BC Min, YH Choi, SM Lee, B Oh, JT Lee, I Park, CC Shin. A compact

narrowband HTS microstrip filter for PCS applications. IEEE Trans Appl Supercond 9:3909–3912, 1999.

111. K Satoh, T Mimura, S Narahasi, T Nojima. The 2 GHz high temperature superconducting receiver equipment for mobile communications. Physica C 357–360: 1495–1502, 2001.
112. JS Hong, MJ Lancaster, RB Greed, D Jadamzik, JC Mage, HJ Chaloupka. A high-temperature superconducting duplexer for cellular base-station applications. IEEE Trans Microwave Theory Tech MTT-48:1336–1342, 2000.

11

High-Temperature Superconducting IR Detectors

John C. Brasunas
NASA's Goddard Space Flight Center, Greenbelt, Maryland, U.S.A.

11.1 INTRODUCTION

It is already over 10 years since the discovery of high-temperature superconductor (HTS) materials with transition temperatures T_c in excess of the liquid nitrogen (LN_2) temperature (77 K at 1 atm). In addition to the continuing mystery of what exactly accounts for their high-T_c, the relative ease of LN_2 cooling versus liquid helium (LHe) cooling promises to make a number of engineering applications practical, ranging from magnetically levitated trains to microelectronics such as SQUID (superconducting quantum interference devices) -based medical imaging devices. In this chapter, we will present an overview of employing HTS materials in thin-film (<1 μm) form for the direct detection of infrared (IR) radiation, spanning the approximate wavelength range of 0.8 μm to 1 mm. Some of the examples, particularly for fast (picosecond) response, will be for HeNe laser sources (0.63 μm). Excluded are heterodyne applications where the HTS material serves the role of mixer (for instance, as a so-called hot electron bolometer) (1) producing a difference frequency between a radio-frequency input and a local oscillator. Also excluded are SQUID approaches in general, as the SQUID is covered in a companion chapter. An excellent review of the detector situation as of 1994 may be found in Ref. 2.

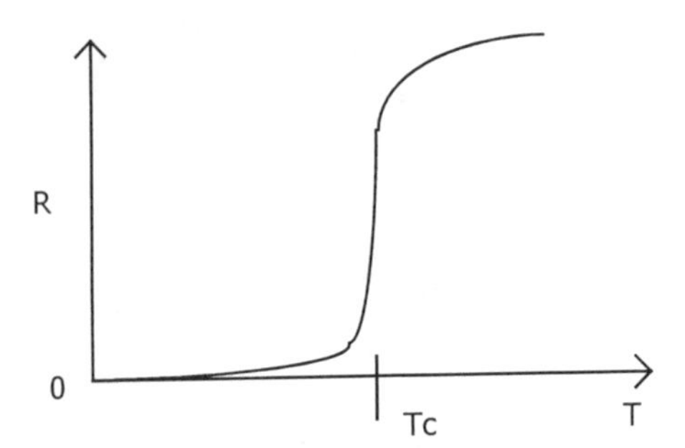

FIGURE 11.1 High-temperature superconductor resistive transition.

A typical resistance curve for an HTS material is shown in Figure 11.1: Typical T_cs include ~90 K for YBCO (yttrium barium copper oxide), 110 K for BSCCO (bismuth Strontium calcium copper oxide), and 125 K for TBCCO (thallium barium calcium copper oxide). Direct detection falls into two main categories: thermal and nonthermal or quantum. In the thermal approach, incoming radiant power W causes a temperature rise in the HTS lattice (phonons are the dominant heat-capacity medium at these temperatures, unlike the electron component for low-T_c materials); this temperature rise then modulate some HTS property such as resistance, which is then detected. The thermal detector is potentially broadband, limited by the spectral properties of the absorber. However it is potentially slow (with a time constant τ), especially due to the thermal inertia of the substrate. An alternate approach is the quantum detector; here, the power W directly interacts with the HTS material with a quantum efficiency η. However, the idea is not to excite the phonons, but rather to directly influence the Cooper pairs. Thus, for example, an incoming signal pulse directly leads to a reduction in the Cooper-pair density, without needing to influence the phonons. The quantum detector is potentially very fast, but at the expense of not being quite as broadband as the thermal (the photons must be energetic enough to break the Cooper pairs). Requiring Cooper pairs, quantum detectors operate at or below T_c; thermal detectors can operate below, at, or above T_c.

Infrared detection in the fully superconducting regime includes quantum detectors operating by pair-breaking, modifying an HTS property such as the kinetic inductance or critical current. The kinetic inductance and critical current can also be changed thermally. Another detector in the superconducting regime is the photofluxonic detector (3), a photon-assisted generation of vortex–antivortex pairs. In the normal region, only thermal detectors are possible. A bolometer is possible based on the metallic property of high-quality HTS materials. Pyroelectric detectors are also possible. In the transition region, both thermal and nonthermal approaches are possible. A very common detection method is thermal modulation of the resistance. Also possible is thermal modulation of the penetration depth, leading to a thermal detector based either on magnetic inductance or kinetic inductance. Detectors have also been made based on thermal modulation of the microwave surface impedance. The penetration depth and surface impedance can also be modified by direct interaction between incoming photons and the Cooper pairs.

11.2 RESISTIVE TRANSITION-EDGE DETECTORS: A FIRST LOOK

One of the most common approaches to HTS detection of IR radiation is the so-called transition-edge (TE) bolometer-type detector. This is a member of the class of thermal detectors for which incoming radiation causes a temperature rise in an absorber. The temperature rise is sensed by a (possibly distinct) thermometer and the deposited heat is eventually transferred to a thermal bath through a thermal link. The HTS material is the thermometer and is held near T_c; incoming radiation causes a rise in temperature that leads to a resistance rise, which is sensed by suitable electronics. The incoming power W can either be directly absorbed into the HTS material (Fig. 11.2a), can be absorbed into a separate but closely coupled absorber (Fig. 11.2b), or it can be coupled by an antenna before being coupled into the HTS material. Although Figure 11.2a may appear the most straightforward, ancillary concerns may make an alternative approach preferable. Figure 11.2a is not suitable, for instance, at long IR wavelengths when the HTS material is fully superconducting; at long enough wavelengths, the Cooper electron pairs are not broken and the HTS absorption goes to zero. For fully superconducting YBCO, the reflectivity can exceed 99% beyond a 25-μm wavelength (4). Also, as one approaches a 1-mm wavelength, the absorber sideway dimension also needs to approach 1 mm for absorption efficiency. When one considers the limited range of substrate materials suited for HTS thin-film growth and the often high specific heats of these candidate substrates, the HTS-absorber detector can be quite slow due to the thermal inertia of the substrate. One can instead use a separate absorber and make the HTS thermometer rather small (Fig. 11.2b), or one can use an antenna to couple W onto a much smaller detector (Fig. 11.2c). The antenna, however, will limit the detector to single-mode operation, whereas the nonantenna approach is multimode (5).

For the resistive TE detector with dc current biasing, the voltage response to steady input power W (in watts) is of the form

$$V_{\text{out}} = \frac{W\eta}{G} I_b \frac{dR}{dT} \tag{1}$$

where η is the absorption efficiency (%), G is the thermal conductance (W/K), I_b is the bias current (A), and dR/dT is the temperature derivative of the resistance

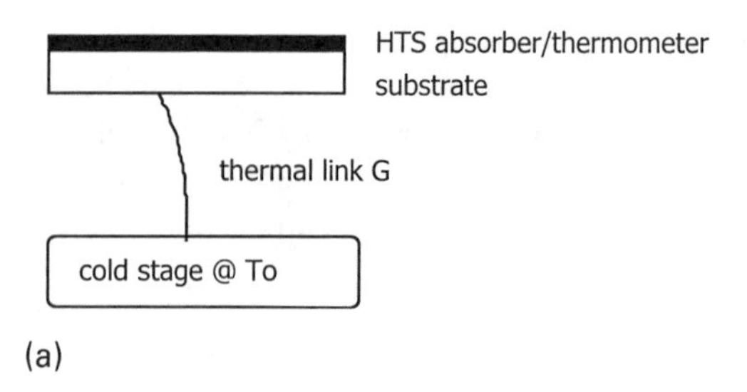

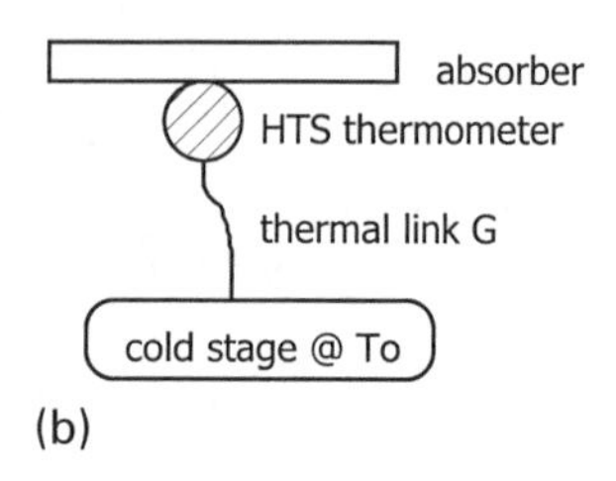

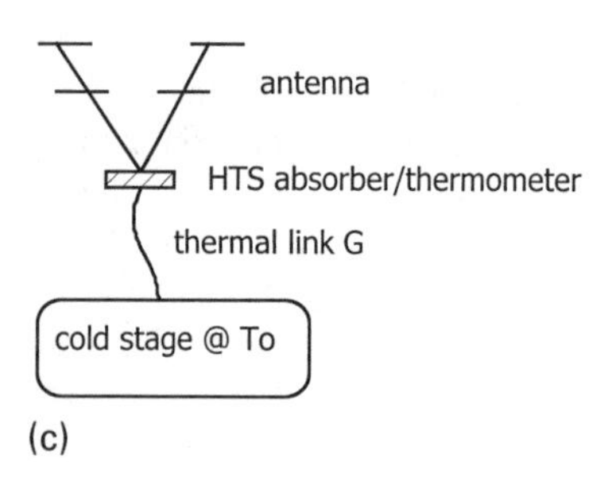

FIGURE 11.2 (a) High-temperature superconductor as absorber and thermometer; (b) HTS as thermometer only; (c) antenna-coupled HTS as absorber and thermometer.

(Ω/K). Defining the detector responsivity as $\Re \equiv V_{out}/W$ (V/W), the responsivity can be factored into volts/watt = (volts/K) (K/watt) as follows:

$$\Re \equiv \Re_{V/W} = \Re_{V/K}\Re_{K/W} = \left(I_b \frac{dR}{dT}\right)\frac{\eta}{G} \tag{2}$$

Consider, now, a body representable as a point with heat capacity or thermal capacity C, connected to a bath with thermal conductance G (6). For non-steady operation, a quantity of heat ΔQ causes a temperature rise ΔT in the body:

$$\Delta Q = C\Delta T \tag{3}$$

After the deposition of this heat, the body will cool at the rate

$$\Delta W = G\Delta T \tag{4}$$

The time response to an input pulse of heat ΔQ is then of the form

$$\Delta W = -\frac{d\Delta Q}{dt} \tag{5}$$

or

$$C\frac{d\Delta T}{dt} + G\Delta T = 0. \tag{6}$$

For an external source of heat with rate W,

$$C\frac{d\Delta T}{dt} + G\Delta T = \eta W \tag{7}$$

For W of the form $W_0 e^{j\omega t}$, the steady-state solution at ω (rads/s) is

$$\Re_{\text{K/W}} = \frac{\Delta T}{W_0} = \frac{\eta}{G + j\omega C} \tag{8}$$

Thus, the ac form of the responsivity is

$$\Re = I_b \frac{dR}{dT} \frac{\eta}{G + j\omega C} \tag{9}$$

Defining the time constant $\tau \equiv C/G$, the responsivity my be re-expressed as

$$\Re = I_b \frac{dR}{dT} \frac{\eta}{G} \frac{1}{1 + j\omega\tau} \tag{10}$$

It is often important to know the signal-to-noise ratio (S/N) of a detector response. In the case of a thermal detector, the minimum noise or noise floor is set by temperature fluctuations in the detector itself. From statistical mechanics, a system with many degrees of freedom satisfies the following equation for the mean square value of temperature fluctuations ΔT (7):

$$\overline{\Delta T^2} = k\frac{T^2}{C} \tag{11}$$

where k is Boltzmann's constant (1.38×10^{-23} J/K).

Considering again the case of a detector with lumped-elements C and G, the spectral decomposition of the mean square temperature fluctuations (mean square fluctuations per hertz) is of the form

$$\overline{\Delta T_f^2} = \frac{4kT^2G}{G^2 + (\omega C)^2} \tag{12}$$

Dividing Eq. (12) by the square of the magnitude of $\Re_{K/W}$ from Eq. (8) (and assuming unity η), temperature fluctuations of the detector thereby correspond to an apparent fluctuation of incoming radiation (units of W^2/Hz) of

$$\overline{\Delta W_f^2} = 4kT^2G \equiv \text{NEP}^2 \tag{13}$$

where NEP is the noise equivalent power ($W/Hz^{1/2}$).

Clearly, it is desirable to make the NEP as small as possible. The minimum possible G, G_{rad}, corresponds to radiative coupling alone. For a detector and background in thermal equilibrium at temperature T (8).

$$G_{rad} = 4\sigma\eta AT^3 \tag{14}$$

where σ is Stefan's constant (5.67×10^{-12} $W/cm^2/K^4$) and A is the detector area. Thus, the minimum NEP_{rad} satisfies the equation

$$\text{NEP}_{rad}^2 = 16Ak\sigma\eta T^5 \tag{15}$$

Because A can vary, NEP has no firm lower limit. Therefore, it is convenient to define the detectivity D^* as

$$D^* = \frac{\sqrt{A}}{\text{NEP}} \tag{16}$$

$$D_{rad}^* = \frac{\sqrt{A}}{\text{NEP}_{rad}} = \frac{1}{4\sqrt{k\sigma\eta T^5}} \tag{17}$$

The best (highest) D^* then corresponds to the lowest NEP. D^* has the nice property that for an array of pixel detectors summed together, the D^* of the array is the same as the D^* of an individual pixel. Assuming unity η, the highest possible D^* is 1.8×10^{10} $cm/Hz^{1/2}/W$ at 300 K (NEP is 5.5×10^{-12} $W/Hz^{1/2}$ for a 1×1-mm detector) and 3.7×10^{11} at 90 K, 20 times higher. This is one of the major drivers to develop an IR detector based on HTS materials, because although numerous, near-optimal thermal detectors have been built for operation at 300 K, there are no more sensitive detectors until one cools to approximately 4 K. Near 90 K, detectors could provide intermediate performance with intermediate cooling demands.

The state of the art for 300 K thermal detectors is 3.6×10^9 for a Golay cell at 6.5 Hz, 3.7×10^9 for a pyroelectric detector at 6.5 Hz, 4×10^9 for a thermocouple detector at a few hertz, and 6.7×10^9 for a thermal-expansion-based detector, at 2–3 Hz (9). None of these detectors is widely available. The best commercially available 300-K detectors are $(1–2) \times 10^9 D^*$ for pyroelectric or thermopile detectors.

11.3 RESISTIVE TRANSITION-EDGE DETECTORS IN DEPTH

11.3.1 Effect of Diffusion and Boundary Resistance on Thermal Isolation; First Look at τ

Consider a transition-edge detector consisting of a HTS thin film on a plate of bulk-type substrate material. Let the plate be mechanically supported and thermally isolated by some insulating fibers, such as Kevlar (Fig. 11.3a). The HTS material is metallized with silver and gold, to which we bond fine gold wires, such as are used in the microelectronics industry for contacting a die within a chip. A practical lower limit for the gold wire diameter is 18 μm (0.8 mil). The thermal isolation is typically adjusted by changing the lengths of the electrical leads, usually four, as in a four-wire connection. However, for moderately cold temperatures such as 90 K and above, attention must be paid to the so-called thermal diffusion length l_0, which satisfies the equation (10)

$$l_0 = \sqrt{\frac{K}{\pi f s \rho}} \equiv \sqrt{\frac{a}{\pi f}} \tag{18}$$

where K is the thermal conductivity (W/cm K), f is the frequency (Hz), s is the specific heat (J/g K), ρ is the density (g/cm^3), and $a \equiv (K/s\rho)$ is the so-called diffusiv-

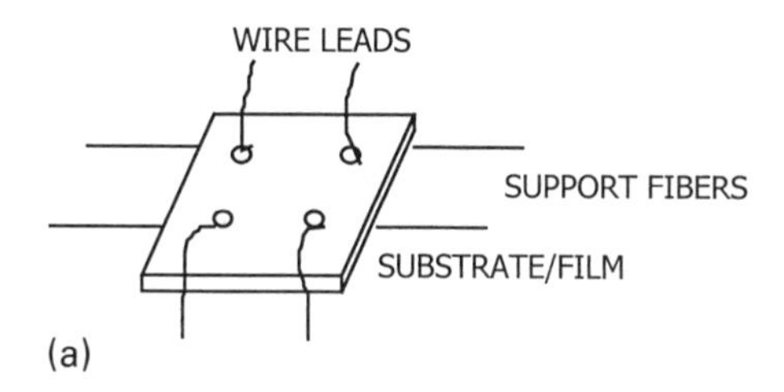

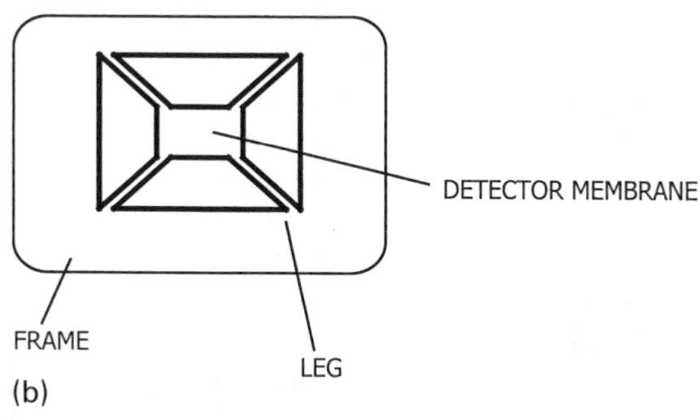

FIGURE 11.3 (a) Hand-crafted detector; (b) monolithic detector pixel, produced by photolithography.

TABLE 11.1 Thermal Properties of IR Detector Materials at 90 K

Material	θ_D	K (W/cm K)	$s\rho$ (J/cm^3 K)	a_{diff} (K/$s\rho$)	l_0 (mm) at 10 Hz
Diamond	1800	100.	0.07	1400.	67.
Sapphire	900	6.4	0.39	16.4	7.2
Silicon	645	10.	0.522	19.2	7.8
MgO	800	3.4	0.53	6.4	4.5
Zirconia	550	0.015	0.7	0.021	0.26
$SrTiO_3$	490	0.2	1.12	0.18	0.76
Gold	180	3	1.93	1.55	2.2
Silver	220	5	1.81	2.76	3.0
Aluminum	375	4	1.25	3.2	3.2
Copper	310	3	1.79	1.67	2.3
YBCO (123)	440		1.13		
In plane		0.1		0.089	0.53
c Axis		0.015		0.013	0.2

ity. Typical values at 90 K for K, $s\rho$, a, l_0, and the Debye temperature θ_D are given in Table 11.1 for YBCO and candidate substrate and wire materials. Note that the diffusion length of gold at 10 Hz and 90 K is fairly short (2.2 mm). Basically, what this means is that the usual formula relating thermal conductance to thermal conductivity for a wire of length l and cross-section area A_w is modified to

$$G = \frac{KA_w}{l}, \quad 0 < l \ll l_0$$
$$G = \frac{KA_w}{l_0}, \quad l_0 \ll l \tag{19}$$

Also, for $l \ll l_0$, the heat-capacity contribution is one-third of the total heat capacity of the wires; for $l \gg l_0$, the contribution is 1.5 times one-third of the contribution of length l_0. For 18-μm-diameter gold wire at 90 K and 10 Hz, four leads, the implied best thermal isolation due to conduction is about 10^{-4} W/K. From Eq. (13), the implied best NEP at 90 K is about 7×10^{-12} and the D^* is 1.5×10^{10} for a 1×1-mm detector. By way of comparison, the unity-absorption radiative coupling [Eq. (14)] is 6×10^{-6} at 300 K and 1.6×10^{-7} at 90 K.

In addition to the thermal isolation between the detector and the bath, there is also some amount of thermal isolation between the HTS film and the substrate. A typical value of the boundary resistance (11) between the HTS film and the substrate is 10^{-3} cm^2 K/W, 80 times larger than the acoustic mismatch model would predict. For a 1×1-mm detector, this implies a G of 10 W/K. A detector based on this thermal isolation has an NEP about 300 times worse than the gold-wire detector, but it is potentially very fast. For a 0.5-μm film of YBCO, the heat capacity is 5.6×10^{-7} J/K and the implied time constant is 57 ns. Because both the heat capacity and thermal conductance scale with area, this time constant should be roughly independent of area.

For the detector configuration of Figure 11.3a, assuming a lumped-elements condition, it is straightforward to calculate the time constant. For quick response, a high Debye temperature is desirable. For temperatures much less than θ_D, the phonon modes freeze out, leading to a T^3 dependence of the heat capacity. From Table 11.1, diamond is the premier candidate, but as will be discussed later, diamond is not well suited as an HTS substrate material. Perhaps the second-best choice is sapphire, which is commercially available in 1-mil (25-μm) thicknesses. A 1×1-mm, 1-mil plate of sapphire has a heat capacity of 10^{-5} J/K; the contribution of the gold wires is about 1.4×10^{-6} J/K, plus a contribution from the gold ball-bond. Together with a thermal isolation of 10^{-4} W/K due to gold wires, the implied time constant is on the order of 100 ms. This is toward the high end of what is desirable for a time constant; the obvious solution is to further thin the sapphire or decrease the area. Decreasing the area, however, will reduce the absorption efficiency toward a 1-mm wavelength, so the wisdom of doing this will depend on the application.

To approach the radiation-limited NEP and D^*, it is necessary to obtain better thermal isolation than is possible with gold-wire bonding; that is, the electrical leads themselves need to be thin films. Consider the detector configuration of Figure 11.3b, a so-called "monolithic" approach. The idea is to start with a fairly thin substrate material, say 1 mil of silicon. Then, most of the material is etched away, leaving a fairly thin frame and much thinner legs and a central portion (membrane) that serves as the substrate for the HTS thin film. The thin-film wire connection and the legs themselves must be thin enough to provide better thermal isolation than gold-wire bonding. The heat capacity must improve even more than the thermal isolation to improve upon the 100-ms time constant, indicating that the membrane needs to approach 1 μm thickness. Some HTS films have even been grown without a substrate (12). As we increase the thermal impedance of the metallic links, by the Wiedemann–Franz law we also increase the electrical impedance. As we will see later, electrical impedance brings with it an extraneous noise term. Berkowitz et al. (13) have presented a superconducting link to ameliorate this noise term.

11.3.2 Bias: Effect of Electrothermal Feedback on Thermal Isolation

As the resistive TE detector is run with an electrical bias, Eq. (7) is modified to

$$C\,\frac{d\Delta T}{dt} + G\Delta T = \eta W + \frac{dW_h}{dT}\,\Delta T \tag{20}$$

where W_h is the heating due to the electrical bias. This may be re-expressed as

$$C\frac{d\Delta T}{dt} + G_e\Delta T = \eta W \tag{21}$$

where the effective thermal conductance satisfies the equation

$$G_e = G - \frac{dW_h}{dT} \tag{22}$$

Two common biasing conditions are fixed current through the detector and fixed voltage across the detector. For fixed current, $W_h = I_b^2 R$ and $dW_h/dT = I_b^2$ dR/dT. Thus, the effective thermal conductance for fixed current is

$$G_{e,i} = G - I_b^2\frac{dR}{dT} \tag{23}$$

As dR/dT is a positive quantity, the effect of constant current bias is to reduce G_e, until the G_e becomes zero (at the destabilization current) and the detector becomes unstable (thermal runaway). Clearly, thermal runaway is most likely at the midpoint of the transition. The reduced G_e is used for the responsivity [Eqs. (8)–(10)], lengthening the effective time constant $\tau_e \equiv C/G_e$. The effect of G_e on the phonon-noise-limited NEP is less clear. The conservative approach (14), which we will use, is to continue to use the low-bias G in Eq. (13). This is not strictly true, as the system is no longer in thermal equilibrium. Indeed, there is evidence that the limiting NEP can be reduced due to electrothermal feedback (15). For constant-voltage biasing, $W_h = V_b^2/R$, $dW_h/dT = -(V_b^2/R^2)\,dR/dT$, and the effective thermal conductance is

$$G_{e,v} = G + \frac{V_b^2}{R^2}\frac{dR}{dT} = G + I_b^2\frac{dR}{dT} \tag{24}$$

Thus, the effect of constant-voltage biasing is to increase G_e, shortening the time constant. Additionally, it will be shown below that a detector is often not operated under optimal (phonon-noise-limited) conditions, particularly at the higher frequencies, and under these circumstances, voltage biasing can improve the NEP. Also, V_b does not appear to be limited (there is no thermal runaway condition), but an arbitrarily high bias can overheat the detector and cause a fuselike destruction mechanism either in the connecting wires/traces or in the HTS film itself.

11.3.3 Effect of Phonon Wave Interference and Mean Free Path on Thermal Isolation

Equation (19) accurately predicts the "bulk" thermal conductance when the dimensions, length and area, are large compared with the mean free path. When a dimension approaches the phonon mean free path, the thermal resistance can be boosted. A possible example is a thin film of HTS material on a substrate: If the film thickness approaches the mean free path, the sideways (parallel to the film–substrate interface) thermal resistance is boosted. For YBCO, the mean free path is 3 nm at 50 K (16). Most IR detectors employ films at least 100 nm thick, so this is not an issue. Films are generally thinner on silicon substrates to avoid microcracking, but the mean free path is probably still not an issue.

We previously discussed the issue of thermal resistance between HTS film and substrate. The thermal isolation is not high, leading to an insensitive but fast detector. Consider the case of Figure 11.2a, where incoming photons are absorbed by the HTS film, launching phonon waves into the HTS material. Due to the property mismatch at the film–substrate interface, the waves will be partly reflected back to the absorber and those waves will be reflected yet again toward the interface. The multiply-reflected waves can lead to destructive and constructive interference, as is the case for optical elements such as Fabry–Perot interferometers. Under the right conditions, the thermal isolation of the film from the substrate can be increased (17). The frequency range of enhanced temperature rise is $\pi D/16d^2$ to $9\pi D/16d^2$, where d is the film thickness and D is the diffusion coefficient (a_{diff} in Table 11.1). For YBCO, the frequency range of increased isolation and thermal response is 6–57 MHz for 200-nm-thick *c*-axis YBCO.

11.3.4 Effect of Bias Current on dR/dT

It is common, given measurements of G and dR/dT, to predict the responsivity of an IR detector using a relation such as Eq. (1). However, care must be taken when the dR/dT is measured with a bias other than the one in use for the detector application, as there are a couple of reasons why dR/dT may depend on the magnitude of the bias current. The first is thermal. R versus T is logically interpreted in terms of T_{film}, the temperature of the HTS film. Yet, if T is measured by a thermometer separate from the detector, the measured T may be more typically T_{stage}, the temperature of a thermally isolated stage that is varied to set the operating temperature of the detector. Consider the case of current biasing. The low-frequency relationship between T_{film} and T_{stage} is

$$T_{\mathrm{film}} = T_{\mathrm{stage}} + \frac{I^2 R}{G} \tag{25}$$

In the fully superconducting regime, $R = 0$ and there is no offset; in the normal regime, there is a positive offset. The transition appears broader versus T_{film} than versus T_{stage}. Equivalently, the transition appears narrower versus T_{stage} than T_{film}. Thus, the film heating effect can make the transition appear narrower by an amount that depends on the bias current. Heating at the contacts will be nearly constant throughout the transition, so contact heating will not change dR/dT but may change the apparent T_c.

The other consideration is that increasing the bias current can broaden the intrinsic width of the transition. This is fairly straightforward to show for a one-dimensional (wire) superconductor (18). Thus, in the general case, there is the possibility for the appearance of both narrowing and broadening, and to determine which, if either, will dominate will depend on the details of a given detector (19).

11.3.5 Considerations on the Absorber Efficiency η

There are numerous possibilities for good absorption efficiency in the infrared regime. Dielectric materials such as paints are hampered, however, by the need to make the absorber about as thick as a wavelength, leading to high heat capacity and long time constants as the operating wavelength approaches 1 mm. Metals are a possibility, in two forms. In one, the so-called "gold black" or "gold soot" (20), the gold is deposited in a poor nitrogen vacuum, forming a frothy layer mostly void filled (about 0.5% of the density of bulk gold). The absorption is near unity in the visible, declining to perhaps 20–40% near 1 mm, depending on the thickness used. Another possibility is the so-called "space-matched" coating (21). Here, an extremely thin metal film is deposited onto a substrate—the film is thin enough so that the resistivity is not the bulk resistivity but, rather, is determined by the film thickness. For the appropriate thickness, giving an impedance of approximately half of free space (377 Ω/square for a substrate index of refraction of 2) radiation incident through the substrate is approximately half absorbed (44% for an index of refraction of 2) at the substrate–film interface, pretty much independent of wavelength; the rest is transmitted and none is reflected. The advantage of the space-matched coating is that it is very broadband and very thin, thereby not increasing the time constant.

Finally, the HTS film itself can be used as the absorber. Fully superconducting, it is mirrorlike at long enough wavelengths, thereby providing low absorption efficiency. Within the transition region, it is metallike; however, relying on it for absorption may impose constraints on the necessary thickness, which may not be compatible with a thickness consistent with good film quality, depending on the substrate. On the other hand, there may be great benefit to depositing the heat directly into the HTS thermometer. We have also mentioned earlier the possibility of antenna coupling the incoming radiation to a smaller-than-usual HTS thermometer, with a limitation to single-mode operation.

11.3.6 Thermal Detector Models with More Complex Thermal Structure

The simplest thermal model consists of a single, lumped heat capacity C and a single, lumped thermal conductance G, a so-called single-node thermal model. The single-node model has the Lorentzian frequency response of Eq. (9), consisting of a flat frequency response and 0° phase shift at low frequencies ($\omega\tau \ll 1$), a high-frequency phase lag of 90°, and a high-frequency amplitude response $\sim 1/\omega$. When the diffusion length becomes comparable to detector dimensions, either a multimode or distributed model is more appropriate. We will now consider a few simple examples of these more involved thermal models.

We have already considered the case of diffusion in the thermal link. At low frequencies, the diffusion length is very long and the thermal conductance is the usual value (call it G_{dc}). For frequencies higher than the value where the diffusion length equals the length of the thermal link (call this ω_D), the thermal conductance depends on frequency as $G_{dc}(\omega/\omega_D)^{1/2}$. Consider the simple case where the heat capacity of the link may be neglected, relative to the detector absorber–thermometer unit. Define the characteristic frequencies $\omega_0\tau_0 = 1$, where $\tau_0 \equiv C/G_{dc}$. There are the two possibilities for the frequency behavior of the responsivity as described by Eq. (9), allowing for G to be a function of frequency through the diffusion length dependence.

Case 1: $\omega_D \gg \omega_0$. In this case, G is not modified by diffusion until the heat capacity becomes the dominant factor in the denominator of Eq. (9), so the frequency response is the typical Lorentzian.

Case 2: $\omega_D \ll \omega_0$. In this case, the responsivity begins falling as $1/\omega^{1/2}$ at ω_D, until

$$G_{dc}\sqrt{\frac{\omega}{\omega_D}} = \omega C \quad \text{or} \quad \omega = \frac{\omega_0^2}{\omega_D} \tag{26}$$

when the responsivity is again dominated by the heat-capacity term, falling off as $1/\omega$.

Another case is the one-dimensional thermal-link diffusion case, but this time the heat capacity is dominated by the thermal link; the heat capacity of the absorber–thermometer may be neglected. In this case, the response is Lorentzian until ω_D, and then it varies as $1/\omega^{1/2}$ for higher frequencies. For both of these cases, for frequencies where the response is non-Lorentzian, the rolloff is *slower* than Lorentzian. Additional considerations concerning the thermal link may be found in Ref. 22.

A further case is an absorber–thermometer–thermal bath (cold-stage) system (23). Let G_1 be the thermal conductance between the bath and the thermometer and G_2 be the conductance between the thermometer and the absorber. Let C_1 be the heat capacity of the thermometer and C_2 be the heat capacity of the absorber. Define $\Omega = C_2/C_1$, $\tau_1 = C_1/G_1$, and $\tau_2 = C_2/G_2$. The unity-emissivity responsivity is terms of Kelvins per watt is of the form

$$\Re_{K/W} = \frac{1}{G_1(1 + j\omega\tau_1')(1 + j\omega\tau_2')} \tag{27}$$

where $\tau_1'\tau_2' = \tau_1\tau_2$ and $\tau_1' + \tau_2' = \tau_1(1 + \Omega) + \tau_2$. In this case, the response is a double Lorenztian, faster than Lorentzian. The high-frequency phase shift is a phase lag of 180°. Thermal modeling of a similar nature may be found in Ref. 24.

11.3.7 Noise Considerations for the Transition-Edge Detector

The basic, irreducible noise floor for any thermal detector is due to the temperature fluctuations of the detector. For a lumped-elements detector, we showed in Eq. (12) that this noise floor, so-called phonon noise, has a Lorentzian spectrum. The phonon noise expressed in terms of voltage is, from Eqs (2) and (12),

$$V_{n,\mathrm{phonon}} = \Re_{\mathrm{V/K}} \sqrt{\Delta T_f^2} = I_b \frac{dR}{dT} \sqrt{\frac{4kT^2G}{G_e^2 + (\omega C)^2}} \tag{28}$$

Because the voltage responsivity is also Lorentzian [Eq. (9)], the resulting NEP is frequency independent [Eq. (13)] as long as we are dominated by phonon noise. The objective in building a detector is to keep all other noise sources small compared with the phonon noise. We will consider three other noise sources: Johnson noise, $1/f$ noise, and signal or photon noise. The Johnson and $1/f$ noise can originate both in the bolometer itself and in following stages of the electronics. The various noise sources are usually uncorrelated and, thus, are added in quadrature. We will first consider these noise terms in the context of current biasing. For current biasing at low frequencies, high bias will make the effective thermal conductance smaller, giving the phonon noise an additional boost, beyond the linear dependence on the bias current.

Johnson noise (25), in units of volts per square root of hertz has the form

$$V_{n,J} = \sqrt{4kTR} \tag{29}$$

At 90 K, typical values are 0.2 nV/Hz$^{1/2}$ for 10 Ω and 0.7 nV/Hz$^{1/2}$ for 100 Ω. Johnson noise is universally present for any resistance or dissipative element and is analogous to blackbody radiation. It arises from the HTS material itself in the transition, from contacts, and from the electronics. It has a flat spectrum versus frequency, unlike the phonon noise, which declines as $1/\omega$ for high frequencies ($\omega\tau \gg 1$). Because the noise remains flat while the response declines, the S/N must ultimately decline at high frequencies, compared to what is predicted by Eq. (13) for the phonon-noise limit. Thus, it is desirable to minimize τ, meanwhile maximizing I_b to boost the phonon voltage noise above the Johnson noise. The maximum bias is limited by thermal runaway conditions, because the detector is typically operated at peak dR/dT. It is also desirable to minimize the contact resistance and noise relative to the detector signal, either by reducing the contact resistance or by patterning the superconductor to make the contact resistance insignificant.

The $1/f$ noise (26) has the approximate form

$$V_{n,\,1/f} = \text{aa}\,\frac{I^x}{f^y} \tag{30}$$

where $x \sim 1$ and $y \sim 1/2$ and aa is a constant depending on the material. Unlike Johnson noise, $1/f$ noise is associated with current flow, I. The approximately linear dependence on current makes $1/f$ noise appear to be resistance fluctuations. Although extremely widespread throughout electronics and across most of natural phenomena, the origin of $1/f$ noise is mysterious. Because of its spectrum, it will tend to dominate the noise budget at low frequencies, thus depressing the S/N at low frequencies. Due to the bias current, there will be $1/f$ noise contributions from the HTS material and from the contacts; $1/f$ noise is generally negligible in wires. Because of its roughly linear dependence on bias current, we are not able to boost the phonon noise from the HTS material above the $1/f$ noise. Using a four-wire/four-contact connection scheme, we are able to minimize the $1/f$ noise due to the contacts because very little current flows through the sensing contacts. The effect of $1/f$ noise from later stages of the electronics, due to different currents, can be lessened by choosing an ac bias for the detector. With an ac bias, the signal is heterodyned to a higher-frequency, where the electronics $1/f$ noise is less. With the combined effects of phonon, Johnson, and HTS $1/f$ noise, the S/N spectrum may be expected to appear similar to Figure 11.4. S/N declines $\sim 1/f$ at high frequencies due to Johnson noise combined with the time constant. S/N theoretically declines $\sim 1/f$ at low frequencies, due to $1/f$ noise. The "sweet spot" is centered at the frequency where the quadrature combination of Johnson and $1/f$ noise is a minimum. The S/N at the sweet spot may or may not approach the level implied by Eq. (13), depending on whether the quadrature combination dominates the phonon noise. The sweet spot depends on the amount of $1/f$ noise, on the amount of Johnson noise, and on the time constant. The sweet spot need not occur at $\omega\tau = 1$. Therefore, the time constant alone can be a misleading discrimina-

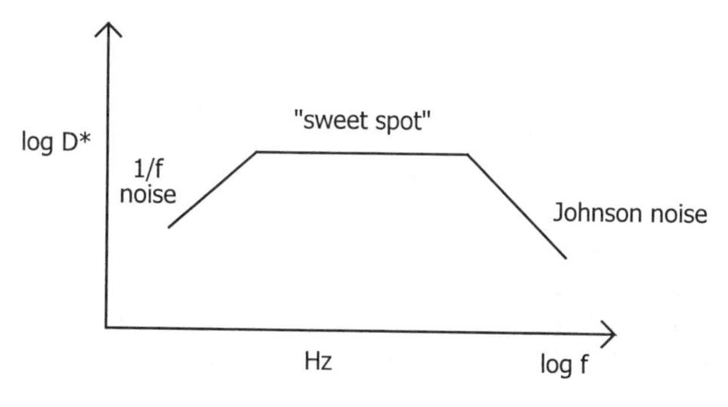

FIGURE 11.4 Detectivity versus operating frequency.

tor if the objective is to choose among a set of candidate detectors for optimal S/N. The best bias is toward the destabilization current, as too low a bias causes the Johnson noise to dominate unduly, whereas $1/f$ noise eliminates the advantage of increasing the bias beyond a certain point.

The target itself can contribute noise, as the emission of photons even from a constant temperature blackbody is a stochastic process. This noise term is called signal or photon noise. Allowing for a detector temperature T and a beam-filling target at temperature T_t, the effect of photon noise (27) is to modify Eq. (15) to

$$\mathrm{NEP}^2_{\mathrm{photon}} = 8Ak\sigma\eta(T^5 + T_t^5) \tag{31}$$

When $T = T_t$, this reduces to Eq. (15); thus, phonon noise can be viewed as photon noise between an isothermal detector–chamber system. Equation (31) indicates that the excellent S/N potential of an HTS detector cannot be maintained when viewing a very bright target. For instance, a 90-K HTS detector viewing a 300-K beam-filling target cannot be more than $\sqrt{2}$ more sensitive (S/N) than an ideal 300 K detector [compare Eqs. (15) and (31)]. Thus, to reach fundamental limits, it is important to introduce cold shields (at or near the detector temperature) to reduce the area and/or solid angle of warm emission reaching the detector.

The above considerations were for the case of current biasing. A different picture emerges for voltage biasing (28). With voltage biasing, there are two major differences: There is no limitation on bias current imposed by thermal runaway; the thermal conductance is increased rather than decreased. Therefore, the potential advantage to using voltage biasing occurs when the detector is Johnson-noise limited. This occurs at high frequencies, where the allowed increase in bias current allows us to further increase the voltage phonon noise compared with the Johnson noise. At low frequencies, the phonon noise increases more slowly than linear with bias [due to the change in effective G, see Eq. (28)], whereas the $1/f$ noise has a roughly linear dependence on bias. Thus, there is no benefit to voltage biasing at low frequencies.

11.4 CURRENT STATE OF THE ART FOR TRANSITION-EDGE BOLOMETERS

As far as competitiveness with room-temperature detectors is concerned, one of the earliest examples was provided by Verghese et al. (29). Using a fairly thin 20-μm sapphire substrate and a YBCO thin film, they achieved an NEP of 2.4×10^{-11} at 10 Hz (about $4 \times 10^9 D^*$). Further progress occurred in 1994, $6 \times 10^9 D^*$ at 4 Hz, employing 1 mm^2 of YBCO on 25-μm-thick sapphire (30). Additional progress occurred with the deposition of YBCO on 12-μm-thick sapphire (31). Broadly speaking sapphire is commercially available as thin as 25 μm, a thickness obtainable by mechanical polishing. Further thinning, necessary for reduced time constants and operating at frequencies above the $1/f$ noise region, is possible by other techniques such as chemical polishing. Additional thinning is key to boosting the detectivity. This was demonstrated in 1997 (32) by using chemical–mechanical polishing techniques to thin sapphire down to 7 μm, producing a bolometer with $1.2 \times 10^{10} D^*$ from 2 to 3.8 Hz; a gold-black layer was used for absorption.

As contrasted with the sapphire approach, which is essentially thinning bulk substrate material, the silicon/membrane route promises thinner substrates and compatibility with semiconductor electronics. Here, the silicon or silicon nitride membrane substrate can be as thin as 1 μm, approximately. The disadvantage is the greater difficulty of depositing HTS films on silicon, as will be discussed. Additionally, silicon is somewhat inferior compared with sapphire in a thermal sense, as can be seen in Table 11.1. Detectivities of 5×10^8 were achieved with YBCO on silicon by Mechin et al. (33). This was further advanced to 2.5×10^9 with a 564-μs time constant (34). The current record for D^* in the few-hertz range is 1.8×10^{10} for 0.2–2 Hz, using GdBaCuO thin films on a silicon nitride membrane (35).

Just as the concept of D^* has an area normalization that is of some use in comparing the NEPs of detectors with different areas, a further normalization of D^* may be useful in comparing detectors with widely different time constants. Bloyet et al. (36) have noted that D^* and the time constant τ roughly follow the empirical relationship

$$D^* \cong 10^{11} \tau^{1/2} \qquad (32)$$

where τ is in seconds. Let us assume that D^* reaches a maximum value from $f_{\min}$ to $f_{\max}$, centered around the sweet spot; then, Eq. (32) may be re-expressed as

$$\sqrt{f_{\max}} D^*_{f=f_{\max}} \cong \text{const} \qquad (33)$$

This constant is evaluated for the above bolometers in Table 11.2. For the detector of Mechin et al., the D^* has been reduced by a factor $\sqrt{2}$ to adjust D^* from its value at low-frequency to its value at the cutoff $f_{\max}$. It is notable that the recent work in sapphire and silicon/silicon nitride is all quite close in terms of the constant of Eq. (33). It will be interesting to see if replacing YBCO with GdBCO on sapphire leads to a yet better D^*.

TABLE 11.2 Detectivity Normalization for Different Operating Frequencies

Reference	f_{up} (Hz)	D^* at f_{up}	$\sqrt{f_{up}} D^*$
Verghese et al. (29)	10	4×10^9	1.26×10^{10}
Brasunas and Lakew (30)	4	6×10^9	1.20×10^{10}
Lakew et al (32)	3.8	1.2×10^{10}	2.34×10^{10}
Mechin et al (34)	254	1.77×10^9	2.82×10^{10}
de Nivelle et al (35)	2	1.8×10^{10}	2.55×10^{10}

The above bolometers have been single-pixel detectors generally. There is great interest in developing HTS arrays for applications such as cameras for thermal sensing. Johnson et al. (37) produced linear arrays with a Si_3N_4 layer for mechanical strength, a yttria-stabilized zirconium (YSZ) buffer layer, YBCO, and Si_3N_4 for a passivation layer. With 85 × 115-μm pixels, the achieved NEP is 9×10^{-12} (D^* of 1.1×10^9) with a 24-ms time constant. Mai et al. (38) fabricated an eight-element YBCO array on YSZ, achieving $2.46 \times 10^9 D^*$ at 10 Hz. The operation of this device was somewhat different, in that it is based on the temperature dependence of the critical current rather than the resistance. Li et al. (39) built single-element and 2 × 2 arrays of GdBCO on 50-μm-thick YSZ substrates; the typical area is 0.8 mm in diameter. They achieved $1.7 \times 10^{10} D^*$ for a single-element array, $0.7 \times 10^{10} D^*$ for the 2 × 2 array, and a time constant of 1–5 ms. However, it is unclear at what frequency the D^* was measured; therefore, it is not possible to evaluate this device in terms of the metric of Eq. (33). More recently, Liu et al. (40) have fabricated 1 × 8 linear arrays of YBCO on a ZrO_2 substrate, achieving 10^8D* to 10^9D* at 10 Hz.

11.5 OTHER TRANSITION-EDGE DETECTORS

In contrast with the transition-edge bolometers, there are a couple of detection techniques that do not require electrical contacts. One advantage of no contacts is the lack of a transport current and, therefore, the lack of $1/f$ noise due to the contacts. Another advantage is the possibility of achieving better thermal isolation, as wire connections are not used. The first detection technique is based on the temperature dependence of the microwave surface resistance (41). In terms of a simple model such as the two-fluid model (42), the real and imaginary parts of the surface impedance increase rapidly as the transition temperature is approached from below the transition temperature T_c. The other detection technique is based on the diamagnetic screening, related to the Meissner expulsion of magnetic flux from the interior of a superconductor except for a surface layer of thickness λ, the magnetic field penetration depth. In terms of the two-fluid model, the penetration depth is finite near $T = 0$ and increases toward infinity as T_c is approached from below (43). This so-called magnetic susceptibility bolometer was first realized by Brasunas et al. (44). Incoming infrared radiation affects the temperature and, thus, the penetration depth of a thin-film superconductor. The penetration depth is monitored by placing the film inside and perpendicular to the axis of a coil. The film thus modifies the self-inductance of the coil, which is monitored by standard electronic means. The response versus temperature becomes sharper as the film becomes thick with respect to the penetration depth. Thus, this detector works via a magnetic induction effect. The other extreme, very thin films, corresponds to the so-called kinetic inductance effect, which is related to the inertia of the electrons.

Additional work with a diamagnetic screening bolometer was done by Zerov et al. (45), who achieved a responsivity of 6.5 V/W at 3.5 mT magnetic field and 88 K.

11.6 ROOM-TEMPERATURE HTS DETECTORS

High-quality HTS films typically have a metallic temperature dependence on the resistance above T_c, therefore, it is a straightforward proposition to make a room-temperature bolometer using HTS films. Neff et al. (46) reported on an HTS bolometer with a D^* of 10^9 in the transition; the D^* is 10^7 between 100 K and 150 K and declines to $(1–2) \times 10^6$ at 300 K. The time constant increases from 0.33 ms in the transition to 1.55 ms at room temperature. HTS films have also been used to make uncooled pyroelectric infrared detectors (47), with a reported D^* of 10^8.

11.7 SUPERCONDUCTING REGIME DETECTORS

We next consider the class of detectors operating below T_c, not necessarily just below T_c as would be the case for the surface impedance and diamagnetic screening detectors. Typical detection methods would be the temperature dependence of the critical current I_c and of the penetration depth λ. It is also possible for incoming photons to influence directly, say, the kinetic inductance by breaking Cooper pairs rather than through the intermediate step of phonons. It has sometimes been somewhat controversial to identify whether a particular detector is exhibiting bolometric or nonbolometric (i.e., nonthermal) behavior. Sometimes fast response will be identified as nonthermal, but care must be taken, as it has been pointed out earlier that nanosecond-scale bolometric response is possible. Sometimes a nonlinear dependence on the bias current will be cited as evidence for nonthermal behavior, but, again, care must be taken because current-biased detectors can exhibit nonlinearity as the destabilization current is approached, modifying the effective thermal conductance G_e. The wavelength dependence for the nonbolometric detector will not be as flat as the bolometric detector, again assuming that the absorber has a flat response versus wavelength. As a general statement, a conclusion that a detector does not behave as one would expect for a thermal detector may, instead, indicate that the assumed thermal model for the detector is inadequate.

Hegmann et al. (48) measured R versus T for a range of ambient magnetic field conditions. They then estimated the effect of magnetic field on bolometric response and compared this with the observed dependence of the optical response on the magnetic field to identify a nonbolometric component below T_c. Schneider et al. (49) examined a nonbolometric response at low-temperature ($dR/dT \sim 0$) and found a photoresponse $\sim\omega^{-2.3}$ for wave numbers from 10 to 1000 cm^{-1}. They identified this as an ac Josephson effect. Semenov et al. (50)

identified a 90-ps response at the bottom of the transition, 78 K, which they identified as Cooper-pair-breaking. Ghis et al. (51) measured at 50 K a 12-ps rise time and a 29-ps width, which they identified as a nonequilibrium phenomenon due to the pair population being temporarily reduced. Hegmann et al. (52) measured a 20-ps response at 77 K, which they attributed to a kinetic inductance effect; whether it is nonequilibrium was unclear. Moix et al. (53) detected a 700-ps response to 35-ps input pulses below T_c. They attributed this to a kinetic inductance effect, noting a linear dependence on the bias current and inferring a 220-nm London depth. Danerud et al. (54) measured a 30-ps response to a 17-ps input, which they identified as electron heating, followed by a slower bolometric decay. Lindgren et al. (55) further investigated the picosecond response, concluding that the incoming photons directly cause a rise in the electron temperature T_e with a rise time of 0.56 ps, followed by a decay time of 1.1 ps as energy is transferred from the electrons to the phonons.

Bhattacharya et al. (56) measured optical responses between 15 K and 35 K and identified the mechanism as a Josephson weak link. The detection mechanism is the temperature dependence of the weak-link critical current I_c, and supporting evidence was the nonlinear dependence on the bias current I_b. The relationship between a voltage response δV and a temperature change δT satisfies the relation

$$\delta V = R\left(\left(\frac{I_b}{I_c}\right)^2 - 1\right)^{-1/2} \frac{dI_c}{dT}\,\delta T \tag{34}$$

Although this detector consisted of a single grain boundary, the mechanism should pertain to granular films in general. Huang et al. (57) measured an NEP of 9×10^{-13} at 10 Hz, below T_c. They ascribe the detection mechanism to the light-induced suppression of the critical current.

11.8 HTS FILM PRACTICAL CONSIDERATIONS

11.8.1 Substrate Selection and Metallization

For thermal detectors in particular, the substrate needs to be thin and of low specific heat. A high Debye temperature is associated both with low heat capacity near 90 K, due to freezing out of phonon modes, and with the strength necessary to enable thinness. Diamond would be the ideal candidate (see Table 11.1), but carbon will poison the YBCO and the diamond will decompose at the high temperatures necessary for HTS film fabrication. Also, the thermal expansion mismatch will lead to cracking when the film and substrate are cycled from deposition temperature to room temperature to operating temperature. Sapphire, the second choice from a thermal viewpoint, is a much better candidate, especially with CeO_2 buffer layers. The highest-quality films are grown on substrates such as $SrTiO_3$, which are less suitable from a thermal viewpoint. An alternative to thinning bulk sapphire is to use grown membranes such as silicon nitride with buffer layers starting with silicon. Films on silicon must be especially thin to avoid problems with microcracking. As for metallization, silver and gold appear to be most benign, in particular for YBCO.

11.8.2 Stability and Passivation

There are a couple of degradation mechanisms of HTS films, in particular YBCO. The first is oxygen loss, which can convert the superconductor to a semiconductor. Rothman et al. (58) noted that the diffusion of oxygen in the *c* direction is 10^6 times lower than diffusion in polycrystals. This is consistent with the observation that high-quality films are more stable and emphasizes the need for *c*-axis-aligned thin films. Michaelis et al. (59) note that H_2O enhances oxygen out-diffusion, pointing out the need for a passivation layer if exposure to high humidity is anticipated. Russek et al. (60) noted that an amorphous surface layer forms on YBCO upon exposure air and to CO_2, with an increase of contact resistivity. Exposure to N_2 also causes a degradation; exposure to vacuum does not. Excess barium in YBCO, especially present at grain boundaries, is also a problem when combining with water and carbon dioxide to form barium carbonate $BaCO_3$. Tatum et al. (61) found that diamondlike carbon and amorphous carbon can protect YBCO against humidity. Aguiar et al. (62) noted that CeO_2 protects YBCO against extreme humidity.

There are particular concerns with stability on silicon due to the microcracking issue, and this is perhaps reflected in the fact that HTS films on silicon substrates do not appear to be widely available commercially. Tiwari et al. (63) found that 60-nm-thick YBCO is crack-free on Si(100) with a CaF_2 buffer layer. Copetti et al. (64) found that 50 nm is the limit for YBCO on YSZ/Si(100) to avoid microcracks. Thinner films are more stable, but still show aging over several weeks. Haakennasen et al. (65) found that 1500–3000 Å of YBCO can be grown with high-quality on a thin silicon substrate, 4000 Å of Si, 300 Å of YSZ, and 80 Å of CeO_2.

11.8.3 Radiation Effects

There is vast literature on the effects of ionizing radiation on HTS thin films. Ionizing radiation comes in various forms (photons, electrons, protons, ions) and energies. The effect of radiation can be beneficial; for instance, high-energy ions can create columnar defects, leading to improved pinning and critical currents. The effects can also be detrimental, say for a transition-edge bolometer, either by broadening the transition, increasing the $1/f$ noise, or lowering T_c. Among others, ionizing radiation is of concern for applications such as operation in outer space, in nuclear reactors, or in military operations.

Brasunas and Lakew (66) studied the effect of 1-MeV γ-rays on the YBCO-resistive transition. Choosing a 100-krad total ionizing dose as being representative of the 12-year NASA/ESA Cassini/Huygens mission launched in 1997 to Saturn and Titan, they determined that the low-frequency (1–25 Hz) noise and transition width were unchanged. Thus, the bolometer S/N is resistant to this level of radiation. Moss et al, (67) made the interesting observation that the adhesion of silver pads to YBCO is actually improved by electron irradiation. However, the contact resistance went up, a deleterious effect.

11.8.4 Temperature Regulation and the Effect of Ambient Magnetic Fields

Consider the case of the transition-edge bolometer, where the required temperature regulation will be necessarily stringent compared with some of the other detection mechanisms. Consider the structure of HTS thin film on the substrate on a temperature-controlled stage (or cold head) (Fig. 11.2a). The thermodynamic fluctuations implied by Eq. (11) for the film and the substrate lead to the phonon-noise limit of Eq. (13). To ensure good detector S/N, there are essentially two requirements: (1) the cold-head temperature must be regulated to the transition region, typically to the midpoint of the transition; (2) the incoming signal must not drive the bolometer outside of the transition (dynamic range consideration). Typically, the cold head is much more massive than the substrate; thus, the thermal fluctuations such as predicted by Eq. (11) would be negligible. However, there can be environmental forcing, such as a diurnal effect or crossing from sunlight to shadow. If the forcing is very small, the cold-head temperature can simply be set within the transition once and for all. However, in general, the forcing will be such that active feedback-type control is necessary to maintain the cold-head temperature within the transition.

This control could be done by a conventional temperature regulator, using, for instance, a diode thermometer. Even more sensitivity is possible if a high-T_c film is used as the thermometer for temperature regulation (68), either an additional HTS film or the detector itself. If the detector itself is used to regulate the temperature, the notion is that the infrared signals are ac (say, 0.1 Hz and above) and the range from dc to, say, 0.01 Hz is used for temperature control. In this case, the feedback attempts to make the observed resistance of the thin-film below 0.01 Hz stable, and the HTS signal is read out in the conventional way. As an alternative, as long as the frequency ranges of the environmental forcing and infrared signals do not overlap, it is possible to increase the frequency range of the feedback control to include both the forcing and the infrared signals. Now, the detector resistance is held truly fixed and the information about the infrared signal is included in the high-frequency (>0.1 Hz) part of the feedback control. The dynamic range can be increased in two ways. First, through the use of feedback, as discussed earlier. Second, by voltage biasing, because voltage biasing decreases the effective thermal isolation, and, thus, it takes a larger signal to drive the HTS film outside the transition.

Because the HTS-resistive transition is shifted and broadened by ambient magnetic field, it is necessary to bound the magnetic excursions as well as the thermal excursions in order to ensure a good S/N. Romani et al. (69) reported that a low-noise laboratory (but not a magnetically shielded room) has an ambient magnetic field noise of about 10^{-9} T/Hz$^{1/2}$ at 1 Hz, and from 10^{-9} to 10^{-10} at 10 Hz. Hegmann et al. (70) show that in the transition midpoint of YBCO at 10 mA bias, a shift of 0.1 K is equivalent to 100 G or 10^{-2} *T*, so 1 *K* corresponds roughly to 0.1 *T*. Now, if we take 10^{-4} W/K as a typical thermal conductance, as implied by Eq. (19) for gold wires, Eq. (13) then gives 7×10^{-12} W/Hz$^{1/2}$ for the NEP (and 1.5×10^{10} D* for a 1×1-mm detector). Given this *G*, this NEP corresponds to a temperature fluctuation of 7×10^{-8} K/Hz$^{1/2}$ and by the above relation about 10^{-8} T/Hz$^{1/2}$. Thus, the typical laboratory magnetic field noise fall short by about a factor of 10 in affecting the detector noise level, but unusually high transients would be of concern.

REFERENCES

1. CT Li, BS Deaver, M Lee, RM Weikle, RA Rao, CB Eom. Gain bandwidth and noise characteristics of millimeter-wave $YBa_2Cu_3O_7$ hot-electron bolometer mixers. Appl Phys Lett 73:1727–1729, 1998.
2. ZM Zhang, A Frenkel. Thermal and nonequilibrium response of superconductors for radiation detectors. J Superconduct 7:871–884, 1994.
3. AM Kadin, M Leung, AD Smith, JM Murduck. Photofluxonic detection: A new mechanism for infrared detection in superconducting thin films. Appl Phys Lett 57:2847–2849, 1990.
4. ZM Zhang, TA Le, MI Flik, EG Cravalho. Infrared optical constants of the high-T_c superconductor $YBa_2Cu_3O_7$. J Heat Transfer 116:253–257, 1994.
5. JP Rice, EN Grossman, DA Rudman. Antenna-coupled high-T_c air-bridge microbolometer on silicon. Appl Phys Lett 65:773–775, 1994.
6. RA Smith, FE Jones, RP Chasmar. The Detection and Measurement of Infra-Red Radiation. 2nd ed. Oxford: Oxford University Press, 1968, pp 49–50.
7. RA Smith, FE Jones, RP Chasmar. The Detection and Measurement of Infra-Red Radiation. 2nd ed. Oxford: Oxford University Press, 1968, pp 211–213.
8. RA Smith, FE Jones, RP Chasmar. The Detection and Measurement of Infra-Red Radiation. 2nd ed. Oxford: Oxford University Press, 1968, p 51.
9. J Brasunas, B Lakew. High-T_c superconductor bolometer with record performance. Appl Phys Lett 64:777–778, 1994.
10. RA Smith, FE Jones, RP Chasmar. The Detection and Measurement of Infra-Red Radiation. 2nd ed. Oxford: Oxford University Press, 1968, pp 52–56.
11. M Nahum, S. Verghese, PL Richards, K Char. Thermal boundary resistance for $YBa_2Cu_3O_{7-\delta}$ films. Appl Phys Lett 59:2034–2036, 1991.

12. LP Lee, MJ Burns, K Char. Free-standing microstructures of $YBa_2Cu_3O_{7-\delta}$: A high-temperature superconducting air bridge. Appl Phys Lett 61:2706–2708, 1992.
13. SJ Berkowitz, AS Hirahara, K Char, EN Grossman. Low-noise high-temperature superconducting bolometers for infrared imaging. Appl Phys Lett 69:2125–2127, 1996.
14. RA Smith, FE Jones, RP Chasmar. The Detection and Measurement of Infra-Red Radiation. 2nd ed. Oxford: Oxford University Press, 1968, pp 258–259.
15. JC Mather. Bolometer noise: nonequilibrium theory. Appl Opt 21:1125–1129, 1982.
16. MI Flik, PE Phelan, CL Tien. Thermal model for the bolometric response of high-T_c superconducting films to optical pulses. Cryogenics 30:1118–1128, 1990.
17. S Bauer, B Ploss, Interference effects of thermal waves and their application to bolometric and pyroelectric detectors. Sensor Actuators A 25–27:417–421, 1991.
18. M Tinkham. Introduction to Superconductivity. Reprint Ed. Malabar, FL: Kreiger, 1980, pp 166–169.
19. M Fardmanesh, K Scoles, A. Rothwarf. DC characteristics of patterned $YBa_2Cu_3O_{7-x}$ superconducting thin-film bolometers: Artifacts related to Joule heating, ambient pressure, and microstructure. IEEE Trans Appl Supercond As-8:69–78, 1998.
20. W Becker, R Fettig, A Gaymann, W Ruppel. Black gold deposits as absorbers for far infrared radiation. Phys. Status Solidi (b) 194:241–255, 1996.
21. M Dragovan, SH Moseley. Gold absorbing film for a composite bolometer. Appl Opt 23:654–656, 1984.
22. M Epifani. Effects of thermal link in bolometric detectors. Appl Opt 34:6327–6331, 1995.
23. JC Brasunas. Measuring and modeling the frequency response of infrared detectors. Inform Phys. Technol 38:69–74, 1997.
24. M Fardmanesh, A. Rothwarf, KJ Scoles. Low and midrange modulation frequency response for YBCO infrared detectors: Interface effects on the amplitude and phase. IEEE Trans Appl Supercond As-5:7–13, 1995.
25. RA Smith, FE Jones, RP Chasmar. The Detection and Measurement of Infra-Red Radiation. 2nd ed. Oxford: Oxford University Press, 1968, pp 181–187.
26. RA Smith, FE Jones, RP Chasmar. The Detection and Measurement of Infra-Red Radiation. 2nd ed. Oxford: Oxford University Press, 1968, pp 196–203.
27. RA Smith, FE Jones, RP Chasmar. The Detection and Measurement of Infra-Red Radiation. 2nd ed. Oxford: Oxford University Press 1968, pp 214–220.
28. AT Lee, JM Gildemeister, SF Lee, PL Richards. Voltage-biased high-T_c superconducting infrared bolometers with strong electrothermal feedback. IEEE Trans Appl Supercond. As-7:2378–2381, 1997.
29. S Verghese, PL Richards, K Char, SA Sachtjen. Fabrication of an infrared bolometer with a high-T_c superconducting thermometer. IEEE Trans Magnetics M-27:3077–3080, 1991.
30. J Brasunas, B Lakew. High-T_c superconductor bolometer with record performance. Appl Phys Lett 64:777–778, 1994.
31. A Pique, KS Harshavardhan, J Moses, M Mathur, T Venkatesan, JC Brasunas, B Lakew. Deposition of high-quality $YBa_2Cu_3O_{7-x}$ films on ultrathin (12 μm thick) sapphire substrates for infrared detector applications. Appl Phys Lett 67:1920–1922, 1995.

32. B Lakew, JC Brasunas, A Pique, R Fettig, B Mott, S Babu, GM Cushman. High-T_c superconducting bolometers on chemically etched 7 μm thick sapphire. Physica C 329:69–74, 2000.
33. L Mechin, JC Villegier, P Langlois, D Robbes, D Bloyet. Sensitive IR bolometers using superconducting YBaCuO air bridges on micromachined silicon substrates. Sensors Actuators A 55:19–23, 1996.
34. L Mechin, JC Villegier, D Bloyet. Suspended epitaxial YBaCuO microbolometers fabricated by silicon micromachining: Modeling and measurements. J Appl Phys 81:7039–7047, 1997.
35. MJME de Nivelle, MP Bruijn, R deVries, JJ Wijnbergen, PAJ deKorte, S Sanchez, M Elwenspoek, T Heidenblut, B Schwierzi, W Michalke, E Steinbeiss. Low-noise high-T_c superconducting bolometers on silicon nitride membranes for far-infrared detection. J Appl Phys 82:4719–4726, 1997.
36. D. Bloyet, D Robbes, P Langlois, A Gilabert, J Aboudihab, A Kreisler, JM Depond, D Pavuna, A Gauzzi, B Dwir, JC Villegier, A Ghis, A Jager, F Pourtier. Studies on radiation detectors using high-T_c superconductors. Physica C 234–240:3391–3392, 1994.
37. BR Johnson, T Ohnstein, CJ Han, R Higashi, PW Kruse, RA Wood, H Marsh, SB Dunham. $YBa_2Cu_3O_7$ superconductor microbolometer arrays fabricated by sillicon micromachining. IEEE Trans Appl Supercond. As-3:2856–2859, 1993.
38. Z Mai, X Zhao, F Zhou, W Song. Infrared radiation detector linear arrays of high-T_c superconducting thin films. Inform Phys Technol 38:13–16, 1997.
39. H Li, R Wang, F Wan, Y Ping, G He, M Yu. High-T_c $GdBa_2Cu_3O_{7-\delta}$ Superconducting thin film bolometers. IEEE Trans Appl. Supercond As-7:2371–2373, 1997.
40. X Liu, B Shi, X Liu, J Chu, D Yang. The fabrication of a 1×8 linear array high-T_c superconductor infrared detector. Inform Phys Technol 40:83–85, 1999.
41. R Kaplan, WE Carlos, EJ Cukauskas, J Ryu. Microwave detected optical response of $YBa_2Cu_3O_{7-x}$ thin films. J Appl Phys 67:4212–4216, 1990.
42. T Van Duzer, CW Turner. Principles of Superconductive Devices and Circuits. New York: Elsevier, 1981, pp 128–131.
43. T Van Duzer, CW Turner. Principles of Superconductive Devices and Circuits. New York: Elsevier, 1981, pp 124–125.
44. J Brasunas, B Lakew, C Lee. High-temperature-superconducting magnetic susceptibility bolometer. J Appl Phys 71:3639–3641, 1992.
45. VY Zerov, VN Leonov, MV Sosnenko, JA Khrebtov, AA Ivanov. A new type of high-temperature superconductor bolometer using the diamagnetic-screening effect. J Opt Technol 65:242–244, 1998.
46. H Neff, J Laukemper, C Hefle, M Burnus, T Heidenblut, W Michalke, E Steinbeiss. Extended function of a high-T_c transition-edge bolometer on a micromachined Si membrane. Appl Phys Lett 67:1917–1919, 1995.
47. DP Butler, Z Celik-Butler, A Jahanzeb, JE Gray, CM Travers. Micromachined YBaCuO capacitor structures as uncooled pyroelectric infrared detectors. J Appl Phys 84:1680–1687, 1998.
48. FA Hegmann, JS Preston. Identification of nonbolometric photoresponse in $YBa_2Cu_3O_{7-\delta}$ thin films based on magnetic field dependence. Appl Phys Lett 62:1158–1160, 1993.

49. G Schneider, PG Huggard, T O'Brien, D Lemoine, W Blau, W Prettl. Spectral dependence of nonbolometric far-infrared detection with thin-films $Bi_2Sr_2CaCu_2O_8$. Appl Phys Lett 60:648–650, 1992.
50. AD Semenov, GN Gol'tsman, IG Gogidze, AV Sergeev, PT Lang, KF Renk. Subnanosecond photoresponse of a YBaCuO thin film to infrared and visible radiation by quasiparaticle induced suppression of superconducitivity. Appl Phys Lett 60:903–905, 1992.
51. A Ghis, JC Villegier, S Pfister, M Nail, Ph. Gibert. Electrical picosecond measurements of the photoresponse in $YBa_2Cu_3O_{7-x}$. Appl Phys Lett 63:551–552, 1993.
52. FA Hegmann, RA Hughes, JS Preston. Picosecond photoresponse of epitaxial $YBa_2Cu_3O_{7-\delta}$ thin films. Appl Phys Lett 64:3172–3174, 1994.
53. DB Moix, DP Scherrer, FK Kneubuhl. Photoresponse of the high-temperature superconductor $YBa_2Cu_3O_{7-\delta}$ to ultrashort 10 μm CO_2 laser pulses. Inform Phys Technol 37:403–426, 1996.
54. M Danerud, D Winkler, M Lindgren, M Zorin, V Trifonov, BS Karasik, GN Gol'tsman, EM Gershenzon. Nonequilibrium and bolometric photoresponse in patterned $YBa_2Cu_3O_{7-\delta}$ thin films. J Appl Phys 76:1902–1909, 1994.
55. M Lindgren, M Currie, C Williams, TY Hsiang, PM Fauchet, R Sobolewski, SH Moffat, RA Hughes, JS Preston, FA Hegmann. Intrinsic picosecond response times of Y–Ba–Cu–O superconducting photodetectors. Appl Phys Lett 74:853–855, 1999.
56. S Bhattacharya, M Rajeswari, I Takeuchi, XX Xi, SN Mao, C Kwan, Q Li, T Venkatesan. Low-temperature optical response of a single grain boundary in superconducting $YBa_2Cu_3O_{7-\delta}$ thin films. Appl Phys Lett 63:2279–2281, 1993.
57. MQ Huang, L Chen, T Yang, JC Nie, PJ Wu, GR Liu, ZX Zhao. Infrared detector fabricated by YBCO Josephson junctions in series. Physica C 282–287:2545–2546, 1997.
58. SJ Rothman, JL Routburt, U Welp, JE Baker. Anisotropy of oxygen tracer diffusion in single-crystal $YBa_2Cu_3O_{7-\delta}$. Phys Rev B 44:2326–2333, 1991.
59. A Michaelis, EA Irene, O Auciello, AR Krauss. A study of oxygen diffusion in and out of $YBa_2Cu_3O_{7-\delta}$ thin films. J Appl Phys 83:7736–7743, 1998.
60. SE Russek, SC Sanders, A Roshko, JW Ekin. Surface degradation of superconducting $YBa_2Cu_3O_{7-\delta}$ thin films. Appl Phys Lett 64:3649–3651, 1994.
61. JD Tatum, JWH Tsai, M Chopra, S Chen, JM Phillips, SY Hau. Use of carbon films for passivation and environmental protection of superconducting $YBa_2Cu_3O_{7-x}$. J Appl Phys 77:6370–6376, 1995.
62. R Aguiar, F Sanchez, C Ferrater, M Varela. Protective oxide coatings for superconducting $YBa_2Cu_3O_{7-x}$ thin films. Thin Solid Films 306:74–77, 1997.
63. AN Tiwari, S Blunier, H Zogg, Ph Lerch, F Marcenat, P Martinoli. Epitaxial growth of superconducting $YBa_2Cu_3O_{7-x}$ on Si(100) with CaF_2 as intermediate buffer. J Appl Phys 71:5095–5098, 1992.
64. CA Copetti, J Schubert, W Zander, H Soltner, U Poppe, Ch Burchal. Aging of superconducting $Y_1Ba_2Cu_3O_{7-x}$ structures on silicon. J Appl Phys 73:1339–1342, 1993.
65. R Haakennasen, DK Fork, JA Golovchenko. High-quality crystalline $YBa_2Cu_3O_{7-\delta}$ films on thin silicon substrates. Appl Phys Lett 64:1573–1575, 1994.

66. JC Brasunas, B Lakew. Transition-edge noise in $YBa_2Cu_3O_{7-x}$ thin films before and after exposure to ionizing radiation. J Appl Phys 75:7565–7566, 1994.
67. SD Moss, RA O'Sullivan, PJK Paterson, IK Snook, AJ Russo, A Katsaros, N Savvides. Modification of the adhesion and contact resistance of the $Ag/YBa_2Cu_3O_7$ interface with KeV electron irradiation. J Appl Phys 78:5782–5786, 1995.
68. E Lesquey, C Gunther, S Flament, R Desfeux, JF Homet, D Robbes. Progress toward a low-noise temperature regulator using a superconductive high-T_c microbridge. IEEE Trans Appl Supercond As-5:2427–2430, 1995.
69. GL Romani, SJ Williamson, L Kaufman. Biomagnetic instrumentation. Rev Sci Instrum 53:1815–1845, 1982.
70. FA Hegmann, JS Preston. Identification of nonbolometric photoresponse in $YBa_2Cu_3O_{7-\delta}$ thin films based on magnetic field dependence. Appl Phys Lett 62:1158–1160, 1993.

12

Cryocoolers and High-T_c Devices

Ray Radebaugh

National Institute of Standards and Technology, Boulder, Colorado, U.S.A.

12.1 INTRODUCTION

12.1.1 Cooling Requirements for High-T_c Superconducting Electronic Devices

A long-range goal in the study of superconductivity is to find a new material with a superconducting transition temperature (T_c) significantly above room temperature so that there would be no need to cool the superconductor. Such a breakthrough would be of profound significance, because it would then free superconductivity of the problems imposed upon it by the need for cooling. Operating temperatures of less than about two-thirds of the transition temperature are required to significantly reduce the temperature dependence of the critical current and to achieve satisfactory performance of superconductors in practical applications. At present, temperatures below 80 K are needed for practical use of even the highest-temperature superconductors. Although liquid nitrogen is often used for laboratory studies of high-T_c devices, it is rarely satisfactory for commercial applications. This dependence on cryocooling then adds another set of problems that must be overcome in moving a superconducting device into the marketplace. In terms of any marketable product, the superconducting device and the cryocooler must be considered an inseparable pair. There are many problems associated with cryocooling the superconductor, and it is these problems that often prevent the superconductor from making it into the marketplace. Studies to improve the performance of superconducting devices and systems should be coupled with studies to improve the performance and lower the cost of cryocoolers.

The purpose of this chapter is to discuss the various methods available for cooling high-T_c superconducting electronic devices. The differing requirements of various superconducting devices often lead to different cooling methods being employed. There is no one method that is best for all applications. Various cooling methods have been review briefly by the author (1). There are many problems associated with all types of cryocoolers, and these will be discussed here. The operating principles of each cryocooler type will be explained in this chapter, along with their advantages and disadvantages, to aid in the selection of the optimum cryocooler for a particular application. The refrigeration power required for high-T_c electronic devices is usually less than a few watts at temperatures between 60 K and 80 K. Thus, the discussion here of the different types of cryocoolers will focus on this requirement for small cooling loads as opposed to the need for much larger cooling loads in bulk superconductor applications usually involving magnets. With such small cooling loads, efficiency is seldom a concern with regard to the cost of the input power, unless it is to be used for satellite applications. However, the dissipation of this power in the form of heat in confined areas can sometimes be a problem, so efficiency then becomes important.

Other requirements of cooling systems for superconducting electronic devices often vary depending on the application. Because electronic devices deal with low-level electromagnetic signals, they are easily disturbed by electromagnetic interference (EMI) from nearby motors. That is a particularly serious problem when cooling a superconducting quantum interference device (SQUID) that can sense magnetic fields as small as a few femtotesla. The SQUID is also sensitive to vibration in the Earth's magnetic field. Excessive vibration of superconductor–insulator–superconductor (SIS) junctions for microwave receivers can lead to distortions of the signal. Because there are no moving parts in most electronic devices, their lifetimes are extremely long. Such hi-tech devices would usually become obsolete after 5–10 years rather than fail. Such long lifetimes are difficult to achieve in cryocoolers. Cost is always an important consideration when attempting to market a superconducting device. Often the cost of the superconductor/cryocooler system is dominated by the cost of the cryocooler. Table 12.1 summarizes the cooling requirements for most high-T_c superconducting electronic devices.

12.1.2 Cooling Systems: General Thermodynamic Introduction

Cooling systems can be either open thermodynamic systems, as shown in Figure 12.1, where mass crosses the system boundary, or closed thermodynamic systems,

TABLE 12.1 Requirements for Cryocooling High-T_c Superconducting Electronic Devices

Requirement	Comment
Low cost	Should not dominate cost of system
High reliability	At least 3-year lifetime (5 years preferred) with little maintenance
High efficiency	Needed for low heat rejection
Low EMI	Should not degrade performance of superconductor
Low vibration	Quality perception as well as required for some applications
Small size	Should not dominate size of complete system

where no mass crosses the system boundary. The open cooling system is represented by a liquefier, which produces some cryogen, such as liquid nitrogen or liquid helium. This cryogen leaves the system and is transported by some means to the site where it is to provide cooling. After evaporation, the gas can be returned to the liquefier or vented to the atmosphere. The use of open systems is also used when analyzing a portion of a complete cryocooler (e.g., a heat exchanger). In that case, mass also crosses the system boundary.

The first and second laws of thermodynamics are used in analyzing both the open and closed cooling systems. The first law of thermodynamics is simply an

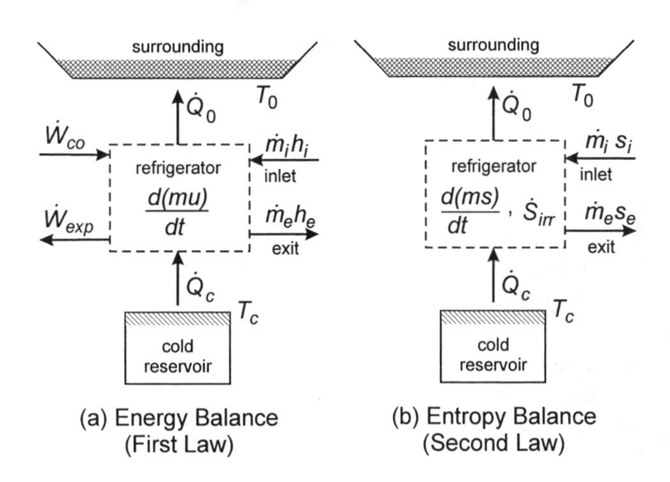

FIGURE 12.1 Open thermodynamic system showing energy and entropy terms appropriate to the analysis by the (a) first and (b) second laws of thermodynamics.

energy balance on the system. Figure 12.1a shows the energy terms appropriate for an open system. In a closed system, the mass flows are zero. For open thermodynamic systems, the first law of thermodynamics is expressed by

$$\dot{Q}_c - \dot{Q}_0 + \dot{m}_i h_i - \dot{m}_e h_e = \dot{W}_{\text{exp}} - \dot{W}_{\text{co}} + \frac{d(mu)}{dt} \tag{1}$$

where $\dot{Q}_c$ is the heat absorbed, or refrigeration power, at the temperature T_c, $\dot{Q}_0$ is the heat rejected to the surrounding at the temperature T_0, $\dot{m}_i$ and $\dot{m}_e$ are the mass flow rates at the system inlet and exit, respectively, h_i and h_e are the specific enthaplies of the fluids crossing the boundary at the system inlet and exit, respectively, $\dot{W}_{\text{exp}}$ is the power produced by the system from any expansion process, $\dot{W}_{\text{co}}$ is the power delivered to the system, such as in a compressor, and mu is the system internal energy. Usually, this compressor power is in the form of electrical power to the compressor motor or possibly the electrical power to the conditioning electronics before the compressor. When dealing with the thermodynamics of the working fluid, the input and expansion powers are expressed as mechanical power, or PV power of the device. The last term in Eq. (1) is the time rate of change of the internal energy of the system, which is zero under steady-state conditions. Many refrigeration systems either recover the expansion work internally or do not recover any expansion work. Thus, $\dot{W}_{\text{exp}}$ is often zero for a complete refrigeration system, which, for a closed refrigeration system in steady-state conditions, leads to the simple energy balance of

$$\dot{Q}_0 = \dot{W}_{\text{co}} + \dot{Q}_c \tag{2}$$

The significance of Eq. (2) is that it tells us that all of the power input to the refrigerator must be rejected to the surroundings in the form of heat. In addition, the heat absorbed at the cold end also must be rejected to ambient, but this is always a small fraction of the power input for cryogenic refrigerators, as we shall see later. The heat rejection to ambient is often a problem associated with closed-cycle cryocoolers, particularly if the efficiency is low and large power inputs are required. In small systems, there may not be a problem with providing several hundred watts or even a kilowatt of power if the power comes from a standard wall circuit. However, if the system is to be made compact and is to rely only on air cooling, it may be more of a problem in rejecting that much heat to ambient. A higher-efficiency cryocooler would then reject less heat to ambient for the same refrigeration power. The first law of thermodynamics says nothing about the relative size of $\dot{W}_{\text{co}}$ and $\dot{Q}_{\text{c}}$ in Eq. (2). For that relationship, we must rely on the second law of thermodynamics.

The entropy balance given by the second law of thermodynamics for open refrigeration systems is represented in Figure 12.1b. For an open system, the entropy change of the refrigerator is given by

$$\frac{d(ms)}{dt} = \frac{\dot{Q}_c}{T_c} - \frac{\dot{Q}_0}{T_0} + \dot{m}_i s_i - \dot{m}_e s_e + \dot{S}_{\text{irr}} \quad (3)$$

where s_i and s_e are the specific entropies of the fluid at the inlet and exit to the system, respectively, and $\dot{S}_{\text{irr}}$ is the entropy production rate (≥ 0) inside the system boundary associated with any irreversible process. For a closed system, the flow-rate terms are zero. For an ideal closed system at steady state where all processes are reversible, $\dot{S}_{\text{irr}} = 0$, the second law of thermodynamics from Eq. (3) becomes

$$\frac{\dot{Q}_c}{T_c} = \frac{\dot{Q}_0}{T_0} \quad \text{(ideal)} \quad (4)$$

Thus, the second law of thermodynamics is used to give the relative size of the heat-flow terms at different temperatures. For a closed ideal system at steady state, the combined first and second laws of thermodynamics, Eqs. (2) and (4), gives

$$\dot{W}_{\text{co}} = \left(\frac{T_0}{T_c} - 1\right)\dot{Q}_c \quad \text{(ideal)} \quad (5)$$

The coefficient of performance (COP) of any system is given by the ratio of the desired power to the actual power required to drive the system, which, for a refrigerator, becomes

$$\text{COP} = \frac{\dot{Q}_c}{\dot{W}_{\text{co}}} \quad (6)$$

In comparing the COP of various practical refrigerators, it is important to understand what conditions were used in arriving at the COP; that is, what are the two temperature levels and what power was used for the input power? Was it the compressor mechanical PV power, the electrical power to the compressor, or the electrical power to some power conditioning electronics? For the ideal reversible refrigerator, the COP from Eq. (5) becomes

$$\text{COP}_{\text{Carnot}} = \frac{T_c}{T_0 - T_c} \quad \text{(ideal)} \quad (7)$$

which is the COP of the Carnot cycle as well as any ideal cycle operating between T_0 and T_c with no irreversible processes. Any real cycle will have a COP less than the Carnot value. The relative COP of an actual refrigerator, often referred to as the second-law efficiency, relative efficiency, or, simply, the efficiency, is expressed as

$$\eta = \frac{\text{COP}_{\text{actual}}}{\text{COP}_{\text{Carnot}}} \quad (8)$$

This efficiency is less dependent on the high- and low-temperature values than is the absolute COP. The inverse of the COP is called the specific power and represents the watts of input power per watt of refrigeration. For a low-temperature of

80 K and an ambient of 300 K, the Carnot COP from Eq. (7) is 0.364 and the specific power is 2.75 W/W. Typically, small cryocoolers may have a COP of about 10% of Carnot, so the specific power would be 27.5 W/W. Small cryocoolers have COP values that range from about 1% to 25% of the Carnot value. Large helium-liquefaction plants may operate at about 30% of Carnot and large air-liquefaction plants operate at about 50% of Carnot.

12.1.3 Open Systems Versus Closed Systems

The cryocooling of superconducting devices is often carried out in the laboratory by the use of liquid nitrogen or liquid helium. The user is seldom aware of any problems associated with the remote liquefier. However, the cost of the cryogen is influenced by such problems. Because most research laboratories are located in or near large metropolitan areas, liquid nitrogen or liquid helium can be obtained within a few days after an appropriate phone call, which is, the primary advantage of relying on an open cryocooling system. Researchers have the technical background that makes dealing with the cryogenic liquids, even liquid helium, a trivial matter. In practical applications with a large market, the end user most often will not have the technical background to be comfortable with the use of cryogenic liquids. Often the location may not allow for easy access to a reliable supply of cryogenic liquids. Only limited applications that are restricted to high-technology facilities would find the use of cryogenic liquids for cryocooling of interest. However, because much of the research on superconducting devices is carried out using liquid nitrogen or liquid helium, the next section will be devoted to their use for cooling superconducting devices.

Closed cryocooler systems generally operate with electrical input power. However, in some large systems, which will not be considered here, a high-temperature heat source, such as gas combustion, may be the power input. That power would either be converted to electrical power via a heat engine and generator or be converted directly to PV power via thermoacoustic drivers. However, these thermally driven systems do not scale down well to the small sizes needed for most superconducting device application. In the case of electrically driven systems, their operation merely requires the flipping of a switch to turn them on. This simple operation is ideal for widespread applications of superconducting devices. They can be incorporated easily into any superconducting system, and, ideally, remain invisible to the user. Unfortunately, closed-cycle cryocoolers still have their own set of problems that keep them from being truly invisible to the user.

12.2 COOLING WITH CRYOGENIC FLUIDS

12.2.1 Properties of Cryogenic Fluids

In most cases, liquid nitrogen is used for cooling high-temperature superconductors and liquid helium is used for the cooling of low-temperature superconductors.

However, liquid helium, or perhaps liquid neon, may be used in the study of high-T_c devices at temperatures below 63 K. The properties of several cryogenic fluids (cryogens) are listed in Table 12.2. In most cases, these properties are taken from NIST Standard Reference Data (2). Liquid neon may occasionally be used for achieving temperatures around 25–30 K, although it is much more expensive even than liquid helium. In the United States, liquid nitrogen typically costs somewhat less than milk, liquid helium costs about the same as inexpensive wine, and liquid neon costs about 10 times that of liquid helium. In the simplest of experiments, a sample to be cooled is immersed in the cryogen at atmospheric pressure. Cooling comes from the heat of vaporization of the liquid. This heat of vaporization increases with the normal boiling point. Temperatures below the normal boiling point can be achieved by pumping on the liquid to reduce its vapor pressure. Usually, the formation of the solid at the triple point determines the

TABLE 12.2 Properties of Several Cryogenic Fluids

Property	He^3	He^4	Para H_2	Normal H_2	Ne	N_2	Ar
Molecular weight (g/mol)	3.017	4.003	2.016	2.016	20.18	28.01	39.95
Normal boiling point (NBP) (K)	3.191	4.222	20.28	20.39	27.10	77.36	87.30
Triple-point temperature (K)	—	2.177	13.80	13.96	24.56	63.15	83.81
Triple-point pressure (kPa)	—	5.042	7.042	7.20	43.38	12.46	68.91
Critical temperature (K)	3.324	5.195	32.94	33.19	44.49	126.19	150.69
Critical pressure (MPa)	0.117	0.2275	1.284	1.315	2.679	3.396	4.863
Liquid density at NBP (g/cm^3)	0.05844	0.1249	0.07080	0.0708	1.207	0.8061	1.395
Gas Density at 0°C and 1 atm (kg/m^3)	0.134	0.1785	0.08988	0.08988	0.8998	1.250	1.784
Heat of vaporization at NBP (J/g)	7.714	20.72	445.5	445.6	85.75	199.18	161.14
Sensible heat from NBP to 300 K (J/g)		1543.3	4010.5	3510.8	255.3	234.03	112.28
Heat of fusion (J/g)		—	58.23	58.23	16.6	25.5	27.8
Heat of vaporization per volume of liquid at NBP (J/cm^3)	0.4508	2.588	31.54	31.55	103.5	160.6	224.8
Sensible heat per volume of liquid at NBP (J/cm^3)		192.8	283.9	248.6	308.1	188.7	156.6

*Lambda-point temperature and pressure

lower-temperature limit for that particular cryogen. When temperatures above the normal boiling point are required, the cold vapor is often used for cooling the sample. Table 12.2 also lists the sensible heat, or the enthalpy change, of the vapor in warming from the normal boiling point to 300 K. The heat absorbed by the vapor when warming to any temperature is proportional to the temperature rise for an ideal gas, because for an ideal gas, the specific heat is independent of temperature.

12.2.2 Cryostat Construction

The cooling of small samples to the temperature of the normal boiling point of a cryogen is most conveniently done by using the storage dewar as the cryostat. The sample is simply immersed in the cryogen, as shown by the first example in Figure 12.2. The O-ring seals shown in Figure 12.2 are needed when using liquid helium to keep air from entering the dewar and solidifying. Such seals are usually not needed when using liquid nitrogen, although some crude seal may be desired to eliminate excessive frost buildup over time. For samples larger than those that will fit down the neck of a storage dewar, it becomes necessary to construct a special cryostat. For simple experiments for short periods of time, an inexpensive cryostat can be made of foam insulation when dealing with liquid nitrogen. Of course, the boil-off rate will be much higher than with a vacuum-insulated dewar. Low-cost vacuum-insulated containers made of stainless steel and with capacities ranging from about 0.5 to 2 L are readily available at sporting goods stores. Although intended for use in keeping beverages hot or cold, they also work well with liquid nitrogen, provided the cap is not sealed tightly to allow the boil-off gas to vent. The advantage of the immersion cryostat is its simplicity. Its disadvantage is

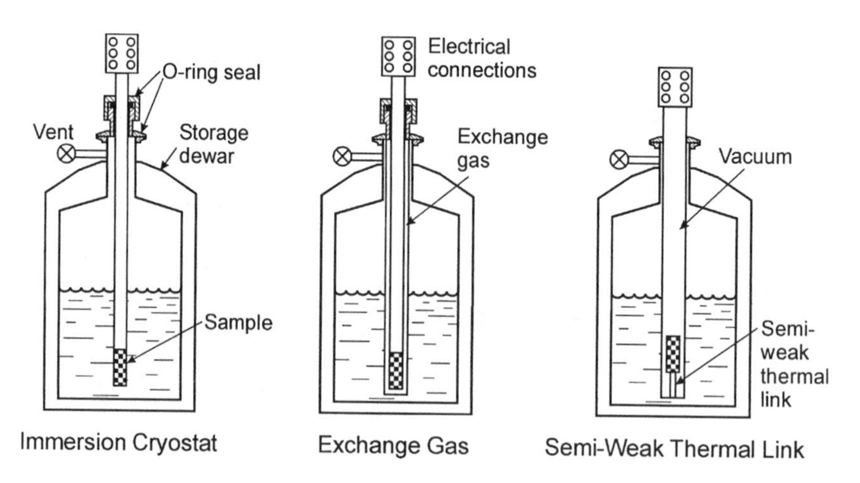

FIGURE 12.2 Use of cryogens for sample cooling.

that it does not allow for much temperature variation of the sample. Temperatures below the normal boiling point can be achieved by pumping on the vapor to reduce the pressure. Good seals are then required.

The center and right illustrations in Figure 12.2 show two methods used to cool samples but allow them to be heated with an electrical heater to some higher temperature than the normal boiling point of the cryogen. Both cases provide a semiweak thermal link between the sample and the cryogenic liquid. The use of a low-pressure exchange gas (usually helium) allows the thermal conductance of the link to be varied by varying the pressure of the exchange gas. For sufficiently low pressures, the thermal conductivity of a gas becomes proportional to the pressure.

12.2.3 Advantages and Disadvantages

The main advantage in using a cryogen like liquid nitrogen is the simplicity of cooling a device by immersion in the liquid. The cost of the liquid nitrogen and the dewar are very low. There is no EMI associated with the use of a cryogen as such, although the dewar may have some magnetic properties that could influence sensitive SQUID devices. Nonmetallic dewars of fiberglass-epoxy are often used with SQUIDs to reduce the magnetic noise of the dewar. The boiling of the liquid cryogen can produce some vibration that could pose a problem in a few cases. In research laboratories, often located in metropolitan areas, liquid nitrogen is readily available. However, even in those cases, there is always the need for human involvement in the transportation and transfer of the liquid. Constant maintenance is not always reliable, especially in remote areas and it can lead to high operating costs. The need for periodic replenishment of the liquid usually becomes a nuisance to the user and keeps the system from being easily marketed. In order to compete with other electronic devices, the user should not even be aware that cooling is required. It should be taken care of automatically with the flip of an "on" switch. That type of cooling is the focus of the rest of the chapter that deals with closed-cycle cryocoolers.

12.3 COOLING WITH CLOSED-CYCLE CRYOCOOLERS

12.3.1 Types of Cryocoolers

Figure 12.3 shows the five types of cryocoolers in common use today. All five are mechanical systems relying on the compression and expansion of a gas. In most cases, the compression is done with moving mechanical parts. Refrigeration with other working fluids, such as electrons (thermoelectric cooler) or photons (laser cooling) that can be driven electronically with no moving mechanical parts, has not advanced to the stage where they can be used to cool any device to cryogenic temperatures. Significant research in thermoelectric or optical materials is required before such systems can ever be used for cooling high-T_c devices.

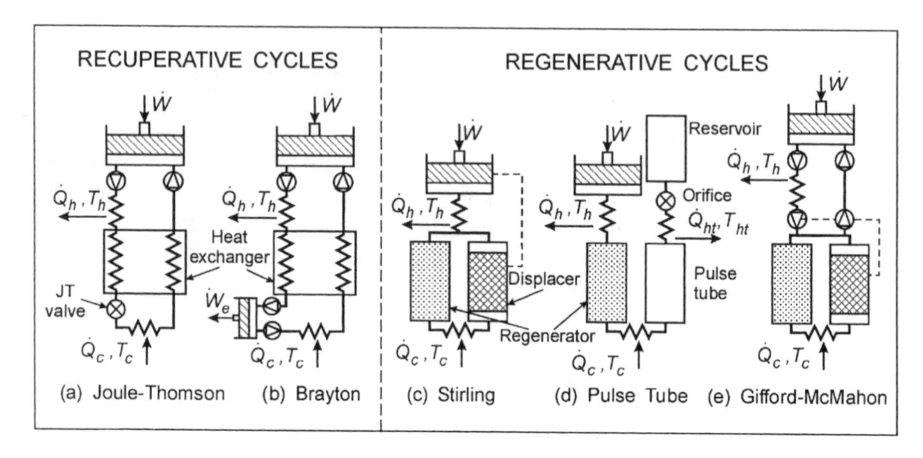

FIGURE 12.3 Schematics of five common types of cryocooler.

The Joule–Thomson (JT) and Brayton cryocoolers, shown in Figure 12.3, are of the recuperative type in which the working fluid flows steadily in one direction, with steady low- and high-pressure lines, analogous to dc electrical systems. The compressor has inlet and outlet valves to maintain the steady flow. The recuperative heat exchangers transfer heat from one flow stream to the other over some distance or across tube walls. Recuperative heat exchangers with the high effectiveness needed for cryocoolers can be expensive to fabricate, especially if they are to be compact. Although not shown here, the Claude cycle is a combination of the Brayton cycle with the addition of a final Joule–Thomson expansion stage for the liquefaction of the working fluid. It is commonly used in air-liquefaction plants and in large helium-liquefaction systems for cooling superconducting magnets and radio-frequency (RF) cavities in accelerators. The three regenerative cycles shown in Figure 12.3 operate with an oscillating flow and an oscillating pressure, analogous to ac electrical systems. Frequencies vary from about 1 Hz for the Gifford–McMahon (GM) and some pulse-tube cryocoolers to about 60 Hz for Stirling and some pulse-tube cryocoolers.

12.3.2 Recuperative Cryocoolers

The steady pressure and the steady flow of gas in these cryocoolers allow them to use large gas volumes anywhere in the system with little adverse effects except for larger radiation heat leaks if the additional volume is at the cold end. Thus, it is possible to “transport cold” to any number of distant locations after the gas has expanded and cooled. In addition, the cold end can be separated from the compressor by a large distance and greatly reduce the EMI and vibration associated with the compressor. Oil-removal equipment with its large gas volume can also be incorporated in these cryocoolers at the warm end of the heat exchanger to remove any traces of compressor oil from the high-pressure working gas before it is cooled in the heat exchanger. Unlike conventional refrigerators operating near ambient temperature, any oil in the working fluid will freeze at cryogenic temperatures and plug the system.

12.3.3 Regenerative Cryocoolers

These cryocoolers operate with oscillating pressures and mass flows in the cold head. The working fluid is almost always helium gas. The oscillating pressure can be generated with a valveless compressor (pressure oscillator) as shown in Figure 12.3 for the Stirling and pulse-tube cryocoolers or with valves that switch the cold head between a low- and high-pressure source, as shown for the Gifford–McMahon cryocooler. In the latter case, a conventional compressor with inlet and outlet valves is used to generate the high- and low-pressure sources. With the Gifford–McMahon cryocooler, an oil-lubricated compressor is usually used and oil-removal equipment can be placed in the high-pressure line, where there is no pressure oscillation. The use of valves greatly reduces the efficiency of the system. Pulse-tube cryocoolers can use either source of pressure oscillations, even though Figure 12.3 indicates the use of a valveless compressor. The valved compressors are modified air conditioning compressors and they are used primarily for commercial applications where low-cost is very important. The amplitude of the oscillating pressure may typically be anywhere from about 10% to as high as 50% of the average pressure. Average pressures are usually in the range of 1.5–3.0 MPa.

The main heat exchanger in regenerative cycles is called a regenerator. In a regenerator, incoming hot gas transfers heat to the matrix of the regenerator, where the heat is stored for a half-cycle in the heat capacity of the matrix. In the second half of the cycle, the returning cold gas, flowing in the opposite direction through the same channel, absorbs heat from the matrix and returns the matrix to its original temperature before the cycle is repeated. Very high surface areas for enhanced heat transfer are easily achieved in regenerators through the use of stacked fine-mesh screen or packed spheres.

12.3.4 Cryocooler Compressors

All of the cryocoolers discussed here are gas systems that rely on the compression and expansion of gas. Many of the problems with cryocoolers, such as cost, reliability, efficiency, EMI, vibration, and size, are associated with the compressor. In the Joule–Thomson and the pulse-tube cryocoolers, the only moving parts are in the compressor. The purpose of the compressor is to convert electrical power into PV power in the gas. In some cases, the compressor may be a thermal or thermoacoustic device that converts heat into PV power. One type of thermal compressor is the sorption compressor, which utilizes either physical adsorption of a gas on the surface of a material like charcoal or the chemical absorption of chemically active gases like hydrogen and oxygen within a material. Heating the sorption bed drives off the gas at a high-pressure, whereas maintaining a bed at ambient temperature causes it to adsorb or absorb the gas at a low-pressure. Once the beds are depleted or saturated, they must be switched, via valves, in order for the flow process to continue. Wade (3) reviewed the use of sorption compressors. Their advantage is the lack of moving parts, except for the check valves that switch once every few minutes. They are easily miniaturized and have been used in a microscale Joule–Thomson cooler operating at 170 K (4). Thermoacoustic drivers convert heat into acoustic waves (oscillating pressures) that can be used to drive pulse-tube refrigerators with no moving parts in the entire system (5–7). Unfortunately, they do not scale down to small sizes needed for cooling electronic devices.

Electromechanical compressors can utilize a wide variety of geometries. The most common geometry is that of oscillating piston driven with a rotary motor and crankshaft as shown in Figure 12.4a or with a linear motor as shown in Figure 12.4b. Linear compressors are generally designed to operate at resonant conditions to maximize their efficiency. The piston compressors provide the oscillating pressures needed for the Stirling and pulse-tube refrigerators. Such a compressor may be called a pressure oscillator or valveless compressor. The ratio of maximum to minimum pressure (pressure ratio) used with these cryocoolers ranges from about 1.3 to 2.0. If inlet and outlet valves (either reed or check valves) are added to the compressor head, a steady flow of gas is generated with a low inlet pressure and a high outlet pressure. These valved compressors are needed for all of the recuperative cryocoolers as well as for the Gifford–McMahon cryocooler and sometimes for the pulse-tube cryocooler. Typical pressure ratios may vary from about 1.4 for some Brayton systems to about 2 for Gifford–McMahon coolers to as high as 200 for some Joule–Thomson coolers. The use of valves on

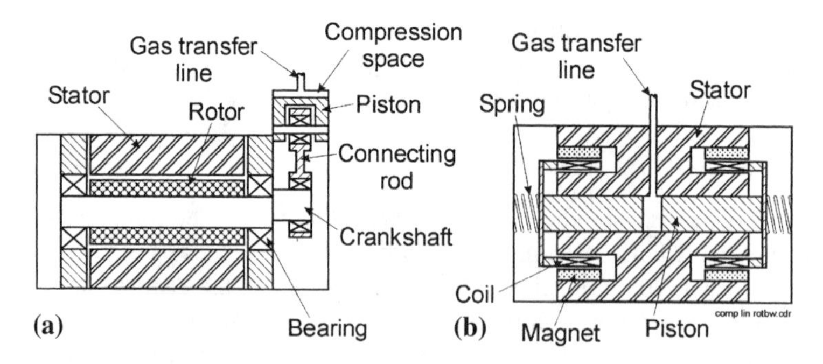

FIGURE 12.4 Schematics of valveless compressors driven with (a) a rotary motor and crankshaft and (b) a linear motor.

these compressors introduces some irreversible losses that are not present in the valveless compressors. In well-designed valveless compressors, the conversion efficiency from electrical to PV power is about 85%, whereas with valved compressors, the efficiency is only about 50% for pressure ratios of about 2. The efficiency is reduced further at larger pressure ratios.

Steady gas flows can also be generated without the use of valves in other compressor geometries, such as scroll, rolling piston, screw, centrifugal, and turbine compressors. Their conversion efficiencies may be higher than the valved piston compressors. For long-life operation, all of the compressors must be sealed, with the motor operating inside the sealed space to eliminate any dynamic seal that would allow the working fluid to escape from the system over time. Refrigeration and air conditioning compressors are typical examples of such sealed compressors. They are readily available in many sizes at low-cost and are very reliable because of the oil lubrication. Adapting them for cryocoolers is not always easy and sometimes not even possible.

In conventional refrigerators, some lubricating oil from the compressor flows with the refrigerant but does not freeze at the cold end. In cryocoolers, the oil would freeze at such low temperatures and plug up orifices or heat-exchanger passageways. Therefore, with cryocoolers, nearly all traces of oil must be removed from the working fluid before it passes to the cold end, otherwise oil-free compressors must be used. Oil-removal equipment adds volume, cost, and maintenance requirements to the compressor system. The large gas volume needed for oil-removal equipment prevents their use with rapidly oscillating pressures, such as with Stirling and some pulse-tube cryocoolers. Oil-free compressors have short lifetimes unless all rubbing contact is eliminated through the use of gas bearings or flexure bearings. Typical lifetimes for oil-free piston compressors where rubbing occurs are about 2500 h for the rotary-motor types shown in Figure 12.4a and 5000 h for linear-motor types shown in Figure 12.4b. There are a few reported cases of such an oil-free compressor lasting as much as 2–3 years, but the results are not repeatable. Further research is required to understand this behavior and make it repeatable.

Rubbing contact between the piston and cylinder can be eliminated through the use of gas bearings or flexure bearings, but usually at additional cost. When the clearance between the piston and the cylinder is reduced to about 15 μm on the radius, the volume of gas that flows through the gap in a half-cycle is much less than the piston-displaced volume. Because the back side of the piston is always sealed, the gas does not leave the system, but returns to the front side of the piston on the return stroke. The pressure on the back side is always at the average pressure. Figure 12.5a shows a cross section of a flexure-bearing compressor and Figure 12.5b shows two examples of the geometry used for flexure bearings. Flexure bearings are flexible in the axial direction but very stiff in the radial direction. The radial stiffness supports the piston inside the cylinder and maintains the

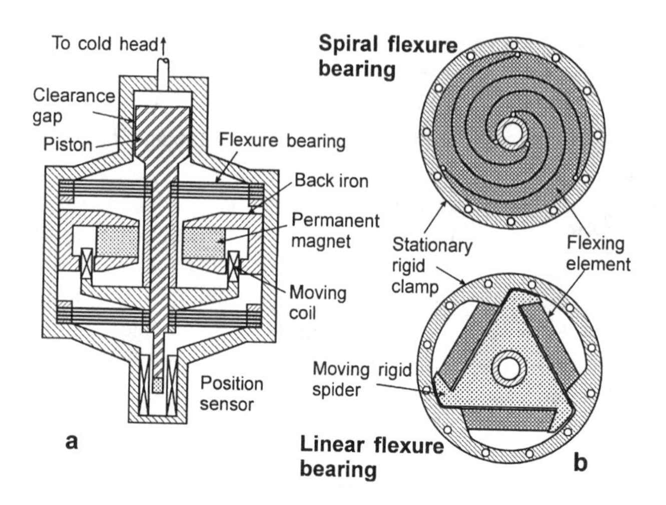

FIGURE 12.5 Schematic of (a) linear compressor with (b) two types of flexure bearing.

clearance gap with no contact. The peak stresses in the flexures are designed to be small enough that the flexures have infinite fatigue lifetimes. These flexures are also used in some cases to support the displacer of Stirling cryocoolers. Some Stirling cryocoolers have now operated in excess of 10 years when using these flexure bearings. These flexure-bearing compressors are also being used on pulse-tube refrigerators when long lifetimes are needed, such as for space applications. The flexure-bearing compressors and displacers were first developed in the mid-1980s for use in space applications of Stirling cryocoolers and were very expensive (8). Progress in flexure-bearing compressors has advanced to the point where the cost has been greatly reduced and such systems are now being developed for commercial applications.

12.3.5 Cryocooler Problems

The problem associated with the use of cryocoolers in most applications is the difficulty in meeting the requirements listed in Table 12.1. For the cooling of high-T_c devices, cost and reliability are usually the most important requirements. Low EMI and vibration are very important in some applications. In some cases, small size and high-efficiency are also important. The need for precision moving parts in cryocoolers leads to high manufacturing costs when there has not been a demand for very large quantities. The compressors for Gifford–McMahon cryocoolers are air conditioning compressors made by the millions but modified for use with helium gas by enhancing the oil flow for increased cooling. Thus, the basic cost of the compressor is very low, but the modification, including the addition of oil-removal equipment, significantly increases the cost. A $1000 cryocooler has been a goal for many years, but studies (9) have shown that even a small $2000 cryocooler would require production rates of at least 10,000 per year. With comparable levels of production of small Stirling cryocoolers for military applications, costs have usually been about $5000. However, the strict military requirements and specifications have often led to increased costs. Costs can be reduced by decreasing the number of moving parts or by making use of innovative fabrication techniques, such as those used in the electronics industry.

Reliability is not always easy to define. Typically, the user in the high-T_c device field is interested in a mean time to failure (MTTF) of at least 3 years and often 5 years. Some maintenance during this time may be tolerated, but it should be minimal (a few hours at the most) and generally not be required more than once per year. Cold moving parts must be oil-free and the need for such long lifetimes under such conditions generally requires noncontact bearings, such as gas bearings or flexure bearings. Such special bearings can greatly increase the cost. The compressor of Stirling or some pulse-tube refrigerators, although operating at room temperature, must still be oil-free. The development of cryocoolers for space applications has led to cryocooler lifetimes of about 10 years with no maintenance, but costs are usually about $1M per cryocooler. Reducing these costs for commercial applications but still maintaining a comparable reliability is the emphasis of much research on cryocoolers. Even a 3-year lifetime for these mechanical coolers is equivalent to operating the engine of an automobile for a distance of 2.1×10^6 km (1.3×10^6 miles) at 80 km/h (50 miles/h).

The use of an electric motor (either rotary or linear) to drive the compressor of a cryocooler leads to the radiation of electromagnetic noise that can be a serious problem when used for cooling-sensitive instruments such as SQUIDs. Moving components, particularly reciprocating elements, can give rise to vibrations that can also affect sensitive devices. Even the oscillating pressures of regenerative cryocoolers can cause vibrations at the cold end from elastic deformations of the tubes.

Efficiencies of small cryocoolers have been quite low for many years. Typically, efficiencies of about 2% of Carnot had been the average for a cryocooler producing 1 W at 80 K (10). However, research in the last 10 years for space applications has led to efficiencies of 10–20% of Carnot and greatly decreased size and weight. Some of these lessons learned from the space applications can be applied to commercial applications with little increase in cost over the older technology.

12.4 JOULE–THOMSON CRYOCOOLERS

12.4.1 Operating Principles

Joule–Thomson (JT) cryocoolers produce cooling when the high-pressure gas expands through a flow impedance (orifice, valve, capillary, porous plug), often referred to as a JT valve. The expansion occurs with no heat input or production of work; thus, according to Eq. (1), the process occurs at a constant enthalpy. The heat input occurs after the expansion and is used to evaporate any liquid formed in the expansion process or to warm up the cold gas to the temperature it was before the expansion occurred. In an ideal gas, the enthalpy is independent of pressure for a constant temperature, but real gases experience an enthalpy change with pressure at constant temperature. Thus, cooling in a JT expansion occurs only with real gases and at temperatures below the inversion temperature. Some heating occurs for expansion at temperatures above the inversion temperature for that particular gas. Typically, nitrogen or argon is used in JT coolers because their inversion temperatures are above room temperature, but pressures of 20 MPa (200 bar) or more on the high-pressure side are needed to achieve reasonable cooling. Such high pressures are difficult to achieve and require special compressors with short lifetimes.

12.4.2 Mixed Gases Versus Pure Gases

Recent advances in JT cryocoolers have been associated with the use of mixed gases as the working fluid rather than pure gases. The use of mixed gases was first proposed in 1936 for the liquefaction of natural gas (see discussion in Ref. 11), but it was not used extensively for this purpose until the last 20 or 30 years. It is commonly referred to as the mixed-refrigerant cascade cycle. The use of small JT coolers with mixed gases for cooling infrared sensors was first developed under classified programs in the Soviet Union during the 1970s and 1980s. Such work was first discussed in the open literature by Little (12). Missimer (13), Radebaugh (1,11), Little (14), and Boiarski et al. (15) reviewed the use of mixed gases in JT cryocoolers. Typically, higher-boiling-point components, such as methane, ethane, and propane, can be added to nitrogen to make the mixture behave more like a real gas over the entire temperature range. The larger enthalpy changes at a constant temperature result in increased cooling powers and efficiencies with pressures (about 2.5 MPa) that can be achieved in conventional compressors used for domestic or commercial refrigeration. The lowest-temperature that can be achieved with mixed-gas JT systems is limited by the freezing point of its components. In general, the freezing point of a mixture is less than that of the pure fluids, so temperatures of 77 K are possible with the nitrogen–hydrocarbon mixtures even though the hydrocarbons freeze in the range of 85–91 K as pure components. The presence of propane also increases the solubility of oil in the mixture at 77 K so that less care is needed in removing oil from the mixture when using an oil-lubricated compressor. Much research is currently underway pertaining to the solubility of oil in various mixtures and the freezing point of mixtures. Temperatures down to 67 K have been achieved in these mixed-gas systems by the addition of neon to the gas mixture (15). Marquardt et al. (16) discussed the optimization of gas mixtures for a given temperature range. The gas mixture in these systems undergoes boiling and condensing heat transfer in the heat exchanger that contributes to its high-efficiency. However, the flowing liquid in the heat exchanger gives rise to increased vibration, which could be a problem in cooling very sensitive SQUID systems.

Mixed-gas JT cryocoolers can be classified into two types. Single-stage systems have one expansion nozzle and allow the refrigerant to flow through the entire system without the use of intermediate expansion stages and phase separators. Sometimes, they are referred to as a throttle-cycle cryocooler. Oil separation from the refrigerant takes place at ambient temperature after the aftercooler. The vapor pressure of the oil at this temperature is high enough that a significant amount of oil can still find its way to the cold end and lead to potential clogging of the cold end. The use of a small conventional vapor-compression refrigerator to precool the gas mixture to about −30°C was used to enhance oil removal at the lower-temperature and to increase the cooling available with the mixed-refrigerant stage (17,18). The second type of mixed-gas system uses multiple expansion nozzles and phase separators with the mixed refrigerant. As with the previous system, it also can be precooled. The intermediate expansion stages allow for cooling at other temperatures and the phase separators at each of these stages tends to block the flow of oil to the coldest stage. This type of system with the phase separators is known as the mixed-refrigerant cascade (MRC) cycle in the liquefied natural gas industry. A schematic of this cycle is shown in Figure 12.6. It also has been referred to as the Kleemenko cycle (14).

12.4.3 Advantages and Disadvantages

The main advantage of JT cryocoolers is the fact that there are no moving parts at the cold end, which makes them inexpensive to fabricate. The cold end can be

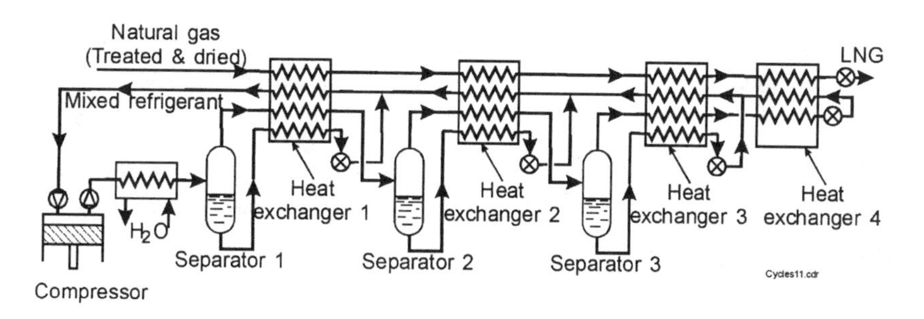

FIGURE 12.6 Schematic of the MRC cycle (also known as the Kleemenko cycle), which is one type of mixed-gas Joule–Thomson cryocooler. It utilizes phase separators in each stage.

miniaturized and provide a very rapid cool-down. This rapid cool-down (a few seconds to reach 77 K) has made them the cooler of choice for cooling infrared sensors used in missile guidance systems. These coolers utilize a small cylinder pressurized to about 45 MPa, with nitrogen or argon as the source of high-pressure gas. In this open-cycle mode, cooling lasts for only a few minutes until the gas is depleted. In the closed-cycle mode, most JT cryocoolers now use mixed gases and operate with a high pressure of only about 2.5 MPa. With such outlet pressures, conventional oil-lubricated refrigeration compressors can be used for a lower cost. These mixed-gas JT coolers now sell for about $2000 to $3000, which is less expensive than other types of cryocoolers at this time. Because only the liquid is flowing at the cold end, there is very little vibration and low levels of EMI as long as the cold end is removed some distance from the compressor.

A major disadvantage of the JT cryocooler is the susceptibility to plugging by moisture or oil in the very small orifice. Continuous operation of over 2 years has been achieved with some of the mixed-gas JT systems using phase separators to prevent oil and water from reaching the cold end. Another disadvantage of JT cryocoolers is the low-efficiency at 80 K and below. Compressor efficiencies are very low when compressing to such high pressures. However, the use of mixed gases has reduced the pressures required and has increased efficiencies, especially with precooling. Still, efficiencies are only about 5% of Carnot at 80 K. The efficiency increases rapidly at temperatures of 90 K and above. Temperatures below about 70 K require the use of a second compressor and a neon or hydrogen working fluid.

12.4.4 Examples of Joule–Thomson Cryocoolers

Figure 12.7 shows a typical JT cryocooler used for missile guidance where cool-down to about 80 K occurs in a few seconds. Miniature finned tubing is used for the heat exchanger. An explosive valve is used to start the flow of gas from the high-pressure cylinder, and after flowing through the cooler, the gas is vented to the atmosphere. Figure 12.8 shows another type of open-cycle miniature JT cryocooler in which the gas-flow paths for the heat exchanger and expansion orifice are etched into a glass substrate and sealed with another layer of glass. These coolers are typically operated with a cylinder of commercial-grade nitrogen gas providing an inlet pressure of about 10 MPa. It can absorb about 0.25 W of cooling at 85 K for about 50 h from a standard nitrogen cylinder. Temperatures down to about 70 K are possible using a vacuum pump on the low-pressure outlet line. The enclosure for the glass microcooler shown in Figure 12.8 is the vacuum vessel. Such coolers are often used for laboratory studies of high-temperature superconductors or in the study of various material properties.

Marquardt et al. (16) showed how a mixed-gas JT cryocooler can be used for a cryogenic catheter only 3 mm in diameter. The heat exchanger was 2.5 mm

FIGURE 12.7 Open-cycle Joule–Thomson cryocooler for missile guidance. (Courtesy of Carleton.)

in diameter and was fabricated by photoetching and diffusion bonding of copper and stainless-steel foils. Figure 12.9 shows the cold tip of this 1-m-long coaxial catheter. Such miniature systems could also be used for cooling superconducting electronic devices and provide spot cooling at several locations with one compressor.

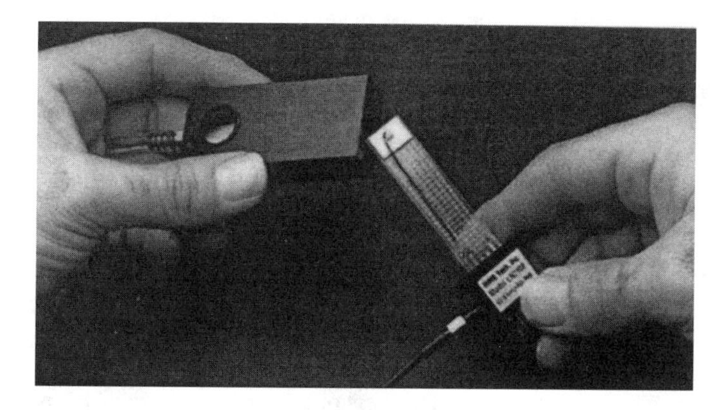

FIGURE 12.8 Open-cycle Joule–Thomson cryocooler with gas-flow channels etched in glass. (Courtesy of MMR.)

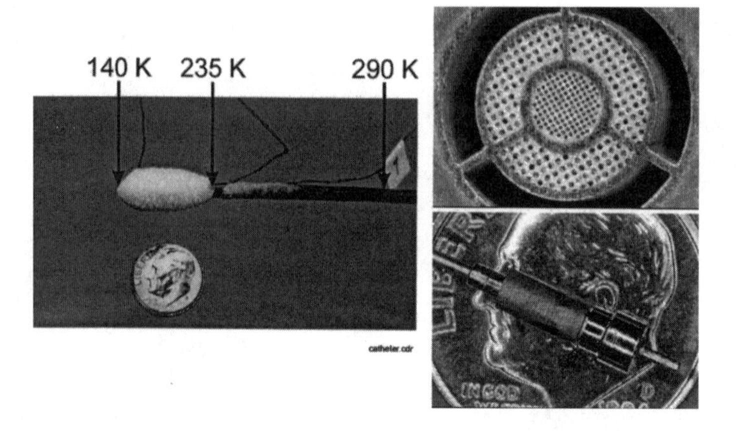

FIGURE 12.9 Cold tip of a 1-m-long cryogenic catheter which utilizes a mixed-gas JT cryocooler and a miniature heat exchanger at the tip. Catheter diameter is 3.0 mm.

12.5 STIRLING CRYOCOOLERS

12.5.1 Operating Principles

The Stirling cycle was invented in 1815 by Robert Stirling for use as a prime mover. Although used in the latter part of that century as a refrigerator, it was not until the middle of the 20th century that it was first used to liquefy air and, soon thereafter, for cooling infrared sensors for tactical military applications. However, they cannot provide the very fast cool-down times of JT cryocoolers, so they are not used on missiles for guidance. The long history of the Stirling cryocooler in cooling infrared equipment has resulted in the development of models tailored specifically to that application that are manufactured by several manufacturers. The refrigeration powers of these models range from 0.15 to 1.75 W, which is also appropriate for many superconducting electronic applications, although problems of reliability and EMI are important issues that must be considered.

A pressure oscillation by itself in a system would simply cause the temperature to oscillate and produce no refrigeration. In the Stirling cryocooler, the second moving component, the displacer, is required to separate the heating and cooling effects by causing motion of the gas in the proper phase relationship with the pressure oscillation. When the displacer in Figure 12.3c is moved downward, the helium gas is displaced to the warm end of the system through the regenerator. The piston in the compressor then moves downward to compress the gas, mostly located at ambient temperature, and the heat of compression is removed by heat exchange with the ambient. Next, the displacer is moved up to displace the gas through the regenerator to the cold end of the system. The piston then moves

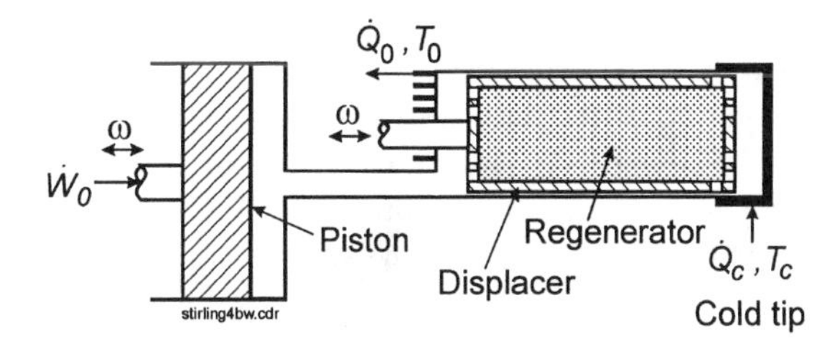

FIGURE 12.10 Piston-displacer Stirling cryocooler with internal regenerator.

up to expand the gas, now located at the cold end, and the cooled gas absorbs heat from the system it is cooling before the displacer forces the gas back to the warm end through the regenerator. Stirling cryocoolers usually have the regenerator inside the displacer, as shown in Figure 12.10, instead of external as shown in Figure 12.3c. The resulting single cylinder provides a convenient cold finger.

In practice, motion of the piston and the displacer are nearly sinusoidal. The correct phasing occurs when the volume variation in the cold expansion space leads the volume variation in the warm compression space by about 90°. With this condition, the mass flow or volume flow through the regenerator is approximately in phase with the pressure. Regenerative cryocoolers are analogous to ac electrical systems, in which voltage is replaced with pressure and current is replaced with volume or mass flow. Real power in electrical systems flows only when the current and voltage are in phase with each other, and for mechanical systems, pressure and flow must be in phase for real power flow. The moving displacer reversibly extracts work from the gas at the cold end and transmits it to the warm end, where it contributes to the compression work. In an ideal system, with isothermal compression and expansion and a perfect regenerator, the entire process is reversible. Thus, the coefficient of performance (COP) for the ideal Stirling refrigerator is the same as the Carnot COP given by Eq. (7).

12.5.2 Advantages and Disadvantages

The main advantages of Stirling cryocoolers are their rather high-efficiency and their commercial availability in several sizes. With efficiencies of about 10% of Carnot, specific powers are about 28 W/W at 80 K. About 140,000 Stirling cryocoolers have been made to date, mostly for the military in cooling infrared sensors onboard tanks, airplanes, helicopters, and handheld systems. They are available in several sizes from many manufacturers. Prices range from about $5000 to $10,000. They can be made small enough to hold in one hand and run with power as low as 3 W (battery operation).

Disadvantages of Stirling cryocoolers include the relatively short lifetime of about 5000 h of continuous operation (about half a year), the vibration from the oscillating displacer, and the EMI generated by the compressor that must be in close proximity to the cold finger. The short lifetimes occur because oil lubrication cannot be used for the rubbing parts. Research on improved materials may lead to lifetimes of 2–3 years in the compressor and displacer. Lifetimes of 3–10 years are possible, but it requires the use of gas or flexure bearings to eliminate rubbing contact (see discussion in Sec. 12.3.4). Much of the vibration of the linear compressor is eliminated by the use of dual-opposed pistons to balance the forces. However, in the cold head, there is only one displacer, and its oscillation causes considerable vibration at the cold tip. The metal screens (usually stainless steel) of the regenerator inside the displacer that moves in the Earth's magnetic field generates enough magnetic noise to greatly interfere with SQUIDs. In the Stirling cryocooler, the cold head must be close to the compressor to reduce void volume that leads to losses. Typically, the separation is kept less than about 100 mm, but distances of up to about 1 m are possible with significant sacrifice in efficiency. The motor in the compressor generates considerable EMI, which requires that it be shielded and/or moved as far away from the cold end as possible whenever this type of cryocooler is used for cooling sensitive instruments such as SQUIDs.

12.5.3 Examples of Stirling Cryocoolers

Figure 12.11 shows four sizes of Stirling cryocoolers that are commonly used for military tactical applications of infrared sensors. All except the smallest cooler in Figure 12.11 are split systems in which the cold finger can be located a short distance from the compressor. In split systems, the oscillating pressure is used to drive the displacer through pneumatic techniques. Operating frequencies for these cool-

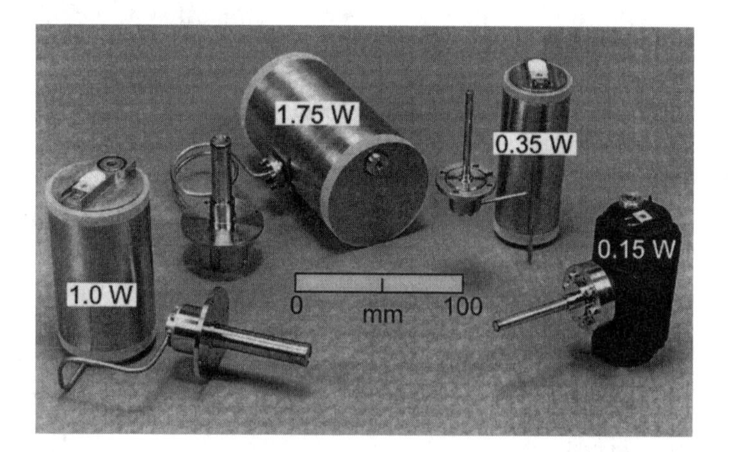

FIGURE 12.11 Four sizes of tactical Stirling cryocoolers with dual-opposed linear compressors. (Courtesy of Texas Instruments.)

ers are typically about 50–60 Hz. Because they are designed for military use, a separate controller is available to provide the required ac voltage from a nominal 24-V dc power source. The refrigeration powers listed for each cooler are for a temperature of about 77–80 K, except the 1.75-W system, which is for a temperature of 67 K. Input powers range from about 10 W for the smallest to about 70 W for the largest. They have efficiencies at 80 K of about 8–10% of Carnot. All of the coolers shown in Figure 12.11 make use of linear-drive motors and dual-opposed pistons to reduce vibration. The single displacer gives rise to considerable vibration unless a passive balancer is used or if two cold fingers are mounted in an opposed fashion. The linear drive reduces side forces between the piston and the cylinder, but the MTTF is still only about 5000 h (half a year of continuous operation). Efforts are currently underway to increase the MTTF of these Stirling cryocoolers because they are the least reliable component in an infrared system. Other linear-drive Stirling cryocoolers are now commercially available with refrigeration powers of about 6 W at 80 K. Figure 12.12 shows the smallest Stirling cryocooler produced to date. It uses a rotary motor and crankshaft to drive the piston, so the lifetime is only a few thousand hours. Refrigeration powers of 0.15 W with 3 W of input power are possible with this small device that is used in handheld infrared camcorders.

The development of cryocoolers for space applications has led to greatly improved reliabilities, and a MTTF of 10 years is now usually specified for these applications. The Stirling cooler was first used in these space applications after flexure bearings were developed (8) for supporting the piston and displacers in their respective cylinders with little or no contact in a clearance gap of about 15 μm. Because these flexure-bearing Stirling cryocoolers were first developed at the University of Oxford, they are sometimes called Oxford coolers. Flexure bearings were discussed in Section 12.3.4. Although originally used for space applications, flexure bearings are now available in some of the tactical Stirling compressors like

FIGURE 12.12 Miniature Stirling cryocooler used for handheld infrared video cameras. (Courtesy of Inframetrics.)

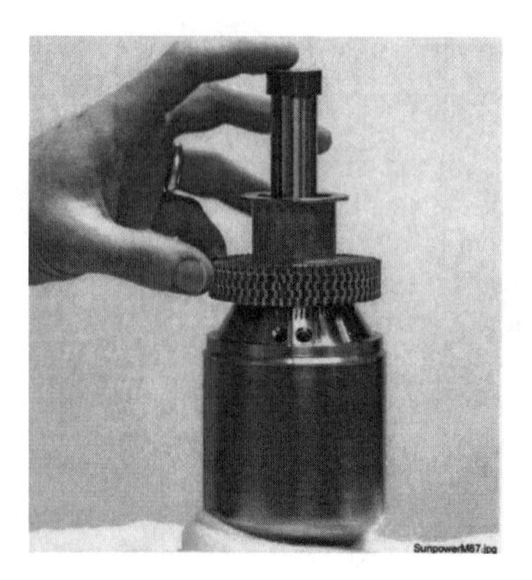

FIGURE 12.13 Gas-bearing Stirling cryocooler with one piston and passive balancer. (Courtesy of Sunpower.)

those shown in Figure 12.11. Such systems are expected to last 3–5 years in commercial applications.

The use of gas bearings is another approach being used to eliminate the contact between the piston and the cylinder. Figure 12.13 shows an example of this type of Stirling cryocooler which is currently being used for cooling high-T_c microwave filters for wireless telecommunication. It produces 7.5 W of cooling at 77 K with 150 W of input power and is expected to have a lifetime of 3–5 years. Such systems are too new to have good statistics on lifetimes.

12.6 GIFFORD–MCMAHON CRYOCOOLERS

12.6.1 Operating Principles

The cold head of the Gifford–McMahon (GM) cryocooler is the same as that of the Stirling cryocooler. They both use a moving displacer, usually with a regenerator matrix on the inside, as shown in Figure 12.10. Thus, the operating principles for the cold head are the same as for the Stirling cryocooler. However, the displacer of a GM cryocooler generally operates at frequencies of about 1–2 Hz as opposed to 30–60 Hz for a Stirling cryocooler. The lower-frequency provides a longer lifetime for the rubbing seal on the displacer. The displacer is driven either with an electric motor synchronously with the valves or pneumatically by the pressure oscillation. The pressure oscillation in GM cryocoolers is provided by valves (rotary, slide, or poppet) located at the warm end of the cold head (or expander unit) that switch between a low- and a high-pressure source. These steady (or dc) pressure sources are provided with a conventional oil-lubricated compressor. Such compressors can be valved piston compressors or scroll compressors. They usually operate at a frequency of 50 or 60 Hz to keep the compressor compact. The compressor package also contains oil-cooling and oil-removal equipment as well as the aftercooler. This compressor package is usually much larger than the cold head, but it can be located quite some distance from the cold head (10 m or even more) and connected by two flexible gas lines and an electrical line for driving the displacer and valves. Figure 12.14 shows a drawing of a typical two-stage GM cryocooler system.

The Gifford–McMahon cycle was first developed (19) in the late 1950s and is described in more detail elsewhere (20,21). These cryocoolers are most commonly used in two-stage 15-K versions for cryopumps in the semiconductor fabrication industry. About 20,000 per year are made worldwide with costs for one unit ranging from about \$10,000 to \$20,000. The GM cryocooler is also used for cooling shields to 10–15 K in magnetic resonance imaging (MRI) systems to reduce the boil-off rate of liquid helium or, in some cases, to reliquefy the helium at 4.2 K. Single-stage units for temperatures above about 30 K are somewhat less expensive. Refrigeration powers at 80 K generally range from about 10 to 300 W, with input powers ranging from about 800 W to 7 kW. The oil-lubricated compressors have lifetimes of at least 5 years, but the adsorber cartridge for oil removal must be replaced once every year or two. Replacement of seals on the displacer must be carried out about once every year.

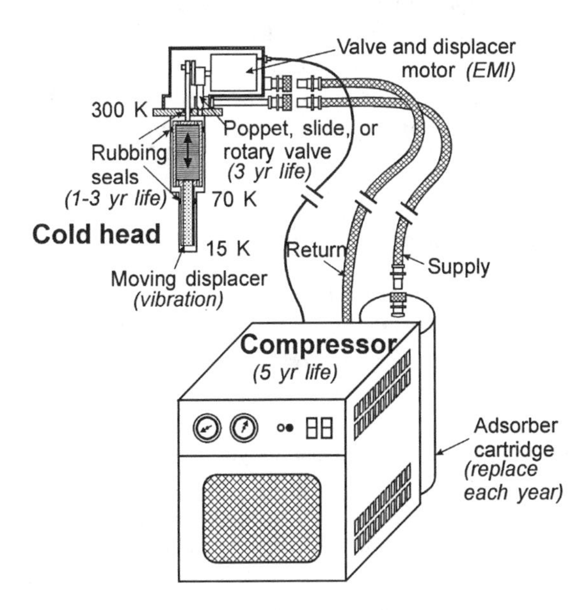

FIGURE 12.14 Drawing of two-stage Gifford–McMahon cryocooler showing the compressor and cold head.

12.6.2 Advantages and Disadvantages

One of the major advantages of the GM cryocooler is the fact that they are commercially available from a large number of manufacturers in several sizes, both in single-stage and two-stage versions. Single-stage versions are useful for temperatures down to about 30 K. With replacement of the displacer seal and the charcoal adsorber on the compressor about once every year of operation, lifetimes of 5 years are possible. Small, single-stage GM cryocoolers are relatively inexpensive and are in the $5000–$10,000 range. Good service is usually available for replacing the displacer seals and for any other problem. The compressor can be located quite some distance away from the cold head (even in another room) to reduce audible and magnetic noise. The flexible gas lines as well as the charcoal adsorber come equipped with self-sealing quick connects to allow them to be disconnected from the system without losing helium gas or letting in air.

One of the major disadvantages of the Gifford–McMahon cryocooler is its low inefficiency. The Carnot efficiencies of single-stage GM cryocoolers at about 80 K are only about 4–8% of Carnot. Thus, they require rather large compressors with input powers that may range from about 800 W to 7 kW. For most high-T_c electronic applications, a system with an input power of 800–1000 W would usually be sufficient because they can provide refrigeration powers at 80 K of about 10–20 W. These smaller systems are generally air-cooled, but dissipating this power in a small, closed environment may sometimes be a challenge. The moving displacer with a frequency of about 1 Hz is a source of considerable vibration and magnetic noise that often interferes with sensitive electronic systems. The audible noise of the moving displacer is sometimes bothersome. The need to replace the seal on the displacer about once every year of continuous operation is often a great disadvantage.

12.6.3 Examples of Gifford–McMahon Cryocoolers

Cooling for most high-T_c electronic applications can be provided with small one-stage GM cryocoolers. Figure 12.15 is a drawing of a cold head or expander unit that can produce about 20 W of cooling at 80 K and has a minimum temperature of about 30 K. The refrigeration power of GM and other regenerative cryocoolers is approximately linear with temperature over a wide temperature range. The cool-down time to 80 K with no additional mass attached is about 30 min. It is powered with a 1.6-kW compressor that weighs about 60 kg. Figure 12.16 shows a photograph of a typical two-stage GM cryocooler in which the second (colder) stage can reach temperatures down to about 10 K. Temperatures down to about 2 K are possible with two-stage GM cryocoolers in which special regenerator materials are used in the second stage that have very high heat capacities below about 15 K due to magnetic transitions.

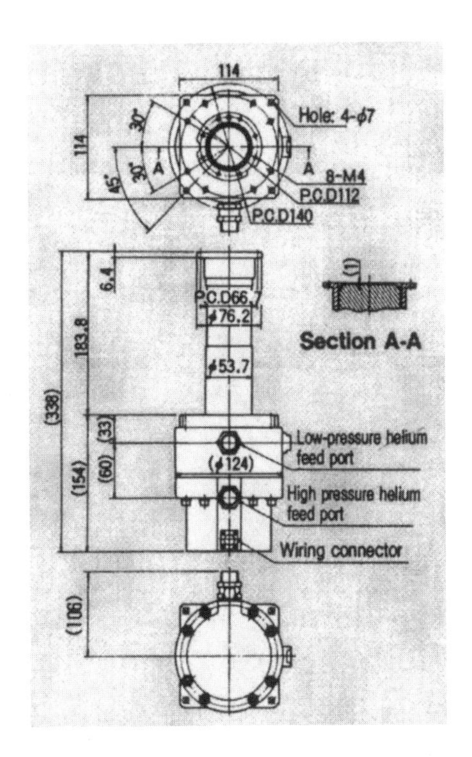

FIGURE 12.15 Drawing of the cold head for a small one-stage Gifford–McMahon cryocooler. (Courtesy of Daikin.)

12.7 PULSE-TUBE CRYOCOOLERS

12.7.1 Operating Principles of Different Types

The pressure oscillation for the cold head of the pulse-tube cryocooler can be provided with a valveless compressor like that of the Stirling cryocooler, as shown in Figure 12.3d. In this case, the pulse tube is often referred to as a Stirling-type pulse tube. If the pressure oscillation is provided with a set of valves like that shown in Figure 12.3e for the Gifford–McMahon cryocooler, the pulse tube is referred to as a GM-type pulse tube. The operating principles of the cold head are the same for both types. However, the Stirling-type pulse tube must operate at the compressor frequency, which usually is as high as about 20–60 Hz to maintain a compact compressor. The GM-type pulse tube generally operates at about 1–2 Hz, even though the compressor operates at 50–60 Hz. The low-frequency is a result of using valve drive motors and electronics from GM cryocoolers, where the low frequency results in longer life for the displacer seals. The low speed used with GM-type pulse tubes ensures a long valve life, which may be as much as 3–5 years.

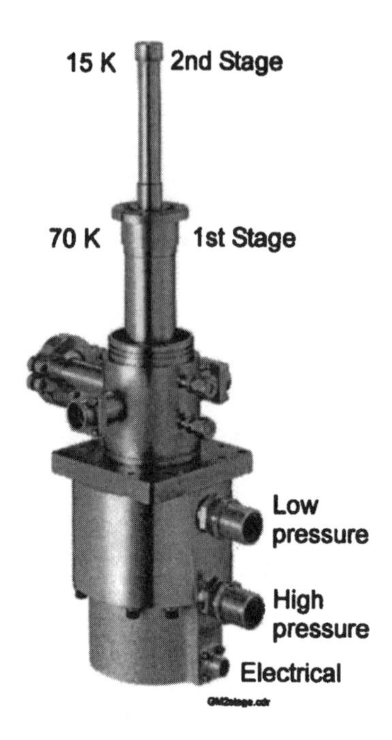

FIGURE 12.16 Two-stage cold head of GM cryocooler. (Courtesy of Daikin.)

The cold end of the pulse tube is different from that of the Stirling and Gifford–McMahon cryocoolers in the fact that there is no moving displacer. The moving displacer has several disadvantages. It is a source of vibration, has a limited lifetime, and contributes to axial heat conduction as well as to a shuttle heat loss. In the pulse-tube cryocooler, shown in Figure 12.3d, the proper gas motion in phase with the pressure is achieved by the use of an orifice, along with a reservoir volume to store the gas during a half-cycle. The reservoir volume, at ambient temperature, is large enough that negligible pressure oscillation occurs in it during the oscillating flow. It maintains a steady average pressure that typically is around 1.5–3.0 MPa. The oscillating flow through the orifice separates the heating and cooling effects just as the displacer does for the Stirling and Gifford–McMahon refrigerators. The orifice pulse-tube refrigerator operates ideally with adiabatic compression and expansion in the pulse tube. Thus, for a given frequency, there is a lower limit on the diameter of the pulse tube in order to maintain adiabatic processes. The four steps in the cycle are as follows. (1) The piston moves down to compress the gas (helium) in the pulse tube. (2) Because this heated, compressed gas is at a higher pressure than the average in the reservoir, it flows through the orifice into the reservoir and exchanges heat with the ambient through the heat exchanger at the warm end of the pulse tube. The flow stops when the pressure in the pulse tube is reduced to the average pressure. (3) The piston moves up and expands the gas adiabatically in the pulse tube. (4) This cold, low-pressure gas in the pulse tube is forced past the cold end by the gas flow from the reservoir into the pulse tube through the orifice. As the cold gas flows through the heat exchanger at the cold end of the pulse tube, it picks up heat from the object being cooled. The flow stops when the pressure in the pulse tube increases to the average pressure. The cycle then repeats. The function of the regenerator is the same as in the Stirling and Gifford–McMahon refrigerators in that it precools the incoming high-pressure gas before it reaches the cold end.

One function of the pulse tube is to insulate the processes at its two ends; that is, it must be large enough that gas flowing from the warm end traverses only part way through the pulse tube before flow is reversed. Likewise, flow in from the cold end never reaches the warm end. Gas in the middle portion of the pulse tube never leaves the pulse tube and forms a temperature gradient that insulates the two ends. Roughly speaking, the gas in the pulse tube is divided into three segments, with the middle segment acting like a displacer but consisting of gas rather than a solid material. For this gas plug to effectively insulate the two ends of the pulse tube, turbulence in the pulse tube must be minimized. Thus, flow straightening at the two ends is crucial to the successful operation of the pulse-tube refrigerator. The pulse tube is the unique component in this refrigerator that appears not to have been used previously in any other system. It could not be any simpler from a mechanical standpoint. It is simply an empty tube. However, the thermohydrodynamic processes involved in it are extremely complex and still not well understood or modeled. The overall function of the pulse tube is to transmit hydrodynamic or acoustic power in an oscillating gas system from one end to the other across a temperature gradient with a minimum of power dissipation and entropy generation.

Pulse-tube refrigerators were invented by Gifford and Longsworth (22) in the mid-1960s, but that type was different than what is shown in Figure 12.3d and only reached a low temperature of 124 K. This type is now referred to as the basic pulse tube and is seldom studied any further. In 1984, Mikulin et al. (23) introduced the concept of an orifice to the original pulse-tube concept and reached 105 K. In 1985, Radebaugh et al. (24) changed the location of the orifice to that shown in Figure 12.3d and reached 60 K. Further improvements since then have led to a low-temperature limit of about 20 K with one stage and 2 K with two stages. This newer type of pulse-tube refrigerator is referred to as the orifice pulse-tube refrigerator (OPTR), although, recently, the use of the word orifice is often omitted and the designation of PTR is now common. The OPTR operates with a completely different principle than does the basic pulse-tube refrigerator (BPTR). The thermoacoustic refrigerator is another related device, similar to the BPTR, but

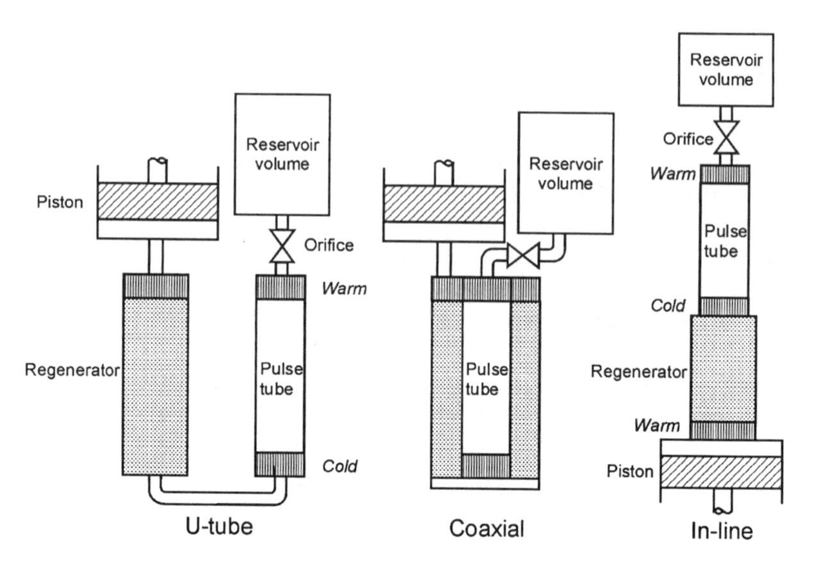

FIGURE 12.17 Three different geometries for pulse-tube cryocoolers.

operating at resonant frequencies of the structure, which may be as high as 500–1000 Hz. That refrigerator, like the BPTR, cannot achieve cryogenic temperatures and is more useful for refrigeration at near-ambient temperatures. The author has recently presented a more extensive review of pulse-tube refrigerators than space permits here (25).

The three different geometries that have been used with pulse-tube cryocoolers are shown in Figure 12.17. The in-line arrangement is the most efficient, because it requires no void space at the cold end to reverse the flow direction nor does it introduce turbulence into the pulse tube from the flow reversal. The disadvantage is the possible awkwardness associated with having the cold region located between the two warm ends. The most compact arrangement and the one most like the geometry of the Stirling cryocooler (see Fig. 12.10) is the coaxial arrangement. That geometry has the potential problem of a mismatch of temperature profiles in the regenerator and in the pulse tube that would lead to steady heat flow between the two components and a reduced efficiency. However, that problem has been minimized, and a coaxial geometry was developed at NIST as an oxygen liquefier for NASA with an efficiency of 17% of Carnot (26). Early pulse-tube cryocoolers were not nearly as efficient as Stirling cryocoolers, but advances in the last 10 years have brought pulse-tube refrigerators to the point of being the most efficient of all cryocoolers. Some details of this rapid progress are given in the following subsection.

12.7.2 Recent Developments

12.7.2.1 Phase-Shift Mechanisms

In regenerative cryocoolers, the optimum phase relationship between the pressure and the mass (or volume) flow is to have them in phase with each other near the center of the regenerator. With such a phase relationship, the magnitude of the average mass flow rate through the regenerator for a given PV power is minimized. Because the regenerator losses (such as pressure drop and imperfect heat transfer) depend mostly on the magnitude of the mass flow rate, the regenerator losses are minimized and the system efficiency is maximized. In a Stirling cryocooler, where the displacer is driven, the optimum phase angle can always be achieved. In the orifice pulse-tube refrigerator, the mass flow rate and the pressure are in phase with each other at the orifice because of the purely resistive nature of the flow impedance in an orifice. In practice, the orifice can be any resistive flow element such as an orifice, a needle valve (for adjustment purposes), a capillary, or a porous plug. Because of the gas volume in the pulse tube and the regenerator, the spatial averaged flow in the regenerator then leads the pressure by as much as 30°–50° and causes greater regenerator losses because of the large amplitude of flow for the same PV power flow. The double-inlet system and the inertance tube are two methods to adjust the phase between the flow and pressure in a beneficial direction.

12.7.2.2 Double Inlet (Secondary Orifice)

In 1990, Zhu et al. (27) introduced the concept of a secondary orifice to the OPTR in which the secondary orifice allows a small fraction (about 10%) of the gas to travel directly between the compressor and the warm end of the pulse tube, thereby bypassing the regenerator, as shown in Figure 12.18. They called this the double-inlet pulse-tube refrigerator. This bypass flow is used to compress and expand the portion of the gas in the warm end of the pulse tube that always remains at the warm temperature. The bypass flow reduces the flow through the regenerator, thereby reducing the regenerator loss. An alternative explanation shows that

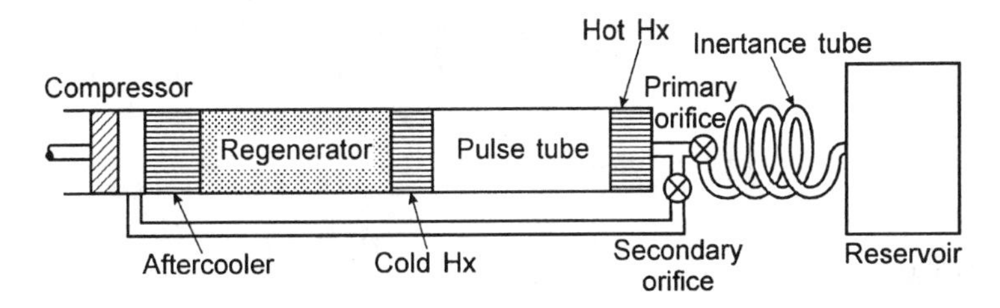

FIGURE 12.18 Schematic of pulse-tube cryocooler with secondary orifice (double inlet) and inertance tube.

the secondary orifice causes the flow at the warm end of the pulse tube to lag the pressure, which is the desired effect. The location of the secondary orifice is shown in Figure 12.18, along with that of an inertance tube described in the next subsection.

Although the introduction of the secondary orifice usually led to increased efficiencies compared to the OPTR, it also introduced a problem. Performance of the double-inlet pulse-tube refrigerator was not always reproducible, and sometimes the cold-end temperature would slowly oscillate by several degrees with periods of several minutes or more. This erratic behavior has been attributed (28) to a steady or dc flow that can occur around the loop formed by the regenerator, pulse tube, and secondary orifice. An asymmetric flow impedance in the secondary is required to cancel such a dc flow, but the amount of asymmetry is very sensitive to operating conditions.

12.7.2.3 Inertance Tube

The inertia of the oscillating mass flow in regenerative systems causes a component of the pressure drop in any element to lead the flow by 90°. This inertance effect is analogous to that of an inductance in electrical systems where the voltage leads the current by 90°. Because of the resistive component of pressure drop and the compliance (analogous to capacitance) due to void volume, the actual phase shift will always be less than 90°. In fact, in the pulse tube and regenerator, the compliance effect dominates at normal operating frequencies of 1–60 Hz. A component with an optimized geometry to maximize the inertance effect is required to bring about the desired phase shift of any significance. This component is known as the inertance tube (1,25) and is shown in Figure 12.18. It is a long thin tube of about 1–5 m in length. In practice, the required resistive component can be incorporated into the inertance tube, so the primary orifice can be eliminated unless some control is desired. With an inertance tube, the flow at the warm end of the pulse tube lags the pressure instead of being in phase as with the primary orifice. If the phase shift is large enough to cause the flow and pressure to be in phase in the regenerator, then no secondary orifice is required and dc flow is eliminated. The maximum phase shift in the inertance tube increases with the frequency and flow in the system. Small pulse tubes of only a few watts of refrigeration at 80 K usually need the addition of the secondary orifice to obtain enough of a phase shift.

12.7.3 Advantages and Disadvantages

The absence of a moving displacer in pulse-tube cryocoolers gives them many potential advantages over Stirling and Gifford–McMahon cryocoolers. These advantages include higher reliability, lower cost, lower vibration, less EMI, and insensitivity to large side forces on the cold region. With some of the latest advances discussed in the previous subsection, the pulse-tube refrigerator has achieve efficiencies as high as 24% of Carnot at 80 K, which makes it the most efficient of all cryocoolers, at least in the temperature range and the small sizes of interest here. These high-efficiencies are only achieved with Stirling-type pulse tubes where there are no valves associated with the compressor. The GM-type pulse tubes have much lower-efficiencies, comparable to those of the GM cryocooler. A comparison of cryocooler efficiencies is given in a later section.

Disadvantages of pulse-tube cryocoolers are the difficulty in scaling to very small sizes (less than about 10 W input power) and the possibility of convective instabilities in the pulse tube when operated with the cold end up or horizontal in a gravity environment. The few experiments that have been performed regarding orientation dependence indicate that for temperatures around 80 K, there is an orientation dependence during operation if the diameter of the pulse tube exceeds some critical value, which may be around 10 mm. Such a pulse-tube diameter corresponds to a refrigeration power around 10 W. Another disadvantage is the relatively short history of about 15 years for the orifice-type pulse tubes, the only type that has achieved cryogenic temperatures. As a result, they are just now beginning to find their way into the marketplace. At this time, there are not many options available regarding commercially available pulse-tube cryocoolers, especially the Stirling-type pulse tubes for high-efficiency.

12.7.4 Examples of Pulse-Tube Cryocoolers

Figure 12.19 shows the first commercial pulse-tube cryocooler, which was first introduced in Japan in 1993, but is now available worldwide. It utilizes the U-tube geometry and is driven with a Gifford–McMahon compressor requiring

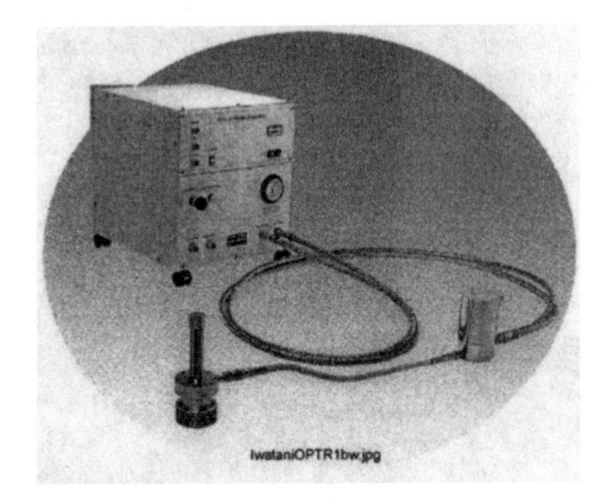

FIGURE 12.19 First commercial pulse-tube cryocooler, showing GM compressor and rotary valve. (Courtesy of Iwatani.)

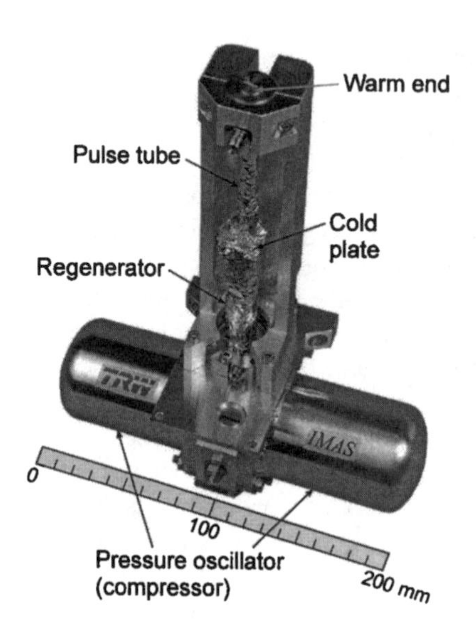

FIGURE 12.20 Light-weight and high-efficiency pulse-tube cryocooler for space applications. (Courtesy of TRW.)

about 800 W of input power. A rotary valve provides the pressure oscillation and is located some distance from the cold head to reduce EMI from the valve motor. Oscillating frequency is about 1 Hz. The reservoir has cooling fins to aid in the heat rejection from the warm end of the pulse tube. This pulse tube provides 2 W of refrigeration at 77 K. A newer, larger version provides 7 W at 77 K with the same compressor and 15 W at 77 K when using a 2.3-kW compressor. Figure 12.20 shows one of the latest space-qualified pulse-tube cryocoolers designed for cooling infrared sensors on the Integrated Multispectral Atmospheric Sounder (IMAS) (29). It provides 1.0 W of cooling at 55 K and 2.5 W at 80 K with only 51 W of input power to the compressor (30) to give an effi-

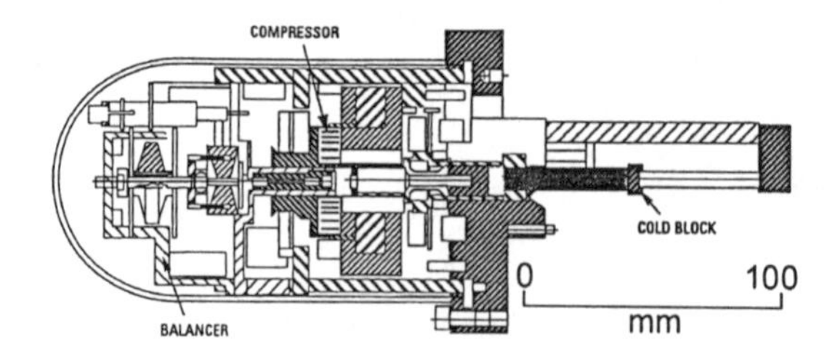

FIGURE 12.21 Cross section of mini-pulse-tube with single piston and passive balancer. (Courtesy of TRW.)

FIGURE 12.22 Mini-pulse-tube cryocooler with U-tube geometry and dual-opposed pistons. (Courtesy of Lockheed Martin.)

ciency of 13.5% of Carnot at 55 K. It uses the latest technology in flexure-bearing compressors to reduce the size and mass of the compressor. The cold head is an in-line arrangement and the support structure also serves as the reservoir. Total mass of the valveless compressor and pulse-tube cold head is only 3.2 kg. A schematic of a mini-pulse-tube cryocooler developed for space applications is shown in Figure 12.21. It is an integral system that produces 0.5 W of cooling at 80 K with 17 W of input power (31). The total length of compressor plus cold end is about 300 mm. Over 20 of these mini-pulse-tubes have been developed for space missions. Another miniature-pulse-tube cryocooler using a U-tube configuration is shown in Figure 12.22.

12.8 CRYOCOOLER OPERATING REGIONS AND COMPARISONS

12.8.1 Operating Regions

The typical operating region for various cryocoolers is shown in Figure 12.23. Some of these boundaries have not been fully explored, especially the larger sizes for pulse-tube refrigerators. Developments on industrial-size pulse-tube refrigerators utilizing flexure-bearing compressors up to 20 kW of input power are just now beginning. These will be useful for the power applications of high-temperature superconductors and for industrial gas liquefaction. The chart in Figure 12.23

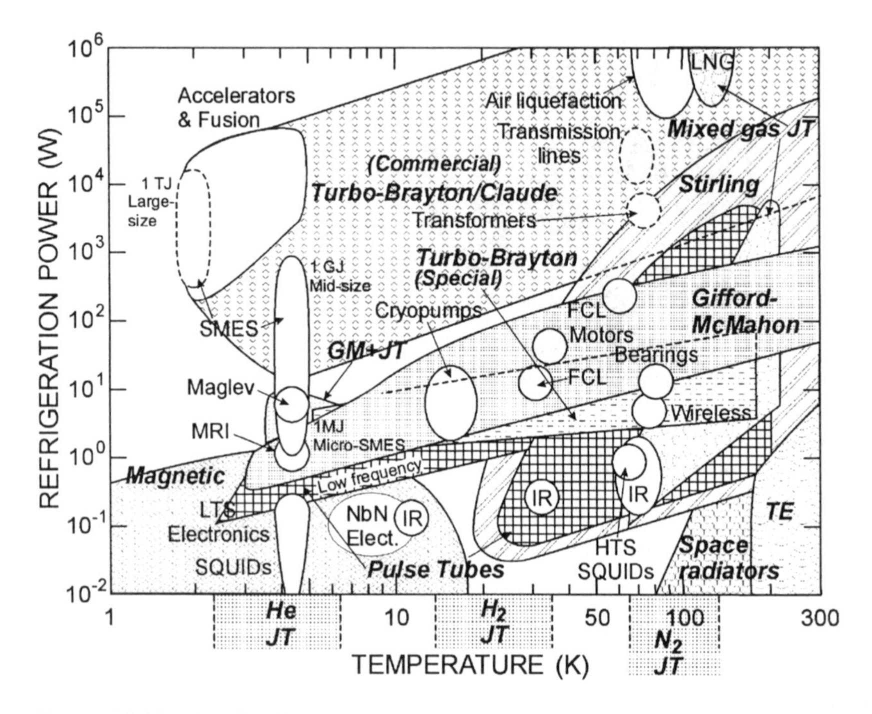

FIGURE 12.23 Applications and operating regions for various cryocoolers.

also shows the operating regions for various applications, including most of the superconducting applications. Electronic applications of high-temperature superconductors generally require refrigeration powers less than about 10 W for temperatures between about 60 and 80 K. Occasionally temperatures down to about 30 K are needed. Figure 12.23 shows that for the 60–80 K region below 10 W, the mixed-gas Joule–Thomson, Stirling, Gifford–McMahon, and pulse-tube cryocoolers would all be useful. It is very important to point out that commercially available cryocoolers often do not meet the specifications required for a successful market introduction of a superconducting device. One or more of the cryocooler problems listed in Section 12.3.5 are often serious enough that special research and development efforts are needed to eliminate or reduce the problem before the complete system becomes marketable.

12.8.2 Cryocooler Efficiencies

Figure 12.24 compares efficiencies of various types of cryocooler operating at 80 K as a function of the input power to the compressor. The data used in

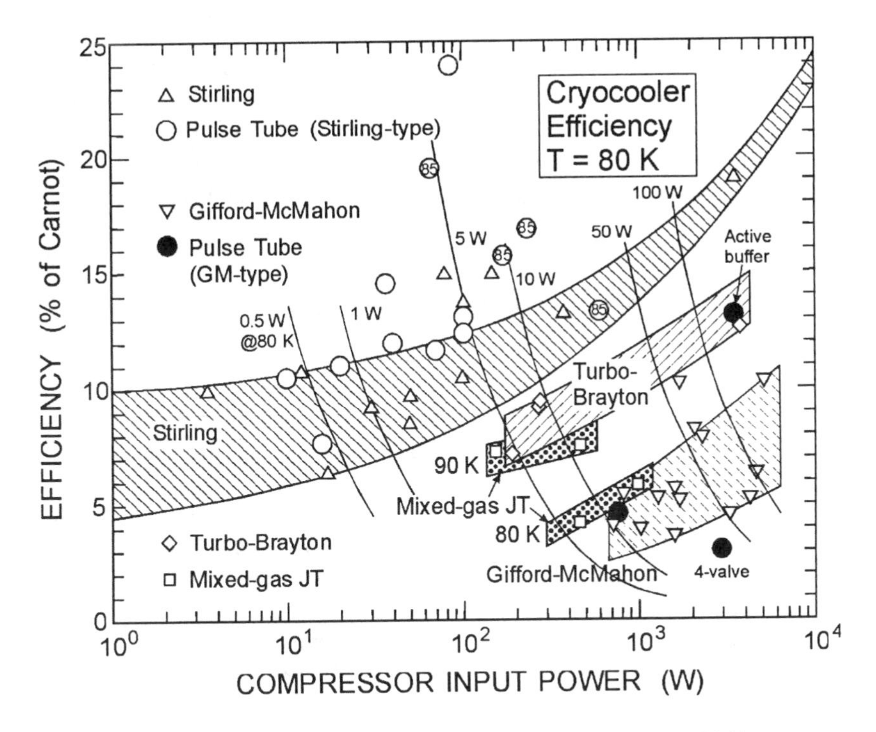

FIGURE 12.24 Efficiencies of various types of cryocoolers at 80 K.

Figure 12.24 are from the last 10 years and include mostly high-efficiency coolers of each type. Two facts emerge from this figure. The first is that efficiencies improve with larger sizes and the second is that pulse-tube refrigerators with Stirling-like (valveless) compressors are the most efficient of all cryocoolers. The highest efficiency of 24% of Carnot at 80 K was recently achieved (32) in a pulse-tube cryocooler similar to that shown in Figure 12.20. The efficiencies of mixed-gas JT cryocoolers increases rapidly as the temperature is increased from 80 to 90 K, whereas the efficiency of the other cryocoolers changes very little with temperature over this temperature range. Recent results using a mixed-gas JT cryocooler with a precooling stage gave an efficiency of 18% of Carnot at 100 K, but it dropped to 6.4% at 90 K (18). The efficiencies of small cryocoolers has increased considerably since 1974, when the survey by Strobridge (10) showed an average efficiency of about 2% of Carnot at a refrigeration power of 1 W at 80 K. A 1998 survey (33) showed the average efficiency for a 1-W cryocooler at 80 K had increased to about 5%. Efficiencies of 10–20% of Carnot for pulse tubes and some Stirling cryocoolers are now being achieved in this size range. Most of these high-efficiencies are with cryocoolers designed for space applications. Lower-cost commercial coolers may not quite achieve these high-efficiencies. Gifford–McMahon cryocoolers and pulse tubes driven with the GM compressors have much lower-efficiencies. These types of cryocooler would normally be used in applications where efficiency is not so important. A recent comparison of efficiencies, mass, and size of all types of space cryocoolers is given by Donabedian et al. (30).

12.8.3 Vibration and Electromagnetic Interference

Vibration and electromagnetic radiation can significantly degrade the performance of sensitive superconducting electronic devices. Acoustic noise, although not necessarily interfering with the electronics, can be a nuisance to personnel when they are near the system for long periods of time. Vibration can be characterized in several ways. One of the most common and easily defined methods is to measure the force transmitted by the cooler into a very rigid and heavy mass. Figure 12.25 compares this transmitted force from the various types of small cryocoolers as well as that from a dewar of boiling liquid nitrogen. Measurement with miniature turbo-brayton cryocoolers have not been able to detect any measurable vibration so far. The cold tip of a pure-nitrogen JT cryocooler (34) has shown lower vibration than that of boiling liquid nitrogen (35), whereas the cold tip of a mixed-gas JT cryocooler shows somewhat more vibration be-

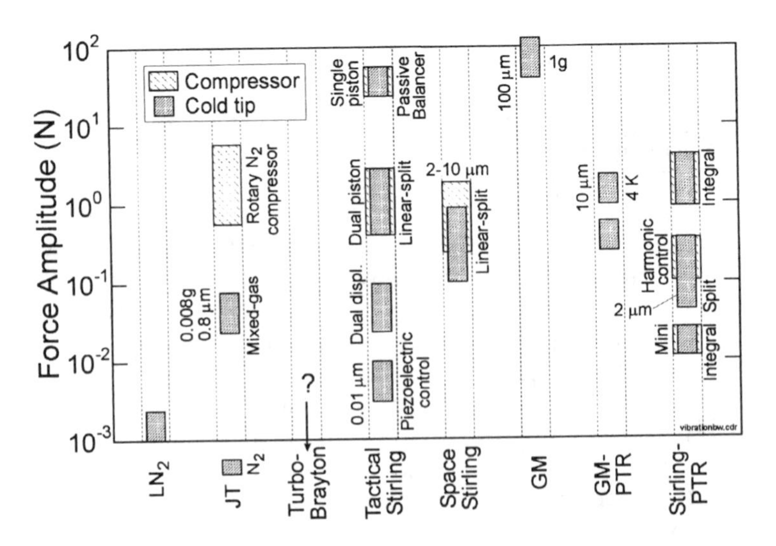

FIGURE 12.25 Vibration force transmitted by various cryocoolers.

cause of the heavier fluids used (35). The size or power of a cryocooler has a large effect on the transmitted force. For example, the mini-pulse-tube cooler (31) has a lower vibration output than the larger integral pulse-tube systems (36). When many of the higher-order harmonics are canceled by electronic feedback control (harmonic waveform suppression), the vibration of these larger systems is reduced further (37). In the case of the tactical Stirling cryocoolers, vibration at the cold end can be reduced by using two opposed displacers (38) or with active suppression of the vibration by piezoelectric actuators (39). The largest vibrations occur with the Gifford–McMahon cryocoolers because of the large displacer required for operation at a low frequency of about 1 Hz. As Figure 12.25 shows, the vibration of even a large 4-K GM-type pulse tube is much less than a GM cryocooler (40). The lower values for the GM-type pulse tube are for smaller 80-K systems. Figure 12.25 also gives some representative amplitudes of motion and acceleration.

The compressors of all the cryocoolers are major sources of radiated magnetic noise because of the large currents required to drive the motors. In some cases, motors are used to drive displacers of Stirling and GM cryocoolers, so they, too, become sources of EMI, but are at somewhat lower levels because of the much lower currents to drive these smaller motors. Figure 12.26 shows a plot of the radiated fields at 7 cm from the small integral pulse-tube cooler illustrated in Figure 12.20 (29). The fields can be reduced by about 20 dB at frequencies below about 200 Hz with the addition of a magnetic shield around the compressor housing. The maximum allowed fields for military requirements are indicated by the

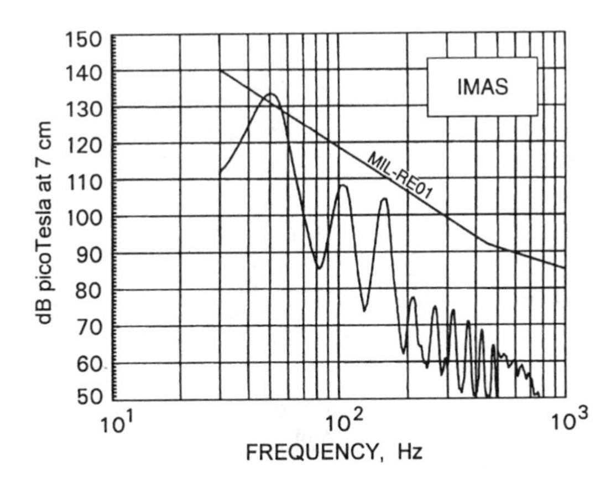

FIGURE 12.26 Radiated magnetic field 7 cm from a small integral pulse-tube cryocooler.

curve labeled MIL-RE01. Most Stirling coolers have very similar radiated fields. Joule–Thomson and GM-type pulse tubes can have lower EMIs because the cold end can be moved quite some distance from the compressor. In the case of the GM-type pulse tube, even the rotary valve needs to be moved some distance from the cold head, but that sacrifices performance.

12.9 APPLICATIONS OF CRYOCOOLERS WITH HTS DEVICES

One of the most challenging problems in cooling superconductors with cryocoolers is that of reducing the associated vibration and EMI caused by the motor and other moving parts. The problem is most serious with SQUIDs, because of their extreme sensitivity to magnetic fields and to vibration in the Earth's magnetic field. Thus, we concentrate this section on these applications.

To reduce noise in a superconducting device caused by the cryocooler, the following points should be considered: (1) selection of cryocooler type, (2) selection of materials, (3) distance between cryocooler and superconductor, (4) mounting platforms, (5) shielding, (6) thermal damping, and (7) signal processing. With regard to cryocooler types, the Joule–Thomson and pulse-tube cryocoolers are good choices because they have no cold moving parts. Because the pulse-tube cryocooler uses oscillating pressures, the temperature of the cold tip will also oscillate slightly at the operating frequency. The first published use of a cryocooler to cool a high-T_c SQUID was in 1994 by Khare and Chaudhari (41) and involved a miniature Stirling cryocooler that provides 150 mW at 77 K and requires 3 W of input power. Figure 12.27 is a photograph of the SQUID package (10 mm × 10 mm) mounted on the cold tip of this miniature cryocooler. At the 43-Hz fundamental frequency, the magnetic noise was about 10^{-9} T/Hz$^{1/2}$, but there was no measurable noise from the cooler in the range 4–100 kHz. The small size of the cooler helps to keep the noise signals low. More recent high-T_c SQUIDs have achieved much lower intrinsic noise levels and special efforts must be made to keep the noise from cryocoolers low enough so as not to significantly affect the SQUID signal.

A combination of techniques such as shielding of the Stirling compressor, use of dual, opposed pistons and displacers, and separation of SQUIDs from the cold finger by flexible copper braids was used for a high-T_c SQUID heart scanner cooled with a pair of tactical Stirling cryocoolers (42). The field noise spectra measure by the SQUID is shown in Figure 12.28 with and without shielding, as well as with the compressor turned off. The noise with the shielded compressor is only slightly above the background level. Figure 12.29 shows a magnetocardiogram recorded with this cryogen-free system.

With a careful selection of materials and a separate support for a high-T_c SQUID magnetometer in a μ-metal shield, Lienerth et al. (43) used a GM-type

FIGURE 12.27 Miniature Stirling cooler with SQUID package mounted on the tip of the cold finger. The SQUID package is approximately 1 × 1 cm. (Reproduced from Ref. 41 with permission.)

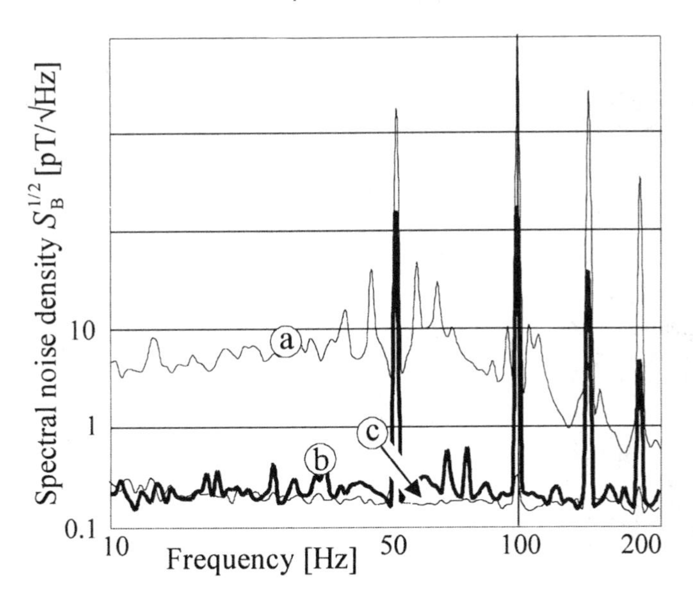

FIGURE 12.28 Magnetic field noise measured by a high-T_c SQUID magnetometer cooled with a Stirling cryocooler: (a) no magnetic shielding; (b) magnetic shielding around the compressors; (c) coolers off. (Reproduced from Ref. 42 with permission of *Cryogenics*.)

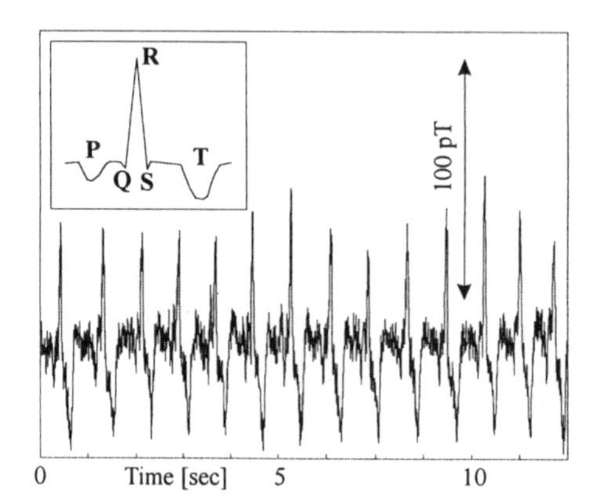

FIGURE 12.29 Magnetocardiogram recorded with high-T_c SQUID magnetometer cooled with a Stirling cryocooler. (Reproduced From ref. 42 with permission of *Cryogenics.*)

pulse-tube cooler for the system and achieved a white-noise level above 1 kHz of 35 fT/Hz$^{1/2}$ compared with 45 fT/Hz$^{1/2}$ for liquid-nitrogen cooling. In fact, the white-noise level for cooling with the pulse tube was less than with liquid-nitrogen cooling for all frequencies above about 2 Hz. However, the pulse-tube produced sharp peaks in the noise spectra at the operating frequency of 4.6 Hz and its harmonics. Earlier measurements in the same laboratory compared vibration and field noise spectra at the cold tips of a mixed-gas JT cooler and a

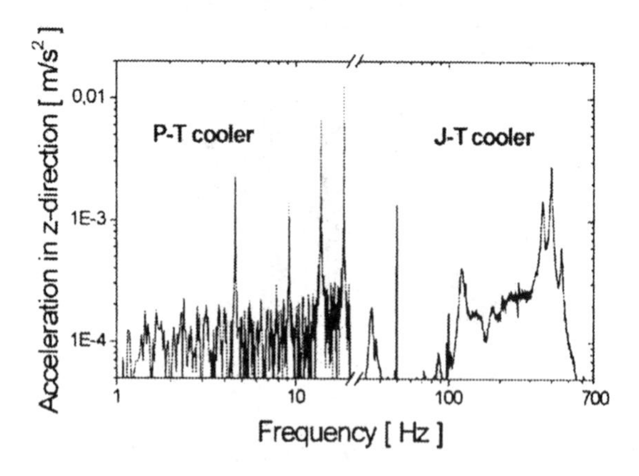

FIGURE 12.30 Vibration at the cold tip of a pulse-tube cryocooler and a mixed-gas JT cryocooler. (Reproduced from Ref. 44 with permission of IEEE.)

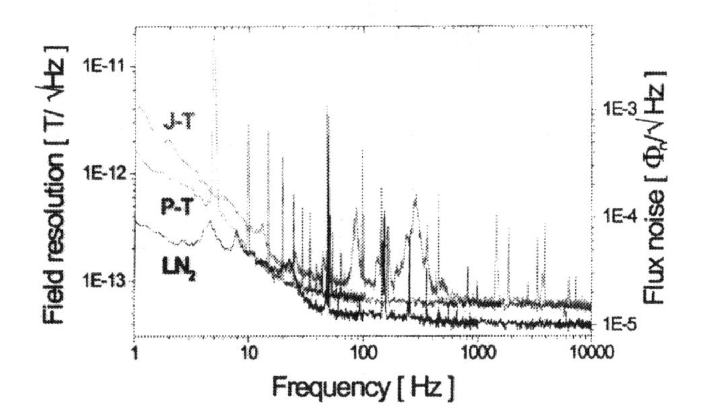

FIGURE 12.31 Field noise in a SQUID caused by various cooling methods. (Reproduced from Ref. 44 with permission of IEEE.)

pulse-tube cooler (44). These comparisons of vibration and field noise are shown in Figures 12.30 and 12.31. Except for the sharp peaks in the noise spectra at the operating frequency and its harmonics, the pulse-tube cooler produced less noise than the JT cooler and less noise than with liquid-nitrogen cooling at frequencies above about 40 Hz. The broad vibration and noise peaks in the frequency range from 100 to 500 Hz with the JT cooler may be caused by turbulent fluid flow. Researchers are still at an early stage in the integration of cryocoolers for use with SQUIDs. Interference problems are less severe when using cryocoolers with other superconducting devices, such as microwave filters for wireless telecommunication, either in ground-based stations (45) or in satellite stations (46).

Considerable research and development is still required to reduce some of the problems associated with cryocoolers in their use with HTS devices. Of particular concern is cost, reliability, EMI, and, in some cases, efficiency, before high-temperature superconducting devices can easily compete with conventional electronics in the marketplace.

REFERENCES

1. R Radebaugh. Advances in Cryocoolers, Proceedings ICEC16/ICMC. Oxford: Elsevier Science, 1997, pp 33–44.
2. NIST Thermodynamic and Transport Properties of Pure Fluids Database, Version 5.0. National Institute of Standards and Technology Standard Reference Database 12, Gaithersburg, MD, September 2000; See www.cryogenics.nist.gov.
3. LA Wade. An overview of the development of sorption refrigeration. In: R W Fast, ed. Advances in Cryogenic Engineering Vol. 37. New York: Plenum Press, 1992, pp 1095–1106.

4. JF Burger, HJ Holland, JH Seppenwoolde, JW Berenschot, HJM ter Brake, JGE Gardeniers, M Elwenspoek, H Rogalla. 165 K microcooler operating with a sorption compressor and a micromachined cold stage. In: R G Ross, Jr., ed. Cryocoolers 11. New York: Plenum Press, 2001, pp 551–560.
5. GW Swift, RA Martin, R Radebaugh. Acoustic cryocooler. U.S. Patent 4,953,366 (1990).
6. R Radebaugh, KM McDermott, GW Swift, RA Martin. Development of a thermoacoustically driven orifice pulse tube refrigerator. Proc. Fourth Interagency Meeting on Cryocoolers, David Taylor Research Center Technical Report DTRC-91/003, 1991, pp 205–220.
7. S Backhaus, GW Swift. A thermoacoustic–Stirling heat engine. Nature 399:335–338, 1999.
8. G Davey. Review of the Oxford Cryocooler. In: R W Fast, ed. Advances in Cryogenic Engineering Vol. 35. New York: Plenum Press, 1990, pp 1423–1430.
9. T Nast, P Champagne, V Kotsubo. Development of a low-cost unlimited-life pulse-tube cryocooler for commercial applications. In: P. Kittel, ed. Advances in Cryogenic Engineering Vol. 43. New York: Plenum Press, 1998, pp 2047–2053.
10. TR Strobridge. Cryogenic Refrigerators—An Updated Survey. National Bureau of Standards Technical Note 655, 1974.
11. R Radebaugh. Recent developments in cryocoolers. Proc. 19th International Congress of Refrigeration, 1995, pp 973–989.
12. WA Little. Recent developments in Joule–Thomson cooling: Gases, coolers, and compressors. Proc. 5th International Conference on Cryocoolers, Monterey, CA, 1988, pp 3–11.
13. DJ Missimer. Auto-refrigerating cascade (ARC) systems—An overview. Tenth Intersociety Cryogenic Symposium, AIChE Spring National Meeting, 1994.
14. WA Little. Kleemenko cycle coolers: Low cost refrigeration at cryogenic temperatures. In: Proceedings of the 17th International Cryogenic Engineering Conference. Bristol: Institute of Physics, 1998, pp 1–9.
15. MJ Boiarski, VM Brodianski, RC Longsworth. Retrospective of mixed-refrigerant technology and modern status of cryocoolers based on one-stage, oil-lubricated compressors. In: P. Kittel, ed. Advances in Cryogenic Engineering Vol. 43. New York: Plenum Press, 1998, pp 1701–1708.
16. ED Marquardt, R Radebaugh, J Dobak. A cryogenic catheter for treating heart arrhythmia. In: P. Kittel, ed. Advances in Cryogenic Engineering Vol. 43. New York: Plenum Press, 1998, pp 903–910.
17. J Dobak, X Yu, K Ghaerzadeh. A novel closed loop cryosurgical device. In: P. Kittel, ed. Advances in Cryogenic Engineering Vol. 43. New York: Plenum Press, 1998, pp 897–902.
18. A Alexeev, C Haberstroh, H Quack. Mixed gas J-T cryocooler with precooling stage. In: R G Ross, Jr., ed. Cryocoolers 10. New York: Plenum Press, 1999, pp 475–479.
19. WE Gifford, HO McMahon. A new refrigeration process. Proc. Tenth International Congress of Refrigeration, 1959, Vol. 1; HO McMahon, WE Gifford. A new low-temperature gas expansion cycle. In: K D Timmerhaus, ed. Advances in Cryogenic Engineering Vol. 5. New York: Plenum Press, 1960, pp 354–372.

20. A Ravex. Small Cryocoolers, B Seeber, ed. In: Handbook of Applied Superconductivity, Volume 1. Bristol: Institute of Physics, 1998, pp 721–746.
21. G Walker, ER Bingham. Low-Capacity Cryogenic Refrigeration. Oxford: Oxford University Press, 1994.
22. WE Gifford, RC Longsworth. Pulse tube refrigeration. Trans. ASME, J Eng Ind August 1964, paper 63-WA-290.
23. EI Mikulin, AA Tarasov, MP Shkrebyonock. Low temperature expansion pulse tubes. In: Advances in Cryogenic Engineering Vol. 29. New York: Plenum Press, 1984, pp 629–637.
24. R Radebaugh, J Zimmerman, DR Smith, B Louie. A comparison of three types of pulse tube refrigerators: New methods for reaching 60 K. In: Advances in Cryogenic Engineering Vol. 31. New York: Plenum Press, 1986, pp 779–789.
25. R Radebaugh. Development of the pulse tube refrigerator as an efficient and reliable cryocooler. Proc Inst Refrig (London) 1999–2000, vol. 96, pp 11–29.
26. ED Marquardt, R Radebaugh. Pulse tube oxygen liquefier. In: Q S Shu, ed. Advances in Cryogenic Engineering Vol. 45. New York: Plenum Press, 2000, pp 457–464.
27. S Zhu, P Wu, Z Chen. Double inlet pulse tube refrigerators: An important improvement. Cryogenics 30:514–520, 1990.
28. D Gedeon. DC gas flows in Stirling and pulse tube refrigerators. In: R G Ross, Jr., ed. Cryocoolers 9. New York: Plenum Press, 1997, pp 385–392.
29. CK Chan, T Nguyen, R Colbert, J Raab, RG Ross Jr, DL Johnson. IMAS pulse tube cooler development and testing. In: R G Ross, Jr., ed. Cryocoolers 10. New York: Plenum Press, 1999, pp 139–147.
30. M Donabedian, DGT Curran, DS Glaister, T Davis, BJ Tomlinson. An overview of the performance and maturity of long-life cryocoolers for space applications. Aerospace Report TOR-98(1057)-3, Revision A, The Aerospace Corporation, El Segundo, CA, 2000; DS Glaister, M Donabedian, DGT Curran, T Davis. An overview of the performance and maturity of long life cryocoolers for space applications. In: R G Ross, Jr., ed. Cryocoolers 10. New York: Plenum Press, 1999, pp 1–19.
31. E Tward, CK Chan, J Raab, R Orsini, C Jaco, M Petach. Miniature long-life space-qualified pulse tube and Stirling cryocoolers. In: R G Ross, Jr., ed. Cryocoolers 8. New York: Plenum Press, 1995, pp 329–336.
32. E Tward, CK Chan, J Raab, T Nguyen, R Colbert, T Davis. High efficiency pulse tube cooler. In: R G Ross, Jr., ed. Cryocoolers 11. New York: Plenum Press, 2001, pp 163–167.
33. JL Bruning, R Torrison, R Radebaugh, M Nisenoff. Survey of cryocoolers for electronic applications (C-SEA). In: R G Ross, Jr., ed. Cryocoolers 10. New York: Plenum Press, 1999, pp 829–835.
34. R Levenduski, R Scarlotti. Joule–Thomson cryocooler development at Ball Aerospace. In: R G Ross, Jr., ed. Cryocoolers 9. New York: Plenum Press, 1997, pp 493–508.
35. DH Hill. Development of a low vibration throttle cycle cooler. In: P. Kittel, ed. Advances in Cryogenic Engineering Vol. 43. New York: Plenum Press, 1998, pp 1685–1692.

36. DL Johnson, SA Collins, MK Heun, RG Ross Jr, C Kalivoda. Performance characterization of the TRW 3503 and 6020 pulse tube coolers. In: R G Ross, Jr., ed. Cryocoolers 9. New York: Plenum Press, 1997, pp 183–193.
37. CK Chan, C Carlson, R Colbert, T Nguyen, J Raab, M Waterman. Performance of the AIRS pulse tube engineering model cryocooler. In: R G Ross, Jr., ed. Cryocoolers 9. New York: Plenum Press, 1997, pp 195–202.
38. AP Rijpma, JFC Verberne, EHR Witbreuk, HJM ter Brake. Vibration reduction in a set-up of two split type Stirling cryocoolers. In: R G Ross, Jr., ed. Cryocoolers 9. New York: Plenum Press, 1997, pp 727–736.
39. CE Byvik, J Stubstad. Summary and results of a space-based active vibration suppression experiment. In: R G Ross, Jr., ed. Cryocoolers 9. New York: Plenum Press, 1997, pp 719–726.
40. C Wang, PE Gifford. Performance characteristics of a 4 K pulse tube in current applications. In: R G Ross, Jr., ed. Cryocoolers 11. New York: Plenum Press, 2001, pp 205–212.
41. N Khare, P Chaudhari. Operation of bicrystal junction high-T_c direct current SQUID in a portable microcooler. Appl Phys Lett 65(18), 2353–2355, 1994.
42. AP Rijpma, CJHA Blom, AP Balena, E de Vries, HJ Holland, HJM ter Brake, H Rogalla. Construction and tests of a heart scanner based on superconducting sensors cooled by small Stirling cryocoolers. In: Cryogenics. Oxford: Elsevier Science, 2000, pp. 821–828.
43. C Lienerth, G Thummes, C Heiden. Progress in low-noise cooling performance of a pulse-tube cooler for HT-SQUID operation. IEEE Trans. on Appl. Superconductivity, Vol. 11, No. 1, 2001, pp 812–815.
44. R Hohmann, C Lienerth, Y Zhang, H Bousack, G Thummes, C Heiden. Comparison of low-noise cooling performance of a Joule–Thomson cooler and a pulse tube cooler using a HT SQUID. IEEE Trans Appl Supercond AS-9:3688–3691, 1999.
45. JL Martin, JA Corey, CM Martin. A pulse tube cryocooler for telecommunications applications. In: R G Ross, Jr., ed. Cryocoolers 10. New York: Plenum Press, 1999, pp 181–189.
46. V Kotsubo, JR Olson, P Champagne, B Williams, B Clappier, TC Nast. Development of pulse tube cryocoolers for HTS satellite communications. In: R G Ross, Jr., ed. Cryocoolers 10. New York: Plenum Press, 1999, pp 171–179.

13

High-Temperature Superconductor Electronics: Status and Perspectives

Shoji Tanaka
Superconductivity Research Laboratory, ISTEC, Tokyo, Japan

13.1 INTRODUCTION

After the discovery in 1986, many kinds of high-T_c superconductor (HTS) have been found, and the critical temperature was raised up to 130 K in mercury-based compounds. In the first 10 years, however, they have been the objects of material sciences primarily, because they are quite peculiar cuprates and it was necessary to understand the physical properties, chemical properties, and so on.

At around 1995, the trends in the applications of HTS became clear. The development of microwave filters and superconducting quantum interference devices (SQUIDs) have been accelerated. Furthermore, the applications of well-known low-T_c superconductors (LTSs) were stimulated by the progress of HTS research and developments, and the application of LTS SQUIDs to medical electronics and study of single flux quantum (SFQ) devices started. As for materials for the LTS devices, only Nb-based compounds are used.

Over the past 20 years, the progress of the Internet has been remarkable; and it covers the whole world and is changing the structure of our society to form the so-called "ubiquitous society." Superconductivity electronics must have a great impact on the progress of the Internet, as will be mentioned in this chapter.

13.2 MICROWAVE FILTERS

High-T_c superconductor microwave filters and low-noise amplifiers for mobile telephone base stations have been the first commercialized devices in HTS electronics. HTS-based systems offer an improved quality in wireless communications, with increased area coverage and reduced interference.

Currently, the mobile telephone system is connected to the Internet, and service is widely improved (e.g., i-mode). Especially in Japan, third-generation (3G) wireless communication began last year in major cites, which is close to becoming a picture phone system. The area of this service is limited at present, but after an extremely large number of users join this system, the HTS microwave filters will be used in order to prevent interference. The number of base stations in Japan is increasing and is expected to become 200 thousand in the year 2010. This may become a large market for the microwave filters.

It is not clear at present when a software-defined radio system will be introduced, but we expect that it will be introduced in the fourth-generation communication system (4G) around the year 2010. However, in order to construct the software-defined radio system, it is necessary to develop a high-quality AD converter of 200 MHz and 16 bits using very high-speed SFQ circuits.

13.3 SQUID

The SQUID is the most well-known superconducting device and it has a high sensitivity for detecting small magnetic fields. Thus, it has many possibilities for applications in various fields. Before 1990, however, real application was limited because LTS SQUIDs must be cooled down to a liquid-helium (He) temperature of 4.2 K.

Around 1990, applications to the medical electronics for magnetic diagnostics of heart and brain activity started in many countries; in Japan, a project for constructing new systems of observing magnetic brain waves by using 256-channel LTS SQUIDs and also of observing magnetic heart waves by using 36-channel HTS SQUIDs was started. The system for detecting human magnetic brain waves is now used for the study of brain activity in universities and the systems for detecting magnetic heart waves will now be commercialized after a license from the government is obtained. Recently, LTS SQUIDs are used even in magnetic heart wave detecting systems to obtain more precise information. It is hoped that the market for these magneto-cardiographic systems expands rapidly and reaches that of magnetic resonance imaging (MRI).

The application of SQUIDs for the precise voltage standard has been performed, and the application to nondestructive evaluation systems in materials of airplanes and other constructions is very hopeful.

Recently, the SQUID microscope has appeared. In this equipment, the LTS SQUID is used to observe the distribution of weak magnetic fields in a sample with high accuracy; thus, it is possible to observe the distribution of magnetic flux quanta in superconducting thin films with a precision of a few microns. There also seems to be a potential market in the application for the observation of biomaterials, where very fine magnetic particles are attached.

13.4 RSFQ LOGIC DEVICES

At present, the developments of rapid-SFQ (RSFQ) logic circuits are the most exciting subject in superconducting electronics, as the RSFQ device has an ultrafast operating speed of several hundred gigahertz and a very small power dissipation of the order of 10 nW. The principle of the RSFQ device is very simple; the quantum flux stored in a superconducting loop is used as an individual bit of information. The quality of Josephson junctions included in the loop is the most important factor in constructing an integrated RSFQ circuit.

The structures of the RSFQ devices of LTS and HTS are shown in Figs. 13.1 and 2. For LTS, the stacked junction of Nb–A1O_x–Nb is used. In the case of HTS, the ramp-edge junction of YBCO–barrier–YBCO is used, and, sometimes, YBCO is used as a counterelectrode to lower deposition temperatures.

The most important factor in both cases is the standard deviation (σ) of the critical current of the junction. At present a nearly 1% deviation in LTS and 6% in 100 junctions of a HTS are obtained at 4.2 K. We expect that a 5% deviation will be obtained soon, with 1000 HTS junctions, which means that the sigma–delta AD converter will be made in the yield of 50% if suitable designs are made, as shown in Fig. 13.3. The road maps of the LTS junctions are now presented by many institutions and one of them is shown in Table 13.1. The road map of a HTS SFQ is not available yet, as the HTS SFQ is only 5 years in use.

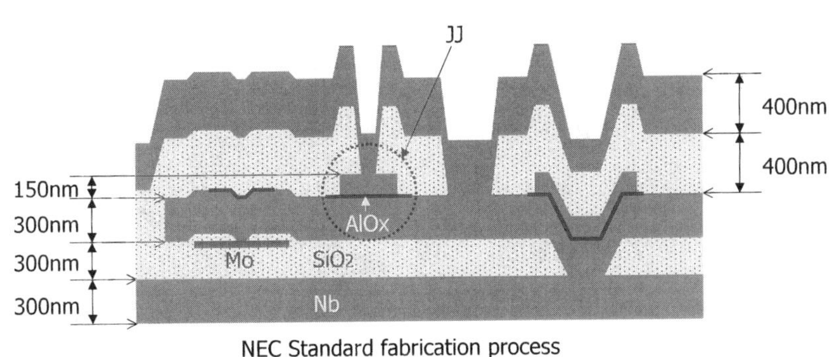

FIGURE 13.1 Structure of a LTS RSFQ device.

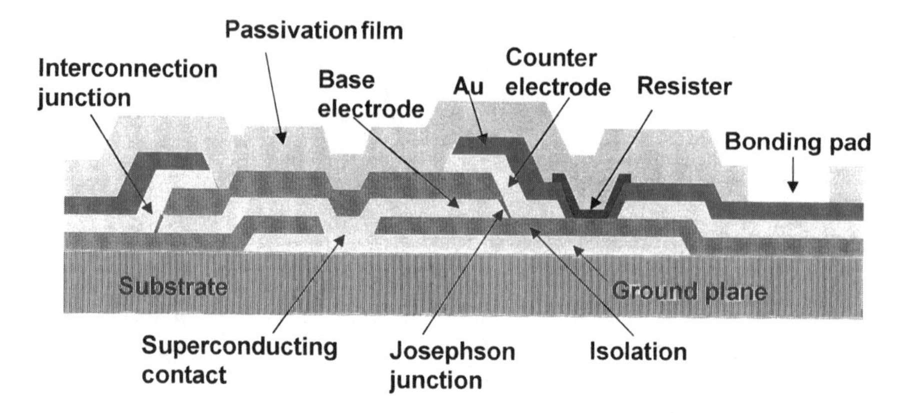

FIGURE 13.2 Structure of a HTS RSFQ device.

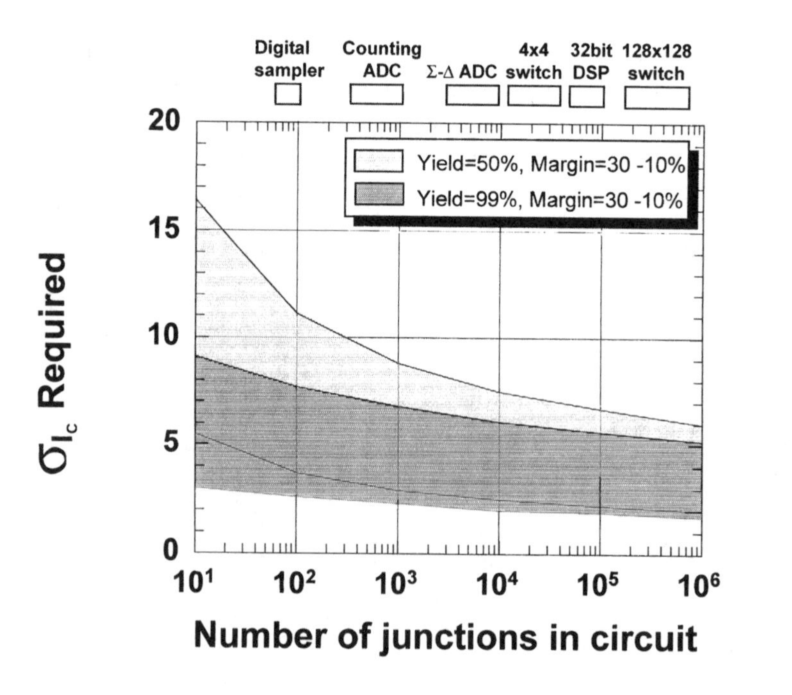

FIGURE 13.3 Relation between the I_c standard deviation and the number of junctions in a circuit.

TABLE 13.1 Superconducting LSI (Large Scale Integration) Technology Road Map

	Year		
	1999	2005	2010
Fabrication process			
Junction size (μm)	2.0	1.0	0.8
Junction current			
Density (kA/cm^2)	2.5	10.0	15.6
Number of junctions	10 K	100 K	1 M
Memory			
Memory density (bits/chip)	256	16 K	1 M
Clock frequency (GHz)	5	10	10
Logic			
Clock frequency (GHz)		50	50
Performance		16-bit MPU	64-bit MPU

The possibility of LTS RSFQ circuits has been discussed in relation to its use in the future peta-flops computers, mainly in the United States. Recently, a new design of the Microprocessor Unit (MPU) of 16 bits and 25 GHz, the Flux Chip, was proposed by the SUNNY group and its production is in progress at TRW.

As for the HTS RSFQ, fundamental circuits for future logic circuits were made already, and they proved to operate in a very high speed of a few hundred gigahertz. These are the toggle flip-flop (9JJs) operated at 270 GHz at 4.2 K (SRL), the ring oscillator (21JJs) at 30 GHz at 30 K (Toshiba), and the sigma-delta modulator (11JJs) at 100 GHz at 20 K (Hitachi). The high-speed sampler is shown in Fig. 13.4, which was developed recently by NEC. It consists of 17 JJs and shows a beautiful wave form of 15 GHz. We expect that it will soon accomplish observa-

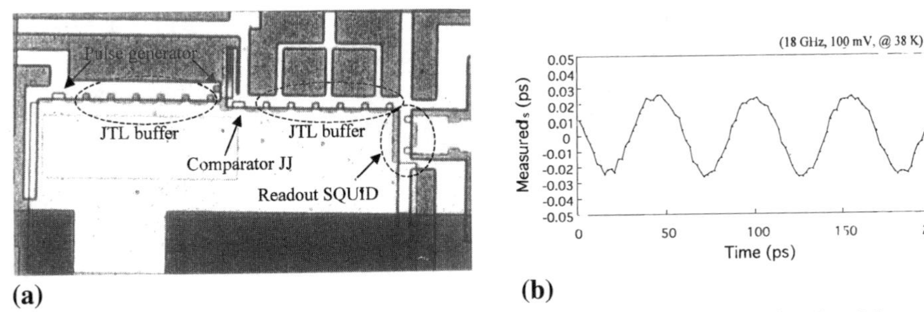

FIGURE 13.4 Sampler circuit with JTL buffers. (a) HTS sampler circuit with 6-stage JTL buffers using 17 Josephson junctions; (b) measured 18 GHz waveform.

tions higher than 50 GHz. These results proved the very fast operations of the HTS RSFQ circuits, and, we hope, circuits with more than 100 JJs will appear very soon.

13.5 IMPACT OF THE INTERNET

The progress of the Internet creates new possibilities for the RSFQ devices. In Japan, optical fibers will be equipped in offices and homes (Fiber To The Homes) this year, and communication at 100 Mbps will become very popular. In such a very fast communication, communication nodes, servers, and routers must be operated in a rate of more than 10 Tbps, which exceeds the operation speed of semiconductor devices. Thus, very fast and very powerful conservative RSFQ circuits are necessary. Furthermore, it is expected that in the mobile communication system of the fourth generation, the communication speed will be 100 Mbps also. Such a "broad band and wireless" technology requires the suitable combinations of optical fibers, semiconductors, and superconductors. Therefore, it is believed that the development of the RSFQ circuits of both LTS and HTS must be accelerated.

13.6 SUMMARY

In the coming 10 years, the primary goal worldwide must be the very fast progress of the communication technology of the Internet. The era of picture communication is coming soon, for which a great amount of information will be exchanged at a very high speed of 100 Mbps. However, the progress of the information processing technology is rather slow compared to that of the communication technology, due to the saturation in the developments of information storage systems and semiconductor devices. This mismatch between the two important technologies will result in the substantial dissipation of electric power in society as a whole. Therefore, the role of the RSFQ circuits of very high-speed operation and very low power dissipation will become very important. The circuits of very high integration of the LTS RSFQ could be used in every node of the Internet. The circuits of medium-scale integration of the HTS RSFQ will be used in the base stations of mobile communication systems in offices and homes. The developments of future RSFQ technologies must be accelerated in order to realize such expectations, and it will also expand the applications of superconductivity electronics in many fields.

Index